Char and Carbon Materials Derived from Biomass

Char and Carbon Materials Derived from Biomass

Production, Characterization and Applications

Mejdi Jeguirim, PhD
Professor
Lionel Limousy, PhD
Professor
University of Strasbourg, University of Upper Alsace, Institute of Materials Science of Mulhouse (IS2M—UMR CNRS 7361), Mulhouse, France

ELSEVIER

Elsevier
Radarweg 29, PO Box 211, 1000 AE Amsterdam, Netherlands
The Boulevard, Langford Lane, Kidlington, Oxford OX5 1GB, United Kingdom
50 Hampshire Street, 5th Floor, Cambridge, MA 02139, United States

Library of Congress Cataloging-in-Publication Data
A catalog record for this book is available from the Library of Congress

British Library Cataloguing-in-Publication Data
A catalogue record for this book is available from the British Library

ISBN: 978-0-12-814893-8

Publisher: Mathew Deans
Acquisition Editor: Christina Gifford
Editorial Project Manager: Joanna Collett
Production Project Manager: Bharatwaj Varatharajan
Cover Designer: Christian J. Bilbow

Typeset by SPi Global, India

Contents

Contributors

Agapios Agapiou University of Cyprus, Department of Chemistry, Nicosia, Cyprus

Ahmed Ahmed Department of Bioresource Engineering, Macdonald Campus, McGill University, Sainte-Anne-de-Bellevue, QC, Canada

Hanene Akrout Wastewaters and Environment Laboratory, Water Research and Technologies Centre (CERTE), Soliman, Tunisia

Ahmed Amine Azzaz Wastewaters and Environment Laboratory, Water Research and Technologies Centre (CERTE), Soliman, Tunisia

Meriem Belhachemi Laboratory of Chemical and Environmental Sciences, University of Tahri Mohamed Bechar, Bechar, Algeria

Simona Bennici University of Strasbourg, University of Upper Alsace, Institute of Materials Science of Mulhouse (IS2M—UMR CNRS 7361), Mulhouse, France

David Bessieres CNRS/Total/University of Pau and Pays de l'Adour/E2S UPPA, Laboratory of Complex Fluids and Their Reservoirs-IPRA, Pau, France

Patrick Brassard Research and Development Institute for the Agri-Environment (IRDA), Quebec City, QC, Canada

Joanna Conder University of Strasbourg, University of Upper Alsace, Institute of Materials Science of Mulhouse (IS2M—UMR CNRS 7361), Mulhouse, France

Claire Courson Univ Strasbourg, CNRS, ICPEES UMR 7515, ECPM, Strasbourg, France

Krzysztof Fic Institute of Chemistry and Technical Electrochemistry, Poznan University of Technology, Poznan, Poland

Imen Ghouma University of Strasbourg, University of Upper Alsace, Institute of Materials Science of Mulhouse (IS2M—UMR CNRS 7361), Mulhouse, France

Stéphane Godbout Research and Development Institute for the Agri-Environment (IRDA), Quebec City, QC, Canada

Richard Hogue Research and Development Institute for the Agri-Environment (IRDA), Quebec City, QC, Canada

Cecile Hort University of Pau and Pays de l'Adour/E2S UPPA, Laboratory of Thermal Engineering, Energy and Processes-IPRA, Pau, France

Thomas Jeanne Research and Development Institute for the Agri-Environment (IRDA), Quebec City, QC, Canada

Mejdi Jeguirim University of Strasbourg, University of Upper Alsace, Institute of Materials Science of Mulhouse (IS2M—UMR CNRS 7361), Mulhouse, France

Salah Jellali Wastewaters and Environment Laboratory, Water Research and Technologies Centre (CERTE), Soliman, Tunisia

Rita Khanna School of Materials Science and Engineering, The University of New South Wales, Sydney, NSW, Australia

Besma Khiari National School of Engineers of Carthage, Tunis, Tunisia

Witold Kwapinski Chemical Sciences, University of Limerick, Limerick, Ireland

Madona Labaki Laboratory of Physical Chemistry of Materials (LCPM/PR2N), Lebanese University, Campus Fanar, Jdeidet El Metn, Lebanon

Vicky Lévesque Agriculture and Agri-Food Canada, Quebec City, QC, Canada

Kejiang Li School of Metallurgical and Ecological Engineering, University of Science and Technology Beijing, Beijing, PR China

Lionel Limousy University of Strasbourg, University of Upper Alsace, Institute of Materials Science of Mulhouse (IS2M—UMR CNRS 7361), Mulhouse, France

Pantelitsa Loizia Open University of Cyprus, Faculty of Pure and Applied Sciences, Environmental Conservation and Management, Laboratory of Chemical Engineering and Engineering Sustainability, Nicosia; I.E.S.T—EnviTech Ltd, (Institute of Environmental Technology and Sustainable Development), Department of Research and Development, Paralimni, Cyprus

Camélia Matei Ghimbeu University of Strasbourg, University of Upper Alsace, Institute of Materials Science of Mulhouse (IS2M—UMR CNRS 7361), Mulhouse; French Research Network on Electrochemical Energy Storage (RS2E), FR CNRS 3459, Amiens, France

Partha Sarathi Mukherjee Institute of Minerals and Materials Technology, Advanced Materials Technology Department, Bhubaneshwar, India

Jose Navarro Pedreno Department of Agrochemistry and Environment, University Miguel Hernández of Elche, Alicante, Spain

Joahnn H. Palacios Research and Development Institute for the Agri-Environment (IRDA), Quebec City, QC, Canada

Deneb Peredo-Mancilla CNRS/Total/University of Pau and Pays de l'Adour/E2S UPPA, Laboratory of Complex Fluids and Their Reservoirs-IPRA; University of Pau and Pays de l'Adour/E2S UPPA, Laboratory of Thermal Engineering, Energy and Processes-IPRA, Pau, France

Vijaya Raghavan Department of Bioresource Engineering, Macdonald Campus, McGill University, Sainte-Anne-de-Bellevue, QC, Canada

Antonio Sepúlveda-Escribano Laboratorio de Materiales Avanzados, Departamento de Quimica Inorganica-Instituto Universitario de Materiales de Alicante, Universidad de Alicante, Alicante, Spain

Marinos Stylianou University of Cyprus, Department of Civil & Environmental Engineering, NIREAS-International Water Research Center, Subsurface Research Laboratory; University of Cyprus, Department of Chemistry, Nicosia, Cyprus

Minmin Sun School of Metallurgical and Ecological Engineering, University of Science and Technology Beijing, Beijing, PR China

Andrius Tamošiūnas Lithuanian Energy Institute, Plasma Processing Laboratory, Kaunas, Lithuania

Mausam Verma Research and Development Institute for the Agri-Environment (IRDA), Quebec City, QC, Canada

Ziming Wang School of Metallurgical and Ecological Engineering, University of Science and Technology Beijing, Beijing, PR China

Jianliang Zhang School of Metallurgical and Ecological Engineering, University of Science and Technology Beijing, Beijing, PR China

Antonis A. Zorpas Open University of Cyprus, Faculty of Pure and Applied Sciences, Environmental Conservation and Management, Laboratory of Chemical Engineering and Engineering Sustainability, Nicosia, Cyprus

Preface

The genesis of this book entitled *Char and Carbon Materials Derived from Biomass—Production, Characterization and Applications* was initiated by the editors Mejdi Jeguirim and Lionel Limousy, researchers at the Material Science Institute of Mulhouse (IS2M—www.is2m.uha.fr/fr/accueil/). These scientists have been working on the valorization of different kinds of biomass (liquid, solid) for more than 15 years, and they have accumulated a lot of experience and knowledge in this field of research. They are involved in different scientific committees of international journals and congresses, have organized several special issues, and managed several international scientific projects in the field of biomass transformation and valorization.

This book is organized into 12 chapters, written by different international experts from 14 countries. The chapters are organized as follows: description of biomass feedstocks, transformation of biomass into chars, characterization of chars, energetic conversion of chars, different applications for chars, and the sustainable assessment approach of char production and valorization. The progress of the book reveals specific information about the material composition, sourcing, transformation, and characterization before describing char elaboration and valorization. This can help nonspecialists to understand the scientific approach as well as experts to find the expected information more easily.

The objective of the editors is to give a specific insight into char elaboration from different feedstocks. Depending on the carbon source and the composition of the raw materials from which they are issued, the chars have to be produced differently; their physicochemical characteristics as well as their fields of valorization are not the same. In fact, chars can be elaborated from different supports, starting with the very conventional "woody biomass" up to unexpected resources like "poultry feather" or "automotive shredder residues." The cost of elaborating chars as well as their impact on the environment are described in this book.

The different chapters of this book show the diversity of research conducted recently on the elaboration, characterization, and use of chars. Agriculture materials, biofuels, adsorbents, catalysts, and supercapacitors are versatile resources that can be used for various applications. Recent developments have shown that the return to soil is also a promising issue since the carbon balance is negative due to carbon sequestration. Other benefits are described below. New developments are still in progress, encouraging the production of this book.

At the beginning of this book (Chapter 1) the authors show an overview of the variety of waste and raw materials that can be converted into chars. The main feedstocks are described in terms of composition, quantities, and location, as well as the experimental conditions in which they are transformed. Among the wide variety of feedstocks, we can mention woody, agricultural, and aquatic biomasses, which are the most used to prepare chars. The elaborating of chars from livestock and industrial and municipal wastes is also described. Chapter 2 presents the required characteristics for char designation. The main criteria correspond to carbon content, H/C_{org} and O/C_{org} ratios (described by the Van Krevelen diagram), remaining volatile organic compounds, pH, and mineral and ash contents. The different standards for measuring these parameters are presented. Thermal conversion via slow and fast pyrolysis as well as by gasification is detailed. The development of biochar production is also detailed (earth pit, kiln, retorts, pyrolyzer, hydrothermal carbonization). Chapter 3 examines the different techniques and methodologies used to characterize the biomass-derived chars. Indeed, chars are high-value materials and their recovery methods depend strongly on their chemical, physical, structural, morphological, and textural properties. All the experimental techniques used to characterize these properties are described; the results are discussed depending on the origin of the biomass and the way the chars were obtained. Chapters 4 and 5 describe the thermochemical conversion of chars by combustion and gasification, respectively. The densification process (pellets and briquettes) is interesting for combustion because it facilitates the management of fuel supply in industrial or domestic devices, e.g., domestic cooking and heating stoves as well as specific industrial applications (iron and steel making, electricity generation, and direct carbon fuel vell). The gasification process is depicted through the different parameters, which influence the reactivity of the chars: char production, catalytic effect of minerals, nature of the gasifying agent (CO_2, H_2O, O_2). The mechanisms and kinetics of solid/gas and gas-phase reactions are necessary to size the gasifiers. The different technologies and controlled parameters are shown (Chapter 5), including the presentation of fixed-bed, updraft, fluidized-bed, and entrained flow gasifiers. Chars can be used for nonenergetic applications (Chapters 6 to 11); their elaboration and characteristics have to be adapted depending on the conditions of use. As an example, biochars can be used as an amendment in agricultural soils due to many interesting characteristics (Chapter 6). The global impact of biochar on various agricultural soils is discussed; it is shown that biochar can improve soil chemical, physical, and biological properties, increasing crop productivity. Carbon sequestration is also highlighted due to its impact on climate change mitigation. Chapter 7 is devoted to the use of chars as adsorbents for gas and liquid pollutants. The importance of surface chemistry and the textural properties of the chars are described through a huge number of examples. Surface activation (physical, chemical) is often necessary to reach high adsorption capacity; the nature of the carbon feedstock is also a key parameter to improve/adapt the adsorption properties of the chars. Chars can be used as support for catalysts but also as catalysts (Chapter 8). The main parameter that has to be controlled corresponds to the surface acid/base properties of the chars, which are the key factors that govern the adsorption ability of solid materials toward the different molecules that can control their performance in catalysis. The elaboration of char-supported catalysts is described and different applications are presented (bio-oils upgrading, tar reforming/cracking, hydrocarbons/oils reforming, methanol and ethanol oxidation). Chapters 9 and 10 focus on energy storage considering chars or activated carbons as adsorbents or electrodes. Chapter 9 presents the recent developments in CH_4,

CO_2, and H_2 storage by activated carbons with a focus on biomass as precursor materials. An analysis of the main carbon properties affecting the activated carbon's adsorption capacity (specific surface area, pore size, and surface chemistry) is discussed in detail. The transformation of biomass into doped activated carbons is presented in Chapter 10 to prepare supercapacitors (electrochemical capacitors). Their elaboration enables their use as capacitor electrode material; the important parameters are the presence of high specific surface area, electronic and ionic conductivities, and sufficient mechanical and chemical stability. The performances of the best supercapacitors are presented and discussed. Chapter 11 presents the specific use of biochars in iron and steel industries. This chapter focuses on the utilization of sustainable biomass-based chars/charcoals to partially replace fossil-based coal and metallurgical coke in various iron- and steelmaking operations (coke making, iron ore sintering, auxiliary fuel injectant, charge carbon, and slag foaming agent in electric arc furnaces). Economic and supply issues are also discussed. The last chapter provides a sustainability assessment for biomass utilization in char production and utilization in various applications, including soil amendment, energy production, as well as the synthesis of green adsorbents, catalysis, and supercapacitors.

The writing of this book was possible thanks to the participation of numerous scientists from various countries. The editors would like to thank warmly all of them for their important contributions to this work.

Biomass feedstocks

1

Mejdi Jeguirim, Besma Khiari†, Lionel Limousy**
*University of Strasbourg, University of Upper Alsace, Institute of Materials Science of Mulhouse (IS2M—UMR CNRS 7361), Mulhouse, France, †National School of Engineers of Carthage, Tunis, Tunisia

Chapter Outline

1.1 Introduction

In view of the rising price and the scarcity of energy and raw materials, as well as the different environmental concerns issued from human activities, substitutes for fossil resources and/or depleting materials used in the management of the environmental are urgently required. There is a growing agreement that biomass could efficiently meet these requirements.

Biomass can be defined as plant and animal resources as well as wastes generated from their transformations. It covers lignocellulosic materials, including wood and agricultural residues, as well as nonlignocellulosic materials like food supply chain residues (farm and food processing residues) and municipal solid wastes (MSW). These different biomass resources could be converted directly or after a pre-treatment/conversion process into high value-added products, including energy, chemicals, and materials.

Among the various thermochemical processes, pyrolysis is a quite simple, low-cost, and robust technology for converting biomass into three different fractions, namely bio-oil, syngas, and biochar. Syngas could be combusted directly using a gas burner to deliver heat for the pyrolyzer. Bio-oil can be collected, stored, and used as fuel in existing industrial boilers. Biochar can be used as an energy source in the existing combustion system as an alternative for solid fuels. In addition, biochar is an

Char and Carbon Materials Derived from Biomass. https://doi.org/10.1016/B978-0-12-814893-8.00001-8

effective adsorbent for both organic and mineral contaminants. It also can be considered as a source of nutrients and therefore its addition to soils leads to significant environmental and agronomic benefits.

The robust nature of the pyrolysis conversion process allows substantial flexibility in both the type and quality of biomass resources. This chapter first provides the basics of the different biomass feedstocks transformed to chars through thermochemical conversion and hydrothermal techniques, and then goes on to classify these various possible feedstocks from their origin to their transformation processes. In addition, their availability, characteristics, conversion operating conditions, and possible applications are discussed.

1.2 Main categories of biomass

Fig. 1 shows the main sources of biomass and pathways used for char production. The factors influencing production and char applications are clearly identified. Biomass can be classified depending on its nature or its composition, but also from the outfalls

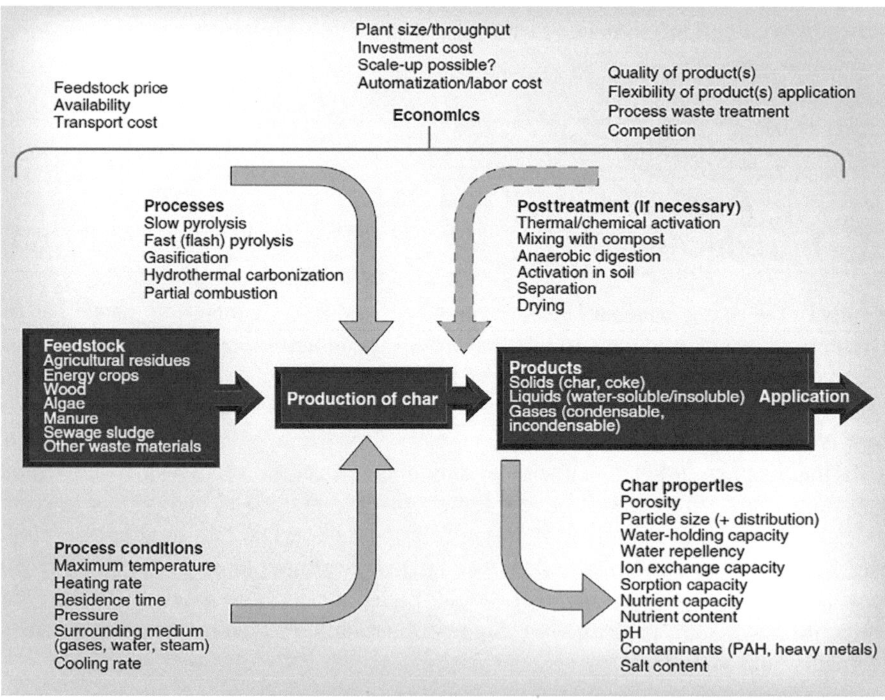

Fig. 1 Biomass feedstocks, char production pathways, and factors influencing potential applications.
Reprinted with permission from Libra, J. A., Ro, K. S., Kammann, C., Funke, A., Berge, N. D., Neubauer, Y., et al. (2011). Hydrothermal carbonization of biomass residuals: a comparative review of the chemistry, processes and applications of wet and dry pyrolysis. *Biofuels*, *2*, 71–106. Copyright 2018, Taylor and Francis.

that can be considered. In this chapter, we have separated the different biomass sources into five categories to better describe their potential from their composition and their characteristics.

The main raw materials converted to char via thermochemical conversion processes were traditionally wood and woody biomass. These woody resources were harvested or collected directly in forestry or generated as by-products of the wood industry. They include stems, branches, foliage, bark, chips, lumps, pellets, briquettes, sawdust, sawmills, and other wastes from various woody species. Biomass volumes associated with these sources are significant.

The second feedstock for char production is agricultural biomass. This may be categorized into agriculture residues and energy crops. Agriculture residues include primary by-products such as corn stover, cereal straw, and secondary by-products from biomass processing such as coffee husk, olive stone, and sugarcane bagasse. Energy crops are grown crops dedicated to biofuel and bioproduct production. The harvests include energy plantations, such as poplars, willows, and eucalyptus, herbaceous crops, such as sugarcane, sorghum, and artichokes, and vegetable oil plants, such as rapeseed and sunflowers.

Aquatic biomass is considered as the third generation of biofuel; it is also used for food processing or as material for wastewater treatment. Most of the work devoted to the energetic valorization of aquatic biomass considers that they are very interesting sources of carbon for bio-oil production by fast or slow pyrolysis. Pyrolysis of this kind of biomass leads to high char yield and also to the production of very interesting and valuable organic products. Biomass-derived chars produced during the process can also be used as biofuel (combustion, gasification), but they present very interesting properties for heavy metal adsorption.

Livestock wastes (cattle, poultry, swine, and turkey manures, chicken litter and feathers, pig slurry, etc.) represent an important problem for industrial countries. The intensification of animal breeding and the massive use of antibiotics and nutritional supplements have led to the classification of these wastes as undesirable. They were spread on farmlands, but the evolution of their composition and their environmental impact have limited their outfalls.

Urban and industrial wastes constitute the last category of potential feedstock for chars considered in this chapter. They can be divided into two major subcategories: MSW and industrial wastes.

MSW are mainly composed of household (paper, glass, plastic, food, and metal) and commercial wastes, which can be eliminated in a classical way (incineration) or by landfilling in an adapted center. The pyrolysis of these wastes leads to the formation of refuse-derived fuel (RDF) mainly in the form of char, oil/wax, and gases. Industrial wastes include sewage sludge, paper sludge, cardboard, textile residues (textile and dyeing sludge), medicinal residues, waste lignin, black liquor, and automotive residues (plastics, rubbers, textiles, foams, fibers, metals, glasses, etc.).

In the following sections we will describe the properties of the different biomass categories (proximate and ultimate analyses, major components) as well as the experimental conditions used to transform these biomasses into chars. Different standard methodologies are necessary to obtain this information. Here is a list with their specific references: moisture content (EN 14774-2, 2009), heating value (EN 14918, 2009), fixed carbon and volatiles content (EN 15148, 2009), ash content (EN 14775, 2009), alkali, carbon, hydrogen and nitrogen, sulfur, chlorine, and fluorine content (EN 15104, EN 15289, 2011), and density and particle size distribution (EN 15103, 2010 and CEN/TS 15149, 2006).

1.3 Woody biomass

Wood is the oldest form of biomass used for the production of charcoal. Woody biomasses are derived from natural forests, natural woodlands, and forestry plantations. These biomass resources include sawdust and other residues from forests (land clearing, logging debris, etc.) and wood processing activities (furniture and building materials manufacture) (Fig. 2). Fuelwood is the principal source for small-scale industries in rural areas of developing countries and for large-scale industries in urban areas. However, while for the former no regular economic circuit is set up, and much of the wood often falls outside the commercial market, the latter are well organized, thus leading to important cost reductions with fewer environment impacts.

Large amounts of woody wastes from different species all over the world and from various industrial processes are generated in the form of sawdust, chips, barks, shavings, etc. Increasingly, these quantities tend to be a potential resource for energy generation and chemical production.

According to their regional availability, they might be rubber wood sawdust (such as in South-East Asia and especially in Malaysia), hornbeams (in the Northern Hemisphere and mainly in Europe for the *Carpinus betulus* L. species), or reed along coastlines in many parts of the world (Estonian and southern Finnish coastal areas). Many other examples may be cited, including varieties of oak, pine, fir, or beech, the

Fig. 2 Woody biomass residues from forest activities.

harvests of which in European countries could easily reach hundreds of millions of tons per year in the form of wood pellets and logs.

Apart from their abundance, woody species' characteristics may come into play. For instance, the height and rapid growth of the European hornbeam, up to 30 meters within only 5–7 years, the deciduous character of maple or cypress trees, or the low weight of the dry reed are key parameters making these materials prime energy resources. As far as the reed feedstock is concerned, it is worth mentioning that its energy production potential is modest; this is why it is often blended with other types of wood.

Several processes at different operating conditions are applied to obtain chars with disparate properties and features. Table 1 reviews these aspects as published by several investigations. An example of each application (as nutrient, fuel, and adsorbent) is detailed hereafter.

Flash carbonization at 800°C and 1 MPa of the upper branches of *Eucalyptus camaldulensis* helped yielding biochar that improved soil properties and corn growth in Thailand (Butnan et al., 2015). Chips from Estonian reed and pellets from Bavarian pine underwent slow pyrolysis at a constant heating rate of 20°C/min until 800°C during 15 min to obtain chars intended for gasifiers (Link et al., 2010).

Before the activation step in the process of activated carbon production, chars are produced from the thermal treatment at 600°C during 4 h of silver fir, holm oak, stone pine, and Pyrenean oak, which are very common trees in Spain.

The woody feedstock has very comparable contents of lignin polymers, whatever its origin. They range from 28% for Turkish hornbeam to 31.2% for Spanish stone pine. The rest of the macromolecules have quite different proportions. Hemicellulose ranges from 11.5% for hornbeam to 25.9% for Pyrenean oak, while cellulose contents vary between 33.9% and 52.1%. Extractive contents usually do not exceed 10% in all selected examples of woody biomasses (Table 2).

The ultimate and proximate analyses (Table 3) as well as ash chemical compositions (Table 4) of the usually used woody biomasses suggest that these latter are excellent candidates to yield chars of good quality, either especially for energy purposes due to highly volatile matter (around 80%) or for interesting contents of carbon (more or less half of the CHONs). The presence of oxygen contributes to the growth of oxygen-containing functional groups, which may be helpful for specific pollutant capture in adsorption processes.

1.4 Agricultural biomass

Huge quantities of agricultural biomasses are produced annually worldwide and their collection and disposal are problematical and costly. Therefore, large amounts are left unused as wastes or burned directly in the agricultural fields creating serious environmental hazards. Currently, the utilization of agricultural wastes as feedstocks for char production is recognized as an attractive route for their management and for the development of high added value products that can meet farming and industry demands.

Table 1 Different wood species and techniques used for char production in the literature

Species	Origin	Form/ part	Conversion process	Operating conditions	Applications	References
Eucalyptus camaldulensis	Khon Kaen province, Thailand	Upper branches	Traditional kiln	350°C, low O_2 concentration, 3 days	Corn growth soil properties	Butnan, Deenik, Toomsan, Antal, and Vityakon (2015)
			Flash carbonization	1 MPa, 800°C, 40 min		
Wood meal, lignin cellulose, xylose	Commercial		Hydrothermal carbonization	3:1 water/biomass ratio (w/w), 225–265°C, 20h	Properties characterization	Kang, Li, Fan, and Chang (2012)
Maple wood	United States	Shavings	Thermal treatment	25°C/min, 300–700°C, 2h	Properties characterization	Cao et al. (2012)
Pine wood	Soria, Spain	Pellets	Thermal treatment	350–450–550°C	Properties characterization	Alburquerque, Sanchez, Mora, and Barron (2016)
Mallee wood	Australia		Slow pyrolysis	5 g, 10°C/min, 700–900°C, 15min	Properties characterization	Asadullah, Zhang, and Li (2010)
			Fast pyrolysis	5 g, 1000°C/s, 700–900°C, 30min		
Jeffrey pine/white fir	Tahoe Forest, United States	Chips	Hydrothermal carbonization	8:1 water/biomass ratio (w/w), 255°C, 30min	Properties characterization	Hoekman, Broch, and Robbins (2011)
Rubber wood	Malaysia	Sawdust	Slow pyrolysis	10g, 5°C/min, 300–700°C, 1–3h	Properties characterization	Shaaban et al. (2014)

Acacia and eucalyptus wood	Varanasi, India	Cubic shape	Slow pyrolysis	5–30°C/min, 270–1200°C, 1 h	Properties characterization	Kumar, Verma, and Gupta (1999)
Aspidosperma quebracho-blanco, *Aspidosperma australe*	Argentina	Sawdust	Slow pyrolysis	5°C/min, 350°C (3 h), 850°C (1 h)	Properties characterization	Della Rocca, Cerrella, Bonelli, and Cukierman (1999)
Hornbeam	Turkey	Sawdust	Slow pyrolysis	15 g, 40°C/min, 400–600°C, 30 min	Properties characterization	Moral, Yavuzel, and Sensöz (2016)
Reed	West Coast, Estonia	Chips	Slow pyrolysis	1 L/min, 20°C/min, 800°C, 15 min	Gasification	Link et al. (2010)
Douglas fir	Bavaria,	Chips				
Pine	Germany	Pellets				
Loblolly pine	Georgia, United States	Wood, bark, residue	Rapid pyrolysis	4 g, 160°C/min, 400–600°C	Properties characterization	Ben, Hao, Liu, and Ragauskas (2017)
Silver fir Holm oak Stone pine Pyrenean oak	Catalonian Pyrenees, Spain	Chips	Thermal treatment	600°C, 4 h	Activated carbon production	López, Centeno, García-Díaz, and Alguacil (2013)
Cypress	Tunisia	Sawdust	Slow pyrolysis	5°C/min, 500°C, 1 h	Biofertilizer	Haddad et al. (2017)
Beech wood	France	Sawdust	Fast pyrolysis	500–1400°C, 2–5 s	Reactivity under O_2	Guizani, Jeguirim, Valin, Limousy, and Salvador (2017)

Table 2 The content of major components in different wood species

	Hemicellulose	Cellulose	Lignin	Extractive	References
Silver fir	15.0	52.1	29.9	2.4	López et al. (2013)
Holm oak	25.9	37.9	27.8	4.8	
Stone pine	20.1	41.0	31.2	6.8	
Pyrenean oak	25.5	33.9	31.2	5.2	
Hornbeam	11.5	51.5	28.0	9.0	Moral et al. (2016)
Pine wood	23.0	45.5	28.0	2.5	Alburquerque et al. (2016)

Table 3 Proximate and ultimate analyses of different wood species

	Proximate analysis (wt%)[a]			Ultimate analysis (wt%)[b]				Reference
	Ash	VM	FC	C	H	O[c]	N	
Mallee wood	0.9	80.9	18.2	48.2	6.1	45.5	0.2	Asadullah et al. (2010)
Maple wood	1.0	*	*	53.4	6.8	39.8	0.0	Cao et al. (2012)
Rubber wood	2.3	72.6	25.1	44.4	7.0	46.3	0.4	Shaaban et al. (2014)
Aspidosperma quebracho	0.4	75.9	16.2	57.7	6.3	35.1	0.9	Della Rocca et al. (1999)
Aspidosperma australe	1.0	72.7	19.3	48.2	5.6	45.3	0.9	
Hornbeam	0.4	78.1	21.5	45.2	6.6	48.2	–	Moral et al. (2016)
Reed	3.2	80.3	16.5	49.1	5.9	44.5	0.5	Link et al. (2010)
Douglas fir	6.7	72.3	21.0	52.0	5.3	42.0	0.7	
Pine	0.2	83.3	16.5	50.6	5.9	43.1	0.4	
Silver fir	0.4	78.7	20.9	51.2	6.4	42.2	0.2	López et al. (2013)
Holm oak	2.3	80.8	16.9	48.0	5.9	45.6	0.5	
Stone pine	0.7	82.1	17.2	50.4	6.0	43.3	0.3	
Pyrenean oak	2.4	80.5	17.1	48.5	5.9	45.1	0.5	
Cypress wood	0.6	80.3	19.1	51.6	6.4	41.8	0.2	Haddad et al. (2017)
Beech wood	0.5	84.3	15.2	49.4	5.7	44.7	0.2	Guizani et al. (2017)

VM, volatile matters; *FC*, fixed carbon.
[a]Dry basis.
[b]Dry ash free.
[c]By difference.
*Not provided.

A significant number of feedstocks have been evaluated for char production. Table 5 presents various agricultural wastes, conversion techniques for char production, and potential applications. These wastes may be generated directly in the field during agricultural operations (crops harvesting and tree pruning) or may be derived from biomass processing (oil extraction, juice extraction).

Table 4 Ash chemical analysis, expressed as oxides for different wood species

	SiO_2	CaO	P_2O_5	Na_2O	MgO	SO_3	Fe_2O_3	K_2O	Al_2O_3	Reference
Reed	73.7	2.9	*	8.4	1.4	*	1.1	5.9	*	Link et al. (2010)
Douglas fir	55.5	10.9	*	1.5	4.9	*	7.2	4.1	11.6	
Pine	11.2	37.3	*	1.3	9.1	*	4.8	7.7	5.0	

*Not provided.

Quantities of rape seeds and stalks as well as sunflower residues are important because their plantations are among the largest source of vegetable oil in the world. These wastes are found in many countries such as Canada, China, India, and Germany for the former and in Russia, Argentina, South Africa, and Turkey for the latter. The European continent is the number one sunflower and rape oil producer with over 400 million tons of seeds, half of which is lost during the crushing process. The trituration by-products can be used for the production of bio-oil, biodiesel, syngas, and biochar fuels.

Another very promising species as an energy resource is *Jatropha curcas* L. Indeed, the extraction process from the seeds of this plant gives oil that can be directly used as fuel for slow-speed diesel engines. Transesterification operation would upgrade this oil to make it suitable for conventional diesel. These last two processes generate agricultural wastes that may be converted into char or into activated carbon through slow pyrolysis.

In countries like Russia, Turkey, Poland, or Ukraine, cherry stones, seeds, and shells may also be candidates for char production because of their high contents of minerals. This is also the case with waste pomegranate peel in subtropical countries like India, Pakistan, and South Africa, where the total quantities are over 1 million ton/year.

Sugarcane bagasses, coco peats, paddy straws, palm kernel shells, empty fruit bunches, wheat, and corn straws (Fig. 3) are fibrous residues that are ideal feedstocks for char production, because their classical ways of management do not lead to high economic added value (litter, livestock feed). Different applications are conducted for the so-obtained chars: adsorption, soil amendment, gasification, etc.

The main component of grasses (such as red canary grass, switchgrass, *Miscanthus*, or even sugarcane), which is used in biofuel production, is the lignocellulosic component. This same part is also of interest while looking for char feedstock, especially in areas where this kind of vegetation is widespread. For example, the top three countries for sugarcane are Brazil, India, and China, while it is in the United States where switchgrass is most found, especially in dedicated crops for agrofuels. Africa and South Asia are the *Miscanthus* native areas.

The lignocellulosic component of the tree branches and trunks is still the reason why these residues are used to produce biomass chars, instead of the ancient methods of burning that have harmful impacts on the environment. According to the climate and the region, they might be discharged from apple trees, palm dates, or cotton shrubs.

Table 5 Different agricultural residues and techniques used for char production in the literature

Agriculture residues	Origin	Conversion process	Operating conditions	Applications	References
Jatropha curcas	Oil extraction process	Slow pyrolysis	20°C/min, 400–800°C, 15–240 min	Activated carbon production	Sricharoenchaikul, Pechyen, Ahtong, and Atong (2008)
Rape Sunflower	Leon, Spain	Slow pyrolysis	30 g, 30°C/min, 550°C, 1 min	Biochar for soil management	Sanchez, Lindao, Margaleff, Martinez, and Moran (2009)
Corn straw	Beijing, China	Pyrolysis	200°C (1 h) followed by 450°C (4 h)	Atrazine adsorption	Zhao et al. (2013)
Cherry pulp	Juice factory, Bursa/Turkey	Pyrolysis	400–700°C, 10–100–200°C/min	Properties characterization	Pehlivan, Ozbay, Yargiç, and Sahin (2017)
Citrus peel waste	Juice factory, Messina, Italy	Pyrolysis	150–650°C, 50°C/min, 30 min	Properties characterization	Volpe, Bermudez Menendez, Ramirez Reina, Messineo, and Millan (2017)
Sugarcane bagasse Coco peat Paddy straw Palm kernel shell	Indonesia	Slow pyrolysis	100–400 g, 10°C/min, 500°C, 1 h	Properties characterization	Lee, Park, et al. (2017)
Waste pomegranate peel	Australia	Slow pyrolysis	15°C/min, 300–600°C; 20–60 min	Properties characterization	Siddiqui et al. (2017)

Grape vine	Pellets, Austria	Pyrolysis	400–500°C	Properties characterization	Colantoni et al. (2016)
Sunflower husk					
Palm kernel shell	Palm oil mill, Selangor, Malaysia	Slow pyrolysis	10°C/min, 500°C, 1 h	Properties characterization	Lee, Lee, Gan, Thangalazhy-Gopakumar, and Ng (2017)
Empty fruit brunch					
Apple tree branches	Weibei, China	Slow pyrolysis	10°C/min, 300–600°C, 2 h	Properties characterization	Zhao, Ta, and Wang (2017)
Olive pomace	Mahdia, Tunisia	Slow pyrolysis	450°C	Gasification	Tamošiūnas et al. (2017)
Red canary grass	United Kingdom	Slow pyrolysis	150–300 g, 350–700°C, 10°C/min, 1 h	Properties characterization	Ryu, Sharifiz, and Swithenbank (2007)

Fig. 3 Agricultural biomass residues. (A) Corn straw, (B) palm kernel shell, (C) sugarcane bagasse, and (D) coco peat.

There is also a growing interest in using grape vine, olive pomace (Fig. 4), and citrus fruit pulps in Mediterranean countries as raw or pretreated materials for char production.

Tables 6, 7, and 8, which give the chemical compounds, explain why all these materials may have a more interesting fate as chars. More precisely, the range of cellulose

Fig. 4 Dried algae, Spirulina sp.
Reprinted with permission from Chaiwong, K., Kiatsiriroat, T., Vorayos, N. Thararax, C. (2013), Study of bio-oil and bio-char production from algae by slow pyrolysis. *Biomass and Bioenergy*, *56*, 600–606. Copyright 2018, Elsevier.

Table 6 The content of major components in different agricultural residues

	Hemicellulose	Cellulose	Lignin	Extractive	References
Jatropha curcas	17.5	56.3	23.9		Sricharoenchaikul et al. (2008)
Cherry pulp	9.1	24.8	49.3	6.4	Pehlivan et al. (2017)
Waste pomegranate peel	8.1	7.9	22.1	57.9	Siddiqui et al. (2017)
Palm kernel shell	14.2	27.5	58.3	7.5	Lee, Lee, et al. (2017), Lee, Park, et al. (2017)
Empty fruit brunch	38.4	24.2	37.3	15.7	
Palm oil sludge	34.7	35.3	30.0	5.6	

content is quite wide, starting from 7.9% for waste pomegranate peel to 56.3% for *J. curcas*. Hemicellulose for the former is still low (8.1%) but extractives are the most registered component (57.9%), as reported by Siddiqui et al. (2017). According to the proximate analysis, a third of the weight would be transformed into char.

All the potential feedstocks reported in Table 5 have carbon contents ranging from two- to three-fifths of all elements contained in the plant. These percentages indicate high output of char when these residues are treated thermally. The ash chemical analysis shows the suitability of straws, husks, and stalks to biochar production and utilization for soil abilities enhancement.

1.5 Aquatic biomass

Aquatic biomass resources are promising feedstocks for terrestrial biomass substitution to avoid competition with arable land and to minimize excessive nutrient use for conventional agriculture. In addition, these resources have high growth rates due to no water limitation, high photosynthetic efficiency (6%–8%) compared to terrestrial biomass (1.8%–2.2%), high productivity, and efficient CO_2 fixation.

Aquatic biomass can be classified according to its origin (marine, freshwater) or its development process (cultivation, eutrophication). Several attempts in the literature were performed to elaborate chars from aquatic biomass, including microalgae, macroalgae, and marine and freshwater plants. The main investigations are summarized in Table 9.

Algae culture and seaweed farming were mainly oriented for human and animal nutrition, medicines, cosmetics, and fertilizers. The recent interest in some of these species as promising feedstock for bioenergy, biorefinery, or biosorption is justified by their high energy density and high uptake of inorganic carbon, notably among other algae, the *Saccharina japonica* and *Undaria pinnatifida* brown algae, the *Spirulina* sp.

Table 7 Proximate and ultimate analyses of different agricultural residues

	Proximate analysis (wt%)[a]			Ultimate analysis (wt %)[b]				
	Ash	VM	FC	C	H	O[c]	N	References
Rapeseed stalk	4.8	85.3	9.9	49.9	6.2	43.2	0.5	Gao et al. (2017)
Jatropha curcas	6.3	55.4	37.2	45.5	7.2	43.3	4.0	Sricharoenchaikul et al. (2008)
Rapeseed	7.3	78.7	14	44.7	5.8	48.1	0.8	Sanchez et al. (2009)
Sunflower	8.3	74.5	17.2	43.6	5.8	49.3	1.0	
Corn straw	5.2	*	*	45.0	5.2	48.9	0.9	Zhao et al. (2013)
Lemon pulp	2.0	73.6	24.4	47.7	5.2	45.7	1.3	Volpe et al. (2017)
Orange pulp	2.5	69.3	28.2	48.5	5.2	45.2	1.1	
Maize stalk	6.0	80.8	13.2	49.3	6.5	42.6	1.6	Fu et al. (2011)
Rice straw	15.0	70.7	14.3	47.5	6.5	44.9	1.1	
Cotton straw	2.9	80.9	16.2	49.4	6.3	43.5	0.8	
Rice husk	18.3	65.6	16.1	50.5	6.9	42.1	0.6	
Sugarcane bagasse	2.4	81.8	15.8	51.7	5.3	42.7	0.3	Lee et al. (2013)
Coco peat	5.8	62.2	32.0	61.6	4.4	33.0	1.0	
Paddy straw	22.5	60.8	16.7	48.8	6.0	43.3	2.0	
Palm kernel shell	3.9	75.8	20.3	55.8	5.6	37.8	0.8	
Pomegranate peel	3.5	68.7	27.8	46.0	5.5	39.0		Siddiqui et al., 2017
Palm kernel shell	7.9	66.0	16.1	55.9	*	41.2	*	Lee, Lee, et al. (2017), Lee, Park, et al. (2017)
Empty fruit brunch	11.5	75.5	13.0	50.9	*	42.4	*	
Palm oil sludge	45.5	29.9	24.6	53.2	*	33.0	*	
Date palm trunk	5.4	70.8	24.8	44.9	6.1	48.6	0.2	Jeguirim, Elmay, Limousy, Lajili, and Said (2014)
Red canary grass	6.0	76.9	17.1	50.0	7.0	40.9	1.3	Ryu et al. (2007)

[a]Dry basis.
[b]Dry ash free.
[c]By difference.
*Not provided.

algae (Fig. 4), the *Posidonia oceanica* sea plant, the Pacific kelp, and the lacustrine freshwater algae.

A very interesting potential in this category comes also from *Ulva prolifera*, which is an invasive marine macroalgae that causes big problems at both environmental and economic levels, as was the case in China's Yellow Sea in 2008 when 10^9 m^3 of this

Table 8 Ash chemical analysis, expressed as oxides for different agriculture residues

	SiO_2	CaO	P_2O_5	Na_2O	MgO	SO_3	Fe_2O_3	K_2O	Al_2O_3	Reference
Maize stalk	57.7	10.3	7.2	0.5	4.0	*	1.4	8.3	6.4	Fu et al. (2011)
Rice straw	66.5	4.4	4.8	0.2	3.6	*	0.3	16.1	0.9	
Cotton straw	6.3	48.8	14.8	0.2	6.8	*	1.3	16.6	0.5	
Rice husk	87.8	1.2	0.9	1.8	0.5	*	0.1	4.7	0.9	

*Not provided.

species proliferated in one summer. Since then, this quantity continues to increase year after year. Lu et al. (2017) investigated its ability to adsorb bisphenol, after a hydrothermal carbonization at 200°C for 4h. The outputs of their study confirmed their interesting potential as biosorbents.

Microalgae are another group of prokaryotic and eukaryotic photosynthetic and heterotrophic microorganisms, the cultivation of which is intensive in different parts of the world. The success of algaculture is due to their composition, which is favorable to produce biochar suitable for soil fertilization and for effluent purification by adsorption. As an example of this last application, pyrolytic char from *Scenedesmus dimorphus* was tested by Bordoloi et al. (2017) in cobalt adsorption. Satisfactory results were achieved, thanks mainly to the microalgae composition rich in protein (53.13%), lipid (10.23%), and carbohydrate (22.90%).

Such high contents, as seen in Table 10 for other biomasses, as well as the high ratios of C/H drawn from Table 11 (between 8 and 10), make them versatile raw materials for the production of different biofuels (solid, liquid, and gas). Furthermore, energy can be recovered even after extracting oils and carbohydrates.

Different configurations are used to exploit the potential of microalgae (closed photoreactors, open ponds, covered basins, etc.) whether in fresh or in very salty waters. Kirtania and Bhattacharya (2015) assessed the *Chlorococcum humicola*-derived char performance for gasification after submitting this green freshwater microorganism to a fast pyrolysis at 1000°C in an entrained flow reactor. For many algae, higher productivity is expected by intensive systems compared to land-use processes, reaching 100ton/ha/year with less allocation and higher payoff (Leng et al., 2015).

The valorization of water hyacinth helps to hit two targets with one bullet. Indeed, this tropical and subtropical perennial and invasive aquatic plant is a potential bioenergy vector thanks to its elemental composition and proximate analysis (Lin et al., 2018), which affix it high reactivity when processed by slow pyrolysis. These authors discovered that, with a 30°C/min heating rate in N_2 for 10–20min, the obtained char becomes a good agent for wastewater treatment because it can effectively adsorb heavy metals such as chromium.

Table 9 Different aquatic biomasses and techniques used for char production in the literature

Species	Origin	Conversion process	Operating conditions	Applications	References
Ulva prolifera	Qingdao City Beach, China	Hydrothermal carbonization	Liquid/solid ratio: 15/1, 200°C, 4h	Bisphenol A adsorption	Lu, Zhang, Wu, and Luo (2017)
Scenedesmus dimorphus	Tezpur, India	Slow pyrolysis	40°C/min, 500°C, 100mL/min	Cobalt adsorption	Bordoloi, Goswami, Kumar, and Katak (2017)
Spirulina, *Platensis*, sp.	Fujian, China	Hydrothermal liquefaction	7.89g SP, 100mL ethanol, 350°C, 20min	Methylene blue adsorption	Leng et al. (2015)
Chlorella spp. *Nannochloropsis*	Ingrepro B.V. (Netherlands)	Slow pyrolysis	100 mg, 15°C/min, 350°C	Properties characterization	Rizzo, Prussi, Bettucci, Libelli, and Chiaramonti (2013)
Saccharina Japonica	Wando Island, Korea	Slow pyrolysis	30g, 430–530°C, 4–10min	Properties characterization	Choi et al. (2016)
Chlorococcum humicola	Monash University, Australia	Fast pyrolysis	1000°C, entrained flow reactor	Gasification	Kirtania and Bhattacharya (2015)
Posidonia oceanica	Italy	Slow pyrolysis	1g, 30°C/min, 400–600°C, 45min	Properties characterization	Chiodo, Zafarana, Maisano, Freni, and Urbani (2016)
Lacustrine alga					
Sargassum sp.	Vietnam	Slow pyrolysis	1g, 360–380°C, 5–20°C/min, 1–4min	Properties characterization	Kim, Bae, and Lee (2013), Kim, Ly, Kim, Choi, and Woo (2013)

Chlorella vulgaris *Scenedesmus obliquus* *Nannochloropsis oculata*	Algae for Future, Pataias, Portugal	Slow pyrolysis	6 mg, 5–20°C/min, 1000°C	Reactivity characterization	Ferreira, Soares Dias, Silva, and Costa (2015)
Water hyacinth	Tongan Lake, Xiamen, China	Slow pyrolysis	15 g, 30°C/min, 300–600°C, 10 min	Chromium adsorption and reactivity	Lin et al. (2018)
Scenedesmus almeriensis *Nannochloropsis gaditana* *Chlorella vulgaris*	Algaenergy, Spain	Slow pyrolysis	40°C/min, 1000°C, 15 min	Reactivity	Lopez-Gonzalez, Fernandez-Lopez, Valverde, and Sanchez-Silva (2014)

Table 10 The content of major components in aquatic biomasses

	Protein	Carbohydrate	Lipid	Fibers	References
Scenedesmus dimorphus	53.1	22.9	10.2		Bordoloi et al. (2017)
Spirulina platensis	63.4	20.5	7.5		Leng et al. (2015)
Chlorella spp.	29.6	15.0	9.0	5.0	Rizzo et al. (2013)
Chlorella vulgaris	58.1	12.4	13.5		Lopez-Gonzalez et al. (2014)
Scenedesmus almeriensis	44.2	25.2	24.6		
Nannochloropsis gaditana	40.5	25.1	26.3		

In general, the macroalgae are lower in C, H, and O than terrestrial biomass and higher in N and S (Table 11). The organic constituents in macroalgae differ chemically from terrestrial biomass. The high protein content (Table 10) is associated with higher nitrogen content. The main component is the carbohydrate fraction, in the absence of lignin and cellulose.

Ash content varies in different species (brown seaweeds have higher contents) and with geographical location. Table 11 shows that ash contents fall in the range 6–40 wt % with larger amounts of macrominerals (Na, K, Ca, P, Si, and Mg) and trace elements (Fe, Zn, Mn, Al, and Cu) than terrestrial biomass. The Na and K contents often dominate the ash components and as such may cause severe problems when used in thermochemical processes.

Finally, differences in amounts of chlorine in the ashes may be attributed to whether the algae were grown in fresh or marine water (Table 12). As an example, the *Nannochloropsis oculata* marine algae have a higher content compared to the freshwater *Chlorella vulgaris* algae.

1.6 Livestock wastes

Excessive animal farming worldwide has led to a huge concentration of animals in small areas. This intensive animal farming generates superfluous amounts of manure that are still unused to insufficient neighboring farmland surface (Fig. 5). Therefore, serious environmental problems are generated, including nitrogen and phosphorous water pollution, bad smells, and methane emissions. Therefore, char production represents an interesting technology for animal manure management, offering the possibility to reduce and valorize these wastes. Table 13 presents the main investigations on char elaboration from different animal manures using thermochemical conversion processes and hydrothermal techniques.

Table 11 Proximate and ultimate analyses of different aquatic biomasses

	Proximate analysis (wt%)[a]			Ultimate analysis (wt%)[b]					
	Ash	VM	FC	C	H	O[c]	N	S	References
Spirulina platensis	37	57.4		43.4	5.9	47.5	3.2	*	Leng et al. (2015)
Chlorella spp.	21.6			58.8	7.8	24.4	8.5	0.5	Rizzo et al. (2013)
Nannochloropsis	13.2			55.5	7.9	27.5	8.6	0.5	
Saccharina Japonica	23.6	72.9	3.5	54.7	5.7	37.1	2.0	0.5	Choi et al. (2016)
Chlorococcum humicola	28.5	57.6	14.9	45.3	7.6	37.2	6.6	3.3	Kirtania and Bhattacharya (2015)
Posidonia oceanica	19.0	59.5	21.5	42.1	4.5	48.9	1.5	0.2	Chiodo et al. (2016)
Lacustrine alga	40.1	27.6	32.3	33.2	4.1	54.9	4.1	0.4	
Sargassum sp.	40.9	49.4	9.7	42.4	6.7	48.5	2.1	0.3	Kim, Bae, et al. (2013), Kim, Ly, et al. (2013)
Chlorella vulgaris	11.6	76.4	12.0	47.1	6.8	35.1	10.3	0.7	Ferreira et al. (2015)
Scenedesmus obliquus	8.2	80.3	11.5	50.3	7.3	33.5	8.3	0.6	
Nannochloropsis oculata	14.4	75.9	9.7	48.1	7.1	36.1	7.8	0.9	
Water hyacinth	8.7	74.9	17.4	43.1	5.9	55.9	4.0	*	Lin et al. (2018)
Scenedesmus almeriensis	20.1	69.9	10.0	41.9	6.7	44.7	5.9	0.8	Lopez-Gonzalez et al. (2014)
Nannochloropsis gaditana	6.6	82.7	10.7	49.4	7.7	34.7	7.0	1.1	
Chlorella vulgaris	16.6	70.4	13.0	44.8	6.8	40.4	7.0	1.0	

[a]Dry basis.
[b]Dry ash free.
[c]By difference.
*Not provided.

Table 12 Ash chemical analysis, expressed as oxides for different aquatic biomasses

	SiO_2	CaO	P_2O_5	Na_2O	MgO	SO_3	Fe_2O_3	K_2O	Al_2O_3	References
Chlorococcum humicola	21.7	21.5	28.9	8.1	8.6	5.4	5.2	1.7	0.3	Kirtania and Bhattacharya (2015)
Chlorella vulgaris	5.8	14.6	40.4	1.4	5.5	1.5	8.9	20.6	0.1	Ferreira et al. (2015)
Scenedesmus obliquus	9.9	11.8	31.3	1.5	5.4	4.1	6.4	26.9	1.8	
Nannochloropsis oculata	0.2	12.4	19.1	25.3	4.9	1.2	1.8	9.9	0.2	
Fucus vesiculosus	1.3	5.7	21.9	15.2	4.9	*	1.3	17.3	0.9	Ross, Jones, Kubacki, and Bridgeman (2008)
Chorda filum	1.0	14.1	16.7	21.0	10.2	*	2.2	6.2	0.4	
Laminaria digitata	0.4	5.0	6.7	19.6	5.2	*	1.0	14.8	0.1	
Fucus serratus	1.5	11.5	12.5	11.8	5.3	*	1.6	11.4	1.1	
Laminaria hyperborea	0.5	16.5	5.4	6.1	40.2	*	7.9	0.5	1.5	
Macrocyctis pyrifera	1.5	10.7	7.0	17.6	4.2	*	1.2	7.6	0.8	

*Not provided.

Fig. 5 Poultry feather waste.

Even when poultry, chicken, broiler, turkey, and hen wastes are used as fertilizers, odorous compounds are still reported, greenhouse gases are again emitted, and more importantly pathogens are spread. The corresponding industries are now unable to properly manage the surplus of litter, manure, meat, and compost. Some authors applied different conversion methods such as hydrothermal carbonization (Cao et al., 2011), hydrothermal liquefaction (Kantarli et al., 2016), slow pyrolysis (Zhang et al., 2016), intermediate pyrolysis (Baniasadi et al., 2016), fast pyrolysis (Kim et al., 2009), gasification (Joseph et al., 2002), and cogasification (Ng et al., 2017) to these kinds of wastes. Adsorption of acesulfame, saccharin, cyclamate, or copper, sorption of microcystin-LR or deisopropylatrazine, reactivity investigation, and product characterization are the opportunities investigated by the above-mentioned authors for yielded chars. Encouraging results were published.

Pig and swine slurry and manure as well as cattle wastes are also of interest when it comes to char feedstocks. Cely et al. (2015) investigated some of these wastes pyrolyzed in a poor oxygen atmosphere with a constant heating rate of 10°C/min until 300–500°C for agronomy applications, whereas Zhang et al. (2016) studied the slow pyrolysis of pig compost for energy purposes.

In general, biochars produced from slow pyrolysis at low temperature (300–600°C) are to be used as fertilizers for their agronomic properties and their impact on microbial growth or as adsorbents (swine wastewater treatment, mercury vapor adsorption, microsystin–tetracycline–atrazine, and *p*-cresol adsorption). Another type of study is mainly focused on the conversion of these wastes in biogas, bio-oil, and biochar. The experimental conditions favor fast pyrolysis and higher temperatures (600–900°C).

Once again, the proximate and ultimate analyses (Table 14) as well as ash composition (Table 15) prove the suitability of livestock wastes for biochar as an amendment or as a fuel. Indeed, the yield of the char is generally high (more than 35 wt% except for poultry meat for which the sum of fixed carbon and ash is 16.1%) and might almost reach 50 wt% as is the case for pig slurry in the work of Cely et al. (2015). No or very little sulfur is reported, which means that no post- or pretreatments are required during the charring process or char usage. The mineral rates, such as those of potassium and phosphor in the studied manures, are interesting for soils

Table 13 Different animal manures and techniques used for char production in the literature

Animal wastes	Origin	Conversion process	Operating conditions	Applications	References
Cattle manure Chicken manure Pig slurry	Avila region, Spain	Pyrolysis	Poor oxygen, 100 g, 10°C/min, 300–500°C	Agronomy applications	Cely, Gascó, Paz-Ferreiro, and Méndez (2015)
Chicken litter	Ireland	Gasification	Gasification, air, 500°C	Characterization	Joseph, Tretsiakova-Mcnally, and Mckenna (2002)
Swine manure	North Carolina, United States	Pyrolysis Hydrothermal carbonization	13°C/min, 620°C, 2 h Swine manure solution (20% solid), 0.7°C/min, 250°C, 20 h	Characterization	Cao, Ro, Chappell, Li, and Mao (2011)
Chicken manure	Chicken farm, Singapore	Cogasification with wood chips	Solid: 10 kg/h, air: 4 L/s, 850–900°C	Adsorption of acesulfame, saccharin, cyclamate	Ng et al. (2017)
Poultry meat Poultry litter	CP Group, Izmir, Turkey	Hydrothermal carbonization Hydrothermal liquefaction Pyrolysis	s/w ratio: 20/80, 5°C/min, 175–200°C, 0–60 min s/w ratio: 20/80, 5°C/min, 200–325°C, 30 min 5°C/min, 250–500°C, 1 h	Products characterization	Kantarli, Kabadayi, Ucar, and Yanik (2016)
Chicken litter Turkey litter	Shenandoah Valley, Virginia	Fast pyrolysis	450–500°C, vapor residence time: 0.5–5 s	Products characterization	Kim, Agblevor, and Lim (2009)

Poultry litter	RES, Ravenna, Italy	Slow/intermediate pyrolysis	30–120°C/min, 400–800°C, 30 min	Products characterization	Baniasadi et al. (2016)
Pig compost Hen compost	Japan	Slow pyrolysis	15°C/min, 700°C, until constant mass	Char reactivity	Zhang et al. (2016)
Broiler cake Broiler litter	Starkville, United States	Slow pyrolysis	700–800°C, 1 h	Copper adsorption	Lima, Boateng, and Klasson (2009)
Chicken manure	China	Slow pyrolysis	300–600°C, 1 h, low O_2	Microcystin-LR sorption	Li et al. (2018)
Broiler litter	Starkville, United States	Slow pyrolysis	350–700°C, 1 h	Deisopropylatrazine sorption	Uchimiya, Wartelle, Lima, and Klasson (2010)

Table 14 Proximate and ultimate analyses of different animal manures

	Proximate analysis (wt%)[a]			Ultimate analysis (wt%)[b]					
	Ash	**VM**	**FC**	**C**	**H**	**O[c]**	**N**	**S**	**References**
Cattle manure	14.6	64.7	20.7						Cely et al. (2015)
Cattle manure mixed with straw	26.5	56.6	17.0						
Chicken manure	20.2	64.6	14.9						
Chicken manure mixed with sawdust	17.4	59.8	22.8						
Pig slurry	47.2	41.7	11.1						
Swine manure	18.5	60.6	8.1	58.0	7.2	28.0	5.6	1.1	Cao et al. (2011)
Chicken manure	28.1	61.3	10.6	37.7	4.7	45.3	10.8	1.5	Ng et al. (2017)
Poultry meal	13.8	83.9	2.3	59.8	8.5	17.8	13.3	0.6	Kantarli et al. (2016)
Poultry litter	16.8	65.6	17.6	47.3	6.4	39.6	6.6	0.1	
Chicken litter	22.8			48.1	6.1	42.2	3.6	*	Kim et al. (2009)
Turkey litter	5.9			48.8	6.5	43.5	1.2	*	
Poultry litter	25.0	67.0	8.0	45.3	7.8	39.7	6.0	1.2	Baniasadi et al. (2016)
Pig compost	20.7	67.9	12.4	44.0	6.1	43.0	5.8	1.1	Zhang et al. (2016)
Hen compost	38.9	55.2	5.9	41.8	4.8	48.7	3.9	0.8	
Poultry litter	13.4			47.5	6.0	46.4	0.1	*	Das, Garcia-Perez, Bibens, and Melear (2008)

[a]Dry basis.
[b]Dry ash free.
[c]By difference.
*Not provided.

Table 15 Ash elemental weight (g/kg_{db}) for different animal manures

	Si	Ca	P	Fe	K	S	Na	Mg	Al	References
Pig compost	3.3	80.6	11.6	5.0	30.0	5.6				Zhang et al. (2016)
Hen compost	3.7	213.6	10.3	1.8	34.2	5.0				
Poultry litter	0.34	17.9	14.2	0.74	19.8	*	4.4	3.7	7.6	Das et al. (2008)

*Not provided.

compared to energy resources. In addition, activation of the resulting char may involve chemical bonds, ion exchange, or electrical attraction forces of these elements in adsorption processes.

1.7 Industrial and urban wastes

The accumulation of urban and industrial wastes in landfill has become a serious environmental and social burden worldwide. Therefore, minimization, material recovery, and energy production have become the priority for several management strategies. Char production from industrial and urban wastes could meet these management strategy objectives. The management of these wastes has been strongly impacted in Europe by the European Union Waste Landfill Directive 1999/31/EC, which defines the objectives of reducing the part of biodegradable wastes for landfilling (from 75% by 1995 to 35% by 2020). Among the various possibilities offered to manage these specific wastes (incineration, composting, anaerobic digestion), pyrolysis has attracted much attention during the last few years due to its low environmental footprint and the possibility of extracting valuable products and biofuels from the thermal conversion of these wastes under inert atmosphere. Pyrolysis of MSW generates refuse-derived char and RDF: gas and liquid) that can be valorized after specific pretreatments.

The amount of MSW generated worldwide represents a fantastic source of renewable energy that has to be considered. In Europe the production of MSW is estimated to be close to 500–700 kg/person/year (Gabbar, Aboughaly, & Ayoub, 2017). Lino and Ismail (2017) reported an estimation of the global generation of MSW given by the World Bank to be close to 1.3×10^9 ton/year in 2012, while this amount is expected to increase to 2.2×10^9 ton/year by 2025. Incineration, recycling, and composting are the main outfalls for developed countries with a homogeneous distribution for Europe, while incineration represents 80% for Japan. While the part of MSW directed to landfill is almost nonexistent for the developed European countries and Japan (0% for Germany and Switzerland), it represents more than 50% in China, the United States, Brazil, and Romania.

Table 16 Proximate and ultimate analyses of different industrial and municipal wastes

	Proximate analysis (wt%)[a]			Ultimate analysis (wt%)[b]						
	Ash	**VM**	**FC**	**C**	**H**	**O[c]**	**N**	**S**	**Cl**	**References**
Refuse-derived fuel	12.9	76.7	10.4	49.9	7.3	41.4	1.3	0.1	0.4	Hwang, Kobayashi, and Kawamoto (2014)
Refuse paper and plastic fuel	7.9	86.7	5.4	56.5	8.7	34.5	0.2	0.1		
Construction and demolition wood	0.5	86.5	13	50.7	6.5	42.7	0.1	0.0	*	
Medicinal residues	33.7	40.0	26.3	49.2	5.0	41.2	3.5	1.1	*	Li et al. (2015)
Municipal solid waste	46.7	17.5	35.8	59.0	6.3	28.7	3.3	0.3	2.4	
Textile dyeing sludge	61.6	37.0	1.4	40.4	8.9	40.8	6.3	3.6	*	Zhang et al. (2017)
Automotive shredder residues	17.0	77.1	5.9	63.7	9.1	21.3	2.4	3.5	2.3	Notarnicola et al. (2017)
Paper	10.9	79.6	9.5	40.4	5.6	53.9	0.04	0.02	*	Berge et al. (2011)
Food waste	7.6	77.6	14.8	45.9	6.3	44.2	3.5	0.02	*	
Digestion waste	35.9	55.9	8.2	50.9	7.5	31.4	8.6	1.4	*	
Municipal waste	28.4	62.0	9.6	39.8	5.3	54.0	0.78	0.07	*	
Used cotton fabrics	2.6	93.9	3.5	53.6	7.3	38.8	0.3	0.01	*	Miranda, Sosa Blanco, Bustos-Martinez, and Vasile (2007)
Municipal solid waste	39.9	49.0	11.1	49.8	7.6	32.5	3.7	2.4	*	Lin et al. (2016)
Sewage sludge	49.8	47.5	2.7	49.9	8.4	30.5	9.7	1.5		Peng et al. (2016)

[a]Dry basis.
[b]Dry ash free.
[c]By difference.
*Not provided.

Table 17 Main metals expressed in g/kg for different industrial and municipal wastes

	Na	K	Ca	Al	Fe	Mg	Mn	P	Zn	Cr	Cu_3	References
Sewage sludge (SS)	18.2	15.4	15.9	33.4	23.0	7.9	0.5	*	0.9	0.3	0.3	Zhang, Mao, Chen, Huang, and Qiu (2013)
Automotive residues	0.17	*	8.6	4.9	26.0	4.1	0.4	*	4.0	0.3	24	Notarnicola et al. (2017)
SS-Datansha	1.4	1.2	4.8	31	*	4.1	*	28.4	*	*	*	Lu et al. (2013)
SS-Liede	1	0.8	6.7	26.2	*	6.3	*	20.1	*	*	*	
SS-Xilang	1.5	1.3	1.5	43.2	*	4.1	*	21.4	*	*	*	
Black liquor	22.3	1.2	*	*	*	*	*	*	*	*	*	Whitty, Kullberg, Sorvari, Backman, and Hupa (2008)
Paper sludge/ wheat husk	0.91	16.8	89.2	*	2.78	4.53	0.127	6.61	*	*	*	Kalderis, Kayan, Akay, Kulakzis, and Gozmen (2017)
Demolition waste	0.68	2.869	*	*	1.58		0.414	0.683	*	*	*	

*Not provided.

Fig. 6 Municipal solid waste.

MSW composition depends on the country where it is issued (Tables 16 and 17). In Europe, the average composition of MSW can be described as 32–42 wt% of organic matter, 24–28 wt% of paper and cardboard, 8–9 wt% of glass, 7 wt% of plastics, 4%–5% of metals, 3 wt% of textiles, and 15 wt% of others (food wastes, etc.) (Fig. 6). From this composition it is estimated that cellulose and lignin represent almost 65 wt%, polymers 15 wt%, while inorganic materials constitute 20 wt% (glass, metals, etc.). The mean composition of Chinese MSW has been described by Lin et al. (2016): it consists of 3 wt% of wood waste, 8.5 wt% of paper, 3.2 wt% of textiles, 11.1 wt% of plastics, 0.9 wt% of rubber, 55.9 wt% of food residue, and 17.5 wt% of others. The proportion of food waste is considerable compared to the one observed in European countries. The situation is globally identical in the United States compared to Europe, while the average composition of MSW can be described as follows: 45.5 wt% of paper, 16.4 wt% of plastics, 9.6 wt% of glass, 10.9 wt% of metals, and 17.6% of food residue (Berge et al., 2011).

The pyrolysis (under nitrogen or CO_2) of MSW was performed between 400°C and 800°C, and gasification of the char (under CO_2 atmosphere) has been studied for a higher temperature (900°C). Few works reported the hydrothermal carbonization of MSW (210–280°C) and the reactivity of the hydrochar (combustion and gasification). The use of char derived from MSW pyrolysis was studied for the adsorption, for instance, of mercury (Li et al., 2015).

RDF is produced from MSW after reducing the noncombustible fraction by different processes. The energetic potential of RDF was widely studied, especially for combustion and gasification applications of its chars, as done, for example, by Cozzani (2000) by applying intermediate pyrolysis to Italian RDF.

Industrial wastes also present a good potential for the elaboration of chars (Table 18). There is a huge variety of wastes generated by the industrial sector; among them we can find municipal sewage sludge, cardboard, textile residues, medicinal

Table 18 Different industrial and municipal wastes and the corresponding techniques used for char production in the literature

Waste	Origin	Conversion process	Operating conditions	Applications	References
Refuse-derived fuel	China	Fast pyrolysis	500–900°C, 60 min	Characterization	Hwang et al. (2014)
Refuse paper/plastic fuel					
Construction wood					
Medicinal residues	China	Slow pyrolysis	200 g, 600°C, 1 h	Mercury removal	Li et al. (2015)
Municipal solid waste					
Sewage sludge	Lijiao WWTP, Guangzhou, China	Slow pyrolysis	300–600°C, 20°C/min, 1–2 h	Pb(II), Cr(VI) sorption	Zhang et al. (2013)
Textile dyeing sludge	Dyeing plant, Jiangsu, China	Pyrolysis	300–800 g, 350–750°C, 20–60 min	Characterization	Zhang et al. (2017)
Automotive shredded residues	Ferramenta Pugliese, Bari, Italy	Pyrolysis	Rotary kiln, 450–650°C	Characterization	Notarnicola et al. (2017)
Paper	United States	Hydrothermal carbonization	Solids concentration of 20% (wt), 250°C, 20 h	Characterization	Berge et al. (2011)
Food waste					
Anaerobic digestion waste					
Mixed municipal waste					

Continued

Table 18 Continued

Used cotton fabrics	T-shirt industry	Pyrolysis	200 g, 600–700°C	Characterization	Miranda et al. (2007)
Refuse-derived fuel	Italy	Slow pyrolysis	30–90°C/min, 500–800°C, 1 h	Combustion, gasification	Cozzani (2000)
Black liquor	Finland	Pressurized pyrolysis	1–20 bar, 650–900°C, 15–20 s	Char yield	Whitty et al. (2008)
Municipal solid waste	China	Hydrothermal carbonization	Solid/water: 1–5; 210–280°C, 30–90 min	Combustion, gasification	Lin et al. (2016)
Municipal solid waste	Australia	Slow pyrolysis	100°C/min, 900°C, 1 h	CO_2 gasification	Hla, Lopes, and Roberts (2016)
Sewage sludge	China	Hydrothermal carbonization	Solid/water: 1–9, 180–300°C, 30–480 min	Char properties, combustion behavior	Peng et al. (2016)
Cotton fiber waste	Textile industry	Pyrolysis	10°C/min, 400–700°C	Alizarin adsorption	Wanassi, Hariz, Ghimbeu, Vaulot, and Jeguirim (2017)

Fig. 7 Industrial residues. (A) Carboard waste, (B) textile residues, (C) medicinal residues, and (D) automotive shredder.

residues, waste lignin, paper sludge, textile dyeing sludge, black liquor, automotive residues, etc. (Fig. 7).

In the industrial sector, recycling and valorization of residues is becoming a priority for several reasons: necessity for the ISO 14000 standard, limitation of costs associated with waste management, production of energy or new derivative products, etc. As energy prices have drastically increased during the last few decades, the transformation of industrial waste into chars has been approached. Industrial wastes are also suitable for feedstock recycling methods using thermal or microwave pyrolysis, steam gasification, or hydrothermal carbonization. The more adapted technology has to be chosen from the composition of the wastes and also from an economical point of view.

Thermal conversion of these wastes into char was investigated through pyrolysis under nitrogen (slow pyrolysis: 350–750°C, fast pyrolysis: 480°C), hydrothermal carbonization (180–300°C), and high-pressurized pyrolysis (650–1100°C, 1–20 bar). The chars have been used for different applications: adsorption (VOCs, dyes, heavy metals, CO_2), energy potential recovery (steam gasification 500–900°C, combustion), and landfilling after pretreatment.

In the following paragraphs, some of the most common industrial wastes that may constitute feedstock materials for char production are presented, that is, sewage sludge, textile dyeing sludge, automotive shredded residues, and black liquor.

Wastewater treatment induces the production of a huge amount of wet sewage sludge, the repartition of which is very disparate. Estimation gives values in China three- to sixfold the quantities in the whole of Europe. The repartition of sewage sludge produced in Europe itself shows that Germany, France, Italy, and Spain are responsible for more than 70 wt%, while Cyprus presents the highest amount of sewage sludge produced per inhabitant (36.8 kg/person/year) (Kelessidis & Stasinakis, 2012). Directive 86/278/EEC (sewage sludge directive) gives the orientation of sewage sludge management, with a focus on the use of this biomass as fertilizer for agriculture with limitations associated with the amount of heavy metals.

Due to its specific composition, its presence in all territories, and the lack of sufficient outfall, the pyrolysis of classical sewage sludge coming from wastewater treatment plants has been studied. The biochar originated from this type of waste is called wastewater sludge-derived biochar or sewage sludge biochar. The main problem with sewage sludge corresponds with their composition. This depends on the composition of the wastewaters, often loaded with hazardous elements such as pathogen microorganisms, heavy metals, and persistent organic pollutants. The disposal of raw sewage sludge as amendment is not simple due to the potential accumulation of organic and mineral pollutants in the soil. Pyrolysis or hydrothermal carbonization processes are good ways to sanitize this biomass and to obtain more valuable products (biofuels, adsorbents, amendment), but the leaching of heavy metals (adsorbents, amendment) or their behavior during combustion (biofuels) needs to be evaluated carefully. In this context, work has been carried out on the elaboration of biochars made of both sewage sludge and other biomasses (Fan et al., 2016; Nascimento et al., 2017).

Textile dyeing sludge is more problematical than sewage sludge since it contains perishable organics, dyeing agents, additives, polycyclic aromatic hydrocarbons, and several heavy metals such as lead and copper. The processing of this sludge to produce char by isothermal 350°C pyrolysis and 750°C pyrolysis for 20 and 60 min was successfully carried out by Zhang et al. (2017) who reported initial ash content more than 61%, which is a very high weight percentage but also a real mine of metals and minerals (Table 17).

Automobile shredder residue management is also a preoccupying issue because of organic (hydrocarbons, polychlorinated biphenyl) and inorganic compounds (heavy metals), the proportions of which vary considerably. A possible way of conversion in a rotary kiln at 450–650°C was tested by Notarnicola et al. (2017) who further characterized the so-obtained tire waste pyrolytic char. The results indicated a possible use in power generation and energy recovery rather than in adsorption or in agronomy.

Black liquor is the spent pulping solution resulting from wood digestion in a solution of sodium hydroxide and sodium sulfide. Conventionally, after a process of concentration, black liquor is burned in a boiler to output energy on the one hand and pulp chemicals on the other hand. To yield char from black liquor, pressurized pyrolysis at 1–20 bar and 650–900°C during 15–20 s was efficiently conducted by Whitty et al. (2008).

1.8 Conclusion

The possibilities for biomass char production are numerous, because the feedstock is diverse and comprises a long list of products and by-products. Potential sources may be found in woody, grassy, fibrous, and herbaceous materials and debris whether they are harvested from dedicated crops or collected from furniture and wood manufacturing. It is still the lignocellulosic component that makes agricultural biomasses excellent candidates for charring. Another significant influx may be assured by livestock solid and liquid wastes. The high potential of aquatic biomass for char production comes from high protein, carbohydrate, and lipid contents, among other reasons. Finally, industrial and urban wastes such as sewage sludge, automotive residues, demolition wastes, plastics, etc. can also ensure long-time availability to pyrolysis and/or gasification processes of char yielding. Provided that operation conditions are mastered, the feedstocks are characterized, and the final applications are defined, the processes could be subdued to output high-quality chars as adsorption agents, energy vectors, soil enhancers, or for carbon sequestration.

References

Alburquerque, J. A., Sanchez, M. E., Mora, M., & Barron, V. (2016). Slow pyrolysis of relevant biomasses in the Mediterranean basin. Part 2. Char characterisation for carbon sequestration and agricultural uses. *Journal of Cleaner Production*, *120*, 191–197.

Asadullah, M., Zhang, S., & Li, C. Z. (2010). Evaluation of structural features of chars from pyrolysis of biomass of different particle sizes. *Fuel Processing Technology*, *91*, 877–881.

Baniasadi, M., Tugnoli, A., Conti, R., Torri, C., Fabbri, D., & Cozzani, V. (2016). Waste to energy valorization of poultry litter by slow pyrolysis. *Renewable Energy*, *90*, 458–468.

Ben, H., Hao, N., Liu, Q., & Ragauskas, A. J. (2017). Solid-state NMR investigation of biochars produced from biomass components and whole biomasses. *Bioenergy Research*, *10*, 1036–1044.

Berge, N. D., Ro, K. S., Mao, J., Flora, J. R. V., Chappell, M. A., & Bae, S. (2011). Hydrothermal carbonization of municipal waste streams. *Environmental Science and Technology*, *45*, 5696–5703.

Bordoloi, N., Goswami, R., Kumar, M., & Katak, R. (2017). Biosorption of Co (II) from aqueous solution using algal biochar: kinetics and isotherm studies. *Bioresource Technology*, *244 p*, 1465–1469.

Butnan, S., Deenik, J. L., Toomsan, B., Antal, M. J., & Vityakon, P. (2015). Biochar characteristics and application rates affecting corn growth and properties of soils contrasting in texture and mineralogy. *Geoderma*, *237–238*, 105–116.

Cao, X., Pignatello, J. J., Li, Y., Lattao, C., Chappell, M. A., Chen, N., et al. (2012). Characterization of wood chars produced at different temperatures using advanced solid-state 13C NMR spectroscopic techniques. *Energy and Fuels*, *26*, 5983–5991.

Cao, X., Ro, K. S., Chappell, M., Li, Y., & Mao, J. (2011). Chemical structures of swine-manure chars produced under different carbonization conditions investigated by advanced solid-state 13C nuclear magnetic resonance (NMR) spectroscopy. *Energy and Fuels*, *25*, 388–397.

Cely, P., Gascó, G., Paz-Ferreiro, J., & Méndez, A. (2015). Agronomic properties of biochars from different manure wastes. *Journal of Analytical and Applied Pyrolysis*, *111*, 173–182.

Chiodo, V., Zafarana, G., Maisano, S., Freni, S., & Urbani, F. (2016). Pyrolysis of different biomass: direct comparison among Posidonia Oceanica, Lacustrine Alga and White-Pine. *Fuel*, *164*, 220–227.

Choi, J. H., Kim, S. S., Suh, D. J., Jang, E. J., Min, K. I., & Woo, H. C. (2016). Characterization of the bio-oil and bio-char produced by fixed bed pyrolysis of the brown alga Saccharina japonica. *Korean Journal of Chemical Engineering*, *33*, 2691–2698.

Colantoni, A., Evic, N., Lord, R., Retschitzegger, S., Proto, A. R., Gallucci, F., et al. (2016). Characterization of biochars produced from pyrolysis of pelletized agricultural residues. *Renewable and Sustainable Energy Reviews*, *64*, 187–194.

Cozzani, V. (2000). Reactivity in oxygen and carbon dioxide of char formed in the pyrolysis of refuse-derived fuel. *Industrial and Engineering Chemistry Research*, *39*, 864–872.

Das, K. C., Garcia-Perez, M., Bibens, B., & Melear, N. (2008). Slow pyrolysis of poultry litter and pine woody biomass: impact of chars and bio-oils on microbial growth. *Journal of Environmental Science and Health Part A*, *43*, 714–724.

Della Rocca, P. A., Cerrella, E. G., Bonelli, P. R., & Cukierman, A. L. (1999). Pyrolysis of hardwoods residues: on kinetics and chars Characterization. *Biomass and Bioenergy*, *16*(7988).

Fan, S., Tang, J., Wang, Y., Li, H., Zhang, H., Tang, J., et al. (2016). Biochar prepared from co-pyrolysis of municipal seawage sludge and tea waste for the adsorption of methylene blue from aqueous solutions: kinetics, isotherm, thermodynamic and mechanism. *Journal of Molecular Liquids*, *220*, 432–441.

Ferreira, A. F., Soares Dias, A. P., Silva, C. M., & Costa, M. (2015). Evaluation of thermochemical properties of raw and extracted microalgae. *Energy*, *92*, 365–372.

Fu, P., Yi, W., Bai, X., Li, Z., Hu, S., & Xiang, J. (2011). Effect of temperature on gas composition and char structural features of pyrolyzed agricultural residues. *Bioresource Technology*, *102*, 8211–8219.

Gabbar, H. A., Aboughaly, M., & Ayoub, N. (**91**, 2018). Comparative study of MSW Heat treatment processes and electricity generation. *Journal of the Energy Institute*. 481–488.

Gao, Y., Wang, X., Chen, Y., Li, P., Liu, H., & Chen, H. (2017). Pyrolysis of rapeseed stalk: influence of temperature on product characteristics and economic costs. *Energy*, *122*, 482–491.

Guizani, C., Jeguirim, M., Valin, S., Limousy, L., & Salvador, S. (2017). Biomass chars: the effects of pyrolysis conditions on their morphology, structure, chemical properties and reactivity. *Energies*, *10*, 796.

Haddad, K., Jeguirim, M., Jerbi, B., Chouchene, A., Dutournié, P., Thevenin, N., et al. (2017). Olive mill wastewater: from a pollutant to green fuels, agricultural water source and biofertilizer. *ACS Sustainable Chemical Engineering*, *5*, 8988–8996.

Hla, S. S., Lopes, R., & Roberts, D. (2016). The CO_2 gasification reactivity of chars produced from Australian municipal solid waste. *Fuel*, *185*, 847–854.

Hoekman, S. K., Broch, A., & Robbins, C. (2011). Hydrothermal carbonization (HTC) of lignocellulosic biomass. *Energy and Fuels*, *25*, 1802–1810.

Hwang, H. I., Kobayashi, J., & Kawamoto, K. (2014). Characterization of products obtained from pyrolysis and steam gasification of wood waste, RDF, and RPF. *Waste Management*, *34*, 402–410.

Jeguirim, M., Elmay, Y., Limousy, L., Lajili, M., & Said, R. (2014). Devolatilization behavior and pyrolysis kinetics of potential Tunisian biomass fuels. *Environmental Progress and Sustainable Energy*, *33*, 1452–1458.

Joseph, P., Tretsiakova-Mcnally, S., & Mckenna, S. (2002). Characterization of cellulosic wastes and gasification products from chicken farms. *Waste Management*, *32*, 701–709.

Kalderis, D., Kayan, B., Akay, S., Kulakzis, E., & Gozmen, B. (2017). Adsorption of 2,4-dichlorophenol on paper sludge/wheat husk biochar: process optimization and comparison with biochars prepared from wood chips, sewage sludge and hog fuel/demolition waste. *Journal of Environmental Chemical Engineering*, *5*, 2222–2231.

Kang, S., Li, X., Fan, J., & Chang, J. (2012). Characterization of hydrochars produced by hydrothermal carbonization of lignin, cellulose, D-xylose, and wood meal. *Industrial and Engineering Chemistry Research*, *51*, 9023–9031.

Kantarli, I. C., Kabadayi, A., Ucar, S., & Yanik, J. (2016). Conversion of poultry wastes into energy feedstocks. *Waste Management*, *56*, 530–539.

Kelessidis, A., & Stasinakis, A. S. (2012). Comparative study of the methods used for treatment and final disposal of sewage sludge in European countries. *Waste Management*, *32*, 1186–1195.

Kim, S. S., Agblevor, F. A., & Lim, J. (2009). Fast pyrolysis of chicken litter and turkey litter in a fluidized bed reactor. *Journal of Industrial and Engineering Chemistry*, *15*, 247–252.

Kim, H. W., Bae, S., & Lee, J. Y. (2013). A Study on the removal of heavy metals in soil by sewage sludge biochar. *Journal of Soil and Groundwater Environment*, *18*(3), 58–64.

Kim, S. S., Ly, H. V., Kim, J., Choi, J. H., & Woo, H. C. (2013). Thermogravimetric characteristics and pyrolysis kinetics of Alga Sagarssum sp. Biomass. *Bioresource Technology*, *139*, 242–248.

Kirtania, K., & Bhattacharya, S. (2015). CO_2 gasification kinetics of algal and woody char procured under different pyrolysis conditions and heating rates. *ACS Sustainable Chemistry and Engineering*, *3*, 365–373.

Kumar, M., Verma, B. B., & Gupta, R. C. (1999). Mechanical properties of acacia and eucalyptus wood chars. *Energy Sources*, *21*, 675–685.

Lee, X. J., Lee, L. Y., Gan, S., Thangalazhy-Gopakumar, S., & Ng, H. K. (2017). Biochar potential evaluation of palm oil wastes through slow pyrolysis: thermochemical characterization and pyrolytic kinetic studies. *Bioresource Technology*, *236*, 155–163.

Lee, Y., Park, J., Ryu, C., Gang, K. S., Yang, W., Park, Y. K., et al. (2017). Comparison of biochar properties from biomass residues produced by slow pyrolysis at 500°C. *Bioresource Technology*, *148*, 196–201.

Leng, L. J., Yuan, X. Z., Huang, H. J., Wang, H., Wu, Z. B., Fu, L. H., et al. (2015). Characterization and application of bio-chars from liquefaction of microalgae, lignocellulosic biomass and sewage sludge. *Fuel Processing Technology*, *129*, 8–14.

Li, J., Cao, L., Yuan, Y., Wang, R., Wen, Y., & Man, J. (2018). Comparative study for microcystin-LR sorption onto biochars produced from various plant- and animal-wastes at different pyrolysis temperatures: Influencing mechanisms of biochar properties. *Bioresource Technology*, *247*, 794–803.

Li, G., Shen, B., Wang, Y., Yue, S., Xi, Y., An, M., et al. (2015). Comparative study of element mercury removal by three bio-chars from various solid wastes. *Fuel*, *145*, 189–195.

Lima, I. M., Boateng, A. A., & Klasson, K. T. (2009). Pyrolysis of broiler manure: char and product gas characterization. *Industrial and Engineering Chemistry Research*, *48*, 1292–1297.

Lin, Y., Ma, X., Peng, X., Yu, Z., Fang, S., Lin, Y., et al. (2016). Combustion, pyrolysis and char CO_2-gasification characteristics of hydrothermal carbonization solid fuel from municipal solid wastes. *Fuel*, *181*, 905–915.

Lin, H. J., Rong, C. X., Jiu, B. B., Li, B. X., Yu, Q. J., Gan, L. H., et al. (2018). Effects of chromium on pyrolysis characteristic of water hyacinth (Eichornia crassipes). *Renewable Energy*, *115*, 676–684.

Link, S., Arvelakis, S., Hupa, M., Yrjas, P., Kulaots, I., & Paist, A. (2010). Reactivity of the biomass chars originating from Reed, Douglas Fir, and Pine. *Energy and Fuels*, *24*, 6533–6539.

Lino, F. A. M., & Ismail, K. A. R. (2017). Incineration and recycling for MSW treatment: case study in Campinas, Brazil. *Sustainable Cities and Society*, *35*, 752–757.

López, F. A., Centeno, T. A., García-Díaz, I., & Alguacil, F. J. (2013). Textural and fuel characteristics of the chars produced by the pyrolysis of waste wood, and the properties of activated carbons prepared from them. *Journal of Analytical and Applied Pyrolysis*, *104*, 551–558.

Lopez-Gonzalez, D., Fernandez-Lopez, M., Valverde, J. L., & Sanchez-Silva, L. (2014). Pyrolysis of three different types of microalgae: kinetic and evolved gas analysis. *Energy*, *73*, 33–43.

Lu, H., Zhang, W., Wang, S., Zhuang, L., Yang, Y., & Qiu, R. (2013). Characterization of sewage sludge-derived biochars from different feedstocks and pyrolysis temperatures. *Journal of Analytical and Applied Pyrolysis*, *102*, 137–143.

Lu, J., Zhang, C., Wu, J., & Luo, Y. (2017). Adsorptive removal of bisphenol A using N-doped biochar made of Ulva prolifera. *Water, Air, and Soil Pollution*, *9*, 228–327.

Miranda, R., Sosa Blanco, C., Bustos-Martinez, D., & Vasile, C. (2007). Pyrolysis of textile wastes I. Kinetics and yields. *Journal of Analytical and Applied Pyrolysis*, *80*, 489–495.

Moral, U., Yavuzel, N., & Sensöz, S. (2016). Pyrolysis of hornbeam (*Carpinus betulus* L.) sawdust: characterization of bio-oil and bio-char. *Bioresource Technology*, *221*, 682–685.

Nascimento, P. F. P., Sousa, J. F., Oliveira, J. A., Possa, R. D., Santos, L. S., Carvalho, F. C., et al. (2017). Wood sawdust and sewage sludge pyrolysis chars for CO_2 adsorption using a magnetic suspension balance. *The Canadian Journal of Chemical Engineering*, *95*, 2148–2155.

Ng, W. C., You, S., Ling, R., Gin, K. Y. H., Dai, Y., & Wang, C. H. (2017). Co-gasification of woody biomass and chicken manure: Syngas production, biochar reutilization, and cost-benefit analysis. *Energy*, *139*, 732–742.

Notarnicola, M., Cornacchia, G., De Gisi, S., Di Canio, F., Freda, C., Garzone, P., et al. (2017). Pyrolysis of automotive shredder residue in a bench scale rotary kiln. *Waste Management*, *65*, 92–103.

Pehlivan, E., Ozbay, N., Yargiç, A. S., & Sahin, R. Z. (2017). Production and characterization of chars from cherry pulp via pyrolysis. *Journal of Environmental Management*, *203 p*, 1017–1025.

Peng, C., Zhai, Y., Zhu, Y., Xu, B., Wang, T., Li, C., et al. (2016). Production of char from sewage sludge employing hydrothermal carbonization: char properties, combustion behavior and thermal characteristics. *Fuel*, *176*, 110–118.

Rizzo, A. M., Prussi, M., Bettucci, L., Libelli, I. M., & Chiaramonti, D. (2013). Characterization of microalga Chlorella as a fuel and its thermogravimetric behavior. *Applied Energy*, *102*, 24–31.

Ross, A. B., Jones, J. M., Kubacki, M. L., & Bridgeman, T. (2008). Classification of macroalgae as fuel and its thermochemical behavior. *Bioresource Technology*, *99*, 6494–6504.

Ryu, C., Sharifiz, V. N., & Swithenbank, J. (2007). Waste pyrolysis and generation of storable char. *International Journal of Energy Research*, *31*, 177–191.

Sanchez, M. E., Lindao, E., Margaleff, D., Martinez, O., & Moran, A. (2009). Pyrolysis of agricultural residues from rape and sunflowers: Production and characterization of bio-fuels and biochar soil management. *Journal of Analytical and Applied Pyrolysis*, *85*, 142–144.

Shaaban, A., Se, S. M., Dimina, M. F., Juoi, J. M., Mohd Husin, M. H., & Mitan, N. M. M. (2014). Influence of heating temperature and holding time on biochars derived from rubber wood sawdust via slow pyrolysis. *Journal of Analytical and Applied Pyrolysis*, *107*, 31–39.

Siddiqui, M. T. H., Nizamuddin, S., Mubarak, N. M., Shirin, K., Aijaz, M., Hussain, M., et al. (2017). Characterization and process optimization of biochar produced using novel biomass, waste pomegranate peel: a response surface methodology approach. *Waste Biomass Valorization*, https://doi.org/10.1007/s12649-017-0091-y.

Sricharoenchaikul, V., Pechyen, C., Ahtong, D., & Atong, D. (2008). Preparation and characterization of activated carbon from the pyrolysis of Physic Nut (*Jatropha curcas* L.). Waste. *Energy and Fuels*, *22*, 31–37.

Tamošiūnas, A., Chouchène, A., Valatkevičius, P., Gimžauskaitė, D., Aikas, M., Uscila, R., et al. (2017). The potential of thermal plasma gasification of olive pomace charcoal. *Energies*, *10*, 710.

Uchimiya, M., Wartelle, L. H., Lima, I. M., & Klasson, K. T. (2010). Sorption of deisopropylatrazine on broiler litter biochars. *Journal of Agricultural and Food Chemistry*, *58*(23), 12350–12356.

Volpe, R., Bermudez Menendez, J. M., Ramirez Reina, T., Messineo, A., & Millan, M. (2017). Evolution of chars during slow pyrolysis of citrus waste. *Fuel Processing Technology*, *158*, 255–263.

Wanassi, B., Hariz, I. B., Ghimbeu, C. M., Vaulot, C., & Jeguirim, M. (2017). Green carbon composite-derived polymer resin and waste cotton fibers for the removal of alizarin Red S Dye. *Energies*, *10*, 1321.

Whitty, K., Kullberg, M., Sorvari, M., Backman, R., & Hupa, M. (2008). Influence of pressure on pyrolysis of black liquor: 2. Char yields and component release. *Bioresource Technology*, *99*, 671–679.

Zhang, S. Y., Cao, J. P., & Takarada, T. (2016). Effect of pretreatment with different washing methods on the reactivity of manure char. *Bioresource Technology*, *101*, 6130–6135.

Zhang, H., Gao, Z., Ao, W., Li, J., Liu, G., Fu, J., et al. (2017). Microwave pyrolysis of textile dyeing sludge in a continuously operated auger reactor: char characterization and analysis. *Journal of Hazardous Materials*, *334*, 112–120.

Zhang, W., Mao, S., Chen, H., Huang, L., & Qiu, R. (2013). Pb(II) and Cr(VI) sorption by biochars pyrolyzed from the municipal wastewater sludge under different heating conditions. *Bioresource Technology*, *147*, 545–552.

Zhao, X., Ouyang, W., Hao, F., Lin, C., Wang, F., Han, S., et al. (2013). Properties comparison of biochars from corn straw with different pretreatment and sorption behaviour of atrazine. *Bioresource Technology*, *147*, 338–344.

Zhao, S. X., Ta, N., & Wang, X. D. (2017). Effect of temperature on the structural and physicochemical properties of biochar with apple tree branches as feedstock material. *Energies*, *10*, 1293.

Further reading

Agrafioti, E., Bouras, G., Kalderis, D., & Diamadopoulos, E. (2013). Biochar production by sewage sludge pyrolysis. *Journal of Analytical and Applied Pyrolysis*, *101*, 72–78.

Chaiwong, K., Kiatsiriroat, T., Vorayos, N., & Thararax, C. (2013). Study of bio-oil and biochar production from algae by slow pyrolysis. *Biomass and Bioenergy*, *56*, 600–606.

Cheng, M., Wu, L., Huang, Y., & Luo, P. (2014). Total concentrations of heavymetals and occurrence of antibiotics in sewage sludges from cities throughout China. *Journal of Soils Sediments: Protection and Risk Assessment and Remediation*, *14*, 1123–1135.

Chouchene, A., Jeguirim, M., Khiari, B., Trouvé, G., & Zagrouba, F. (2010). Study on the emission mechanism during devolatilization/char oxidation and direct oxidation of olive solid waste in a fixed bed reactor. *Journal of Analytical and Applied Pyrolysis*, *87*, 168–174.

Feng, L., Luo, J., & Chen, Y. (2015). Dilemma of sewage sludge treatment and disposal in China. *Environmental Science and Technology*, *49*, 4781–4782.

Ghouma, I., Jeguirim, M., Ouederni, A., & Limousy, L. (2017). Pyrolysis of olive pomace: degradation kinetics, gaseous analysis and char characterization. *Waste and Biomass Valorization*, *8*, 1689–1697.

Guizani, C., Haddad, K., Jeguirim, M., Colin, B., & Limousy, L. (2016). Combustion characteristics and kinetics of torrefied olive pomace. *Energy*, *107*, 453–463.

Guizani, C., Jeguirim, M., Esceduro Sainz, J., Gadiou, R., & Salvador, S. (2016). Biomass char gasification by H_2O, CO_2 and their mixture: evolution of chemical, textural and structural properties of the chars. *Energy*, *112*, 133–145.

Jeguirim, M., Elmay, Y., & Limousy, L. (2016). Thermal degradation kinetics and mechanisms of Posidonia Oceanica under inert and oxidative atmospheres. *International Journal of Green Energy*, *13*, 665–671.

Kraiem, N., Lajili, M., Limousy, L., Said, R., & Jeguirim, M. (2016). Energy recovery from Tunisian Agri-food wastes: evaluation of combustion performance and emissions characteristics of green pellets prepared from tomato residues and grape marc. *Energy*, *107*, 409–418.

Li, H., Hu, J., Menga, Y., Su, J., & Wang, X. (2017). An investigation into the rapid removal of tetracycline usingmultilayered graphene-phase biochar derived from waste chicken feather. *Science of the Total Environment*, *603–604*, 39–48.

Libra, J. A., Ro, K. S., Kammann, C., Funke, A., Berge, N. D., Neubauer, Y., et al. (2011). Hydrothermal carbonization of biomass residuals: a comparative review of the chemistry, processes and applications of wet and dry pyrolysis. *Biofuels*, *2*, 71–106.

Minoriuchimiy, A., Wartelle, L. H., Lima, I. M., & Klasson, K. T. (2010). Sorption of deisopropylatrazine on broiler litter biochars. *Journal of Agricultural and Food Chemistry*, *58*, 12350–12356.

Zambon, I., Colosimo, F., Monarca, D., Cecchini, M., Gallucci, F., Rosario Proto, A., et al. (2016). An innovative agro-forestry supply chain for residual biomass: physicochemical characterisation of biochar from olive and hazelnut pellets. *Energies*, *9*, 526.

Char production technology

2

Witold Kwapinski
Chemical Sciences, University of Limerick, Limerick, Ireland

Chapter Outline

2.1 Introduction

Char is a product of pyrolysis and natural fire of any carbonaceous material. At temperatures above 300°C, polymeric blocks in biomass begin depolymerization and fragmentation to form smaller molecules, which are released as volatiles and can react with residual solids producing condensed structures.

Char has been known from the beginning of human history, and is the first synthetic material ever produced by humans. Before the dawn of recorded history, humans employed shallow pits of char to smelt tin needed for the manufacture of bronze tools (Antal & Gronli, 2003). Now, its main application is in smokeless fuel for cooking and heating, metallurgy for steel making, medicine, chemicals, activated carbon (filters), drugs, electronics, photovoltaic panels, agronomy, and in art for drawing. As a product of carbonization the material can be very rich in carbon. However, its exact composition depends on the initial biomass and on its production conditions. Charcoal can contain over 90 wt.% of carbon but it can also be much less when using mineral-rich feedstock such as animal manure. The material contains fixed and volatile carbon, the proportions of which can greatly vary due to manufacturing process conditions.

Char and Carbon Materials Derived from Biomass. https://doi.org/10.1016/B978-0-12-814893-8.00002-X

In recent years, one more category of pyrolysis product residue has been selected and is known as biochar. International researchers from a group called the International Biochar Initiative (IBI) formulated the Biochar Standards document (IBI, 2015), followed by the European Biochar Certificate (EBC) (EBC, 2017). Biochar is defined as a heterogeneous substance rich in aromatic carbon and minerals. It is produced by pyrolysis of sustainably obtained biomass under controlled conditions with clean technology. Biochar needs to be used for any purpose that does not involve its rapid mineralization and CO_2 generation, and may become a soil amendment. Lehmann and Joseph (2009) stressed that biochar is produced with the intent to be applied to soil as a means of improving soil productivity, carbon storage, or filtration of percolating soil water. Following the EBC, there are several conditions for a material to be called biochar, where they encourage the design of pyrolysis meeting a number of product and process design elements leading to more environmentally friendly production.

Char production was intimately bound up with the beginning of metallurgy approximately 5000 years ago. Attempts to smelt metals using wood fires could never have been successful, since it would have been impossible to achieve sufficiently high temperatures, mainly due to a large quantity of water driven off biomass. Burning char, on the other hand, produces a much higher fire temperature (well over 1000°C), with little smoke. Oxide ores of copper were first reduced with char in about 3000 Before Christ (BC), initiating the era we know as the Bronze Age. Iron is more difficult to smelt than copper, requiring higher temperatures and a greater blast of air, and was first achieved in about 1200 BC, marking the beginning of the Iron Age. (Harris, 1999).

Currently, more than 800×10^3 tons of charcoal are used annually in Europe. Applying traditional technology, between 5 and 12 tons of wood are required to produce just 1 ton of wood-based charcoal; however, modern pyrolyzers are much more efficient. Around 70% of the charcoal used annually in Europe is imported. Nigeria is Europe's biggest charcoal supplier, followed by Namibia, South Africa, Egypt, and Ivory Coast. In 2017, the United Kingdom imported 87×10^3 tonnes of charcoal to meet the demand of the market due to relatively small domestic production (5×10^3 tonnes) (TFT, 2015). Brazil is by far the largest char producer in the world generating 9.9×10^6 tons/year. Other important biochar-producing countries are Thailand (3.9×10^6 tons/year), Ethiopia (3.2×10^6 tons/year), Tanzania (2.5×10^6 tons/year), and India and the Democratic Republic of Congo (1.7×10^6 tons/year each). Despite being the 10th largest char producer in the world (0.9×10^6 tons/year), most of the char consumed in the United States is imported from other countries (Garcia-Perez, Lewis, & Kruger, 2010).

2.2 Char formation

Biomass mainly consists of cellulose, hemicellulose, and lignin, whose ratios vary in different sources. These three major compounds have significantly different thermal stabilities and interact between each other during thermal processing. Hemicellulose, generally represented by xylan, is the easiest polymer to be degraded because its thermal degradation occurs between 220°C and 350°C. Next, cellulose thermal degradation takes place between 315°C and 400°C, while that of lignin covers a whole temperature range from 150°C to 550°C or even 900°C (Stefanidis et al., 2014; Yang, Yan, Chen, Lee, & Zheng, 2007). At low temperatures (<500°C), the pyrolysis of hemicellulose and lignin involves exothermic reactions, while the pyrolysis of cellulose is endothermic. However, at high temperatures (>500°C), the situation changes inversely (Yang et al., 2007). The complexity of biomass pyrolysis arises from the difference in decomposition of the biomass components with varying reaction mechanisms and reaction rates, which also partly depend on the thermal processing conditions and reactor designs (Kan, Strezov, & Evans, 2016). Hemicellulose and lignin interaction increases the production of lignin-derived phenols, while it hinders generation of hydrocarbons (Wang, Guo, Wang, & Luo, 2011). Although cellulose-hemicellulose interaction is not significant, considerable effects on pyrolysis behavior, including gas, tar, and char-forming behavior, is observed (Hosoya, Kawamoto, & Saka, 2007). During biomass pyrolysis, a large number of reactions take place in parallel and series, including dehydration, depolymerization, isomerization, aromatization, decarboxylation, and charring. The simplified reaction pathway base on cellulose decomposition is presented in Fig. 1, where dehydrogenation, depolymerization, and fragmentation are the main competitive reactions dominant at different temperature ranges. Also, it is generally accepted that the pyrolysis of biomass consists of three main stages: (1) initial evaporation of free moisture, (2) primary decomposition, followed by (3) secondary reactions (oil cracking and repolymerization) (Kan et al., 2016; Van de Velden, Baeyens, Brems, Janssens, & Dewil, 2010; White, Catallo, & Legendre, 2011).

In Section 2.3, "char" is used as a general expression that includes "charcoal" and "biochar." Char is a black-colored solid residue after the carbonization process. Its morphology retains the original feedstock. Its composition can vary greatly depending on the raw material elemental content. The main component is carbon, usually over

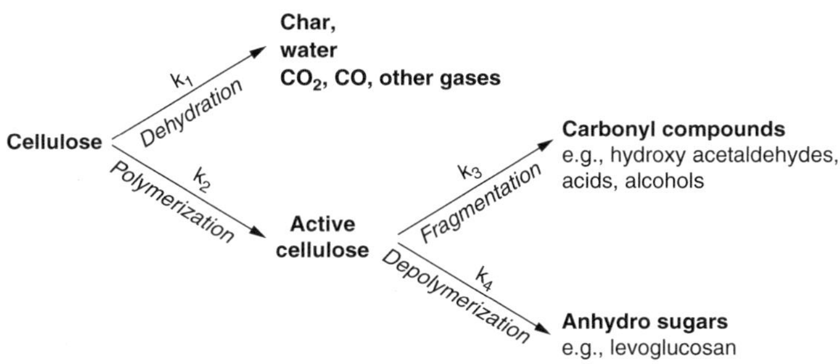

Fig. 1 The primary decomposition of cellulose according to the Waterloo mechanism. Based on Kan, T., Strezov, V., & Evans, T. J. (2016). Lignocellulosic biomass pyrolysis: a review of product properties and effects of pyrolysis parameters. *Renewable & Sustainable Energy Reviews*, *57*, 1126–1140; Van de Velden, M., Baeyens, J., Brems, A., Janssens, B., & Dewil, R. (2010). Fundamentals, kinetics and endothermicity of the biomass pyrolysis reaction. *Renewable Energy*, *35*(1), 232–242.

90 wt.% in which fixed carbon is over 70 wt.%. Char is a product of both primary (char) and secondary (coke) reactions. Increasing char yields require minimizing the carbon losses in the form of gases and liquids and promoting the desired pathways: primary solid-phase dehydration, decarboxylation, and decarbonylation reactions and secondary conversion of pyrolysis vapors to solids (Czernik, 2008). Hydrochar is a somewhat similar product to char but has different physical and chemical properties and is produced in a form of slurry in very different process conditions, described elsewhere in this book.

Approximate stoichiometric Eq. (1) is a representative example of cellulose pyrolysis at 400°C product compositions after the attainment of thermochemical equilibrium (Antal & Gronli, 2003):

$$C_6H_{10}O_5 \rightarrow 3.74C + 2.65H_2O + 1.17CO_2 + 1.08CH_4 \quad (1)$$

The results indicate that the residence time of volatile products within the pyrolyzing cellulose matrix is extremely important in determining conversion. Suggested pathways for cellulose pyrolysis involve primary decomposition to an oxygen-rich intermediate (probably levoglucosan), which then participates in three processes depending on experimental conditions: (1) direct escape of volatiles from the decomposing material into the ambient gas; (2) polymerization, cross-linking, and cracking to form char, and (3) pyrolysis to smaller volatiles, some of which inhibit the char formation in (2) or autocatalyze (3) (Lewellen, Peters, & Howard, 1977).

A number of works have been published in the field of biomass (and mainly cellulose as a model compound) thermal decomposition kinetic studies summarized by Lede (2012). In spite of a considerable amount of published results, there is no clear consensus in the literature for describing kinetics pathways and constants that could be valid in any situation (Lede, 2013). Following Lede (2013), the main reasons are:

- variety of biomass types and compositions, with specific physicochemical properties;
- kinetic behaviors related to the reactor and to the different process conditions, which are individual for each laboratory;
- number of assumptions and simplifications in models and uncertainties made on the values of several physical constants;
- many models issued from results obtained under low-temperature and small heating rates conditions;
- Arrhenius kinetic constants derived from reaction temperature, which is often very difficult to determine and usually changes in time as well as across a reactor.

More research is needed because the pyrolysis process is very complex and it will be very challenging to obtain a universal kinetic model for the process.

It is important not to confuse charcoal with other forms of impure noncrystalline carbon such as coke and soot. Although coke, like charcoal, is produced by solid-phase pyrolysis (usually of bituminous coal), it is distinguished from char in that a fluid phase is formed during carbonization. The structure and properties of cokes and chars are quite different. In the case of soot, it is formed in the gas phase by incomplete combustion rather than by solid-phase pyrolysis, and it has a microstructure quite distinct from either coke or char (Harris, 1999).

In many low-income countries, charcoal remains the primary source of energy for cooking. Chars possess a number of advantages over raw biomass for cooking:

- they have a much higher calorific value (approximately 30 MJ/kg against about 18.5 MJ/kg);
- they burn with less smoke;
- they are hydrophobic with low moisture content;
- they have a much lower bulk density;
- they can be stored for a long time without degradation.

Biomass and major biomolecular components of source materials occupy fairly specific locations on a plot called the Van Krevelen diagram. Fig. 2 illustrates the atomic ratio distribution in material elemental composition. As carbonization temperature increases, H and O are depleted in the char and it becomes rich in carbon.

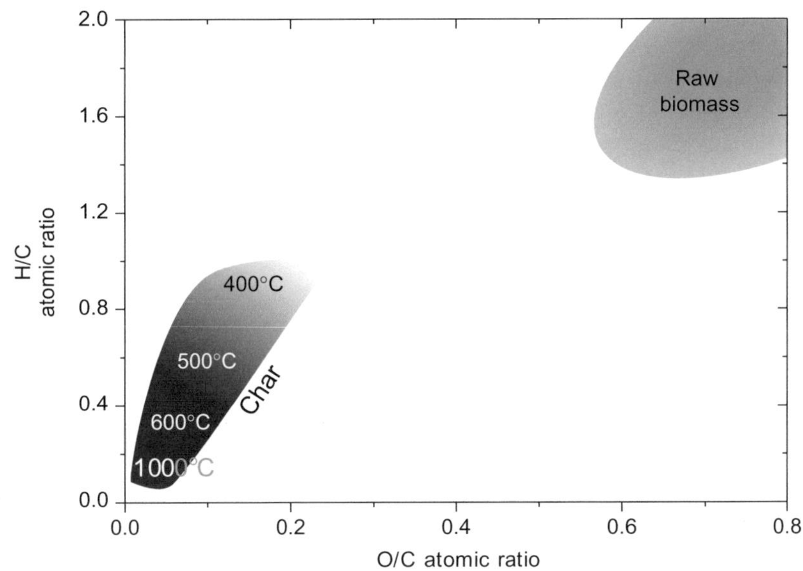

Fig. 2 Van Krevelen diagram of char at different carbonization temperatures.
Based on Budai, A., Wang, L., Gronli, M., Strand, L. T., Antal, M. J., Abiven, S., et al. (2014). Surface properties and chemical composition of corncob and miscanthus biochars: effects of production temperature and method. *Journal of Agricultural and Food Chemistry*, *62*(17), 3791–3799; Lee, Y., Eum, P. R. B., Ryu, C., Park, Y. K., Jung, J. H., & Hyun, S. (2013). Characteristics of biochar produced from slow pyrolysis of Geodae-Uksae 1. *Bioresource Technology*, *130*, 345–350; Schellekens, J., Silva, C. A., Buurman, P., Rittl, T. F., Domingues, R. R., Justi, M., et al. (2018). Molecular characterization of biochar from five Brazilian agricultural residues obtained at different charring temperatures. *Journal of Analytical and Applied Pyrolysis*, *130*, 106–117; Zhao, S. X., Ta, N., & Wang, X. D. (2017). Effect of temperature on the structural and physicochemical properties of biochar with apple tree branches as feedstock material. *Energies*, *10*(9).

Table 1 Typical process conditions.

Process parameters	Slow pyrolysis	Fast pyrolysis	Gasification
Temperature, °C	350–1000	>400	>600
Heating rate, °C/s	<80	100–1000	Few hundreds
Oxygen	Limited <2%	Free	0.2–0.45 stoichiometric amount
Reaction time	Minutes to days	Up to a few seconds	Up to a few seconds

2.3 Thermal conversion processes toward char

Char can be produced by thermal processes with restricted oxygen supply, preferably with slow heating rate (lower than 80°C/min), which allows long vapor residence time and consequently more efficient secondary cracking reactions. Typical values of process operating parameters are presented in Table 1. The indicative values of product yield are presented in Fig. 3 as columns; however, these greatly depend on feedstock composition, heating rate, and residence time (Cha et al., 2016; Zhang, Liu, & Liu, 2015); also particle size, temperature, and deviation can be significant, which is marked by bars in Fig. 3.

The char yield decreases by increasing the process temperature, as presented in Fig. 4. The highest treatment temperature and feedstock selection play an important role in the development of char functional properties, while overall heating rate has no significant effect on pH, stable-C, or labile-C concentrations. Increasing the process temperature reduces labile-C content and increases the yield of stable-C present within biochar (Crombie, Masek, Cross, & Sohi, 2015).

From a practical standpoint, the pyrolysis conditions that favor high char yields are (Lohri, Rajabu, Sweeney, & Zurbrugg, 2016):

- high lignin and nitrogen content in the biomass,
- low moisture content,
- pyrolysis temperatures less than 400°C, but lower temperature also leads to lower fixed carbon content,
- elevated process pressure (1 MPa) because a higher concentration of pyrolysis vapor increases the rate of secondary reactions,
- long vapor residence time because extended vapor-solid contact promotes secondary char formation,
- low heating rate due to slow formation (and escape) of organic vapors from feedstock particles,
- large biomass particle size to reduce heat and mass transfer rate within feedstock particles,
- efficient heat transfer to feedstock to minimize biomass burn-off.

The pressure in the reactor during pyrolysis has no influence on the quantity of the products from slow pyrolysis; however, it does have an influence on the quality of

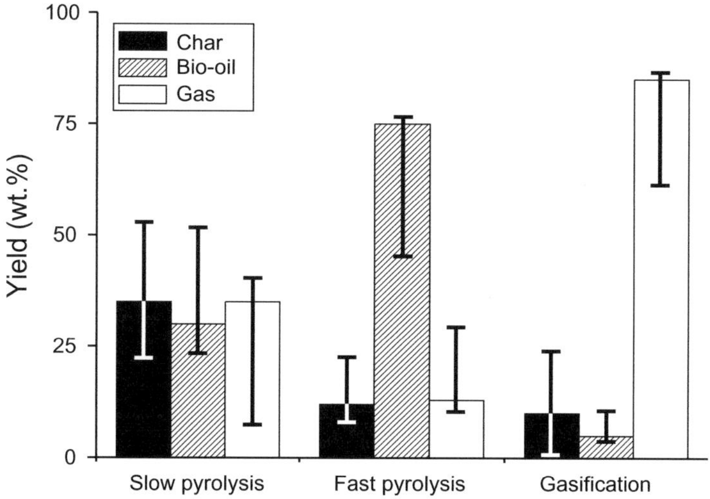

Fig. 3 Indicative value of char yield from thermal processes.
Data from Alvarez, J., Lopez, G., Amutio, M., Bilbao, J., & Olazar, M. (2014). Bio-oil production from rice husk fast pyrolysis in a conical spouted bed reactor. *Fuel*, *128*, 162–169; Bridgwater, A. V., & Peacocke, G. V. C. (2000). Fast pyrolysis processes for biomass. *Renewable & Sustainable Energy Reviews*, *4*(1), 1–73; Gonzalez, J. F., Encinar, J. M., Canito, J. L., Sabio, E., & Chacon, M. (2003). Pyrolysis of cherry stones: energy uses of the different fractions and kinetic study. *Journal of Analytical and Applied Pyrolysis*, *67*(1), 165–190; Isahak, W., Hisham, M. W. M., Yarmo, M. A., & Hin, T. Y. Y. (2012). A review on bio-oil production from biomass by using pyrolysis method. *Renewable & Sustainable Energy Reviews*, *16*(8), 5910–5923; Sohi, S. P., Krull, E., Lopez-Capel, E., & Bol, R. (2010). A review of biochar and its use and function in soil. *Advances in Agronomy*, *105*, 47–82; Wilkomirsky, I., Moreno E., & Berg A. (2014). Bio-oil production from biomass by flash pyrolysis in a three-stage fluidized bed reactors system. *Journal of Materials Science and Chemical Engineering*, *2*, 6–10.)

the products. As pressure is increased, the surface area is decreased, because the pores become clogged with trapped tar, and these also collapse. The nuclear magnetic resonance spectra indicate that chars formed at higher pressures have more extended fused aromatic structures, reflected also in the higher carbon contents, than those formed at lower pressures (Melligan et al., 2011).

Because biochar can play a significant role as solid additive, it is extremely important to conduct a process for its manufacture in a way that improves its quality. The best biochars for soil applications are produced from lignocellulosic materials and energy crops with well-defined cell structures. There is abundant evidence to suggest prolific microbial associations with soil biochar, and the biomass from these will be transformed into soil humic substances whose nutrient holding capacities are well recognized. Experimental results have shown that, depending on the preparation conditions, biochar can inhibit as well as promote plant growth (Kwapinski et al., 2010). Field testing of selected biochar is required first to validate laboratory assessed

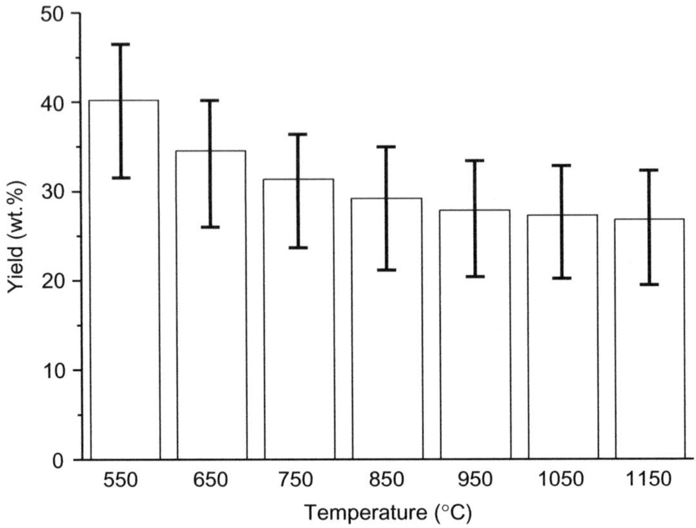

Fig. 4 Yield of char at various process temperatures during slow pyrolysis.
Data from Demirbas, A. (2001). Carbonization ranking of selected biomass for charcoal, liquid and gaseous products. *Energy Conversion and Management*, *42*(10), 1229–1238.

functions to behavior in soil and observe the development of functional properties with time (Crombie et al., 2015).

In "fast" pyrolysis the same products are obtained but in different proportions, the heat rate is a few hundreds of °C/s, and the bio-oil yield can be much larger than other products. Bio-oil is derived from the rapid condensation of vapors released during the pyrolysis of biomass. It is a complex mixture of many organic molecules having diverse molar masses and structures. The chemical compounds found in bio-oils are derived from the breakdown of cellulose, lignin, hemicellulose, and biomass extractives (Garcia-Perez, Adams, Goodrum, Geller, & Das, 2007). It is very important that vapor residence time in a reactor is as short as possible to minimize unwanted secondary cracking to lower hydrocarbons that can easily remain in the gas phase after cooling to ambient temperature (Bridgwater, 2012; Czernik & Bridgwater, 2004; Ronsse, 2016). Additionally, the vapor phase passing by a hot filter for small particle removal should be rapidly quenched (Wilkomirsky, Moreno, & Berg, 2014). The pyrolytic oil (bio-oil) is very different from petroleum-derived fuels and cannot be used directly as a fuel because of its physicochemical properties, such as low calorific value, high oxygen and water contents, low pH, high viscosity, and low stability (Liu, Li, Leahy, & Kwapinski, 2015).

In the gasification process, substoichiometric amounts of oxidants (pure O_2 or in air, water steam) are used to produce syngas (a mixture of CO and H_2), and if the gas mixture contains light hydrocarbons it is called producer gas. Pressurized syngas can be used for further catalytic synthesis of more complex chemicals. Depending on producer gas purity it can be used for combustion in boilers, engines, or in gas turbines if the level of tar content is very low. During gasification, char is converted into the gas and usually the process is optimized toward its production. In a new type of recirculated fluidized bed gasifier, char production is reduced to a few percentage points in the form of small particles caught by hot filters (Kirubakaran et al., 2009; Kumar, Jones, & Hanna, 2009; Kwapinska et al., 2015).

2.4 Char and biochar characterization

The following restrictions regarding properties do not necessary apply to char or charcoal; however, analytical methods for properties investigations, as pointed out below, can be applied for analyses of every biomass pyrolysis residue material.

Biochar is produced by biomass pyrolysis, a process whereby organic substances are broken down at temperatures ranging from 350°C to 1000°C in a low-oxygen thermal process. Torrefaction, hydrothermal carbonization, and coke production are further carbonization processes whose end products cannot be called biochar. Most often, the reason is that the process temperature and carbon content are too low in the final solid residue. However, the scientific literature is not consistent and a solid residue after thermal processes of biomass or biowaste with restricted oxygen access is very often called biochar, despite its origin, properties, carbon content, or production method.

Because there are personal and environmental health and safety risks inherent in producing biochar, the IBI has developed guidelines to assist in the safe and effective development and testing of biochar production technologies (Lynch & Joseph, 2010). The top concerns are to:

- ensure the safety of equipment operators and the general public;
- minimize emissions of atmospheric contaminants;
- produce biochar that is suitable for soil application.

Following the IBI and EBC, there are specific, required, physicochemical properties that have to be fulfilled to call a material biochar:

- The biochar's carbon content must be higher than 50% of the dry mass (according to the DIN 51732 test method). Pyrolyzed organic matter with carbon content lower than 50% is classified as pyrogenic carbonaceous material.
- The molar $H/C_{organic}$ ratio must be less than 0.7 (according to the DIN 51732 test method), and the molar $O/C_{organic}$ ratio must be less than 0.4 (according to the DIN 51733 test method). The ratio is an indicator of the degree of carbonization and therefore of the biochar's stability. Because a biochar is a carbon storage material the ratio should be as low as possible.
- The quantity of volatile organic compounds that condensate from gases into biochar surfaces and in pores during pyrolysis must be measured (by thermogravimetric analyses). Condensed volatile compounds can significantly change a biochar's functionality.
- The biochar nutrient contents with regard to nitrogen, phosphorus, potassium, magnesium, and calcium are subject to major fluctuations (according to the DIN EN ISO 17294-2 (E29) test method). Not every form of nutrient in biochar is equally available for plants, e.g., for phosphor (P) it is essential to distinguish between inorganic P fraction, which is mainly

labile P (weakly bound to the sample matrix), organic P fraction with the apatite P fraction, which is a stable form of P and assumed to be associated with Ca, and finally nonapatite inorganic P fraction, which is a moderately labile P and assumed to be associated with oxides and hydroxides of Al, Fe, and Mn (Pardo, Lopez-Sanchez, & Rauret, 2003).

- There are two groups that follow different thresholds for heavy metals (according to the DIN EN ISO17294-2 (E29) and for mercury the DIN EN1483 (E12) test methods):
  - The lower quality group follows Germany's Federal Soil Protection Act (Bundes-Bodenschutzverordnung or BBodSchV): Pb < 150 g/t dry mass (DM); Cd < 1.5 g/t DM; Cu < 100 g/t DM; Ni < 50 g/t DM; Hg < 1 g/t DM; Zn < 400 g/t DM; Cr < 90 g/t DM; As < 13 g/t DM;
  - The higher quality group follows Switzerland's Chemical Risk Reduction Act (Schweizerische Chemikalien-Risikoreduktions-Verordnung or ChemRRV): Pb < 120 g/t DM; Cd < 1 g/t DM; Cu < 100 g/t DM; Ni < 30 g/t DM; Hg < 1 g/t DM; Zn < 400 g/t DM; Cr < 80 g/t DM; As < 13 g/t DM.

According to the EBC, toxic accumulation of heavy metals could practically be ruled out, even when thresholds are higher, because the amount of heavy metals contained in the original feedstock will remain in the final product, and biochar is able to very effectively bind a number of heavy metals, thereby immobilizing them for a long period of time. However, the above statement should be verified, as at this stage there is not enough proof in the literature.

- Certified biochar must have a specific pH value, specific surface area, and ash content. The biochar's pH value is usually basic and influences its sorption capacity of nutrients. Specific surface area should be high since it influences not only water holding capacity, but also biochar and can be a perfect shelter for microorganisms that support plant growth. Values of Brunauer-Emmett-Teller surface recommended for biochar by the EBC are over 150 m^2/g, which might be very difficult to reach. Values reported in the literature for various carbonized biomaterials are often below 50 m^2/g (Sigmund, Huffer, Hofmann, & Kah, 2017; Zhao et al., 2018; Zhao, Ta, & Wang, 2017).
- The biochar's polyaromatic hydrocarbon (PAH) content (sum of the Environmental Protection Agency's 16 priority pollutants) must be under 12 mg/kg for lower grade and under 4 mg/kg for higher grade biochar (according to the DIN EN 15527: 2008–09 (with toluene extraction) test method). Advanced methods of biochar production allow for significant reduction of PAH content.
- Polychlorinated biphenyl content must be below 0.2 mg/kg and levels of dioxins and furans must be below 20 ng/kg (according to the AIR DF 100, HRMS test method). Levels of those contaminants are usually very low in advanced installations. In cases where feedstock contains larger amounts of chlorine, a higher concentration of dioxins can occur.

It is very important that biomass feedstock for biochar production comes primarily from agricultural products, is grown in a sustainable manner, and ideally should not be transported more than 80 km for carbonization. The EBC also regulates the pyrolysis process with regard to external energy used: indeed, energy used for operating the reactor must not exceed 8% or 4% (respectively to a biochar grade) of the calorific value of the biomass pyrolyzed in the same period, in steady-state process conditions. The pyrolysis gases produced during pyrolysis must be stored and used later for subsequent energy purposes or burned. Its combustion must comply with national emission thresholds for such furnaces. The heat produced by the reactor must be recycled. At least 70% of this must be used for drying biomass, for heating, for generating electricity, or for similar sustainable purposes. A small-scale biochar production unit with an annual output of less than 50 tons is highly recommended, and is exempt from heat recovery.

Sustainability principles of biochar production formulated by the Australian and New Zealand Biochar Researchers Network (2009), and also applied by Lynch and Joseph (2010) in their guideline for biochar production, are:

- Soil quality should always be maintained or improved, but never degraded. The proposed application of biochar to soil/land should demonstrate, on the balance of probabilities and after seeking appropriate scientific advice and opinions, that the receiving soils and the proposed land use will benefit as anticipated by the application (quantum and characteristics) of the particular biochar materials presented. As an absolute minimum it should be demonstrated that soil quality will not be degraded by the activity.
- Biochar production should always be able to demonstrate the sustainability of the biomass (and mineral) resources supply.
- Biochar production processes should always be able to demonstrate a genuine "community licence to operate" in addition to any statutory approvals necessary from the prevailing jurisdictions.

For every biochar production installation it is necessary to conduct a life-cycle assessment to check that the system emits fewer greenhouse gases than when it is sequestered in biochar and estimate energy balance for the system. Processes should be summarized by overall mass and energy balance. Hazard and operability concerns should also be addressed. The industrial-scale biochar production system boundaries are presented in Fig. 5 (Roberts, Gloy, Joseph, Scott, & Lehmann, 2010). Once the biomass is collected, it is transported to the pyrolysis facility where it is reduced in

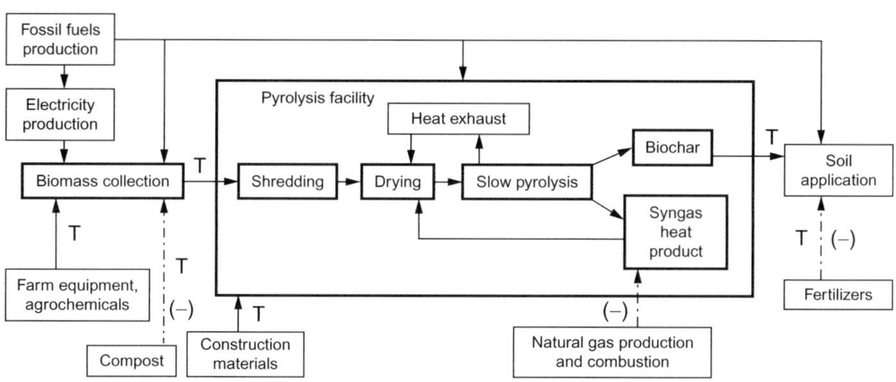

Fig. 5 Industrial biochar system. Dashed arrows with (-) indicate avoided processes. The "T" represents transportation.
Reproduced with permission from Roberts, K. G., Gloy, B. A., Joseph, S., Scott, N. R., & Lehmann, J. (2010). Life cycle assessment of biochar systems: estimating the energetic, economic, and climate change potential. *Environmental Science & Technology*, *44*(2), 827–833. Copyright 2018, ACS.

size and dried; this is called preprocessing. This is followed by materials handling (by transportation and/or storage) before it is transported by feeding equipment to the pyrolysis reactor. The produced syngas and oils are combusted on-site for heat applications. The solid residue (biochar) is transported to a farm and applied to annual crop fields.

2.5 Char production development

A key issue in developing countries is the use of traditional energy in inefficient kilns that need a lot of wood to produce charcoal and that release greenhouse gases into the atmosphere. Once made, charcoal is often burnt in closed atmospheres in houses with inefficient stoves, raising public health issues. Charcoal production can also sometimes be linked with poor working conditions, child labor, human rights abuses, and land rights conflicts (TFT, 2015).

The first documented technique that used biochar to improve soil fertility was discovered in South America in the Amazon region. The biochar was probably produced in a pit by pyrolysis of biowaste and biomass at least 2500 years ago. The area that the carbonized material was applied to is known in Brazil as Terra Preta, and is highly fertile even today, while the soil in the region was known as highly infertile (Glaser, Lehmann, & Zech, 2002; Laird, 2008).

Through the centuries, the process of char production evolved; however, in some undeveloped regions it has not changed much. There are a number of monographs and handbooks that provide information on various types of reactors for char, charcoal, or biochar production (Emrich, 1985; FAO, 1983; Garcia-Perez, Garcia-Nunez, Lewis, Kruger, & Kantor, 2012; Garcia-Perez et al., 2010).

2.5.1 Earth pit

A small or large pit was dug into a soil, filled with wood, and after setting fire to a stack and sufficient heat was generated, the pit was covered with earth as a shield against oxygen and heat loss (Fig. 6). For some period of time a small air inlet could be left to allow partial burning, which is a heat donor for pyrolysis. Carbonization may take a couple of days before the temperature inside is low enough that the char will not burn exposed to air, and the pit can be uncovered. Because the distance varies from a source of heat to various parts of a pit, there is a temperature distribution and different heat rates, and the char will be heterogeneous having different properties. This batch process is impossible to control, and if temperature in some parts is too low, the material might not be fully carbonized.

A further problem with pits is reabsorption of pyroligneous acid through rain falling on the pit. The pyroligneous acids tend to condense in the foliage and earth used to cover the pit. When heavy rain falls they are washed back down and are absorbed by the charcoal. They cause jute bags to rot and, on burning, the charcoal produces unpleasant smoke and greenhouse gases. The emissions, particulate matter (black carbon), and toxic volatile organic compounds affect the local environment and pose a health hazard

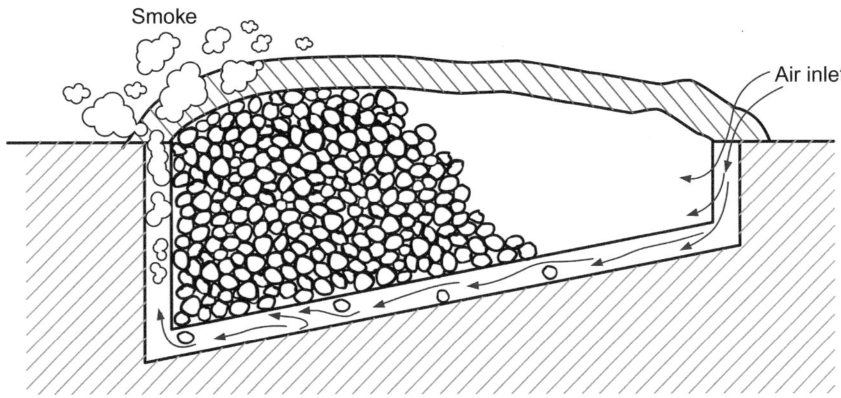

Fig. 6 Earth pit.
Based on FAO. (1983). *Simple technologies for charcoal making*. Food and Agriculture Organization of the United Nations.

to the workers operating in these pits (Pennise et al., 2001). The life expectancy of such workers is significantly shorter. Nevertheless, skilled operators using pits that are not too big can make excellent quality char. Earth pit is the oldest method but is still used nowadays for producing charcoal because it is simple and cheap. This method is ideal where the soil is well drained, deep, and loamy (FAO, 1983).

2.5.2 Kiln

A kiln consists of mounds or silos covered with soil, bricks, concrete, or metal, constructed over a kind of pit.

The simplest kiln to build is a *mound kiln* (Fig. 7). Most often, a rectangular pile of evenly cross-cut logs is stacked on a grid of crossed logs to allow volatiles to circulate. Inlets for air are opened at the base of the pile and are used to control the rate of burning. Attempts have been made to operate a modified form of this system on a larger scale using earth-moving equipment. Large logs are rolled into a shallow excavation and more logs are rolled and piled on top, using a bulldozer. Foliage is spread over the pile and earth is bulldozed over the heap to cover the fuelwood. A fire is lit at one or more points and, when burning well, the fire points are scaled with earth. The system is successful where no air leaks occur in the cover. In practice, poor yields of charcoal are common because the process is difficult when large logs are rolled into place to get a well-packed stack, gas circulation is erratic, and large amounts of uncarbonized wood result. Sealing of the piles is difficult and at times dangerous for the operator to repair. The result is that air leaks are not controlled and the charcoal is reduced to ashes in some parts of the pile before the remainder has been carbonized properly (FAO, 1983).

Brick kilns must be simple to construct, relatively unaffected by thermal stresses in heating up and cooling, and strong enough to withstand the mechanical stresses of loading and unloading. The kiln must allow controlled air entry at all times, and during

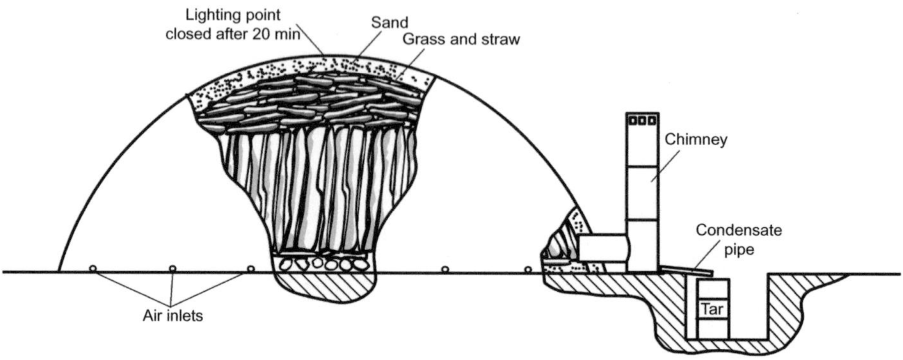

Fig. 7 Mound kiln.
Based on FAO. (1983). *Simple technologies for charcoal making*. Food and Agriculture Organization of the United Nations.

the cooling phase must be able to be readily sealed hermetically to prevent the entry of air. It must be of reasonably light weight construction to allow cooling to take place fairly easily. Good thermal insulation for the wood undergoing carbonization is crucial, otherwise cold spots appear due to wind impact on the kiln walls, which prevents proper burning of the charcoal and can lead to excessive production of partially carbonized wood pieces ("brands") and low yields (FAO, 1983). Usually, all those conditions are possible to fulfill because brick kilns are very popular and easily recognized, for example, the Missouri kiln (Wolk, Lux, Gelber, & Holcomb, 2007), the Brazilian beehive kiln (Bustamante-Garcia et al., 2013), the Argentine half-orange kiln (Rueda, Baldi, Gasparri, & Jobbagy, 2015), and the Schwartz kiln (FAO, 1983). The main difference between the Schwartz kiln and the others is that it uses an external grate, passed through the kiln, while the others burn some of the carbonized material to supply heat. The main drawback of Argentine and Brazilian kilns is that they are very environmentally unfriendly since they do not allow the burning of by-products (volatiles including tar), and are consequently thermally ineffective. In those kilns, the temperature can be monitored by thermocouples and the process can be controlled by regulating the air flow inlet.

Metal kilns can have a number of various constructions where an air inlet is placed at the bottom and produced volatiles move inside the chamber up and down to be evacuated by an exhaust chimney. One of the main advantages of metal kilns is the possibility of moving them when different seasons dictate. However, the raw material for carbonization needs to be in a form of relatively small chips (about $5 \times 30\,\text{cm}$), which requires a significant amount of additional labor.

Kilns operate in batch mode, or in a number of variations in a system of periodic reactors that supply relatively stable flow of pyrolytic gases for efficient burning. The periodic mode should allow environmental criteria of biochar production to be fulfilled.

The concept of a smokeless fire was the starting point for the *Kon-Tiki flame curtain kiln* (Schmidt & Taylor, 2014) design for char making. There are two concepts: a steel, deep-cone bowl (Fig. 8) or a soil pit, consisting of a conically shaped hole in the

Fig. 8 The Kon-Tiki kiln.

ground. Both are cost-effective and very easy to operate. They can easily be built and used by farmers in both the developed and developing world. It was shown that the quality of biochar produced from various feedstocks complies with international quality standards required by the IBI and EBC. Gas and aerosol emissions were very low compared with all other low-cost and traditional charcoal and biochar production devices (Cornelissen et al., 2016). Smebye, Sparrevik, Schmidt, and Cornelissen (2017) concluded during the life-cycle analyses that flame curtain kilns have more positive potential environmental impacts than traditional earth-mound and retort kilns due to their smaller production emissions. The use of earth-mound kilns should not be advocated. Even though they do not require any material or investment, they are slow, laborious, and have negative potential impacts for the environment.

2.5.3 Retorts

In kilns, part of the biomass is combusted to generate energy for drying and pyrolysis, and sometimes the produced gas is burned. However, when the heat is generated by those pyrolysis gases and vapors and used as an energy source the reactors should be called *retorts* (Fig. 9). Lack of burning, smoke, and wood gases produced during pyrolysis in traditional kilns generates pollution. The retort process is divided into two phases of operation. During the first phase, which takes about 4 h, the hot volatiles from the fire box are directly channeled to the wood chamber and the wood is dried. Once the smoke from the chimney, which is mainly steam, becomes more yellow in color, this is an indication that the second phase, the retort operation, can begin. The smoke and wood gases are redirected toward the fire box, the wood gases are flared, and the heat generated is recycled (Adam, 2009). The walls of a retort are heated by direct contact with flue gases. Heat transfer to the material is not rapid and the process

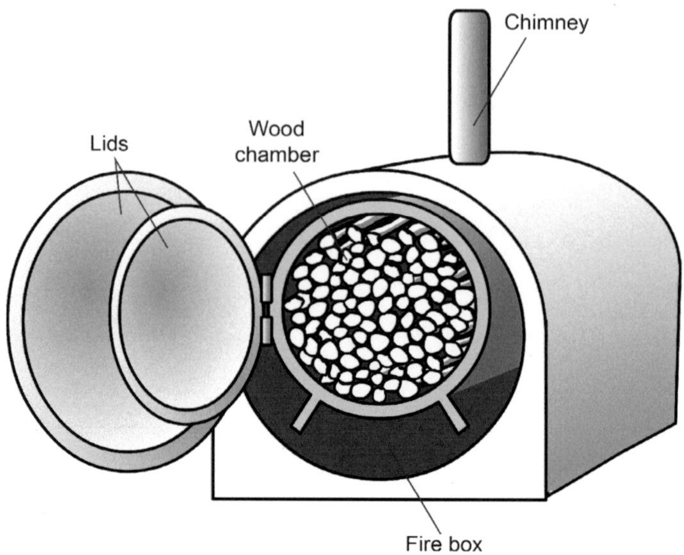

Fig. 9 Exeter retort loaded with short rotational wood.

is called "slow" pyrolysis. There are a number of available designs and some offer the possibility of moving the retorts from one place to another.

Hawaii Natural Energy Institute (HNEI) researchers have patented a Flash Carbonization process for charcoal production with a thermal afterburner that destroys smoke. A packed bed of moist biomass contained in a pressure vessel can be ignited easily under an air pressure of 1 MPa. When the bottom of the bed is ignited, and air is delivered to the top of the bed, a flash fire quickly migrates up the beds against the flow of air triggering the transformation of the moist biomass to carbon (i.e., charcoal) and gas (i.e., steam, carbon dioxide, carbon monoxide, methane, and hydrogen). Under these circumstances, the fixed-carbon yield of charcoal throughout the bed can reach the thermochemical equilibrium "limit" in less than 30 min of reaction time (Antal, Mochidzuki, & Paredes, 2003).

2.5.4 Pyrolyzer (drum reactor)

Nowadays, a more common construction can offer a continuous process, which consists of a feeding system for biomass from one side of a retort (often called a pyrolyzer), and with the help of a screw the biomass is slowly moved to the other side and converted to biochar and vapor (Fig. 10). This solution offers the possibility of making a continuous process. One of the newest pilot-scale solutions is being tested at the University of Edinburgh (United Kingdom) (Masek et al., 2018). Often for transportation, a rotary drum is used with a screw or buffers on its internal surface to ensure a good mixing and proper movement of biomass; alternatively, rotary screw/s can be used. The drum/reactor is directly heated by combustion gases at the start of the process or if the biomass is very wet natural gas burners are used.

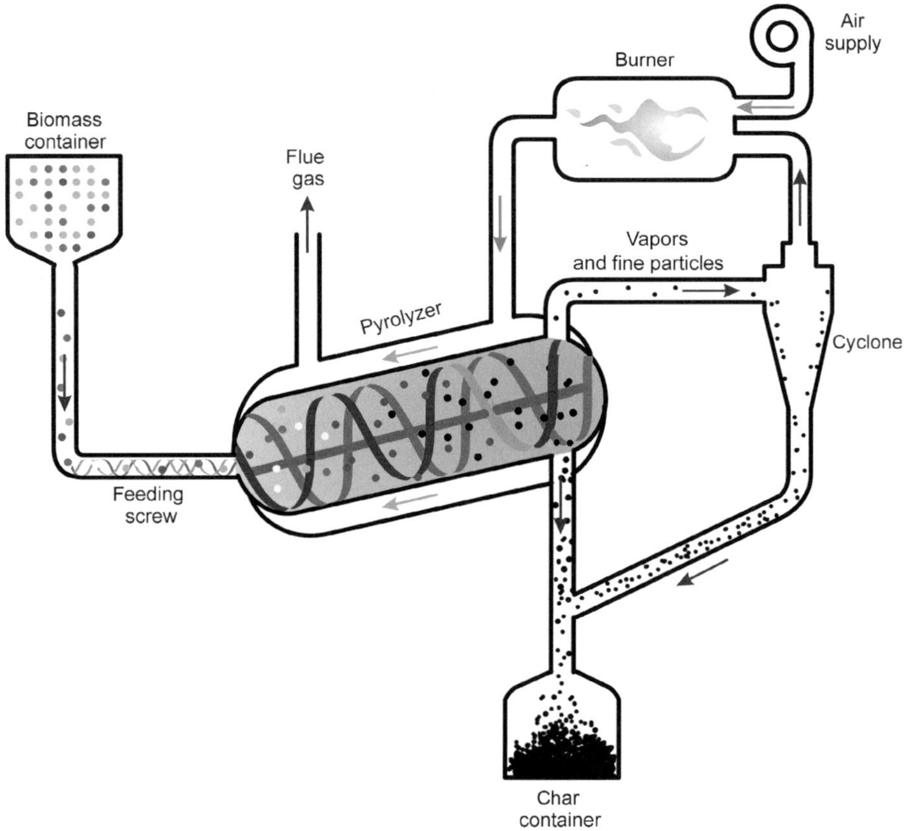

Fig. 10 Slow pyrolysis continuous screw reactor, indirectly heated with the combusted pyrolysis gases and vapors.

Industrial-scale installations are also available and are in operation; various solutions are proposed by 3RAgrocarbon (Hungary), Premier Green Energy (Ireland), Pyreg (Germany), and Biogreen (France).

Usually, a high temperature is introduced to the original predried biomass material; pyrolysis occurs and heats the biomass to a high temperature in the absence of oxygen. The source of the energy for pyrolysis can be heat from burning the produced gases and vapors generated during pyrolysis. Biomass predrying reduces the moisture content in biomass, which increases the pace of the process in the pyrolyzer. Predried and preheated biomass is sent to the main part of the reactor and the following process occurs. All moisture accumulated in a raw biomass has to evaporate to initiate a carbonization process. It can take from a few seconds to a few minutes, and its efficiency depends on the temperature, structure of the moving bed, and size of the particles. In the next step, the biomass is degasified at high temperature and volatiles are evacuated from the biomass. The material remains in the hot zone and the carbonization process begins. The process can be controlled by biomass dosing speed, the biomass moisture content, and residence time of the biomass, which moves forward in the reactor by changing the screw rotation and the temperature of the reactor walls. The biochar leaving the pyrolyzer needs to cool down in the absence of oxygen to prevent autoignition. The vapors leaving the reactor have a high temperature and depend on the technical solution that will be adopted: they can be burnt to produce heat for pyrolysis and drying processes, they can go directly to a separator, e.g., a flash condenser, to produce bio-oil, or a noncondensable gas can be used as a fuel or can be further cleaned for other applications.

The process whereby heat is transferred to a biomass rapidly, the heating rate is over 100°C/s (Amutio et al., 2012; Demirbas, 2004), and vapor residence time is short (<2 s) is called "fast" pyrolysis. As already mentioned, in fast pyrolysis the main product is a dark brown homogeneous liquid called bio-oil with a heating value about half of diesel oil. There are a number of constructions and processes that can fulfill those conditions, including a bubbling fluidized bed, circulating fluid bed, rotating cone reactor, and vacuum moving bed (Agblevor, Besler, & Wiselogel, 1995).

Bubbling fluid beds have the advantages of a well-understood technology that is not very advanced in construction; however, they are not simple to operate, but have good temperature control and very efficient heat transfer to biomass particles arising from the high solids density. Fig. 10 shows a typical configuration using electrostatic precipitators for coalescence and collection of what are referred to as aerosols char. Vapor and solid residence time is controlled by the fluidizing gas flow rate and is higher for char than for vapors. Because char acts as an effective vapor cracking catalyst at fast pyrolysis reaction temperatures, rapid and effective char separation is important. This is usually achieved by ejection and entrainment followed by separation in one or more cyclones; so, careful design of sand and biomass/char hydrodynamics is important. The by-product char is typically about 15 wt.% of the products but about 25% of the energy of the biomass feed (Bridgwater, 2012). For large throughput, a circulating fluidized bed can be used, which is a modification of the installation from Fig. 11. The main amendment is a combustor where the char from cyclones is directed to produce additional energy and hot sand is recirculated to a fluidized bed. This technology is used in the large bio-oil production plant operated by Fortum Otso in Finland. Produced bio-oil is combusted for energy generation, and the greatest benefit is that it reduces CO_2 emissions by ~90% compared to fossil fuels.

Ablative pyrolysis can be considered as a possible alternative to a fluidized bed. In ablative pyrolysis the surface, heated by hot flue gas, is rotating and biomass is pressed onto the hot surface (approximately 600°C). This type of process is called "flash" pyrolysis, and the process is characterized by much higher heating rates of 10^4°C/s and shorter residence times (<0.5 s), resulting in very high bio-oil yields up to 80 wt.%. A rotating cone reactor is a modification of ablative pyrolysis investigated by Aston University (Birmingham, United Kingdom) and developed by BTG (the Netherlands). It operates as a transported bed reactor, but with transport caused by centrifugal forces in a rotating cone rather than gas (Bridgwater, 2012; Venderbosch & Prins, 2010).

Other techniques such as microwave pyrolysis where heat derives inside a particle by radiation (Huang, Chiueh, Kuan, & Lo, 2016; Li et al., 2016; Salema, Afzal, &

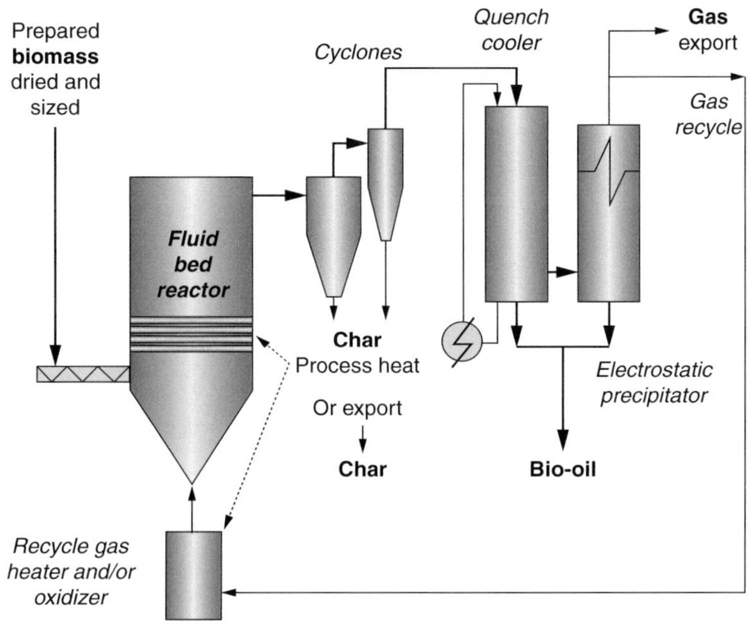

Fig. 11 Bubbling fluid bed reactor with electrostatic precipitator.
Reproduced with permission from Bridgwater, A. V. (2012). Review of fast pyrolysis of biomass and product upgrading. *Biomass & Bioenergy*, *38*, 68–94. Copyright 2018, Elsevier.

Bennamoun, 2017), or induction heating applying a high-frequency magnetic field (Lee, Tsai, Tsai, & Lin, 2010; Muley, Henkel, Abdollahi, Marculescu, & Boldor, 2016), can also qualify as fast pyrolysis with reduced char production and bio-oil as the main product.

A classification of the available carbonization technology ranging from small-scale, low-cost pyrolyzers to more modern systems is presented in Table 2. This classification is based on the reactor type, operation type, scale, construction material, conversion efficiency, emissions, and auxiliary requirements.

2.6 Hydrothermal carbonization

Biomass and biowaste contain a high percentage of moisture, which has a negative impact because a drying step is necessary. The hydrothermal carbonization (HTC) process is considered to be a promising treatment option due to its ability to convert wet material into valuable products with less volume and mass (Cao et al., 2013; Ghanim, Kwapinski, & Leahy, 2017; Ramzan, Ashraf, Naveed, & Malik, 2011). HTC, also known as wet torrefaction, is a thermochemical conversion process; it mimics the natural coalification process and was first proposed by Friedrich Bergius in the early part of the 19th century (Kambo & Dutta, 2015). HTC can be defined as a

Table 2 Classification and important characteristics of carbonization technologies.

Reactor type	Process type	Capacity	Construction materials	Conversion efficiency (wt.%)	Energy source	Residence time	Emissions (g/kg char)	Auxiliary	Portability	Capital cost
Earth pit mound	Batch	10–330 m^3	Soil, sod	90 kg char/ m^3 wood	Partial oxidation	20 days/ 180 m^3	CO_2: 1058–3027; CO: 143–333; CH_4: 32–62; TSP: 13–411	None	Impermanent	€24/ton charcoal
Brick kilns Brazilian Beehive	Batch	50–330 m^3	Brick, mortar	90 kg char/ m^3 wood	Partial oxidation	20–30 days/ 270 m^3	CO_2: 1533; CO: 373; CH_4: 57	None	Stationary	€130–1300
Metal kilns Missouri	Batch	350 m^3	Steel, brick/ concrete	25–36	Partial oxidation	80 h	CO_2: 543–560; CO: 140–162; CH_4: 37–54; TSP: 160	Tar recovery	Stationary	€13,000
Drum reactors ARTI	Batch	200 dm^3	Mild steel	19–30	Partial oxidation	~1 day	CO_2: 1517; CO: 336; CH_4: 58	None	Portable	€11–55/ ton charcoal
Low-tech retorts Adam	Batch	3 m^3	Brick or earth blocks	30–42	Partial oxidation and volatile combustion	12 h	n/a	None	Stationary and portable version	€300
High-tech retorts Carbo twin retort	Batch	2×5 m^3	Steel	30–33	External heat and volatile combustion	32–36 h	Complies with Dutch emission standards	Oil burner, fork lift, hoist and rail, sand lock	Stationary	€1 million

Pyrolyzer PYREG	Continuous	1400 tons p.a.	Steel	60	500 kW and volatile combustion	n/a	n/a	None	Stationary	n/a
Pyrolyzer 3RAgrocarbon	Continuous	28,500 tons p.a.	Steel	50–80	n/a	>20 min	n/a	None	Stationary	n/a
Flash carbonization HNEI	Batch	594 tons p. a.	Steel vessel and piping components	30–50	Partial oxidation	20 min	n/a	Compressed air source, electric ignition	Stationary	€180/ton charcoal
Hydrothermal carbonization AVA-CO_2	Batch	2664 tons p.a. char produced	Steel vessel and piping components	37–60	Steam	5–10 h	n/a	Mixing tank high pressure reactors, buffer tank	Stationary	€10–12 million

n/a, not available.
Modified from Lohri, C. R., Rajabu, H. M., Sweeney, D. J., & Zurbrugg, C. (2016). Char fuel production in developing countries—a review of urban biowaste carbonization. *Renewable & Sustainable Energy Reviews*, *59*, 1514–1530 and other references mentioned in the text.

pressurized thermochemical process for conversion of biomass and/or biowaste with high moisture content and/or in the presence of water at relatively moderate temperatures (180–260°C) under self-generated pressure with residence times from a few minutes to several hours (Ghanim, Pandey, Kwapinski, & Leahy, 2016; Gu et al., 2017; Reza, Coronella, Holtman, Franqui-Villanueva, & Poulson, 2016; Wikberg, Ohra-aho, Pileidis, & Titirici, 2015). The degradation of feedstock material is the main role of the HTC process, which is performed in the liquid state, making it unique among thermochemical conversion processes for biomass and/or biowaste. During HTC, water can act as a mild acid and mild base catalyst because its ionic product is maximized at high treatment temperatures (Lynam, Coronella, Yan, Reza, & Vasquez, 2011); this can facilitate the degradation process. Thus, the carbonization process undergoes a complex process of reactions such as hydrolysis, dehydration, decarboxylation, and other reactions. However, depolymerization and polymerization are considered the major steps in hydrochar production (Reza, Rottler, Herklotz, & Wirth, 2015). The key variables that determine the HTC process and its products are treatment temperature, residence time, and initial pH (chemical addition), while the raw material/water ratio, heating/cooling rate, and mixing/stirring characteristics also contribute, but to a lesser extent. The residue material obtained after HTC is called hydrochar and significantly differs from biochar having higher H/C and O/C ratios, reduced higher heating value, and lower porosity. However, due to the simplicity of the process, its low-energy intensity, and possible agriculture applicability of the hydrochar, it has gained a lot of attention.

2.7 Sustainability assessment

To assess the sustainability of char production, different technologies were evaluated by Lohri et al. (2016) using technical, financial, and environmental criteria. The technical criterion is based on biomass feedstock suitability (moisture, particle size), feedstock pretreatment (drying, grinding), carbonization capacity, system portability, labor intensity, carbonization controllability (temperature, residence time, energy consumption), lifespan, char yield, and demonstrated use (technical functionality validation). The financial criterion is based on the capital cost, operating cost, and gas recovery (heat or other applications). The environmental criterion is based on pollutants emission, tar recovery, and water requirement. Further details on these criteria are found in Lohri et al. (2016). These authors have assessed the different carbonization technologies based on the literature review and their experience. They have proposed a five-point scale (+2: much better than, +1: better than, 0: equal to, −1: worse than, −2: much worse than) to evaluate the different technologies with the baseline. Drum reactors were chosen as the baseline technology since they are widely used for agricultural waste carbonization. Weights were ascribed to designate the importance of each criterion for overall technology sustainability (3: high importance, 2: medium importance, 1: low importance). Total scores were obtained by multiplying weights and scores. The obtained values allowed ranking of the different technologies. Table 3 shows the overall sustainability assessment matrix.

Table 3 Technology assessment matrix (drum reactor as baseline technology).

Assessment criteria	***Weight***	**Reactor type**							
		Earthen pit/ mound	**Brick kiln**	**Metal kiln**	**Drum reactor (baseline)**	**Low-tech retort**	**High-tech retort**	**Flash carbonizer**	**HTC reactor**
Technical aspects									
Suitability for biowaste	*3*	−1	−1	−1	0	0	0	0	+2
Feedstock pretreatment	*2*	−1	−1	−1	0	0	−1	−1	+2
Throughput	*2*	−2	−1	−1	0	0	+2	+1	+1
Portability	*2*	−2	−2	−1	0	−1	−2	−1	−1
Labor intensity	*2*	−1	+1	+1	0	0	−1	−1	−1
Controllability	*2*	−2	−1	−1	0	+1	+2	+2	+2
Conversion efficiency	*2*	0	0	0	0	+1	+1	+2	+2
Lifespan	*2*	−2	+2	0	0	0	+2	+2	+1
Demonstrated use	*2*	+1	0	0	0	−1	−1	−2	−2
Financial aspects									
Capital cost	*3*	+1	−1	−1	0	−1	−2	−2	−2
Operating cost	*3*	+1	0	0	0	0	−2	−2	−2
Gas recovery	*2*	0	0	0	0	+1	+1	+1	+1
Environmental and health aspects									
Pollutant emissions	*3*	0	0	0	0	+1	+1	+1	+1

Continued

Table 3 Continued

Tar recovery	*1*	−1	0	0	0	+1	+1	+2	+2
Water requirement	*2*	0	0	0	0	0	0	0	−2
Safety	*3*	−1	+1	0	0	+1	+1	−2	−2
Total weighted score		**−19**	**−7**	**−12**	**0**	**+6**	**+1**	**−7**	**−1**
(Overall ranking)		*(7)*	*(5)*	*(8)*	*(3)*	*(1)*	*(2)*	*(5)*	*(4)*

Note: The italic numbers in the second column are weights (not scores). The italic numbers in brackets in the last row are ranks (not scores). All other (nonitalic) numbers are scores. Reproduced with permission from Lohri, C. R., Rajabu, H. M., Sweeney, D. J., & Zurbrugg, C. (2016). Char fuel production in developing countries—a review of urban biowaste carbonization. *Renewable & Sustainable Energy Reviews*, *59*, 1514–1530. Copyright 2018, Elsevier.

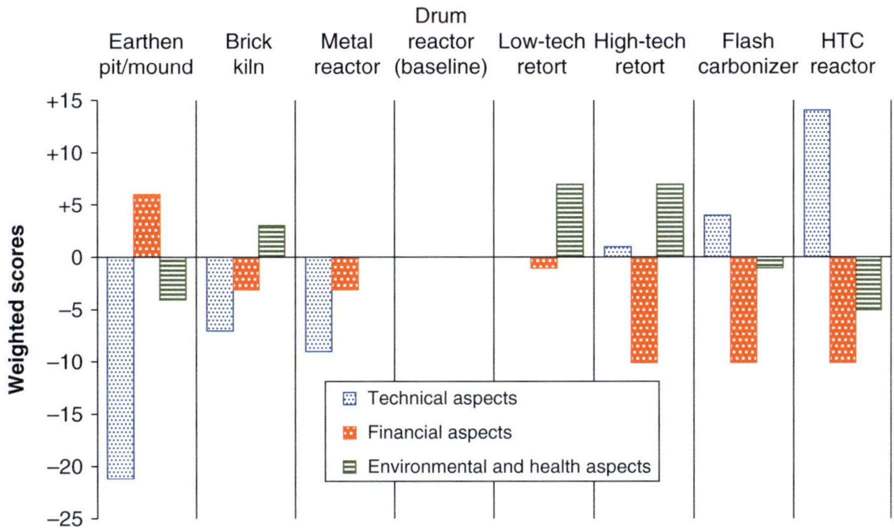

Fig. 12 Assessment results by criteria categories: Technology suitability compared to drum reactor. *HTC*, hydrothermal carbonization.
Reproduced with permission from Lohri, C. R., Rajabu, H. M., Sweeney, D. J., & Zurbrugg, C. (2016). Char fuel production in developing countries—a review of urban biowaste carbonization. *Renewable & Sustainable Energy Reviews*, *59*, 1514–1530. Copyright 2018, Elsevier.)

The low-tech retorts obtained the highest weighted score. Therefore, this technology was ranked first in terms of overall sustainability for biomass feedstock carbonization, followed by high-tech retorts. Drum reactors were ranked third, while metal kilns and earthen pit/mounds obtained the lowest weighted score. Furthermore, Fig. 12 shows the sustainability of the different technologies from the technical, financial, and environmental criteria. The summed results of each criterion are expressed as weighted score differences compared to the baseline (drum reactor). This comparison indicates that the high-tech systems (retorts, flash pyrolysis, HTC) score positively from a technical point of view. However, these technologies receive remarkably negative scores in the financial criterion. In contrast, earthen pit/mounds have low-tech retorts followed by metal reactors.

2.8 Conclusions

Char is the solid product obtained during the pyrolysis process, and can be a product in the biorefinery concept. Char is the general name for any carbonized material, charcoal refers to the product of pyrolysis of vegetable or animal matter in kilns, and biochar refers to a carbonized biomass that has a rigor production method in regard to sustainability. Chars' unique properties allow for broad application, and its production methods are still modified and upgraded depending on needs or regulations.

Biochar can play a significant role as a soil additive and further biochar field testing is required to validate laboratory-assessed behavior in various soils and observe the development of functional properties during time.

The process of material formation is very complex and was tackled by a number of researchers for a more universal process model description. There is no clear consensus for describing kinetics pathways and constants that could be valid in large numbers of situations.

Char can be produced in advanced high-tech retorts or pyrolyzers that allow product separation and heat recovery as well as in simple methods, e.g., Kon-Tiki flame curtain kiln. This simple kiln does not allow for energy recovery from vapors but can produce good-quality char. In the case of traditional technology (pit or kiln), skilled operators are the most important element because proper condition adjustment can produce char with high yield and excellent quality. However, traditional pits and kilns have very high gas, vapor, and condensate emissions and pollute the atmosphere and soil.

References

Adam, J. C. (2009). Improved and more environmentally friendly charcoal production system using a low-cost retort-kiln (Eco-charcoal). *Renewable Energy*, *34*(8), 1923–1925.

Agblevor, F. A., Besler, S., & Wiselogel, A. E. (1995). Fast pyrolysis of stored biomass feedstocks. *Energy & Fuels*, *9*(4), 635–640.

Amutio, M., Lopez, G., Artetxe, M., Elordi, G., Olazar, M., & Bilbao, J. (2012). Influence of temperature on biomass pyrolysis in a conical spouted bed reactor. *Resources Conservation and Recycling*, *59*, 23–31.

Antal, M. J., & Gronli, M. (2003). The art, science, and technology of charcoal production. *Industrial & Engineering Chemistry Research*, *42*(8), 1619–1640.

Antal, M. J., Mochidzuki, K., & Paredes, L. S. (2003). Flash carbonization of biomass. *Industrial & Engineering Chemistry Research*, *42*(16), 3690–3699.

Australian and New Zealand Biochar Researchers Network. (2009). *Guidelines for a safe biochar industry*. .

Bridgwater, A. V. (2012). Review of fast pyrolysis of biomass and product upgrading. *Biomass & Bioenergy*, *38*, 68–94.

Bustamante-Garcia, V., Carrillo-Parra, A., Gonzalez-Rodriguez, H., Ramirez-Lozano, R. G., Corral-Rivas, J. J., & Garza-Ocanas, F. (2013). Evaluation of a charcoal production process from forest residues of *Quercus sideroxyla* Humb., & Bonpl. in a Brazilian beehive kiln. *Industrial Crops and Products*, *42*, 169–174.

Cao, X. Y., Ro, K. S., Libra, J. A., Kammann, C. I., Lima, I., Berge, N., et al. (2013). Effects of biomass types and carbonization conditions on the chemical characteristics of hydrochars. *Journal of Agricultural and Food Chemistry*, *61*(39), 9401–9411.

Cha, J. S., Park, S. H., Jung, S. C., Ryu, C., Jeon, J. K., Shin, M. C., et al. (2016). Production and utilization of biochar: a review. *Journal of Industrial and Engineering Chemistry*, *40*, 1–15.

Cornelissen, G., Pandit, N. R., Taylor, P., Pandit, B. H., Sparrevik, M., & Schmidt, H. P. (2016). Emissions and char quality of flame-curtain "Kon Tiki" kilns for farmer-scale charcoal/biochar production. *PLoS ONE*, *11*(5).

Crombie, K., Masek, O., Cross, A., & Sohi, S. (2015). Biochar—synergies and trade-offs between soil enhancing properties and C sequestration potential. *Global Change Biology Bioenergy*, *7*(5), 1161–1175.

Czernik, S. (2008). Fundamentals of charcoal production. In: *Proceedings of the IBI conference on biochar, sustainability and security in a changing climate, 8-10 September, Newcastle, UK.*

Czernik, S., & Bridgwater, A. V. (2004). Overview of applications of biomass fast pyrolysis oil. *Energy & Fuels, 18*(2), 590–598.

Demirbas, A. (2004). Pyrolysis of municipal plastic wastes for recovery of gasoline-range hydrocarbons. *Journal of Analytical and Applied Pyrolysis, 72*(1), 97–102.

EBC. (2017). *Guidelines for a sustainable production of biochar*. Version 6.3E of 14th August 2017.

Emrich, W. (1985). *Handbook of biochar making. The traditional and industrial methods.* D. Reidel Publishing Company.

FAO. (1983). *Simple technologies for charcoal making*. Food and Agriculture Organization of the United Nations.

Garcia-Perez, M., Adams, T. T., Goodrum, J. W., Geller, D. P., & Das, K. C. (2007). Production and fuel properties of pine chip bio-oil/biodiesel blends. *Energy & Fuels, 21*(4), 2363–2372.

Garcia-Perez, M., Garcia-Nunez, J. A., Lewis, T., Kruger, C., & Kantor, S. (2012). *Methods for producing biochar and advanced bio-fuels in Washington state part 3: Literature review.*

Garcia-Perez, M., Lewis, T., & Kruger, C. E. (2010). *Methods for producing biochar and advanced biofuels in Washington State, part 1: Literature review of pyrolysis reactors.* First Project Report Pullman, WA: Department of Biological Systems Engineering and the Center for Sustaining Agriculture and Natural Resources. Washington State University. 137 pp.

Ghanim, B. M., Kwapinski, W., & Leahy, J. J. (2017). Hydrothermal carbonisation of poultry litter: effects of initial pH on yields and chemical properties of hydrochars. *Bioresource Technology, 238*, 78–85.

Ghanim, B. M., Pandey, D. S., Kwapinski, W., & Leahy, J. J. (2016). Hydrothermal carbonisation of poultry litter: effects of treatment temperature and residence time on yields and chemical properties of hydrochars. *Bioresource Technology, 216*, 373–380.

Glaser, B., Lehmann, J., & Zech, W. (2002). Ameliorating physical and chemical properties of highly weathered soils in the tropics with charcoal—a review. *Biology and Fertility of Soils, 35*(4), 219–230.

Gu, L., Wu, S. F., Li, B. L., Wen, H. F., Zhang, D. F., Ye, J. F., et al. (2017). Persulfate oxidation assisted hydrochar production from Platanus Orientalis Leaves: physiochemical and combustion characteristics. *Bioresource Technology, 244*, 517–524.

Harris, P. (1999). On charcoal. *Interdisciplinary Science Reviews, 24*(4), 301–306.

Hosoya, T., Kawamoto, H., & Saka, S. (2007). Cellulose-hemicellulose and cellulose-lignin interactions in wood pyrolysis at gasification temperature. *Journal of Analytical and Applied Pyrolysis, 80*(1), 118–125.

Huang, Y. F., Chiueh, P. T., Kuan, W. H., & Lo, S. L. (2016). Microwave pyrolysis of lignocellulosic biomass: heating performance and reaction kinetics. *Energy, 100*, 137–144.

IBI (2015). *Standardized product definition and product testing guidelines for biochar that is used in soil.* International Biochar Initiative.

Kambo, H. S., & Dutta, A. (2015). A comparative review of biochar and hydrochar in terms of production, physico-chemical properties and applications. *Renewable & Sustainable Energy Reviews, 45*, 359–378.

Kan, T., Strezov, V., & Evans, T. J. (2016). Lignocellulosic biomass pyrolysis: a review of product properties and effects of pyrolysis parameters. *Renewable & Sustainable Energy Reviews, 57*, 1126–1140.

Kirubakaran, V., Sivaramakrishnan, V., Nalini, R., Sekar, T., Premalatha, M., & Subramanian, P. (2009). A review on gasification of biomass. *Renewable & Sustainable Energy Reviews, 13*(1), 179–186.

Kumar, A., Jones, D. D., & Hanna, M. A. (2009). Thermochemical biomass gasification: a review of the current status of the technology. *Energies*, *2*(3), 556–581.

Kwapinska, M., Xue, G., Horvat, A., Rabou, L., Dooley, S., Kwapinski, W., et al. (2015). Fluidized bed gasification of torrefied and raw grassy biomass (miscanthus x gigantenus). The effect of operating conditions on process performance. *Energy & Fuels*, *29*(11), 7290–7300.

Kwapinski, W., Byrne, C. M. P., Kryachko, E., Wolfram, P., Adley, C., Leahy, J. J., et al. (2010). Biochar from biomass and waste. *Waste and Biomass Valorization*, *1*(2), 177–189.

Laird, D. A. (2008). The charcoal vision: a win-win-win scenario for simultaneously producing bioenergy, permanently sequestering carbon, while improving soil and water quality. *Agronomy Journal*, *100*(1), 178–181.

Lede, J. (2012). Cellulose pyrolysis kinetics: an historical review on the existence and role of intermediate active cellulose. *Journal of Analytical and Applied Pyrolysis*, *94*, 17–32.

Lede, J. (2013). Biomass fast pyrolysis reactors: a review of a few scientific challenges and of related recommended research topics. *Oil & Gas Science and Technology—Revue d'IFP Energies Nouvelles*, *68*(5), 801–814.

Lee, M. K., Tsai, W. T., Tsai, Y. L., & Lin, S. H. (2010). Pyrolysis of napier grass in an induction-heating reactor. *Journal of Analytical and Applied Pyrolysis*, *88*(2), 110–116.

Lehmann, J., & Joseph, S. (Eds.), (2009). Biochar for environmental management: Science and technology. London, Sterling, VA. Earthscan.

Lewellen, P. C., Peters, W. A., & Howard, J. B. (1977). Cellulose pyrolysis kinetics and char formation mechanism. *Symposium (International) on Combustion*, *16*(1), 1471–1480.

Li, J., Dai, J. J., Liu, G. Q., Zhang, H. D., Gao, Z. P., Fu, J., et al. (2016). Biochar from microwave pyrolysis of biomass: a review. *Biomass & Bioenergy*, *94*, 228–244.

Liu, Y., Li, Z., Leahy, J. J., & Kwapinski, W. (2015). Catalytically upgrading bio-oil via esterification. *Energy and Fuels*, *29*, 3691–3698.

Lohri, C. R., Rajabu, H. M., Sweeney, D. J., & Zurbrugg, C. (2016). Char fuel production in developing countries—a review of urban biowaste carbonization. *Renewable & Sustainable Energy Reviews*, *59*, 1514–1530.

Lynam, J. G., Coronella, C. J., Yan, W., Reza, M. T., & Vasquez, V. R. (2011). Acetic acid and lithium chloride effects on hydrothermal carbonization of lignocellulosic biomass. *Bioresource Technology*, *102*(10), 6192–6199.

Lynch, J., & Joseph, S. (2010). *Guidelines for the development and testing of pyrolysis plants to produce biochar.* .

Masek, O., Buss, W., Roy-Poirier, A., Lowe, W., Peters, C., Brownsort, P., et al. (2018). Consistency of biochar properties over time and production scales: a characterisation of standard materials. *Journal of Analytical and Applied Pyrolysis*, *132*, 200–210.

Melligan, F., Auccaise, R., Novotny, E. H., Leahy, J. J., Hayes, M. H. B., & Kwapinski, W. (2011). Pressurised pyrolysis of Miscanthus using a fixed bed reactor. *Bioresource Technology*, *102*(3), 3466–3470.

Muley, P. D., Henkel, C., Abdollahi, K. K., Marculescu, C., & Boldor, D. (2016). A critical comparison of pyrolysis of cellulose, lignin, and pine sawdust using an induction heating reactor. *Energy Conversion and Management*, *117*, 273–280.

Pardo, P., Lopez-Sanchez, J. F., & Rauret, G. (2003). Relationships between phosphorus fractionation and major components in sediments using the SMT harmonised extraction procedure. *Analytical and Bioanalytical Chemistry*, *376*(2), 248–254.

Pennise, D. M., Smith, K. R., Kithinji, J. P., Rezende, M. E., Raad, T. J., Zhang, J. F., et al. (2001). Emissions of greenhouse gases and other airborne pollutants from charcoal making in Kenya and Brazil. *Journal of Geophysical Research-Atmospheres*, *106*(D20), 24143–24155.

Ramzan, N., Ashraf, A., Naveed, S., & Malik, A. (2011). Simulation of hybrid biomass gasification using Aspen plus: a comparative performance analysis for food, municipal solid and poultry waste. *Biomass & Bioenergy*, *35*(9), 3962–3969.

Reza, M. T., Coronella, C., Holtman, K. M., Franqui-Villanueva, D., & Poulson, S. R. (2016). Hydrothermal carbonization of autoclaved municipal solid waste pulp and anaerobically treated pulp digestate. *ACS Sustainable Chemistry & Engineering*, *4*(7), 3649–3658.

Reza, M. T., Rottler, E., Herklotz, L., & Wirth, B. (2015). Hydrothermal carbonization (HTC) of wheat straw: influence of feedwater pH prepared by acetic acid and potassium hydroxide. *Bioresource Technology*, *182*, 336–344.

Roberts, K. G., Gloy, B. A., Joseph, S., Scott, N. R., & Lehmann, J. (2010). Life cycle assessment of biochar systems: estimating the energetic, economic, and climate change potential. *Environmental Science & Technology*, *44*(2), 827–833.

Ronsse, F. (2016). *Biochar production.* Cambridge University Press.

Rueda, C. V., Baldi, G., Gasparri, I., & Jobbagy, E. G. (2015). Charcoal production in the Argentine Dry Chaco: where, how and who? *Energy for Sustainable Development*, *27*, 46–53.

Salema, A. A., Afzal, M. T., & Bennamoun, L. (2017). Pyrolysis of corn stalk biomass briquettes in a scaled-up microwave technology. *Bioresource Technology*, *233*, 353–362.

Schmidt, H. P., & Taylor, P. (2014). Kon-Tiki flame curtain pyrolysis for the democratization of biochar production. *The Biochar Journal*, *1*, 14–24.

Sigmund, G., Huffer, T., Hofmann, T., & Kah, M. (2017). Biochar total surface area and total pore volume determined by N-2 and CO2 physisorption are strongly influenced by degassing temperature. *Science of the Total Environment*, *580*, 770–775.

Smebye, A. B., Sparrevik, M., Schmidt, H. P., & Cornelissen, G. (2017). Life-cycle assessment of biochar production systems in tropical rural areas: comparing flame curtain kilns to other production methods. *Biomass & Bioenergy*, *101*, 35–43.

Stefanidis, S. D., Kalogiannis, K. G., Iliopoulou, E. F., Michailof, C. M., Pilavachi, P. A., & Lappas, A. A. (2014). A study of lignocellulosic biomass pyrolysis via the pyrolysis of cellulose, hemicellulose and lignin. *Journal of Analytical and Applied Pyrolysis*, *105*, 143–150.

TFT. (2015). *Charcoal.* The Forest Trust. *http://www.tft-earth.org/.*

Van de Velden, M., Baeyens, J., Brems, A., Janssens, B., & Dewil, R. (2010). Fundamentals, kinetics and endothermicity of the biomass pyrolysis reaction. *Renewable Energy*, *35*(1), 232–242.

Venderbosch, R. H., & Prins, W. (2010). Fast pyrolysis technology development. *Biofuels Bioproducts & Biorefining-Biofpr*, *4*(2), 178–208.

Wang, S. R., Guo, X. J., Wang, K. G., & Luo, Z. Y. (2011). Influence of the interaction of components on the pyrolysis behavior of biomass. *Journal of Analytical and Applied Pyrolysis*, *91*(1), 183–189.

White, J. E., Catallo, W. J., & Legendre, B. L. (2011). Biomass pyrolysis kinetics: a comparative critical review with relevant agricultural residue case studies. *Journal of Analytical and Applied Pyrolysis*, *91*(1), 1–33.

Wikberg, H., Ohra-aho, T., Pileidis, F., & Titirici, M. M. (2015). Structural and morphological changes in kraft lignin during hydrothermal carbonization. *ACS Sustainable Chemistry & Engineering*, *3*(11), 2737–2745.

Wilkomirsky, I., Moreno, E., & Berg, A. (2014). Bio-oil production from biomass by flash pyrolysis in a three-stage fluidized bed reactors system. *Journal of Materials Science and Chemical Engineering*, *2*, 6–10.

Wolk, R. H., Lux, S., Gelber, S., & Holcomb, F. H. (2007). *Direct carbon fuel cells: Converting waste to electricity.*

Yang, H. P., Yan, R., Chen, H. P., Lee, D. H., & Zheng, C. G. (2007). Characteristics of hemicellulose, cellulose and lignin pyrolysis. *Fuel*, *86*(12–13), 1781–1788.

Zhang, J., Liu, J., & Liu, R. L. (2015). Effects of pyrolysis temperature and heating time on biochar obtained from the pyrolysis of straw and lignosulfonate. *Bioresource Technology*, *176*, 288–291.

Zhao, B., O'Connor, D., Zhang, J. L., Peng, T. Y., Shen, Z. T., Tsang, D. C. W., et al. (2018). Effect of pyrolysis temperature, heating rate, and residence time on rapeseed stem derived biochar. *Journal of Cleaner Production*, *174*, 977–987.

Zhao, S. X., Ta, N., & Wang, X. D. (2017). Effect of temperature on the structural and physicochemical properties of biochar with apple tree branches as feedstock material. *Energies*, *10*(9), 1–15.

Further reading

Alvarez, J., Lopez, G., Amutio, M., Bilbao, J., & Olazar, M. (2014). Bio-oil production from rice husk fast pyrolysis in a conical spouted bed reactor. *Fuel*, *128*, 162–169.

Bridgwater, A. V., & Peacocke, G. V. C. (2000). Fast pyrolysis processes for biomass. *Renewable & Sustainable Energy Reviews*, *4*(1), 1–73.

Budai, A., Wang, L., Gronli, M., Strand, L. T., Antal, M. J., Abiven, S., et al. (2014). Surface properties and chemical composition of corncob and miscanthus biochars: effects of production temperature and method. *Journal of Agricultural and Food Chemistry*, *62*(17), 3791–3799.

Demirbas, A. (2001). Carbonization ranking of selected biomass for charcoal, liquid and gaseous products. *Energy Conversion and Management*, *42*(10), 1229–1238.

Gonzalez, J. F., Encinar, J. M., Canito, J. L., Sabio, E., & Chacon, M. (2003). Pyrolysis of cherry stones: energy uses of the different fractions and kinetic study. *Journal of Analytical and Applied Pyrolysis*, *67*(1), 165–190.

Isahak, W., Hisham, M. W. M., Yarmo, M. A., & Hin, T. Y. Y. (2012). A review on bio-oil production from biomass by using pyrolysis method. *Renewable & Sustainable Energy Reviews*, *16*(8), 5910–5923.

Lee, Y., Eum, P. R. B., Ryu, C., Park, Y. K., Jung, J. H., & Hyun, S. (2013). Characteristics of biochar produced from slow pyrolysis of Geodae-Uksae 1. *Bioresource Technology*, *130*, 345–350.

Schellekens, J., Silva, C. A., Buurman, P., Rittl, T. F., Domingues, R. R., Justi, M., et al. (2018). Molecular characterization of biochar from five Brazilian agricultural residues obtained at different charring temperatures. *Journal of Analytical and Applied Pyrolysis*, *130*, 106–117.

Sohi, S. P., Krull, E., Lopez-Capel, E., & Bol, R. (2010). A review of biochar and its use and function in soil. *Advances in Agronomy*, *105*, 47–82.

Characterization of biomass-derived chars

3

Meriem Belhachemi, Besma Khiari†, Mejdi Jeguirim‡, Antonio Sepúlveda-Escribano§*

*Laboratory of Chemical and Environmental Sciences, University of Tahri Mohamed Bechar, Bechar, Algeria, †National School of Engineers of Carthage, Tunis, Tunisia, ‡University of Strasbourg, University of Upper Alsace, Institute of Materials Science of Mulhouse (IS2M—UMR CNRS 7361), Mulhouse, France, §Laboratorio de Materiales Avanzados, Departamento de Quimica Inorganica-Instituto Universitario de Materiales de Alicante, Universidad de Alicante, Alicante, Spain

Chapter Outline

Char and Carbon Materials Derived from Biomass. https://doi.org/10.1016/B978-0-12-814893-8.00003-1

3.1 Introduction

Thermal decomposition of an organic material under inert atmosphere at temperatures between 400°C and 1000°C produces chars. Heteroatoms (O, H, N, etc.) are removed by heating and the material becomes richer in carbon atoms, which polymerize in the form of aromatic sheets having a planar structure arranged in an irregular manner creating interstices. Examination of the structure, texture, and chemical composition of chars is very important to fully comprehend and exploit this by-product for high-value materials. This chapter describes methodologies and techniques used to characterize chars. Sometimes one has to combine two or three methods to obtain the complete information. The choice of the appropriate technique to characterize this material depends on the final objective.

This chapter is divided into four main sections. The first section examines the physical and thermal properties of chars. The second section reports the techniques used to study the morphology and texture of chars. The third section describes methods used to detect the chemical composition and surface functional groups present in chars. The last section is devoted to the techniques used to characterize the structure of chars at larger and chemical bond levels. Several techniques are presented in this chapter, including: thermogravimetric analysis/differential thermal analysis (TGA/TDA), scanning electron microscopy (SEM), gas adsorption, Fourier transform infrared spectroscopy (FTIR), X-ray photoelectron spectroscopy (XPS), chemical composition analysis, X-ray diffraction (XRD), and Raman spectroscopy. Furthermore, a set of data for char characteristics is given in this chapter. These data could be useful for researchers, engineers, and industry for the selection of suitable feedstock to design chars for their appropriate applications.

3.2 Physical and chemical properties

Physical and chemical properties of chars may be characterized through the determination of pH, electrical conductivity (EC), elemental analysis, and cation exchange capacity (CEC). Thermogravimetric analysis is also used to characterize the proximate analysis of chars and to investigate their thermal degradation behavior under inert and oxidative atmospheres.

3.2.1 pH and electrical conductivity

EC gives total water-soluble ions, whereas pH is a measure of the hydrogen ion concentration of a solution. Since chars do not settle from suspension, their pH and EC are obtained with the help of special protocols. For instance, pH can be determined using a saturated paste approach. During these analyses, an amount of char is mixed with a volume of distilled water. The pH is then measured with a probe submerged in the paste. EC is usually measured by preparing a 1% (w/w) carbon/water suspension, which is stirred for a given period of time (Suliman et al., 2016b). In general, chars act as buffers toward pH changes. After acid or basic addition, the pH values rebound to their original levels within 60 min and stabilize at their new values within 120 min. Thus, a 2-h equilibrium period is used prior to other analyses. The pH value depends on the original biomass species and preparation conditions. An example is given in Fig. 1 for chars from pine, oak, and grass for different production temperatures (Mukherjee, Zimmerman, & Harris, 2011).

Most biomass-derived chars are alkaline (pH > 7) (Xu et al., 2017). This may be due to the presence of alkali metals such as sodium and potassium, and alkaline earth metals such as calcium and magnesium in the form of carbonates (Hass et al., 2012; Yuan, Xu, & Zhang, 2011). Char ash contents are positively correlated to their alkalinity, but weakly linked to pH as pointed out by Xu, Kan, Zhao, and Cao (2016). These authors also found that, although the pH of sludge char was lower, its mineral content was higher than char prepared from wheat straw. The authors explained this behavior by the abundance of Fe in the sludge char, which causes hydrolysis and generates H^+ in the solution. Consequently, it was concluded that the mineral composition has a greater influence on the char pH than the mineral content. Functional surface groups such as —COOH and —OH also contribute to the acidity of the chars, mainly for those obtained at low temperatures (Yuan et al., 2011). Low pH values were found for biomasses carbonized at low temperature (300°C). According to González et al. (2013), the char produced at this temperature is rich in acids and phenolic compounds resulting from the decomposition of cellulose and hemicelluloses.

The pH values of the char increase with increasing pyrolysis temperature. This last statement is attributed to the removal of acid functional groups (—COOH, —OH) and to the enhancing of mineral (KOH, NaOH, $MgCO_3$, $CaCO_3$) separation from the char, which causes higher pH values (Meng et al., 2013; Mohan, Sarswat, Ok, & Pittman, 2014).

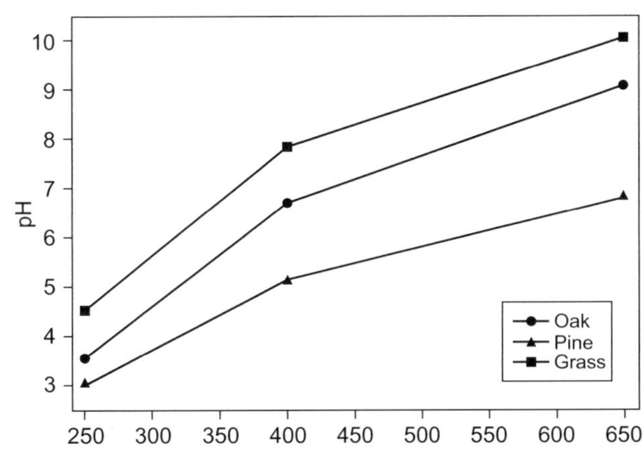

Fig. 1 pH of chars from different biomasses at 250°C under air, and at 400°C and 650°C under N_2. Reprinted with permission from Mukherjee, A., Zimmerman, A. R., Harris, W. (2011). Surface chemistry variations among a series of laboratory-produced biochars. *Geoderma*, *163*(3), 247–255. Copyright 2018, Elsevier.

Furthermore, these alkaline minerals are the reason for the enhanced buffering capacity of char (Xu & Chen, 2015).

The EC of chars is a measure of their salinity. It depends mainly on the feedstock composition and is important to avoid creating unwanted salt effects in soils. Typical EC values vary from tens to thousands of μS/cm. Cantrell, Hunt, Uchimiya, Novak, and Ro (2012), for example, found ECs ranging between 194 μS/cm for char from swine solids and 2217 μS/cm for poultry litter char at the same conditions of pyrolysis (Cantrell et al., 2012). In fact, poultry litter is usually rich in EC-influencing elements, issued from incomplete assimilation of nutrients by the animals. Their pyrolysis, especially at high temperature, increases their EC, probably due to the loss of volatiles resulting in concentration of elements in the ash fraction. Alburquerque, Sánchez, Mora, and Barrón (2016) noted a positive correlation between EC and mineral contents. The authors reported that the quantity and hydrolysis reactions of salts such as Ca, K, Na, and Mg influence mainly the salinity and alkalinity of chars (Alburquerque et al., 2016).

Similarly to pH values, electrical conductivity increases linearly with the pyrolysis temperature, especially at low temperatures (Fig. 2), and varies very significantly from one char to another. This result might be related to the ash fraction in the chars (Suliman et al., 2016b).

Table 1 shows the properties of chars together with proximate analysis reported in the literature.

3.2.2 Density

Solid or skeletal density is the mass per unit volume of the char at the molecular level, whereas bulk density or apparent density is the mass per unit volume of a collection of char particles. It includes the macroporosity within the particles and the interparticle voids.

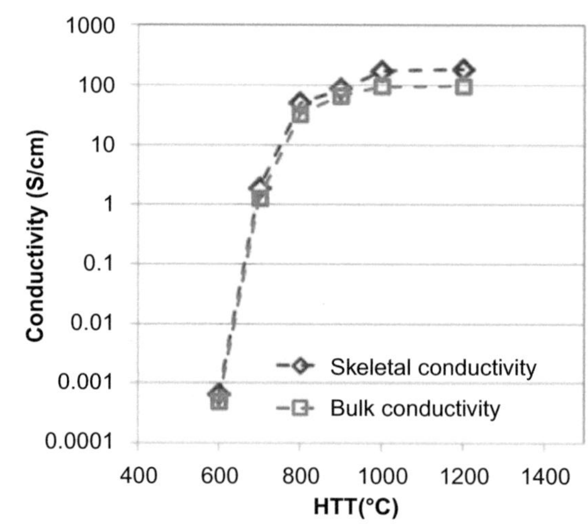

Fig. 2 Skeletal conductivity as a function of heating treatment temperature (HTT) for char from cellulose.
Reprinted with permission from Shao, Y., Guizani, C., Grosseau, P., Chaussy, D., Beneventi, D. (2018). Biocarbons from microfibrillated cellulose/lignosulfonate precursors: a study of electrical conductivity development during slow pyrolysis. *Carbon*, *129*, 357–366. Copyright 2018, Elsevier.

Table 1 Proximate analysis and some properties of chars

Char from	Operating conditions	pH	EC (μs/cm)	VM (%)	Cfixe (%)	Ash (%)	HHV (MJ/kg)	References
Cellulose	Hydrothermal carbonization 200°C 30min	–	–	84	14.7	1.3	23	Kim, Yoshikawa, and Park (2015)
Pea pod	Pyrolysis 200°C 60min	8.84	589	–	–	–	–	Stella Mary et al. (2016)
Cauliflower leaves		9.84	1310	–	–	–	–	
Orange peel		9.43	231	–	–	–	–	
Poultry litter	Pyrolysis 350°C 60min	8.7	1405	42.3 ± 1.2	27 ± 0.6	30.7 ± 0.6	21.53	Cantrell et al. (2012)
Swine solids		7.8	216	49.8 ± 0.6	17.7 ± 1.2	32.5 ± 0.9	23.63	
Wood	Torrefaction 300°C 60min	–	–	42.33	56.89	0.78	26.6	Lu, Lee, Chen, and Lin (2013)
Poplar wood	Pyrolysis 400°C 30min 190K/min	10.1	0.14	–	–	3.8 ± 1	–	Suliman et al. (2016b)
Algae	Pyrolysis 450°C 10min 17K/min	9.1	–	27.5	72.5	68.6	9.22	Frederik, Sven, Dane, and Wolter (2013)

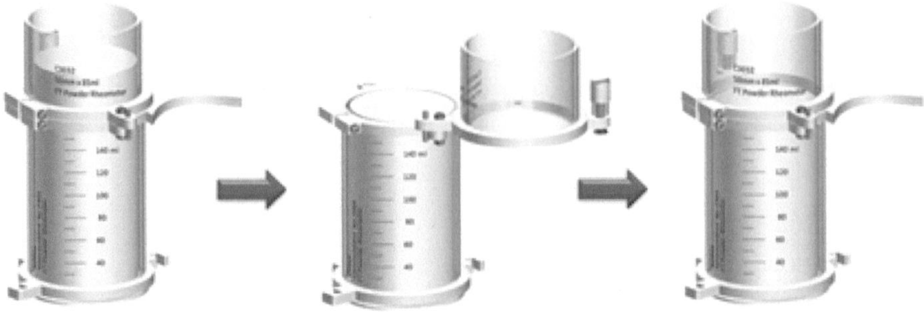

$$\text{Conditioned bulk density} = \frac{\text{Split mass after conditioning}}{\text{Sample volume}}$$

Fig. 3 Bulk density measurement.

To measure bulk density, a glass cylinder is usually filled to a specified volume with mesh powder char and dried in an oven at 80°C overnight. The cylinder is tapped for a few minutes to compact the char and the bulk density is then calculated by dividing the weight of dry material by the volume of packed dry materials (Fig. 3) (Stella Mary et al., 2016).

Several ways are proposed to attain solid density. For instance, Brewer et al. (2014) measured char skeletal density by helium pycnometry and envelope density by displacement of a dry granular suspension (Brewer et al., 2014). Using this method, char skeletal density ranged between 1.34 and 1.96 g/cm^3, whereas the envelope density varied from 0.25 to 0.60 g/cm^3. The highest values were recorded for wood chars whose plant cell structures were preserved during pyrolysis, which was not the case for grass chars, for example (Brewer et al., 2014). The typical dry solid density of char is 1.5–1.7 g/cm^3, whereas typical dry bulk density ranges from 0.25 to 0.75 g/cm^3 (Jin et al., 2018; Weber & Quicker, 2018).

3.2.3 Cation exchange capacity

CEC is a measure of the negative charge of a material that can be neutralized by exchangeable cations. This measure indicates the applicability of the char as a soil amendment and carbon sequestering agent (Nair, 2002). The Brazilian archeological chars, for instance, give CEC values very close to those of humus (Lee et al., 2010).

Typically, solutions are used to replace all surface ions with Ca^{2+}, Mg^{2+}, K^+, Na^+, and NH_4^+, which are then replaced by mass action with ions of another salt. CEC is deduced from the cations released, accounting for entrained salt. Depending on pH, hydroxyl, carboxylate, and carbonyl groups are able to chelate metals with different binding intensities (strong for Fe^{3+} or Al^{3+}, weak for Mg^{2+}, and medium for Ni^{2+}, Pb^{2+}, Ca^{2+}, and Zn^{2+}). The CECs of char are quite variable, ranging from mmol/kg (Cheng, Lehmann, & Engelhard, 2008; Gundale & DeLuca, 2006) to cmol$_c$/kg (centimoles of charges per kg) and depending on temperature, pH, and the original biomass. For instance, in the work of Mukherjee et al. (2011), the mean CEC of char obtained at 250°C was higher than that at 400 or 650°C (51.9 ± 15.3, 16.2 ± 6.0, and 21.0 ± 17.2 cmol$_c$/kg, respectively) at near neutral pH (Mukherjee et al., 2011). The CEC of char obtained at 250°C from grass, pine, and oak increased by a factor of 4–7 when pH varied from 1.5 to 7. However, among 400°C and 650°C chars, only grass 650°C char showed CEC dependency on pH, with an increase from 10.2 to 40.8 cmol$_c$/kg. Liang et al. (2006) reported that the incorporation of carboxylic oxygen groups and other functionalities surface with negative charge improve the CEC (Liang et al., 2006). By increasing the CEC in the soil, its fertility can be improved by reducing the leaching of nutrients.

3.2.4 Proximate analysis

It is widely admitted that fixed carbon, volatile matter (VM), and ash influence the properties and applications of chars (Buss, Graham, Shepherd, & Mašek, 2016; Yuan et al., 2011; Yuan et al., 2015). Some values are reported in Table 1. They are determined by using a high-temperature furnace and a TGA according to specific protocols. During this characterization, a determined mass of char is kept in a drying oven for at least 2 h at 100°C. VM and ash contents are obtained as weight losses after combustion at 900°C for 6 min and at 750°C for 6 h, respectively. All weighing are proceeded after cooling in a desiccator for 1 h (Crombie, Mašek, Sohi Saran, Brownsort, & Cross, 2012; Mitchell, Dalley, & Helleur, 2013; Suliman et al., 2016a).

Minerals are often detected by X-ray fluorescence (XRF) and XRD. The XRF technique is a nondestructive analysis that is used to identify and quantify the elemental composition of chars. X-rays are used to excite the atoms in the sample, causing them to emit energy X-rays characteristic of each element present. The intensity and energy of these X-rays are then measured. On the other hand, the XRD tool is described in Section 3.5 (char structure).

Ash is composed of mineral components, mainly alkali metals such as K and Na and alkaline earth metals such as Ca or Mg, which are present either as discrete mineral phases or are associated with the surface functional groups of chars (Buss et al., 2016; Gunes et al., 2015; Wang et al., 2015; Yuan et al., 2011) in the form of carbonates, phosphates, nitrates, or oxides (Cao & Harris, 2010; Xu, Cao, Zhao, Zhou, & Luo, 2014). A wide range of ash content values is reported in the literature. First, the ash content depends of the feedstock nature; it can vary from 2% to 43% of the total mass char. As an example, poplar biomass pyrolyzed at 525°C yielded 6.8% of ash in the resulting char (Kloss et al., 2012). It may reach 90% for sludge char, between 20% and 80% for manure char and below 20% for plant residues (Zhao & Brooks, 2013). Mineral species such as K, Ca, Na, and Ni act as catalysts in char gasification and tar cracking.

Xu et al. (2017) showed that initial biomass had no significant effect on volatiles, but ash contents were 3–4 times greater for grass chars compared with those of oak or pine, probably because of higher K, Ca, and Mg in grass (Xu et al., 2017). The volatile and ash contents ranged from 25.2% to 66% and from 0.3% to 15.9%, respectively. While the former decreased with the pyrolysis temperature, the latter increased.

Similarly, Song and Guo (2012) reported that the content of P, K, Ca, and Mg in poultry manure chars increased by 34.4%, 32%, 30.9%, and 30.1%, respectively, when production temperature was raised from 300°C to 600°C (Song & Guo, 2012). This may be explained (1) by organic compounds being converted to gas while the less volatile mineral elements remain and are thus concentrated in char (Zhang & Wang, 2016) and (2) by minerals in chars becoming more crystallized and less soluble with increasing temperature, especially above 500°C (Cao & Harris, 2010; Yuan et al., 2011). Colantoni et al. (2016) reported that the increasing pyrolysis temperature of sunflower husk pellets and grape vine pellets increased the concentration of minerals such as potassium and phosphorus, which are potential fertilizers (Colantoni et al., 2016). The resulting minerals in chars, after high temperature pyrolysis, increase the EC and influence the pH value (Lee et al., 2013; Narzari et al., 2017).

As for carbon content in chars, the International Biochar Initiative classifies these latter into three categories according to their organic carbon content (Fu et al., 2011): Class 1 ($C > 60\%$), Class 2 ($30\% < C < 60\%$), and Class 3 ($10\% < C < 30\%$).

For Class 1 chars, mainly issued from plant residues, no relationship between minerals and carbon is evidenced. This behavior is probably due to the low mineral content, the effects of which can be masked by those of other constituents such as hydrogen, oxygen, or nitrogen. However, in Classes 2 and 3, mineral content increases, and a strong negative correlation between carbon and ash content is noticed. Singh, Singh, and Cowie (2010) found that the carbon content in high-ash char decreased with the final pyrolysis temperature, which is the opposite situation in low-ash chars (Singh et al., 2010; Zhang & Wang, 2016). The reason was mainly the different contents of minerals, which were not volatilized, thus enriching the ashes with increasing temperature (Zhang & Wang, 2016).

Minerals influence the carbon content but also the carbon distribution in the surface and bulk of chars (Fig. 4). Sun et al. (2013) declared that after de-ashing manure-derived chars, the surface carbon content decreased, while the total carbon increased (Sun et al., 2013).

This behavior was not observed for char derived from plant residues, in which minerals had no effect on the spatial distribution of carbon. One can then conclude that the nature and distribution of minerals in biomass play important roles in the content and distribution of carbon in the resulting char.

Char VM contains easily decomposable substrates, which can support plant growth and may play a critical role in determining its agronomic value as a soil amendment (Jindo, Mizumoto, Sawada, Sanchez-Monedero, & Sonoki, 2014). More precisely, high VM content may stimulate microbial activity, reduce N uptake, and may not be suitable for soil amendment in the short term.

As far as heating rate influence is concerned, and considering a wide selection of chars (Enders, Hanley, Whitman, Joseph, & Lehmann, 2012), volatiles range from 10% to 70%. Fast pyrolysis chars yield 28%–57% VM. Contrarily to ash, VM depends less on feedstock than on temperature. The VM of chars made from woody materials such as hazelnut, oak, and pine vary greatly, for example, from 28% to 61% for oak. At the same time, chars derived from animal manures and organic wastes such as food and paper vary slightly (maximum 3%).

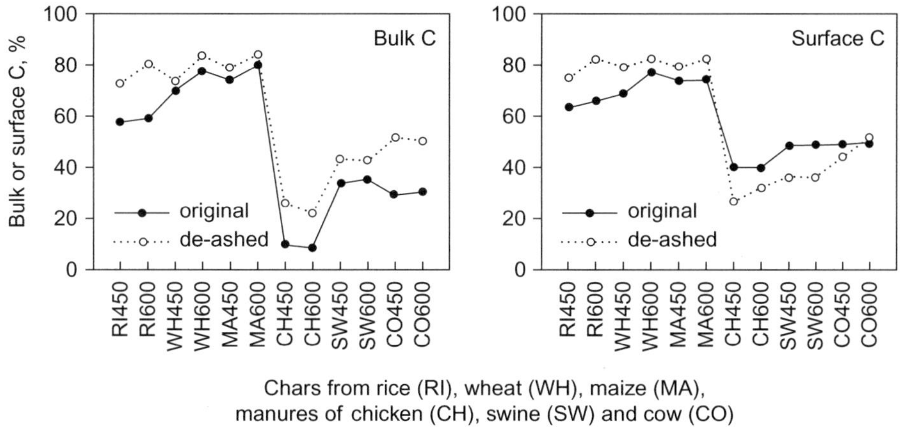

Fig. 4 Comparison of bulk or surface carbon before and after de-ashing.
Reprinted with permission from Sun, K., Kang, M., Zhang, Z., Jin, J., Wang, Z., Pan, Z., et al. (2013). Impact of deashing treatment on biochar structural properties and potential sorption mechanisms of phenanthrene. *Environmental Science and Technology*, *47*(20), 11473–11481. Copyright 2018, American Chemical Society.

3.2.5 Ultimate analysis

Total C, H, N, and S contents are usually determined using an elemental analyzer following the ASTM method. Oxygen mass fraction is obtained either by subtracting the ash and the earlier mentioned elements from the total mass of the sample or directly by dry combustion using a CHNS/O analyzer. These results are used to calculate atomic H/C, O/C, and C/N ratios, which are indicative of the bonding (aromaticity) arrangement and polarity (Cantrell et al., 2012).

Břendováa, Pavel, Jiřina, and Jan (2012) found that during the characterization of different types of char (wood maize and meadow grass) nitrogen content is below 2 wt%, hydrogen content is more or less 2%, whereas oxygen content varies between 13% and 23% and carbon content varies between 44% and 64% (Břendováa et al., 2012). Gaskin, Steiner, Harris, Das, and Bibens (2008) reported 3.4% of nitrogen in poultry litter char with conservation of about 24% at 400–500°C. No or very little sulfur is generally detected in different chars. It is thought that it escapes during pyrolysis (Gaskin et al., 2008).

The lowest C/N ratio was recorded in char from maize, while the highest ratio was observed in 550°C wood mixture char. Knicker, Totsche, Almendros, and González-Vila (2005) suggested that if the H/C ratio is around 1.3, carbon is directly bonded to a proton or is connected through —OH (Knicker et al., 2005). At the same time, higher H/C, O/C, and (O + N)/C ratios indicate chars with more interactive polar compounds (Wang, Cook, Tao, & Xing, 2007). Molar ratios decrease with increasing temperature, because of oxygen and hydrogen losses and chars tend to be more polar (high O/C and O + N/C ratios) at lower pyrolysis temperatures. It was suggested that if the H/C ratio ranges from 0.4 to 0.6 in the aromatic portion of chars, then every second to third carbon is connected to a proton, unlike soot and lignite, which often have H/C ratios less than 0.1, indicating a more graphite-like structure (Hammes et al., 2006; Knicker et al., 2005).

Increasing pyrolysis temperature results in higher carbon content. The changes in carbon content occurred concurrently with hydrogen and oxygen losses (Cantrell et al., 2012).

Table 2 reports the properties of chars prepared at different temperatures. When temperature increases from 450°C to 550°C, carbon content increases up to 6%, and oxygen and hydrogen decrease by about 4% and less than 1%, respectively (Narzari et al., 2017). Increasing heating treatment temperature (HTT) leads to lower H/C ratios, revealing an increased aromaticity and, as a consequence, a higher stability (Van Zwieten et al., 2010). Similar results can be found for hydrochars prepared with wet torrefaction.

3.2.6 Thermal analysis: TGA/TDA

TGA measures the change in mass of a material as a function of temperature and time in a controlled atmosphere. In pyrolysis conditions, this technique is used to estimate pyrolysis temperature and evaluate volatile content, thermal stability, degradation characteristics, and the kinetics of chemical pyrolysis reactions. Indeed, parameters such as VM, fixed carbon, and ashes are important in combustion applications. For example, if char has a low content in volatiles, then it would not ignite. At the same time, if char has a high percentage of volatiles, then combustion would release smoke and bad odors. Characterization may also be continued by TGA carried out in an oxidant atmosphere, or in a CO_2/H_2O atmosphere. These aspects are treated in other chapters.

TGA is often coupled with DTA, which is an experimental method consisting of heating the sample abreast a reference material in the same furnace, and measuring the difference in temperature between sample and reference (inert material) as a function of temperature or time. When a transformation occurs, a peak appears in the curve of temperature difference as a function of temperature or time. Generally, phase transitions and evaporation of solvents result in endothermic peaks. On the other hand, crystallization, oxidation, and some decomposition reactions are characterized by exothermic peaks.

Joseph, Tretsiakova-McNally, and McKenna (2012) conducted TGA analysis of char obtained from chicken litter. The weight loss from 120°C to 900°C was attributed to the heterogeneous nature of the material (Joseph et al., 2012). From Fig. 5, and from comparable TGAs, one can determine the temperature of char elaboration, which corresponds to the inflection point (500°C). Leng et al. (2015) studied the production of char from three different feedstocks: spirulina (SP), rice straw (RS), and sewage sludge (SS) (Leng et al., 2015). TGA (Fig. 6) shows that the decomposition of RS, SS, and SP mainly occurred from 200°C to 400°C; the samples were thermally stable up to the temperature of 400°C, but more significant decomposition was noticed afterward. This was indeed the temperature at which the chars were prepared.

Table 2 Properties of chars prepared from different feedstocks

Feedstock	PT (° C)	C (%)	H (%)	O (%)	N (%)	O/C	H/C	pH	References
Maple wood[a]	400	74.8	4.0	18.6	<0.5	0.64	0.19	n/a	Cao et al. (2012)
	500	83.2	3.2	10.6	<0.5	0.47	0.10	n/a	
Apple tree branches[a]	400	71.13	4.03	15.05	1.94	≈0.7	≈0.15	11.4	Zhao, Na, and Wang (2017)
	500	84.44	2.88	11.67	1.0	≈0.45	≈0.10	11.6	
Noxious weed[a]	400	51.76	1.65	44.41	2.18	0.64	0.38	7.83	Narzari et al. (2017)
	500	57.26	1.01	40.34	1.43	0.53	0.21	8.67	
Jatropha curcas seed cover[a]	400	67.48	2.01	28.36	2.15	0.32	0.36	8.9	Narzari et al. (2017)
	500	68.04	1.75	28.87	1.32	0.32	0.31	9.8	
Duckweed[b]	160	44.4	6.1	36.8	3.7	n/a	n/a	n/a	Zhang, Chen, Li, Dong, and Xiong (2016)
	250	48.3	5.2	23.4	2.4	n/a	n/a	n/a	

PT, pyrolysis temperature; *n/a*, not available.
[a]Pyrolyzed.
[b]Wet torrefaction.

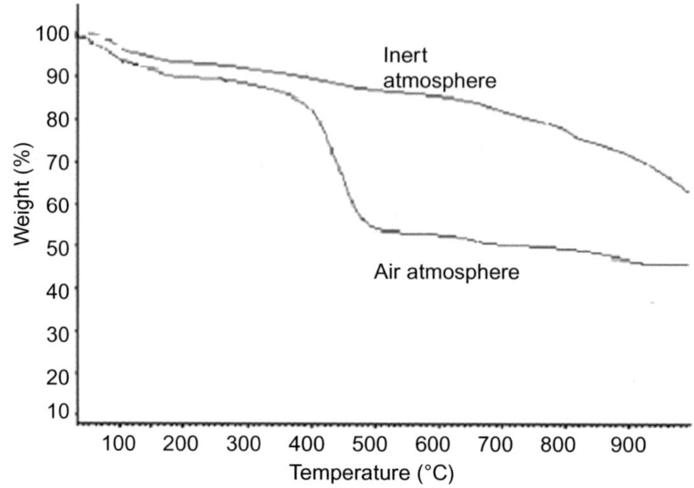

Fig. 5 Thermogravimetric analysis of chicken litter char recorded in nitrogen and air atmospheres.
Adapted with permission from Joseph, P., Tretsiakova-McNally, S., McKenna, S. (2012). Characterization of cellulosic wastes and gasification products from chicken farms. *Waste Management*, *32*(4), 701–709. Copyright 2018, Elsevier.

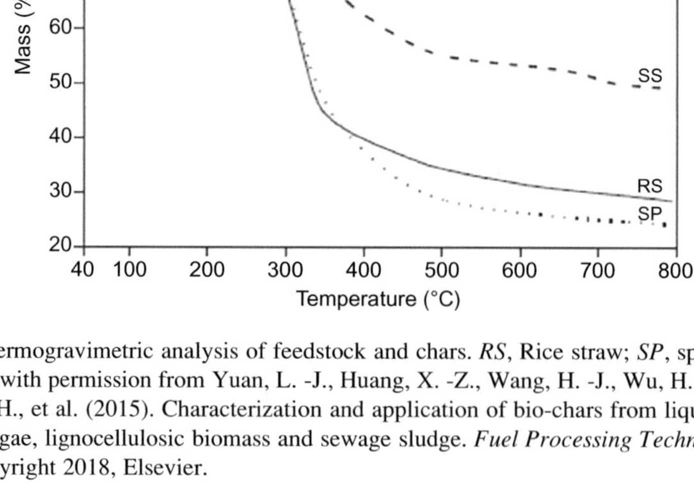

Fig. 6 Thermogravimetric analysis of feedstock and chars. *RS*, Rice straw; *SP*, spirulina.
Reprinted with permission from Yuan, L. -J., Huang, X. -Z., Wang, H. -J., Wu, H., Fu, Z. -B., Peng, L. -H., et al. (2015). Characterization and application of bio-chars from liquefaction of microalgae, lignocellulosic biomass and sewage sludge. *Fuel Processing Technology*, *129*, 8–14, Copyright 2018, Elsevier.

According to Yang et al. (2006), the whole process generally proceeds through a series of complex reactions (Yang et al., 2006) but, in general, char pyrolysis exhibits different stages of thermal degradation detected by TG/DTG analysis (Fig. 7) (Leng et al., 2015; Lu et al., 2013; Tamošiūnas et al., 2017). The first stage is attributed to the loss of entrapped water molecules (Shak & Wu, 2014); it is in the range between 2% and 10% at 150°C. In the second stage, loss occurs in the form of CO_2 from organic matter from 150°C to 750°C. Between 400°C and 700°C, organic vapors leave the samples (Teh, Wu, & Juan, 2014).

3.3 Morphological and textural properties of chars

Chars have different porous structures that depend on the feedstock and the pyrolysis or carbonization conditions. Thus, the arrangement of aromatic layers during thermal treatment is the basis of pore formation in chars. The elimination of VM also allows important cavities on the surface of chars. The morphological texture of chars can be well characterized by SEM and adsorption of gases (nitrogen and carbon dioxide) techniques.

3.3.1 Scanning electron microscopy

The SEM technique is usually used to study the morphology of material surfaces, and it is also used to follow the development of macroporosity and large mesoporosity. Several works have shown the change of char surface by SEM micrographs. These changes were reported for different types of chars as well as for chars prepared under different experimental conditions (temperature, heating rate, residence time, etc.). Narzari et al. (2017) examined the heterogeneity and porosity of chars by using SEM micrographs (Narzari et al., 2017). The samples were prepared from three different biowastes prepared at different temperatures (305–650°C) under nitrogen stream. The authors showed that the porosity was more developed at high temperature and that char exhibited a honeycomb structure, which originates from the nature of the feedstock structure (Fig. 8).

A similar trend was observed by Wang et al. (2018) during the pyrolysis of corncobs. The development of pores occurs as a result of the release of VM, and mineral species distributed on the surface of the char appear luminous (Fig. 9). The original structure of corncob was preserved in chars.

The effect of pyrolysis conditions has been examined in the literature. Fig. 10 shows SEM micrographs of hardwood residue (*Aspidosperma australe*) and hardwood chars prepared by pyrolysis at 350°C and 850°C for two residence times, 3 h and 1 h, respectively (Rocca, Cerrella, Bonelli, & Cukierman, 1999). These images exhibit evidence of surface erosion creating extensive macroporosity, although the chars maintained the cellular structure of the raw material.

A different trend was observed for chars prepared by hydrothermal carbonization. Hydrochars do not present cracks and holes like chars prepared by pyrolysis. The surface of hydrochars shows formation of microspheres (Fig. 11). According to Kang, Li,

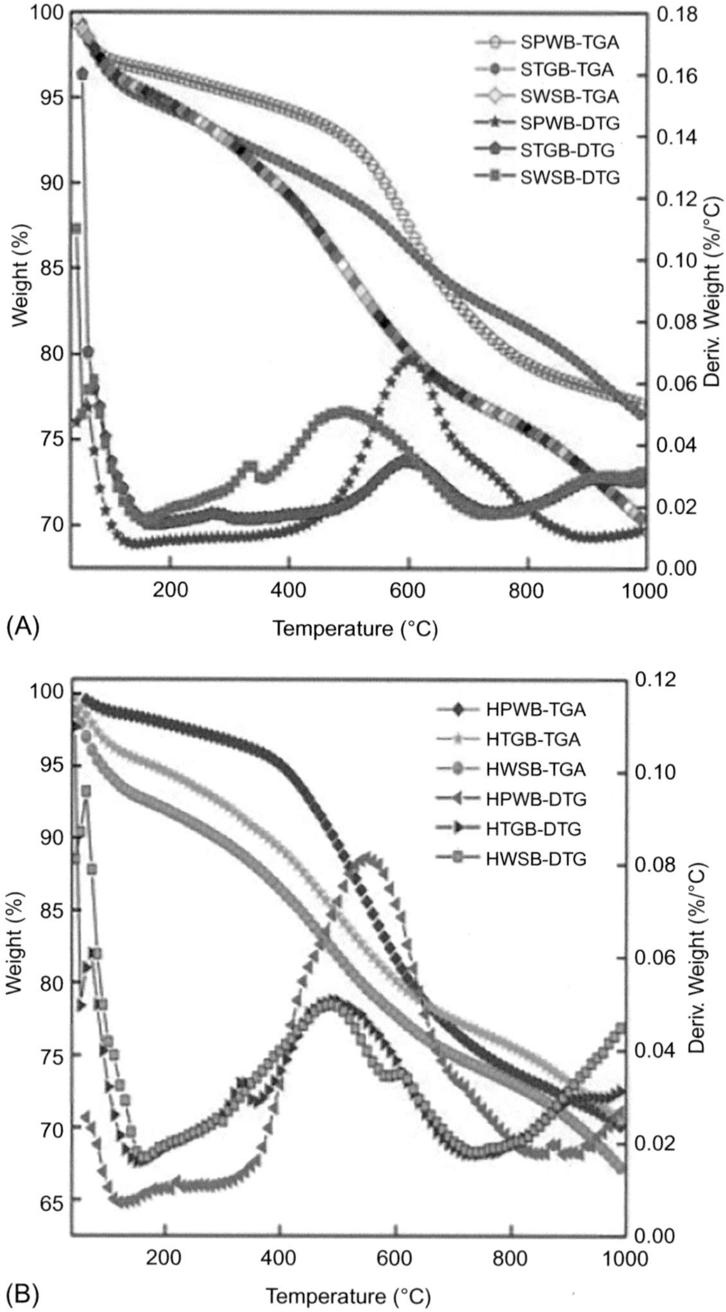

Fig. 7 Thermogravimetric and differential thermogravimetric analysis of different chars (wheat straw, timothy grass, and pinewood) at (A) high heating rate and (B) slow heating rate. Reprinted with permission from Mohanty, P., Nanda, S., Pant, K. K., Naik, S., Kozinski, J. A., Dalai, A. K. (2013). Evaluation of the physiochemical development of biochars obtained from pyrolysis of wheat straw, timothy grass and pinewood: effects of heating rate, *Journal of Analytical and Applied Pyrolysis*, *104*, 485–493. Copyright 2018, Elsevier.

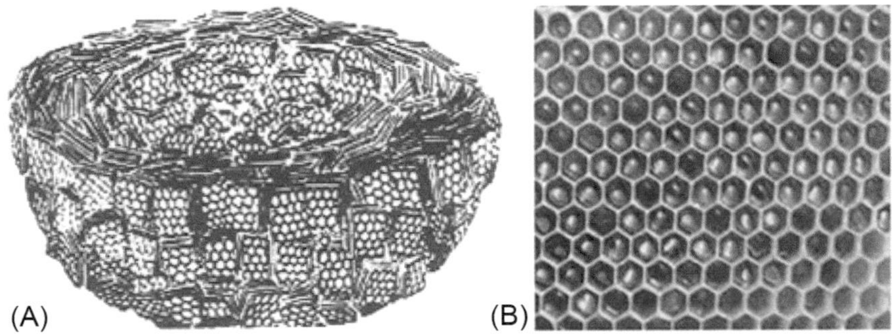

Fig. 8 (A) Structure model of carbon; (B) honeycomb texture.

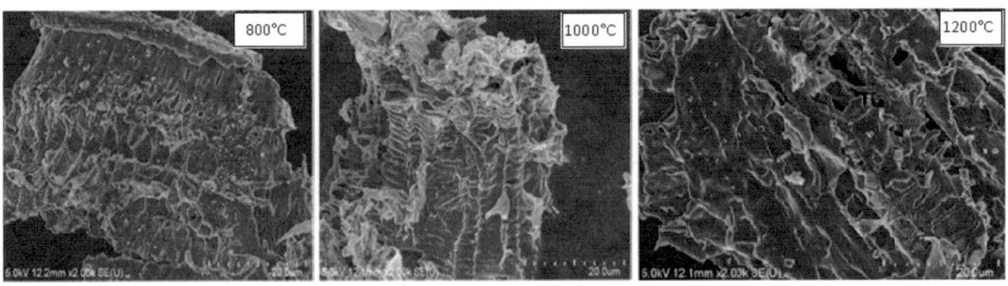

Fig. 9 Scanning electron microscopy images of corncob chars obtained at different temperatures of pyrolysis.
Adapted with the permission from Wang, G., Zhang, J., Chang, W., Li, R., Li, Y., Wang, C. (2018). Structural features and gasification reactivity of biomass chars pyrolyzed in different atmospheres at high temperature. *Energy*, *147*, 25–35. Copyright 2018, Elsevier.

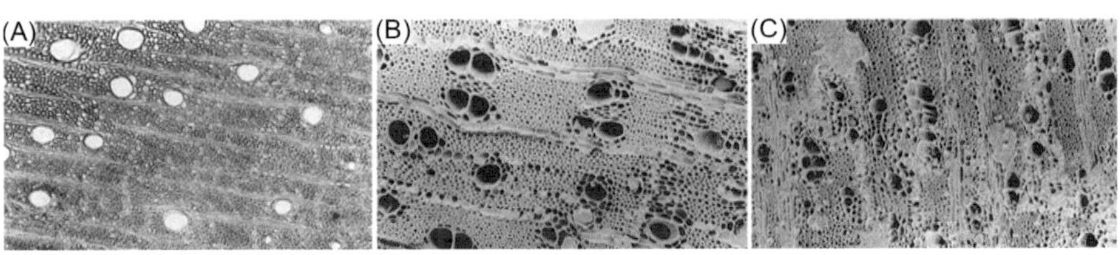

Fig. 10 (A) Optical micrograph of hardwood; (B) scanning electron microscopy (SEM) micrograph of hardwood prepared at 350°C; and (C) SEM micrograph of hardwood prepared at 850°C.
Reprinted with the permission from Rocca, P. A. D., Cerrella, E. G., Bonelli, P. R., Cukierman, A. L. (1999). Pyrolysis of hardwoods residues: on kinetics and chars characterization. *Biomass and Bioenergy*, *16*(1), 79–88, Copyright 2018, Elsevier.

Fan, and Chang (2012) the microspheres were formed from the polymerization of soluble matter in water. The authors prepared hydrochars from D-xylose and lignin at 225°C, and they reported that the polymerization of furfural from D-xylose and phenolics from dissolved lignin covered the surface of chars.

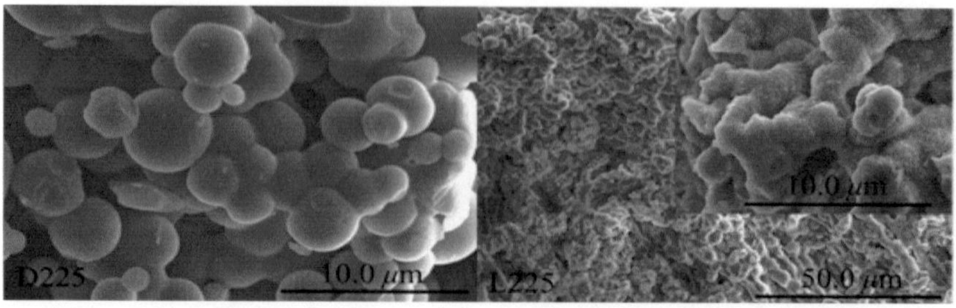

Fig. 11 Scanning electron microscopy images of D-xylose and lignin hydrochars. Reprinted with permission from Kang, S., Li, X., Fan, J., Chang, J. (2012). Characterization of hydrochars produced by hydrothermal carbonization of lignin, cellulose, d-Xylose, and wood meal. *Industrial and Engineering Chemistry Research, 51*(26), 9023–9031, Copyright 2018, American Chemistry Society.

3.3.2 Adsorption of gases analysis

The adsorption of gases is a usual and suitable technique used to characterize the microporosity and small mesoporosity in carbonaceous materials. Chars are porous materials and their texture is composed of pores with different sizes, the distribution of which varies according to the material nature and the carbonization conditions. The classification of pores adopted by the International Union of Pure and Applied Chemistry is based on their size, and according to this classification there are three categories of pores:

- micropores with a size less than 2 nm;
- mesopores with a diameter average between 2 and 50 nm;
- macropores with a size higher than 50 nm.

The micropores are divided themselves into three types: supermicropores (1.4–2 nm), micropores (0.5–1.4 nm), and ultramicropores (<0.5 nm).

It is known that micropores are mainly present in the internal area of chars, and that the mesopores act as transitional pores and are filled by capillary condensation. On the other hand, the macropores allow the routing of a fluid inside the material, acting as transporting pores.

The adsorption of gases on the disordered structure of the microporous chars, whose pore size is comparable to that of the adsorbate molecules, makes the interpretation of the physical adsorption on these materials very delicate. It is now well established that the increase of the adsorption potential in ultramicropores leads to a strong condensation of the adsorbate molecules, starting from very low relative pressures (Seaton, Walton, & Quirke, 1989). On the other hand, the constrictions in the microporous network are at the origin of the activated diffusion phenomenon at low adsorption temperature when the adsorbate does not have sufficient energy to penetrate through the microporous space; this behavior has been found in the adsorption of nitrogen at 77 K (Rodríguez-Reinoso, Garrido, Martín-Martínez, Molina-Sabio, & Torregrosa, 1989). Chars can also show selectivity toward the adsorbate molecules according to their pore shape, with preferential adsorption of the planar molecules.

The adsorption of gases on carbon materials is quantified by the adsorption isotherm, which is the expression of the amount of gas adsorbed as a function of the relative pressure:

$$n_{ads} = f(P/P_0) \tag{1}$$

where P_0 is the saturation pressure at the working temperature and P is the equilibrium pressure.

The study of the adsorption of a gas by a solid is generally intended to provide information on its specific surface and porous texture. The amount of gas retained by a given solid depends on its nature, the adsorption temperature, the nature of the gas to be adsorbed, and its vapor pressure. According to the literature, the combined use of nitrogen and carbon dioxide adsorption can provide useful information regarding the size distribution of the micropores and the degree of access of the porous structure (Belhachemi et al., 2009).

Fig. 12 shows three types of gas adsorption isotherms on chars. When the isotherm has a narrow knee with a well-defined plateau, it indicates the presence of narrow microporosity (Fig. 12a). The isotherm with a wide knee (Fig. 12b) corresponds to char with larger micropore size distribution, while Fig. 12c characterizes the presence of mesopores in char, according to the presence of a hysteresis.

As an example, Fig. 13 shows the adsorption/desorption isotherms of N_2 at −196°C for a series of chars and activated carbons issued from different waste food (Ghouma et al., 2015; Limousy, Ghouma, Ouederni, & Jeguirim, 2017), silver fir (*Abies alba*), holm oak (*Quercus ilex*), stone pine (*Pinus pinea*), and Pyrenean oak (López, Centeno, García-Díaz, & Alguacil, 2013). The shape of the N_2 adsorption isotherms of the chars is type I, with a fast rise of the initial branch followed by a well-defined plateau. This type of isotherm corresponds to microporous materials with narrow micropore size distribution.

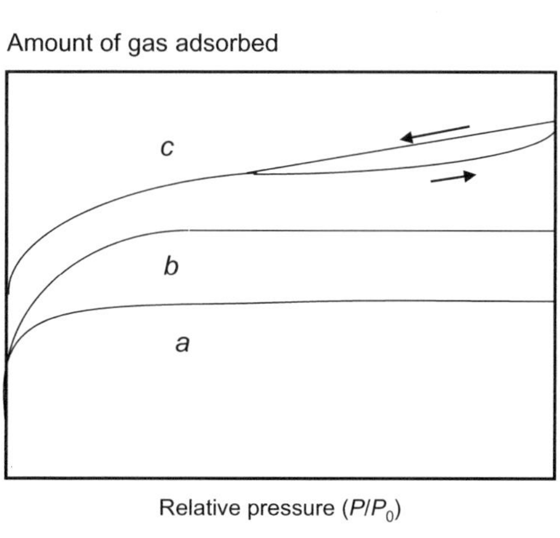

Fig. 12 Schematic representations of gas adsorption isotherms of chars.

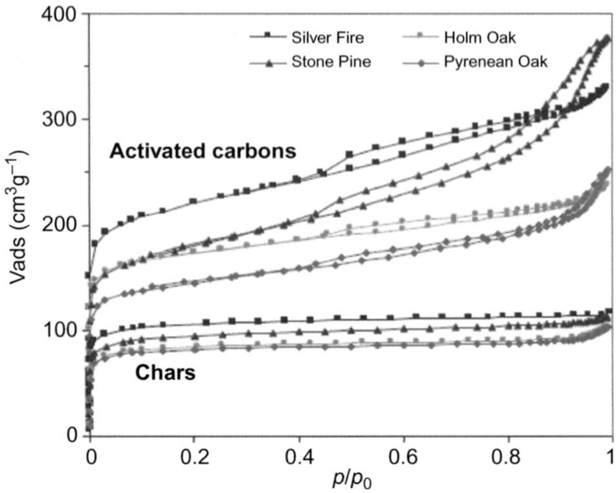

Fig. 13 N_2 adsorption/desorption isotherms at 77 K for chars and activated carbons. Reprinted with permission from López, F. A., Centeno, T. A., García-Díaz, I., Alguacil, F. J. (2013). Textural and fuel characteristics of the chars produced by the pyrolysis of waste wood, and the properties of activated carbons prepared from them. *Journal of Analytical and Applied Pyrolysis*, *104*, 551–558. Copyright 2018, Elsevier.

Table 3 Textural characteristics of chars obtained from N_2 adsorption isotherms

Samples	HTT (°C)	S_{BET} (m^2/g)	V_T (cm^3/g)	V_{meso} (cm^3/g)
Silver fir char	600	405	0.17	0.01
Holm oak char	600	316	0.13	~0.00
Stone pine char	600	361	0.16	0.02
Pyrenean oak char	600	314	0.15	0.01
Wood chips char	650	352	0.24	nd
Pellets char	700	128	0.18	nd

The analysis of the obtained isotherms provides the adsorption capacity. Physical gas adsorption also allows char micropore size distribution to be achieved by using density functional theory.

Table 3 summarizes the specific surface area and pore volumes of chars prepared from different feedstocks (Benedetti, Patuzzi, & Baratieri, 2017). These raw materials generate porous chars. In general, pyrolysis of biomass at temperatures higher than 400°C gives rise to microporous materials with important specific surface areas (Wanassi et al., 2017). Thus, the pyrolysis of important wood species at 550°C leads to chars with surface areas greater than 350 m^2/g (Keiluweit, Nico, Johnson, & Kleber, 2010). The removal of VM causes the development of pores of different sizes. By increasing the pyrolysis temperature, the micropore volume and surface develop (Bonelli, Della Rocca, Cerrella, & Cukierman, 2001).

Fu et al. (2011) examined the effect of temperature pyrolysis on the porosity (surface area, micropore and mesopore volumes) of chars prepared from maize stalk, rice straw, and cotton straw at different temperatures ranging from 600°C to 1000°C under a nitrogen stream. Their results show the increasing of Brunauer-Emmett-Teller (BET) surface area, and micropore and mesopore volumes obtained from nitrogen adsorption with temperatures from 600°C to 900°C. However, at 1000°C these parameters may decrease because of increasing structure order and pore widening. The maximum BET surface areas attained by these chars are 81.6, 30.9, and 66.5 m^2/g for maize stalk, rice, and cotton straws, respectively (Fu et al., 2011).

In general, the textural characterization of chars is mainly carried out by using nitrogen adsorption at –196°C. However, CO_2 adsorption is less used in characterizing chars. We commented above that CO_2 adsorption characterizes better the narrow micropores in chars. Indeed, the problem of restricted diffusion of nitrogen at –196°C in the micropores allows the use of CO_2 adsorption at 0°C. Bonelli et al. (2001) reported the textural analysis of chars prepared from virgin shells using N_2 and CO_2 at 196°C and 25°C, respectively. BET and Dubinin-Radushkevich (DR) equations are applied to calculate the specific surface areas and micropore volumes from the N_2 and CO_2 isotherms. Fig. 14 shows N_2 adsorption isotherms of shells and chars prepared at 350°C, 600°C, and 850°C. The isotherms of chars obtained at 600°C and 800°C indicate higher adsorption capacity than the biomass and the sample produced at 350°C.

The CO_2 adsorption data are well fitted by the DR equation in the low relative pressure range (Fig. 15). Surface areas calculated from CO_2 adsorption are higher than

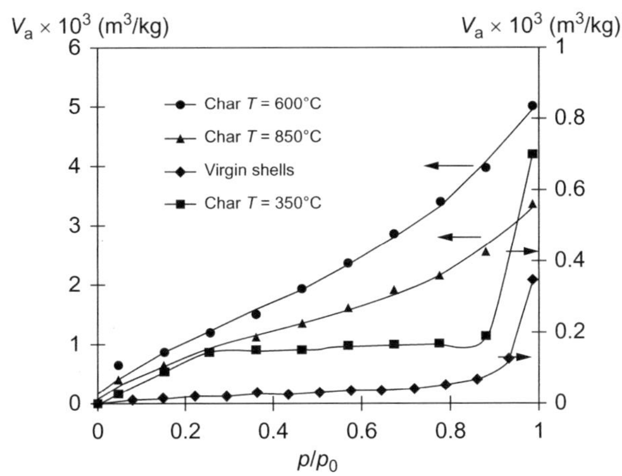

Fig. 14 N_2 adsorption isotherms at –196°C for chars.
Reprinted with the permission from Bonelli, P. R., Della Rocca, P. A., Cerrella, E. G., Cukierman, A. L. (2001). Effect of pyrolysis temperature on composition, surface properties and thermal degradation rates of Brazil Nut shells. *Bioresource Technology*, *76*(1), 15–22. Copyright 2018, Elsevier.

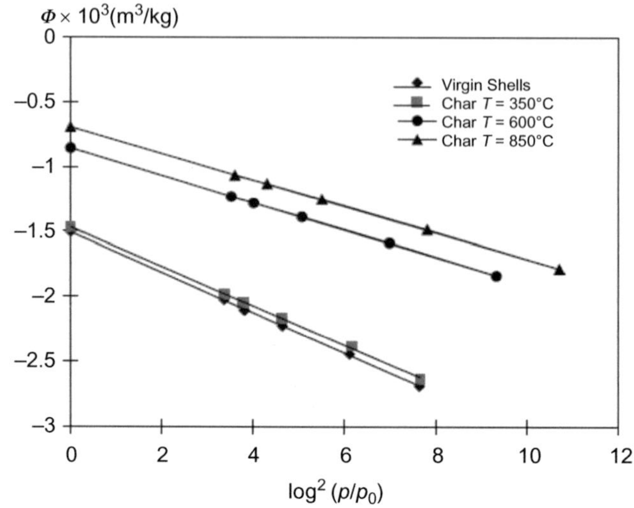

Fig. 15 Dubinin-Radushkevich fits of CO_2 adsorption data of chars.
Reprinted with the permission from Bonelli, P. R., Della Rocca, P. A., Cerrella, E. G., Cukierman, A. L. (2001). Effect of pyrolysis temperature on composition, surface properties and thermal degradation rates of Brazil Nut shells. *Bioresource Technology*, *76*(1), 15–22. Copyright 2018, Elsevier.

those calculated from N_2 adsorption; this indicates the presence of very narrow micropores where nitrogen diffusion is restricted at the low adsorption temperature. The maximum S_{CO2} obtained is 624 m^2/g for the sample prepared at 850°C. However, S_{N2} does not exceed 4.2 m^2/g.

3.4 Char surface chemistry

The chemical properties of chars can be examined by studying the chemical composition and surface functional groups of these materials. The chars are constituted by carbon, hydrogen, oxygen, nitrogen, and mineral components that originate from feedstock. On the other hand, the surface functional groups can also be introduced during the preparation of chars. They mainly depend on the composition of the feedstock and the conditions of char preparation. The active site surface composition influences the applicability of these chars, because they confer acidic, neutral, or basic properties to the carbon surface. The surface of carbons may comprise different types of oxygen- and nitrogen-containing groups, as shown in Fig. 16. Indeed, carboxylic acids and anhydrides, lactones, lactols, and phenols are acidic, while carbonyl and ether groups are neutral, and basic functionalities are presented by quinone, chromene, pyrone, and nitrogen groups (Shafeeyan, Daud, Houshmand, & Shamiri, 2010). A great deal of work has been devoted to highlighting the chemical surface groups of the various types of carbons (chars, activated carbons, carbon blacks, carbon fibers,

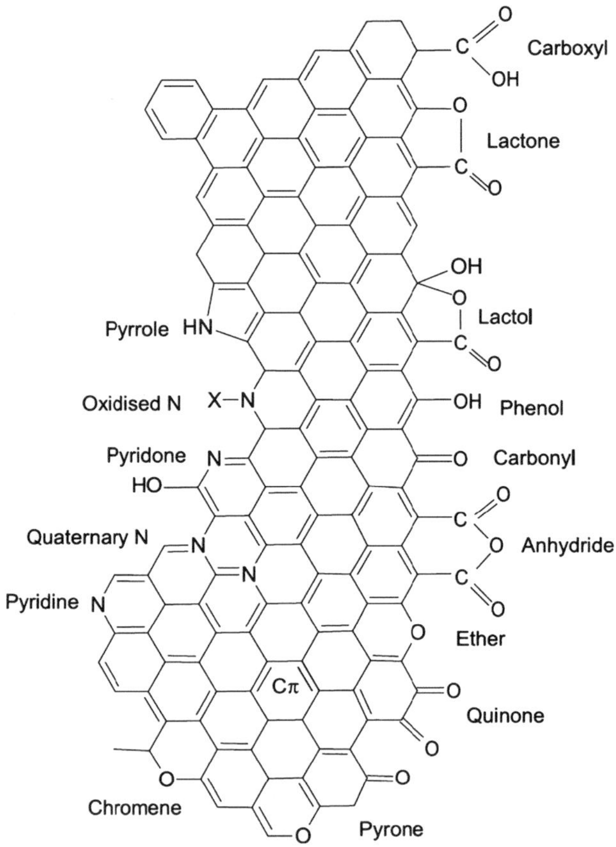

Fig. 16 Nitrogen and oxygen surface functional groups on carbon.
Reprinted with permission from Figueiredo, J. L., Pereira, M. F. R. (2010). The role of surface chemistry in catalysis with carbons. *Catalysis Today*, *150*(1), 2–7. Copyright 2018, Elsevier.

etc.). In particular, one can refer to articles by Boehm (1994) and Puri, Kaistha, Vardhan, and Mahajan (1973), as well as the monographs by Lahaye (1998) and Figueiredo and Pereira (2010).

The composition of surface functional groups can be analyzed by several techniques such as FTIR, XPS, Boehm titration, and temperature-programmed desorption (TPD). However, FTIR and XPS techniques have shown the best applicability for this type of analysis.

3.4.1 Fourier transform infrared spectroscopy

The FTIR technique is used to identify surface functional groups on chars. An infrared radiation source (wave number ranging from 400 to 4000 cm^{-1}) interacts with the sample. If the energy of the electromagnetic wave is close to the vibrational energy of a

Table 4 Fourier transform infrared spectroscopy band assignments of functional groups on chars

Functional group	Assignment range (cm^{-1})
O—H stretching vibration in:	3500–3200
water	
hydrogen bonding hydroxyl	
C—H stretching vibration in:	
Aromatic	3070–3000
Aliphatic	2980–2820
C=O stretching vibration in:	
Esters (acetyl esters in wood)	1740–1730
Carboxylic groups, aldehydes, ketones	1700
Conjugated ketones and quinones	1600
C=C stretching in aromatic	1610–1580
C=C stretching vibration (lignin carbohydrate)	1440
C—OH stretching in phenolics, phenols, ligneous syringyl	1325
Symmetric C—O stretching vibration in carbohydrate (cellulose, hemicelluloses, and lignin)	1100–1030
Aromatic C—H out-of-plane deformation	875; 885
Pyridine in pyridine ring vibration	781

molecule, it absorbs the radiation; thus, the intensity of the transmitted or reflected light decreases. This technique is easy to use and is nondestructive.

Chars contain raised quantities of oxygen, hydrogen, nitrogen, and sulfur. FTIR analysis provides evidence of the presence of oxygen groups and nitrogen functional groups (Table 4). Chars are devoid of oxygen groups when they are treated at high temperature (>950°C) in vacuum or hydrogen and cooled down at room temperature under vacuum. The exposure of chars to air introduces oxygen at their surface; this leads to the formation of oxygen surface groups (Marsh & Rodríguez-Reinoso, 2006a).

Several researchers have used the FTIR technique to analyze surface functionalities during char preparation. The FTIR spectra of beech wood chars prepared at different temperatures are shown in Fig. 17. In particular, the effect of temperature on the presence of heteroatoms on chars is examined (Guizani et al., 2017). Fig. 17 illustrates three categories of materials: raw samples with important kinds and quantities of surface functional groups, chars prepared in the range of 500–800°C with a low amount of heteroatoms, and chars pyrolyzed above 800°C that have lost almost completely the surface complexes. Fig. 18 shows that high extent pyrolysis removes sulfur compounds at 950°C, and oxygen groups that correspond to O—H stretching (3300–3400 cm^{-1}) disappear above 600°C as a result of the cracking of phenolic/aromatic structures and elimination of oxygen groups (Fanning & Vannice, 1993; Narzari et al., 2017). The same behavior was reported for nitrogen groups that volatilize between 400°C and 800°C, while important quantities are protected in heterocyclic compounds (Knicker, Scaroni, & Hatcher, 1996).

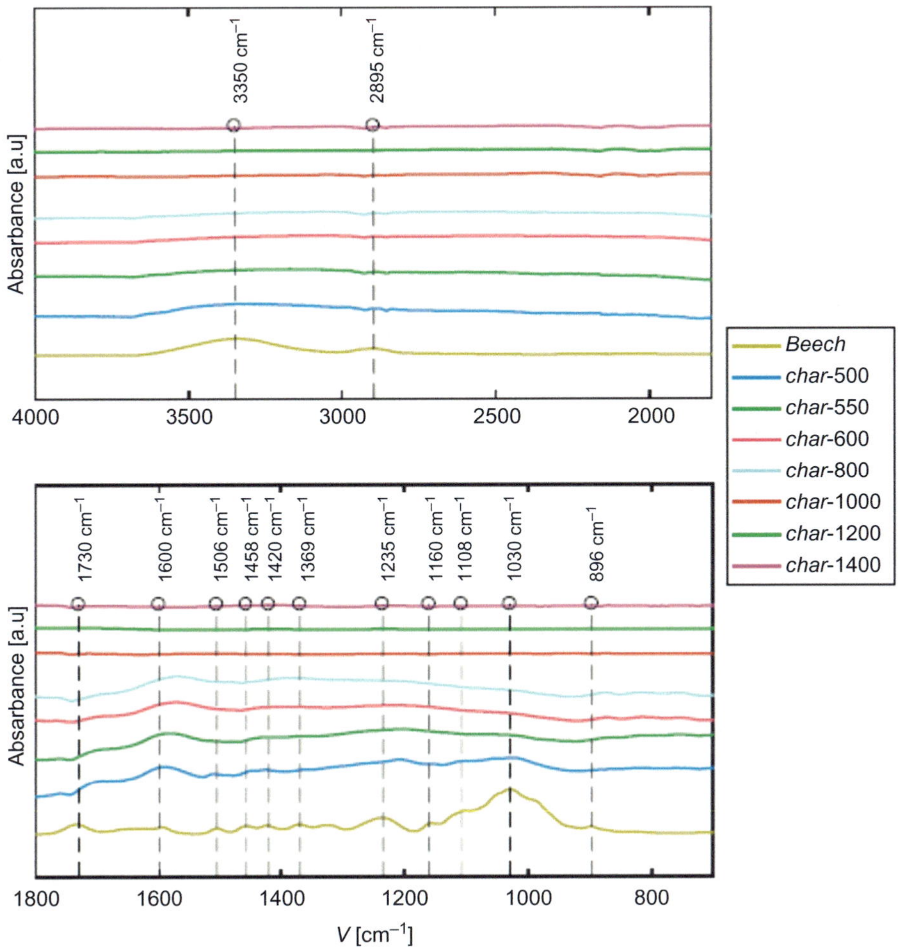

Fig. 17 Fourier transform infrared spectroscopy spectra (1800–500 cm^{-1}) of chars produced from beech wood at different temperatures (500–1400°C) (Guizani, Jeguirim, Valin, Limousy, & Salvador, 2017).

Budai et al. (2017) analyzed the evolution of surface chemistry of corncob and *Miscanthus* during the pyrolysis process. These authors reported that lignocellulosic biomass shows similar FTIR spectra (Fig. 18) due to the presence of carbohydrate polymers (cellulose, hemicelluloses) and an aromatic polymer (lignin) (Budai et al., 2017). The bands at 1732, 1375, 1240, 1165, 1062, and 1030 cm^{-1} were assigned to C=O, C—H, C—O—C, and C—O stretching deformations of functional complexes in carbohydrate. The 1100–1000 cm^{-1} bands were attributed to C—O—C and C—O stretching vibrations of aliphatic and alcohol groups. The wide band at 3600–3320 was assigned to different O—H stretchings in carboxylic and hydroxyl groups. The authors also showed that the high extent pyrolysis of the feedstock

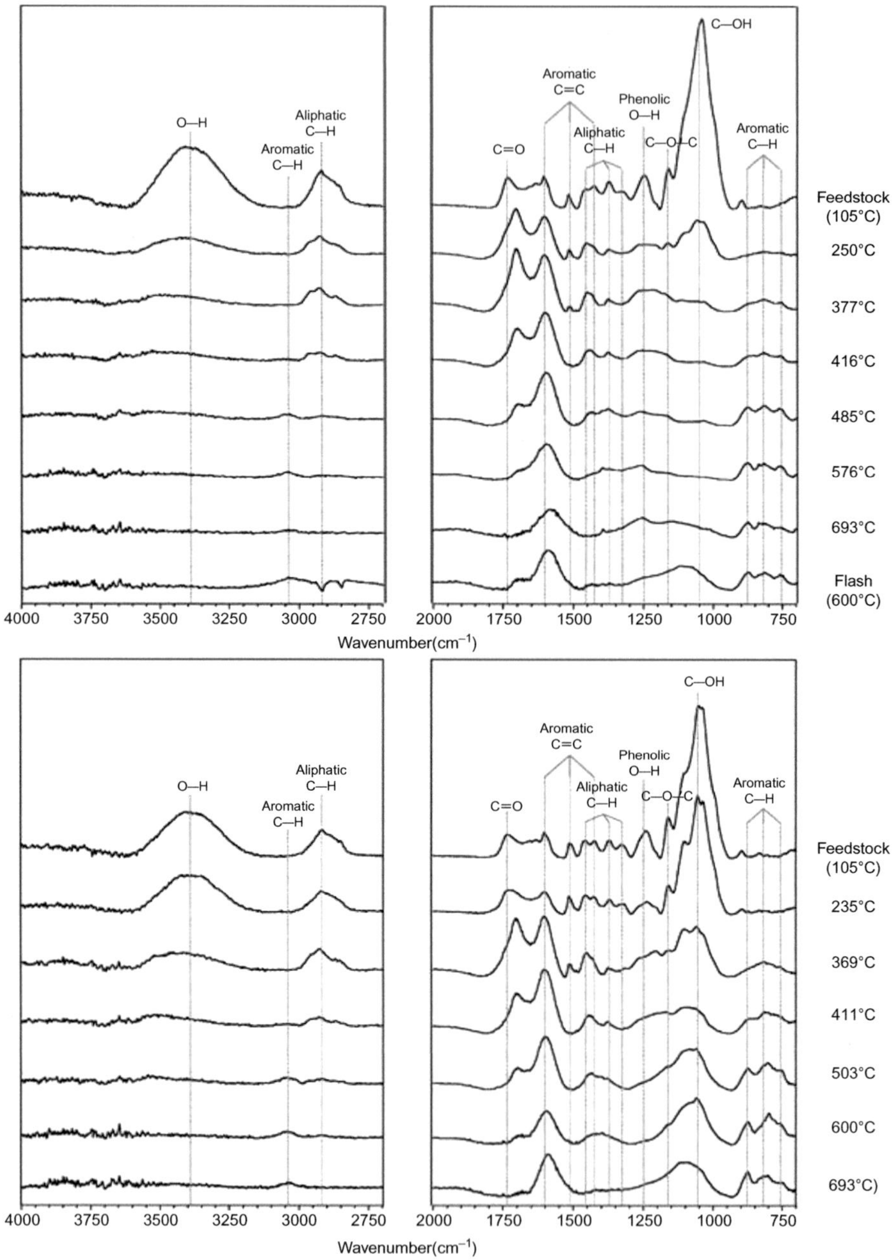

Fig. 18 Fourier transform infrared spectroscopy spectra of corncob and *Miscanthus* biomasses and chars produced at temperatures between 235°C and 693°C.

Reprinted with permission from Budai, A., Calucci, L., Rasse, D. P., Strand, L. T., Pengerud, A., Wiedemeier, D., et al. (2017). Effects of pyrolysis conditions on Miscanthus and corncob chars: characterization by IR, solid state NMR and BPCA analysis. *Journal of Analytical and Applied Pyrolysis*, *128*, 335–345. Copyright 2018, Elsevier.

(corncob and *Miscanthus*) at 800°C produced chars free from surface functional complexes. This trend is explained by the entire aromatization of the char.

3.4.2 X-ray photoelectron spectroscopy

XPS (or ESCA for electron spectroscopy for chemical analysis) is a nondestructive analytical technique that determines the chemical composition of material surfaces. The technique is based on the bombardment of the sample with X-rays of a certain wavelength.

From previous XPS studies of chars the following assignments have been made: aliphatic and aromatic C—C, C=C or C—H at 285 eV, —C—OR at 286.5 eV, C=O at 288 eV, and C—OOR at 289.2 eV (Joseph et al., 2015; Taherymoosavi, Joseph, & Munroe, 2016).

Nitrogen surface groups have also been successfully studied by XPS with N1s spectra. Using the XPS technique, several works have studied the amine groups present on char (Jones & Sammann, 1990; Tomlinson, Freeman, & Theocharis, 1993). XPS band assignments of functional groups on chars are amide 399.5–400.8 eV, pyrrole 400.3 eV, aromatic amine (—NH_2) 401 eV, nitrile 399.4 and 401.1 eV, and quaternary nitrogen 402.6 eV (Kloss et al., 2012; Zhao et al., 2017).

Taherymoosavi et al. (2016) examined the surface functionalities of chars as a function of temperature. They used a mixture of wheat straw and chicken litter (1/1) to produce chars at different pyrolysis temperatures in the range 450–650°C. The authors investigated the change in organic species as a function of pyrolysis temperature. Their results show that the quantity of C—C, C=C, and C—H increases with the increase in pyrolysis temperature. However, the concentration of C—O and C=O decreases due to the loss of VM. A higher amount of nitrogen groups, i.e., pyridine (N=C) and pyrrole (N—C=O) groups at 399 and 400.4 eV, respectively, was noted for the char prepared at 450°C (Taherymoosavi et al., 2016).

3.4.3 Temperature-programmed desorption

TPD experiments evaluate the amount and nature of oxygen groups. This technique is based on the decomposition of oxygen surface groups on chars upon heating in an inert atmosphere, the emitted gases consisting mainly of CO_2 at lower temperatures and principally CO at higher temperatures. It is well known that CO_2 evolution results from the decomposition of carboxylic acids at low temperatures (<450°C) or lactones at higher temperatures. Finally, carboxylic anhydrides originate from both CO and CO_2 at high temperatures; carbonyl, ether, quinone, and phenol groups decompose as CO at temperatures up to 1000°C (Table 5 and Fig. 19).

In a recent study, Hervy et al. (2018) used TPD analysis to evaluate the amount of acidic and basic oxygen groups present on the char's surface (Hervy et al., 2018). The materials were prepared from used wood pellets (UWPs) and a mixture (50/50) of food waste (FW) and coagulation-flocculation sludge (CFS). To increase the amount of oxygen functionalities on the char surface, the prepared samples were oxidized with oxygen at 280°C. The results show that FW/CFS chars exhibit higher amounts of basic

Table 5 Decomposition of oxygen surface functionalities on char by TPD

Functional group	Evolved gas	Peak temperature
Carboxylic H—C—OH	CO_2	250°C 100–500°C 200–250°C
Lactone O—C=O	CO_2	400°C 627°C 190–650°C
Anhydride O=C—O—C=O	CO_2+CO	350–400°C 627°C 600°C
Phenol —OH	CO	600–700°C
Carbonyl C=O	CO	800–900°C 700–1000°C
Ether —O—	CO	700°C
Quinone	CO_2	700–980°C 800–900°C

oxygen-containing surface groups than UWP chars. This trend has been explained by the presence of higher quantities of basic minerals coupled to oxygen.

3.4.4 ^{13}C nuclear magnetic resonance spectroscopy

^{13}C NMR spectroscopy is an important technique for the characterization of chars. It is used to evaluate the carbon chemistry of chars. ^{13}C NMR spectroscopy is carried out by using the direct polarization (DP) and cross-polarization techniques. The latter requires fewer achievement times, despite more production scans than DP. However, the (DP) NMR spectra evaluate with precision the amount of the char structure. It is easy to evaluate the aromaticity of chars than the degree of aromatic condensation (Amin et al., 2016). ^{13}C NMR spectroscopy of chars exhibits the following signals: carbohydrate signals are 74–110 ppm, hemicellulosic signals are 10–173 ppm, and lignin resonates between 115 and 160 ppm. Methyl and carboxylic carbon peaks are assigned at 21 and 172 ppm, respectively. Protonated aromatic carbons resonate at 115.4 and 127 ppm; ortho- and para-aromatic carbons resonate at 120 ppm. CH_2OH group signals are around 62 ppm and CHOH group signals are around 72 ppm (Budai et al., 2017; Cao, Ro, Chappell, Li, & Mao, 2011).

Cao et al. (2012) used solid-state ^{13}C NMR techniques to characterize the chemical structure of chars (Cao et al., 2012). These authors prepared chars from maple wood at different pyrolysis temperatures ranging from 300 to 700°C and they examined the structural change of chars with HTT. They reported that the char obtained by pyrolysis at 300°C kept the most lignocellulosic characteristics (Fig. 20A–C). They noted larger signals of alkyl carbon, increased signals from methylene, OCH_3, and aromatic carbon, and new signals of aldehyde/ketone carbons. At 350°C, the aromatic carbon peak

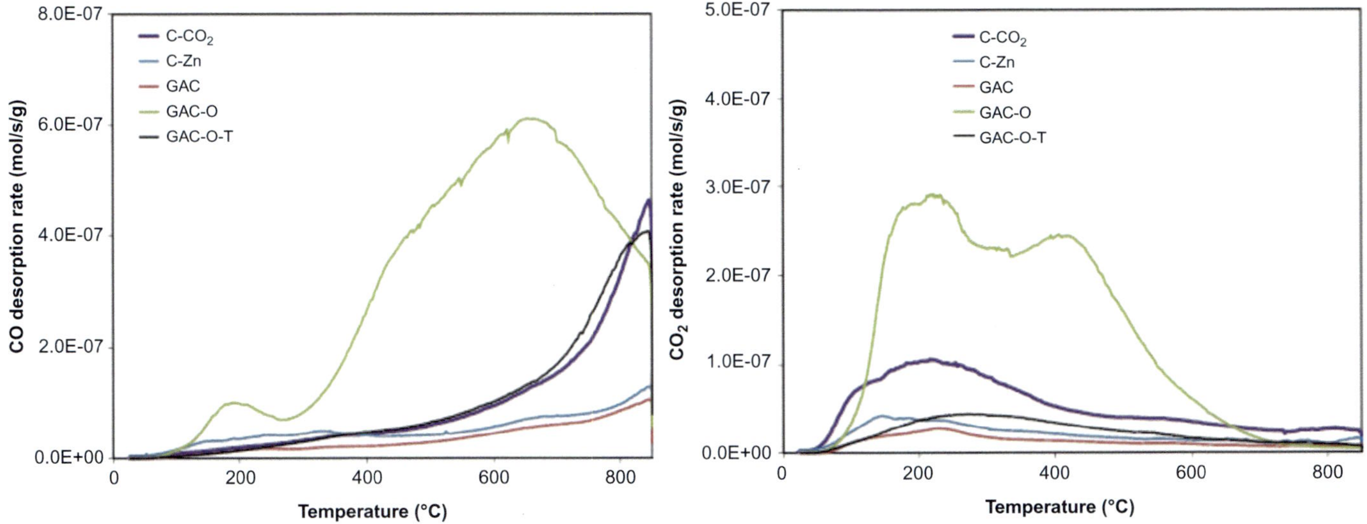

Fig. 19 CO and CO_2 temperature-programmed desorption mass spectrometry profiles of different carbons.
Reprinted with permission from Belhachemi, M., Jeguirim, M., Limousy, L., Addoun, F. (2014). Comparison of NO_2 removal using date pits activated carbon and modified commercialized activated carbon via different preparation methods: effect of porosity and surface chemistry. *Chemical Engineering Journal*, *253*, 121–129. Copyright 2018, Elsevier.

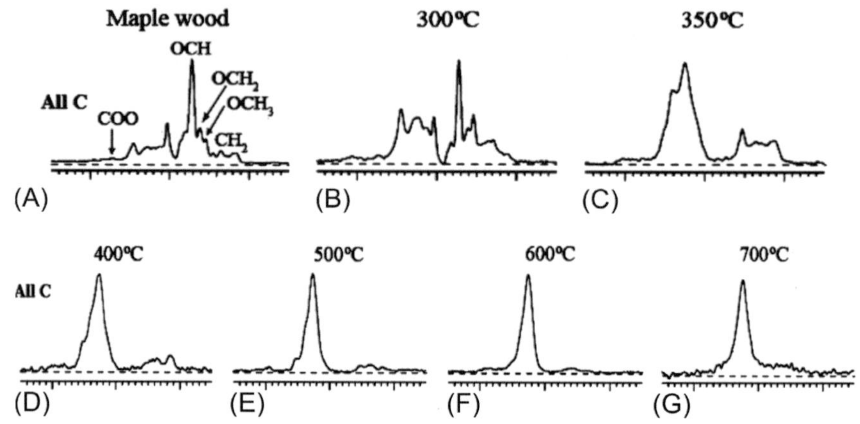

Fig. 20 ^{13}C nuclear magnetic resonance spectra of chars produced from maple wood at different temperatures.
Reprinted with the permission from Cao, X., Pignatello, J. J., Li, Y., Lattao, C., Chappell, M. A., Chen, N., et al. (2012). Characterization of wood chars produced at different temperatures using advanced solid-state ^{13}C NMR spectroscopic techniques. *Energy and Fuels*, *26*(9), 5983–5991. Copyright 2018, American Chemical Society.)

Fig. 21 Nuclear magnetic resonance spectra of wood char. Reprinted with permission from Joseph, P., Tretsiakova-McNally, S., McKenna, S. (2012). Characterization of cellulosic wastes and gasification products from chicken farms. *Waste Management*, *32*(4), 701–709. Copyright 2018, Elsevier.

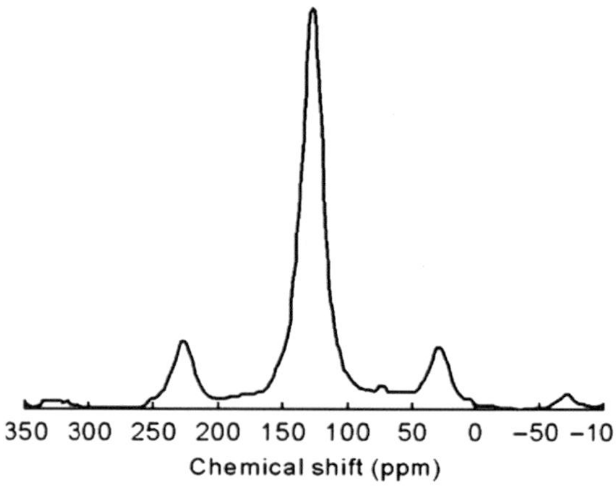

intensity increased and at 400°C the aromatic C—O groups (about 148 ppm) decreased. Above the latter temperature, the lingocellulosic structure disappeared and the chars became dominated by aromatic structure (Fig. 21D–G). The protonated aromatic carbons increased by increasing HTT, while the nonprotonated aromatic carbons rose up to 400°C and are reduced for higher temperatures.

Fig. 21 represents NMR spectra of jarrah wood char prepared at 600°C with a residence time of 12 h (Joseph et al., 2015). The image exhibits a high amount of aromatic C (110–145 ppm) and considerable content of amorphous C associated with oxygen surface groups. Little quantity of alkyl (0–45 ppm), O-alkyl (60–95 ppm), and ketone (190–215 ppm) are observed.

3.5 Structure of chars

Chars are described as carbonaceous materials with disordered and complex structures. However, the carbon matrix of chars is composed of ordered structures with short distances cross-linked to amorphous structures. The ordered structures consist mainly of graphene sheets that involve a two-dimensional network of benzene rings. A wide variety of disordered structures intertwine with leaflets of graphene in the carbon matrix of the chars. This disorder can come from defects in the graphene sheets, such as gaps (absence of one or more carbon atoms), the insertion of impurities (mainly heteroatoms O or H), or the presence of a cycle of approximately six carbons.

Carbonaceous materials are generally formed of well-ordered domains. Carbons do not have the hexagonal arrangement of graphite (Fig. 22). The structure is still lamellar but the long-distance order is no longer respected and the spacing of the aromatic planes is greater than 3.36 Å. Most of these layouts are disoriented with respect to each other, which gives the chars a turbostratic structure (Fig. 22). It was reported that most nongraphitizable carbons come from lignocellulosic materials (Rodríguez-Reinoso & Sepúlveda-Escribano, 2001). The biomass conserves its polymeric structure during the high-temperature treatment and the resulting char structures are mainly due to the elimination of VM.

3.5.1 X-ray diffraction

XRD analysis detects the amorphous phase given by the carbon structure and the crystalline phase due to mineral compositions such as calcite, quartz, lime, and portlandite present in chars. Generally, the XRD spectra of carbonaceous materials exhibits two peaks located at $2\theta = 22.5$ and 44 degrees, which correspond to the diffuse graphite planes of index (0 0 2) and (1 0 0), respectively (Fig. 23). The intensity of the background of the XRD pattern is mainly related to the presence of amorphous carbon in the char. The peak (1 0 0) is attributed to graphite structures in the plane and therefore reflects the size of the aromatic layers (Lu, Kong, Sahajwalla, & Harris, 2002). The narrower the peak indicates the higher degree of condensation of the aromatic rings. The peak (0 0 2) is attributed to the distance between the graphitic planes of

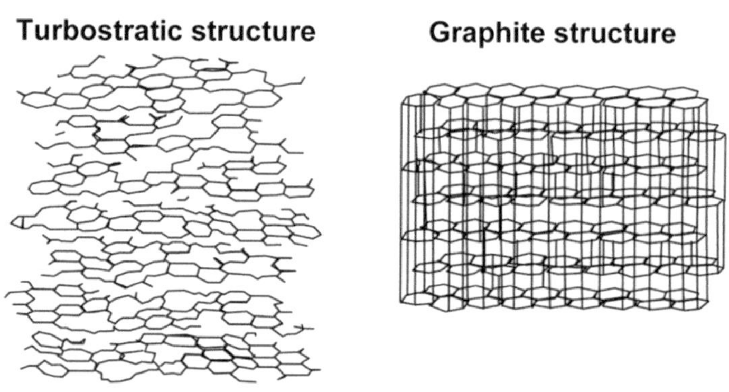

Fig. 22 Turbostratic and graphite structures.

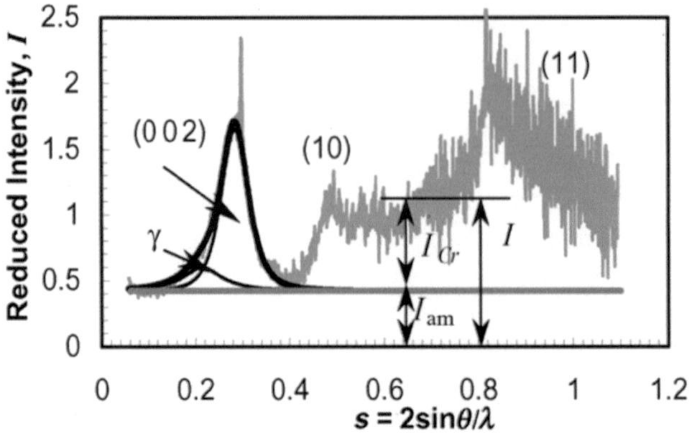

Fig. 23 X-ray diffraction spectra of char prepared at 1200°C (λ is the wavelength of the X-ray; I_{am} and I_{Cr} are the intensities of amorphous and crystalline structures).
Reprinted with permission from Lu, L., Kong, C., Sahajwalla, V., Harris, D. (2002). Char structural ordering during pyrolysis and combustion and its influence on char reactivity. *Fuel*, *81*(9), 1215–1225. Copyright 2018, Elsevier.

the crystallites present in the char (Lu et al., 2002). In theory, this peak is symmetrical. However, several studies have shown that an asymmetry characterized by the appearance of a shoulder on the left side of the peak can reflect the presence of a stack of saturated structures such as aliphatic chains. Increasing pyrolysis temperature leads to sharper and less wide (0 0 2) bands, which correspond to an increase in crystallite size. However, background intensity decreases with increasing HTT of pyrolysis indicating the formation of ordered and aromatic structures of chars.

Fu et al. (2011) reported the X-ray spectra of chars prepared from maize stalk, rice straw, cotton straw, and rice husk by fast pyrolysis at different temperatures (600–1000°C). The X-ray diffraction spectra of the three chars show a large (0 0 2) peak at 23 degrees that indicates a highly disordered structure of chars (Fu et al., 2011). By increasing the temperature of treatment this peak becomes sharper and symmetric (Fig. 24).

XRD analysis is used to detect mineral components such as sylvite, calcite, apatite, quartz, and whewellite. It was reported that the diffraction patterns of pine bark chars samples prepared by pyrolysis at 300°C and 500°C show similar crystalline phases (González et al., 2013). Quartz, calcite, anorthite, and sylvite were detected. The char prepared at 500°C showed more peaks. This trend may be due to the increase in mineral concentration.

By increasing the pyrolysis temperature, the chars loose organic matter and the crystalline mineral phase decreases leading to an increase in metal and salts contents in chars.

3.5.2 Raman spectroscopy

Raman spectroscopy is a powerful technique widely used to characterize chars. It examines the ordered and disordered structure of carbon. The Raman spectra of graphite and graphene are distinguished by bands G at 1560 cm^{-1} and 2D at 2710 cm^{-1}.

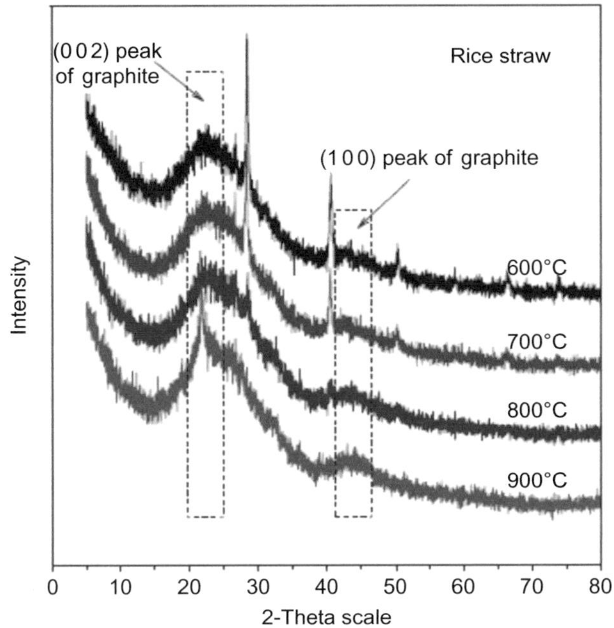

Fig. 24 X-ray diffraction spectra of chars prepared at high temperature of pyrolysis. Reprinted with permission from Fu, P., Yi, W., Bai, X., Li, Z., Hu, S., Xiang, J. (2011). Effect of temperature on gas composition and char structural features of pyrolyzed agricultural residues. *Bioresource Technology*, *102*(17), 8211–8219. Copyright 2018, Elsevier.

The 2D band is related to a second vibrational order. The presence of amorphous structures in these carbonaceous materials leads to a shifting and widening of the G band and the appearance of the D band at 1350 cm^{-1} (Ferrari, 2007). These bands are the main peaks present in chars, indicating the dual composition of chars (i.e., the presence of ordered and disordered carbon rings). The G band represents the graphitic structures, while the D band describes the defects of the carbon structure. Raman spectra can contain 4, 5, or 10 visible peaks. Fig. 25 shows the deconvolution of a Raman spectrum showing 10 bands. The assignments of these 10 bands are recapitulated in Table 6.

These methods are in accordance to give bands G and D the presence of ordered graphene-type leaflets and defects in graphene structures, respectively. The valley separating these two major bands corresponds to the amorphous carbon structures. At low Raman (position) displacement values, the signal reflects the presence of C—H bonds and aliphatic structures.

Guizani et al. (2017) examined the effect of increasing pyrolysis temperature on the structure of beech wood chars (Guizani et al., 2017). The Raman spectrum of the samples prepared at temperatures ranging from 500°C to 1400°C showed two overlapping peaks that can be distinguished at 1350 and 1600 cm^{-1}. These two peaks correspond to the D and G bands, respectively. The D peak is higher than the G peak, suggesting a low amount of graphene-like structures. The authors reported that by increasing the pyrolysis temperature, the structural order of the char increases. As can be seen,

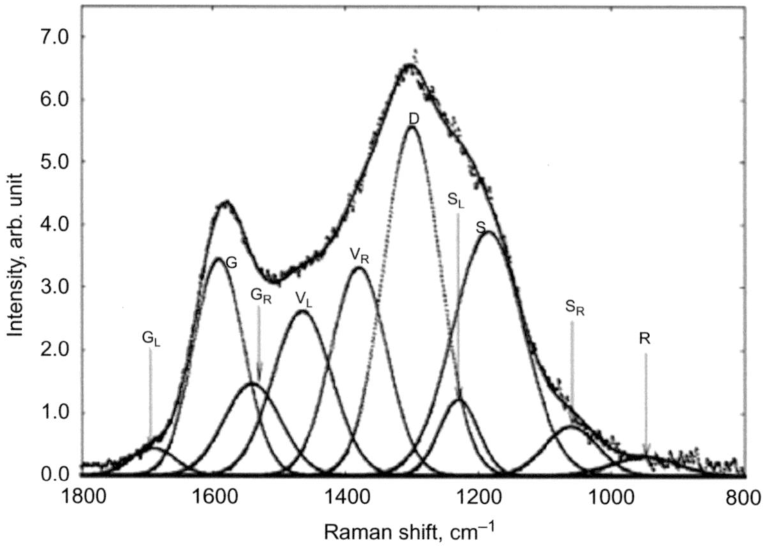

Fig. 25 Raman spectra deconvolution in 10 peaks.
Reprinted with permission from Li, X., Hayashi, J. -I., Li, C. -Z. (2006). FT-Raman spectroscopic study of the evolution of char structure during the pyrolysis of a Victorian brown coal. *Fuel*, *85*(12), 1700–1707. Copyright 2018, Elsevier.

Table 6 Assignment of Raman peaks

Peak name	**Peak position (cm^{-1})**	**Description**
G_l	1700	Carbonyl group (C=O)
G	1585	Graphene structure, aromatic cycle, alkene C=C
G_R	1540	Amorphous carbon structures, aromatics (3–5 rings)
V_l	1465	Methyl groups or methylene, amorphous carbon structures
V_R	1380	Methyl groups, amorphous carbon structures
D	1350	Aromatic rings (≥6), graphene structure type
S_l	1230	Aryl-alkyl-ether, para-aromatic
S	1185	Aromatic ether (aliphatic), C—C on aromatic rings, C—H on aromatic rings
S_R	1060	C—H on aromatic rings, benzene ring
R	960–800	C—C on alkane and cyclic alkane, C—H on aromatic rings

the D peak becomes sharper and less broad for chars pyrolyzed at high temperatures of 1200°C and 1400°C (Fig. 26), suggesting the increase of order in chars' skeletons. A similar trend has been reported by Asadullah, Zhang, and Li (2010). Increasing pyrolysis temperature increases the condensed aromatic rings (>6 rings) and

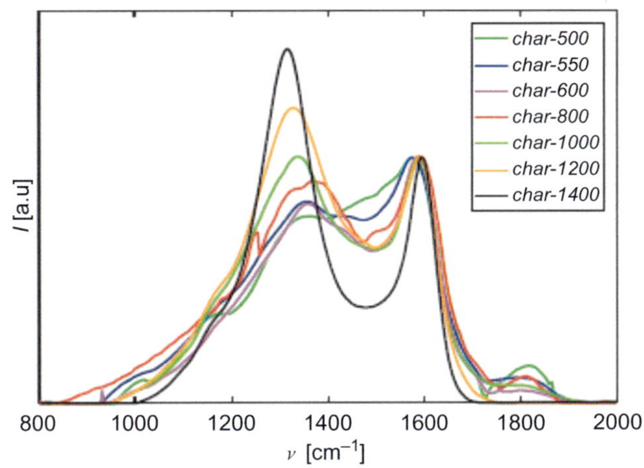

Fig. 26 Raman spectra of chars prepared at different temperatures (Guizani et al., 2017).

amorphous chars' structure (3–5 rings), as a result of the aromatization of the chars and elimination of oxygen groups.

Raman spectroscopy can be related to other techniques such as XRD and transmission electron microscopy (TEM) to complete and confirm the results obtained. Indeed, XRD makes it possible to characterize the structure of a material on a larger scale (diffraction on several planes), whereas Raman spectroscopy highlights the disorder in a structure on the scale of the chemical bonds.

3.5.3 *Transmission electron microscopy*

TEM is a very useful technique to characterize the nanostructure, porosity, and local composition of the carbon matrix of char. The images produced by TEM show different dimensions of the defective graphene layers. The spaces are generated between these graphene layers from the microporosity of carbon (Marsh & Rodríguez-Reinoso, 2006b). Rouzaud and Clinard (2002) directly imaged the defective graphene layers that form the skeleton of two different carbon materials using high-resolution transmission electron microscopy (Rouzaud & Clinard, 2002). In this way, Fig. 27 shows the TEM images of these two carbons: soot nanoparticles with a concentric microstructure and microporous-activated saccharose-based carbon. The images indicate labyrinth diagrams related to the nonlinear graphene layers of carbons.

3.6 Conclusions

Physical, chemical, textural, and morphological properties of a char have impacts on its mobility in the environment and its suitability as an ecological niche for amendments or for thermal valorization. However, the wide range of char characteristics makes it challenging to find appropriate methods to correlate all aspects. Nevertheless, some findings could be made. For example, for almost all chars, surface area

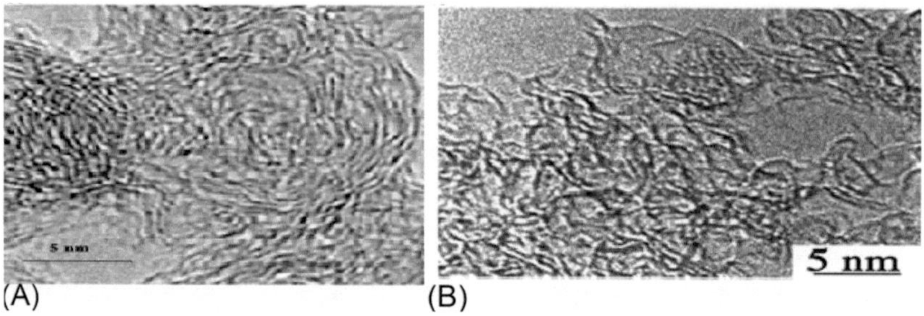

Fig. 27 Transmission electron microscopy images of carbons: (A) soot nanoparticles with a concentric microstructure and (B) microporous-activated saccharose-based carbon. Reprinted with the permission from Rouzaud, J. -N., Clinard, C. (2002). Quantitative high-resolution transmission electron microscopy: a promising tool for carbon materials characterization. *Fuel Processing Technology*, *77–78*, 229–235. Copyright 2018, Elsevier.

and pH increase with increasing pyrolysis temperature, while volatiles and CEC decrease. This is explained by the progressive loss of acidic surface functional groups such as aliphatic carboxylic acids, leading to pH increase. These conclusions might be confirmed by FTIR or SEM images. As another example, TEM and Raman analysis of chars show that the number of condensed aromatic rings increases and amorphous structure changes as a result of the aromatization of the chars and elimination of oxygen groups.

References

Alburquerque, J. A., Sánchez, M. E., Mora, M., & Barrón, V. (2016). Slow pyrolysis of relevant biomasses in the Mediterranean basin. Part 2. Char characterisation for carbon sequestration and agricultural uses. *Journal of Cleaner Production*, *120*, 191–197.

Amin, F. R., Huang, Y., He, Y., Zhang, R., Liu, G., & Chen, C. (2016). Biochar applications and modern techniques for characterization. *Clean Technologies and Environmental Policy*, *18* (5), 1457–1473.

Asadullah, M., Zhang, S., & Li, C.-Z. (2010). Evaluation of structural features of chars from pyrolysis of biomass of different particle sizes. *Fuel Processing Technology*, *91*(8), 877–881.

Belhachemi, M., Rios, R. V. R. A., Addoun, F., Silvestre-Albero, J., Sepúlveda-Escribano, A., & Rodríguez-Reinoso, F. (2009). Preparation of activated carbon from date pits: effect of the activation agent and liquid phase oxidation. *Journal of Analytical and Applied Pyrolysis*, *86*(1), 168–172.

Benedetti, V., Patuzzi, F., & Baratieri, M. (2017). Characterization of char from biomass gasification and its similarities with activated carbon in adsorption applications. *Applied Energy*, *227*, 92–99.

Boehm, H. P. (1994). Some aspects of the surface chemistry of carbon blacks and other carbons. *Carbon*, *32*(5), 759–769.

Bonelli, P. R., Della Rocca, P. A., Cerrella, E. G., & Cukierman, A. L. (2001). Effect of pyrolysis temperature on composition, surface properties and thermal degradation rates of Brazil Nut shells. *Bioresource Technology*, *76*(1), 15–22.

Břendováa, K., Pavel, T., Jiřina, S., & Jan, H. (2012). Biochar properties from different materials of plant origin *Eur. Chemical Bulletin (Beijing, China)*, *1*(12), 535–539.

Brewer, C. E., Chuang, V. J., Masiello, C. A., Gonnermann, H., Gao, X., Dugan, B., et al. (2014). New approaches to measuring biochar density and porosity. *Biomass and Bioenergy*, *66*, 176–185.

Budai, A., Calucci, L., Rasse, D. P., Strand, L. T., Pengerud, A., Wiedemeier, D., et al. (2017). Effects of pyrolysis conditions on Miscanthus and corncob chars: Characterization by IR, solid state NMR and BPCA analysis. *Journal of Analytical and Applied Pyrolysis*, *128*, 335–345.

Buss, W., Graham, M. C., Shepherd, J. G., & Mašek, O. (2016). Suitability of marginal biomass-derived biochars for soil amendment. *Science of the Total Environment*, *547*, 314–322.

Cantrell, K. B., Hunt, P. G., Uchimiya, M., Novak, J. M., & Ro, K. S. (2012). Impact of pyrolysis temperature and manure source on physicochemical characteristics of biochar. *Bioresource Technology*, *107*, 419–428.

Cao, X., & Harris, W. (2010). Properties of dairy-manure-derived biochar pertinent to its potential use in remediation. *Bioresource Technology*, *101*(14), 5222–5228.

Cao, X., Pignatello, J. J., Li, Y., Lattao, C., Chappell, M. A., Chen, N., et al. (2012). Characterization of wood chars produced at different temperatures using advanced solid-state 13C NMR spectroscopic techniques. *Energy & Fuels*, *26*(9), 5983–5991.

Cao, X., Ro, K. S., Chappell, M., Li, Y., & Mao, J. (2011). Chemical structures of swine-manure chars produced under different carbonization conditions investigated by advanced solid-state 13C nuclear magnetic resonance (NMR) spectroscopy. *Energy & Fuels*, *25*(1), 388–397.

Cheng, C.-H., Lehmann, J., & Engelhard, M. H. (2008). Natural oxidation of black carbon in soils: Changes in molecular form and surface charge along a climosequence. *Geochimica et Cosmochimica Acta*, *72*(6), 1598–1610.

Colantoni, A., Evic, N., Lord, R., Retschitzegger, S., Proto, A. R., Gallucci, F., et al. (2016). Characterization of biochars produced from pyrolysis of pelletized agricultural residues. *Renewable and Sustainable Energy Reviews*, *64*, 187–194.

Crombie, K., Mašek, O., Sohi Saran, P., Brownsort, P., & Cross, A. (2012). The effect of pyrolysis conditions on biochar stability as determined by three methods. *GCB Bioenergy*, *5*(2), 122–131.

Enders, A., Hanley, K., Whitman, T., Joseph, S., & Lehmann, J. (2012). Characterization of biochars to evaluate recalcitrance and agronomic performance. *Bioresource Technology*, *114*, 644–653.

Fanning, P. E., & Vannice, M. A. (1993). A DRIFTS study of the formation of surface groups on carbon by oxidation. *Carbon*, *31*(5), 721–730.

Ferrari, A. C. (2007). Raman spectroscopy of graphene and graphite: Disorder, electron–phonon coupling, doping and nonadiabatic effects. *Solid State Communications*, *143*(1), 47–57.

Figueiredo, J. L., & Pereira, M. F. R. (2010). The role of surface chemistry in catalysis with carbons. *Catalysis Today*, *150*(1), 2–7.

Frederik, R., Sven, v. H., Dane, D., & Wolter, P. (2013). Production and characterization of slow pyrolysis biochar: influence of feedstock type and pyrolysis conditions. *GCB Bioenergy*, *5*(2), 104–115.

Fu, P., Yi, W., Bai, X., Li, Z., Hu, S., & Xiang, J. (2011). Effect of temperature on gas composition and char structural features of pyrolyzed agricultural residues. *Bioresource Technology*, *102*(17), 8211–8219.

Gaskin, J., Steiner, C., Harris, K., Das, K. C., & Bibens, B. (2008). Effect of low-temperature pyrolysis conditions on biochar for agricultural use. *Transactions of the ASABE*, *51*, 2061.

Ghouma, I., Jeguirim, M., Dorge, S., Limousy, L., Matei Ghimbeu, C., & Ouederni, A. (2015). Activated carbon prepared by physical activation of olive stones for the removal of NO2 at ambient temperature. *Comptes Rendus Chimie*, *18*(1), 63–74.

González, M. E., Cea, M., Sangaletti, N., González, A., Toro, C., Diez, M. C., et al. (2013). Biochar derived from agricultural and forestry residual biomass: characterization and potential application for enzymes immobilization. *Journal of Biobased Materials and Bioenergy*, *7*(6), 724–732.

Guizani, C., Jeguirim, M., Valin, S., Limousy, L., & Salvador, S. (2017). Biomass chars: the effects of pyrolysis conditions on their morphology, structure, chemical properties and reactivity. *Energies*, *10*(6), 796.

Gundale, M. J., & DeLuca, T. H. (2006). Temperature and source material influence ecological attributes of ponderosa pine and Douglas-fir charcoal. *Forest Ecology and Management*, *231*(1), 86–93.

Gunes, A., Inal, A., Sahin, O., Taskin, M. B., Atakol, O., Yilmaz, N., et al. (2015). Variations in mineral element concentrations of poultry manure biochar obtained at different pyrolysis temperatures, and their effects on crop growth and mineral nutrition. *Soil Use and Management*, *31*(4), 429–437.

Hammes, K., Smernik, R. J., Skjemstad, J. O., Herzog, A., Vogt, U. F., & Schmidt, M. W. I. (2006). Synthesis and characterisation of laboratory-charred grass straw (Oryza sativa) and chestnut wood (Castanea sativa) as reference materials for black carbon quantification. *Organic Geochemistry*, *37*(11), 1629–1633.

Hass, A., Gonzalez, J. M., Lima, I. M., Godwin, H. W., Halvorson, J. J., & Boyer, D. G. (2012). Chicken manure biochar as liming and nutrient source for acid Appalachian soil. *Journal of Environmental Quality*, *41*(4), 1096–1106.

Hervy, M., Pham Minh, D., Gérente, C., Weiss-Hortala, E., Nzihou, A., Villot, A., et al. (2018). H2S removal from syngas using wastes pyrolysis chars. *Chemical Engineering Journal*, *334*, 2179–2189.

Jin, J., Sun, K., Liu, W., Li, S., Peng, X., Yang, Y., et al. (2018). Isolation and characterization of biochar-derived organic matter fractions and their phenanthrene sorption. *Environmental Pollution*, *236*, 745–753.

Jindo, K., Mizumoto, H., Sawada, Y., Sanchez-Monedero, M. A., & Sonoki, T. (2014). Physical and chemical characterization of biochars derived from different agricultural residues. *Biogeosciences*, *11*(23), 6613–6621.

Jones, C., & Sammann, E. (1990). The effect of low power plasmas on carbon fibre surfaces. *Carbon*, *28*(4), 509–514.

Joseph, P., Tretsiakova-McNally, S., & McKenna, S. (2012). Characterization of cellulosic wastes and gasification products from chicken farms. *Waste Management*, *32*(4), 701–709.

Joseph, S., Pow, D., Dawson, K., Mitchell, D. R. G., Rawal, A., Hook, J., et al. (2015). Feeding biochar to cows: an innovative solution for improving soil fertility and farm productivity. *Pedosphere*, *25*(5), 666–679.

Kang, S., Li, X., Fan, J., & Chang, J. (2012). Characterization of hydrochars produced by hydrothermal carbonization of lignin, cellulose, d-Xylose, and wood meal. *Industrial & Engineering Chemistry Research*, *51*(26), 9023–9031.

Keiluweit, M., Nico, P. S., Johnson, M. G., & Kleber, M. (2010). Dynamic molecular structure of plant biomass-derived black carbon (biochar). *Environmental Science & Technology*, *44*(4), 1247–1253.

Kim, D., Yoshikawa, K., & Park, K. (2015). Characteristics of biochar obtained by hydrothermal carbonization of cellulose for renewable energy. *Energies*, *8*(12), 12412.

Kloss, S., Zehetner, F., Dellantonio, A., Hamid, R., Ottner, F., Liedtke, V., et al. (2012). Characterization of slow pyrolysis biochars: effects of feedstocks and pyrolysis temperature on biochar properties. *Journal of Environmental Quality*, *41*(4), 990–1000.

Knicker, H., Scaroni, A. W., & Hatcher, P. G. (1996). 13C and 15N NMR spectroscopic investigation on the formation of fossil algal residues. *Organic Geochemistry*, *24*(6), 661–669.

Knicker, H., Totsche, K. U., Almendros, G., & González-Vila, F. J. (2005). Condensation degree of burnt peat and plant residues and the reliability of solid-state VACP MAS 13C NMR spectra obtained from pyrogenic humic material. *Organic Geochemistry*, *36*(10), 1359–1377.

Lahaye, J. (1998). The chemistry of carbon surfaces. *Fuel*, *77*(6), 543–547.

Lee, J. W., Kidder, M., Evans, B. R., Paik, S., Buchanan Iii, A. C., Garten, C. T., et al. (2010). Characterization of biochars produced from cornstovers for soil amendment. *Environmental Science & Technology*, *44*(20), 7970–7974.

Lee, Y., Park, J., Ryu, C., Gang, K. S., Yang, W., Park, Y.-K., et al. (2013). Comparison of biochar properties from biomass residues produced by slow pyrolysis at 500°C. *Bioresource Technology*, *148*, 196–201.

Leng, L.-j., Yuan, X.-z., Huang, H.-j., Wang, H., Wu, Z.-b., Fu, L.-h., et al. (2015). Characterization and application of bio-chars from liquefaction of microalgae, lignocellulosic biomass and sewage sludge. *Fuel Processing Technology*, *129*, 8–14.

Liang, B., Lehmann, J., Solomon, D., Kinyangi, J., Grossman, J., O'Neill, B., et al. (2006). Black carbon increases cation exchange capacity in soils. *Soil Science Society of America Journal*, *70*(5), 1719–1730.

Limousy, L., Ghouma, I., Ouederni, A., & Jeguirim, M. (2017). Amoxicillin removal from aqueous solution using activated carbon prepared by chemical activation of olive stone. *Environmental Science and Pollution Research International*, *24*(11), 9993–10004.

López, F. A., Centeno, T. A., García-Díaz, I., & Alguacil, F. J. (2013). Textural and fuel characteristics of the chars produced by the pyrolysis of waste wood, and the properties of activated carbons prepared from them. *Journal of Analytical and Applied Pyrolysis*, *104*, 551–558.

Lu, K.-M., Lee, W.-J., Chen, W.-H., & Lin, T.-C. (2013). Thermogravimetric analysis and kinetics of co-pyrolysis of raw/torrefied wood and coal blends. *Applied Energy*, *105*, 57–65.

Lu, L., Kong, C., Sahajwalla, V., & Harris, D. (2002). Char structural ordering during pyrolysis and combustion and its influence on char reactivity. *Fuel*, *81*(9), 1215–1225.

Marsh, H., & Rodríguez-Reinoso, F. (2006a). Characterization of activated carbon. In *Activated Carbon* (pp. 143–242). Oxford: Elsevier Science Ltd. (chapter 4).

Marsh, H., & Rodríguez-Reinoso, F. (2006b). SEM and TEM Images of Structures in Activated Carbons. In *Activated Carbon* (pp. 366–382). Oxford: Elsevier Science Ltd. (chapter 7).

Meng, J., Wang, L., Liu, X., Wu, J., Brookes, P. C., & Xu, J. (2013). Physicochemical properties of biochar produced from aerobically composted swine manure and its potential use as an environmental amendment. *Bioresource Technology*, *142*, 641–646.

Mitchell, P. J., Dalley, T. S. L., & Helleur, R. J. (2013). Preliminary laboratory production and characterization of biochars from lignocellulosic municipal waste. *Journal of Analytical and Applied Pyrolysis*, *99*, 71–78.

Mohan, D., Sarswat, A., Ok, Y. S., & Pittman, C. U. (2014). Organic and inorganic contaminants removal from water with biochar, a renewable, low cost and sustainable adsorbent—a critical review. *Bioresource Technology*, *160*, 191–202.

Mukherjee, A., Zimmerman, A. R., & Harris, W. (2011). Surface chemistry variations among a series of laboratory-produced biochars. *Geoderma*, *163*(3), 247–255.

Nair, P. K. R. (2002). The Nature and Properties of Soils, 13th Edition. By N. C. Brady and R. R. Weil. *Agroforestry Systems*, *54*(3), 249.

Narzari, R., Bordoloi, N., Sarma, B., Gogoi, L., Gogoi, N., Borkotoki, B., et al. (2017). Fabrication of biochars obtained from valorization of biowaste and evaluation of its physicochemical properties. *Bioresource Technology*, *242*, 324–328.

Puri, B. R., Kaistha, B. C., Vardhan, Y., & Mahajan, O. P. (1973). Studies in surface chemistry of carbon blacks-vi. adsorption isotherms of benzene on carbons associated with different surface oxygen complexes. *Carbon*, *11*(4), 329–336.

Rocca, P. A. D., Cerrella, E. G., Bonelli, P. R., & Cukierman, A. L. (1999). Pyrolysis of hardwoods residues: on kinetics and chars characterization. *Biomass and Bioenergy*, *16*(1), 79–88.

Rodríguez-Reinoso, F., Garrido, J., Martín-Martínez, J. M., Molina-Sabio, M., & Torregrosa, R. (1989). The combined use of different approaches in the characterization of microporous carbons. *Carbon*, *27*(1), 23–32.

Rodríguez-Reinoso, F., & Sepúlveda-Escribano, A. (2001). Porous carbons in adsorption and catalysis-A2. In H. S. Nalwa (Ed.), *Handbook of Surfaces and Interfaces of Materials* (pp. 309–355). Burlington: Academic Press (chapter 9).

Rouzaud, J.-N., & Clinard, C. (2002). Quantitative high-resolution transmission electron microscopy: a promising tool for carbon materials characterization. *Fuel Processing Technology*, *77–78*, 229–235.

Seaton, N. A., Walton, J. P. R. B., & Quirke, N. (1989). A new analysis method for the determination of the pore size distribution of porous carbons from nitrogen adsorption measurements. *Carbon*, *27*(6), 853–861.

Shafeeyan, M. S., Daud, W. M. A. W., Houshmand, A., & Shamiri, A. (2010). A review on surface modification of activated carbon for carbon dioxide adsorption. *Journal of Analytical and Applied Pyrolysis*, *89*(2), 143–151.

Shak, K. P. Y., & Wu, T. Y. (2014). Coagulation–flocculation treatment of high-strength agro-industrial wastewater using natural Cassia obtusifolia seed gum: Treatment efficiencies and flocs characterization. *Chemical Engineering Journal*, *256*, 293–305.

Singh, B., Singh, B. P., & Cowie, A. L. (2010). Characterisation and evaluation of biochars for their application as a soil amendment. *Soil Research*, *48*(7), 516–525.

Song, W., & Guo, M. (2012). Quality variations of poultry litter biochar generated at different pyrolysis temperatures. *Journal of Analytical and Applied Pyrolysis*, *94*, 138–145.

Stella Mary, G., Sugumaran, P., Niveditha, S., Ramalakshmi, B., Ravichandran, P., & Seshadri, S. (2016). Production, characterization and evaluation of biochar from pod (Pisum sativum), leaf (Brassica oleracea) and peel (Citrus sinensis) wastes. *International Journal of Recycling of Organic Waste in Agriculture*, *5*(1), 43–53.

Suliman, W., Harsh, J. B., Abu-Lail, N. I., Fortuna, A.-M., Dallmeyer, I., & Garcia-Perez, M. (2016a). Influence of feedstock source and pyrolysis temperature on biochar bulk and surface properties. *Biomass and Bioenergy*, *84*, 37–48.

Suliman, W., Harsh, J. B., Abu-Lail, N. I., Fortuna, A.-M., Dallmeyer, I., & Garcia-Perez, M. (2016b). Modification of biochar surface by air oxidation: role of pyrolysis temperature. *Biomass and Bioenergy*, *85*, 1–11.

Sun, K., Kang, M., Zhang, Z., Jin, J., Wang, Z., Pan, Z., et al. (2013). Impact of deashing treatment on biochar structural properties and potential sorption mechanisms of phenanthrene. *Environmental Science & Technology*, *47*(20), 11473–11481.

Taherymoosavi, S., Joseph, S., & Munroe, P. (2016). Characterization of organic compounds in a mixed feedstock biochar generated from Australian agricultural residues. *Journal of Analytical and Applied Pyrolysis*, *120*, 441–449.

Tamošiūnas, A., Chouchène, A., Valatkevičius, P., Gimžauskaitė, D., Aikas, M., Uscila, R., et al. (2017). The potential of thermal plasma gasification of olive pomace charcoal. *Energies*, *10*(5), 710.

Teh, C. Y., Wu, T. Y., & Juan, J. C. (2014). Potential use of rice starch in coagulation–flocculation process of agro-industrial wastewater: treatment performance and flocs characterization. *Ecological Engineering*, *71*, 509–519.

Tomlinson, J. B., Freeman, J. J., & Theocharis, C. R. (1993). The preparation and adsorptive properties of ammonia-activated viscose rayon chars. *Carbon*, *31*(1), 13–20.

Van Zwieten, L., Kimber, S., Morris, S., Chan, K. Y., Downie, A., Rust, J., et al. (2010). Effects of biochar from slow pyrolysis of papermill waste on agronomic performance and soil fertility. *Plant and Soil*, *327*(1), 235–246.

Wanassi, B., Hariz, I. B., Ghimbeu, C. M., Vaulot, C., Hassen, M. B., & Jeguirim, M. (2017). Carbonaceous adsorbents derived from textile cotton waste for the removal of Alizarin S dye from aqueous effluent: kinetic and equilibrium studies. *Environmental Science and Pollution Research International*, *24*(11), 10041–10055.

Wang, X., Cook, R., Tao, S., & Xing, B. (2007). Sorption of organic contaminants by biopolymers: role of polarity, structure and domain spatial arrangement. *Chemosphere*, *66*(8), 1476–1484.

Wang, Z., Liu, G., Zheng, H., Li, F., Ngo, H. H., Guo, W., et al. (2015). Investigating the mechanisms of biochar's removal of lead from solution. *Bioresource Technology*, *177*, 308–317.

Wang, G., Zhang, J., Chang, W., Li, R., Li, Y., & Wang, C. (2018). Structural features and gasification reactivity of biomass chars pyrolyzed in different atmospheres at high temperature. *Energy*, *147*, 25–35.

Weber, K., & Quicker, P. (2018). Properties of biochar. *Fuel*, *217*, 240–261.

Xu, X., Cao, X., Zhao, L., Zhou, H., & Luo, Q. (2014). Interaction of organic and inorganic fractions of biochar with Pb(ii) ion: further elucidation of mechanisms for Pb(ii) removal by biochar. *RSC Advances*, *4*(85), 44930–44937.

Xu, X., Kan, Y., Zhao, L., & Cao, X. (2016). Chemical transformation of CO2 during its capture by waste biomass derived biochars. *Environmental Pollution*, *213*, 533–540.

Xu, X., Zhao, Y., Sima, J., Zhao, L., Mašek, O., & Cao, X. (2017). Indispensable role of biochar-inherent mineral constituents in its environmental applications: a review. *Bioresource Technology*, *241*, 887–899.

Xu, Y., & Chen, B. (2015). Organic carbon and inorganic silicon speciation in rice-bran-derived biochars affect its capacity to adsorb cadmium in solution. *Journal of Soils and Sediments*, *15*(1), 60–70.

Yang, H., Yan, R., Chen, H., Zheng, C., Lee, D. H., & Liang, D. T. (2006). Influence of mineral matter on pyrolysis of palm oil wastes. *Combustion and Flame*, *146*(4), 605–611.

Yuan, H., Lu, T., Huang, H., Zhao, D., Kobayashi, N., & Chen, Y. (2015). Influence of pyrolysis temperature on physical and chemical properties of biochar made from sewage sludge. *Journal of Analytical and Applied Pyrolysis*, *112*, 284–289.

Yuan, J.-H., Xu, R.-K., & Zhang, H. (2011). The forms of alkalis in the biochar produced from crop residues at different temperatures. *Bioresource Technology*, *102*(3), 3488–3497.

Zhang, J., & Wang, Q. (2016). Sustainable mechanisms of biochar derived from brewers' spent grain and sewage sludge for ammonia–nitrogen capture. *Journal of Cleaner Production*, *112*, 3927–3934.

Zhang, S., Chen, T., Li, W., Dong, Q., & Xiong, Y. (2016). Physicochemical properties and combustion behavior of duckweed during wet torrefaction. *Bioresource Technology*, *218*, 1157–1162.

Zhao, L., & Brooks, R. R. (2013). Tailoring trusted semantic information. In: Proceedings of the Eighth Annual Cyber Security and Information Intelligence Research Workshop, *ACM. Oak Ridge, Tennessee, USA*, pp. 1–4.

Zhao, S.-X., Na, T., & Wang, X.-D. (2017). Effect of temperature on the structural and physicochemical properties of biochar with apple tree branches as feedstock material. *Energies*, *10*, 1293.

Biochar for soil amendment

4

Patrick Brassard, Stéphane Godbout*, Vicky Lévesque†,*
Joahn H. Palacios, Vijaya Raghavan‡, Ahmed Ahmed‡, Richard Hogue*,*
Thomas Jeanne, Mausam Verma**
*Research and Development Institute for the Agri-Environment (IRDA), Quebec City, QC, Canada, †Agriculture and Agri-Food Canada, Quebec City, QC, Canada, ‡Department of Bioresource Engineering, Macdonald Campus, McGill University, Sainte-Anne-de-Bellevue, QC, Canada

Chapter Outline

Char and Carbon Materials Derived from Biomass. https://doi.org/10.1016/B978-0-12-814893-8.00004-3

4.1 Introduction

Interest in biochar (also known as char) as a soil amendment originates from studies of Amazonian soils that were amended with carbonaceous material (Glaser & Birk, 2012), known as *terra preta* (Fig. 1). The fertility of these soils was improved as compared to typical Amazonian jungle soils, and crop yields were increased for thousands of years. Thus, restoration of degraded soil to increase crop yields could be the main motivation for farmers to amend their soils with biochar.

Over the last few years, several investigations were carried out to assess the global impact of biochar in various agricultural soils. Many studies reported positive effects on crop productivity (Asai et al., 2009; Hossain, Strezov, Yin Chan, & Nelson, 2010; Wang, Pan, Liu, Zhang, & Xiong, 2012). In fact, biochar amendment alters the biological and physicochemical properties of soils, which affect soil processes such as organic matter degradation, nutrients cycling, and water holding capacities. However, the mechanisms involved in soil following biochar amendment are not always well understood. In addition to the increase in soil health and crops yield, it was demonstrated that biochar can contribute to mitigate climate change by sequestrating carbon (C) in soil for thousands of years, and by mitigating soil nitrous oxide (N_2O) emissions. Moreover, because of its high porosity and surface area, biochar can be used as a remediation tool for polluted agricultural soil, reducing the bioavailability of heavy metals and organic compounds such as pesticides. Finally, the production of biochar from residual biomass could be a solution for farmers to manage their wastes.

Fig. 1 Typical profiles of terra preta (A) and Oxisol (B) sites (Glaser, Haumaier, Guggenberger, & Zech, 2001).

The physicochemical properties of biochars depend strongly on both production operating parameters and biomass feedstock. Thus, specific biochars have to be produced for their use in specific applications. This chapter gives an overview of the potential benefits of biochar used as soil amendment, summarizing the results of recent research studies and giving input regarding the use of biochar in soil.

4.2 Biochar's uses and impacts on agricultural soil properties and agronomic performance

4.2.1 Impacts on soil physical properties

Biochar has significant effects on the physical nature of agricultural soils (Table 1). These changes affect plant growth by altering root penetration depth and availability of water. The density (ρ) of biochar particles is lower than the density of soil particles (Sharma, Wang, Pareek, Yang, & Zhang, 2014). The density of soil particles is $2.6\,\mathrm{t\,m^{-3}}$ regardless of particle size, whereas the density of wood-derived biochar (WBC) with a particle size of $<70\,\mu\mathrm{m}$ ranges from 0.6 to $1.6\,\mathrm{t\,m^{-3}}$ and depends on the type of wood (Hu et al., 2016). Soils amended with biochar show an increase in surface area and porosity, and a reduction in ρ (Eastman, 2011; Hardie, Clothier, Bound, Oliver, & Close, 2014). For example, a study from Reddy, Yaghoubi, and Yukselen-Aksoy (2015) revealed that the specific gravity of silty clay amended with 5%, 10%, and 20% WBC (particle sizes $<420\,\mu\mathrm{m}$; $\rho = 0.81$) decreased from 2.6 to 2.1, 2.0, and $1.8\,\mathrm{t\,m^{-3}}$, respectively.

Soil compaction is defined as the densification of soil whereby air-filled porosity is reduced (Harris, 1971). Modifications of soil physical properties lead to changes in soil state under compaction (Chen & Weil, 2011). Soils with a lower ρ have lower energy requirements of mechanical tillage (Carter, 1990). Because compaction increases the soil mechanical impedance, the elongation and proliferation of roots are affected (Boone & Veen, 1994). The degree of compaction-induced damage could be minimized if the behavior of the soil/biochar mix is understood. Soil compressibility decreases with an increase in the WBC amendment and with a decrease in WBC particle size (Reddy et al., 2015). Busscher et al. (2010) showed that the application of $44\,\mathrm{t\,ha^{-1}}$ of pecan shell-derived biochar (2% *w*/w) decreased the soil penetration resistance (PR; measured at a water content (θ) of 10%) of a fine loamy sand from 2.90 to 1.18 MPa and raised the soil ρ from 1.45 to $1.52\,\mathrm{t\,m^{-3}}$. It was also demonstrated by Bekele, Roy, and Young (2015) that the PR of a loamy soil was lowered following WBC amendment. However, Eastman (2011) found that a 25-$\mathrm{t\,ha^{-1}}$ WBC amendment did not have a significant effect on the PR of a silt loam 1 year after incorporation.

Soil consistency limits can provide an indicator to describe the degree and kind of cohesion and adhesion between the soil particles and biochar with respect to the resistance of the soil to deformation or rupture. As the plasticity of soil increases, the range of moisture over which the soil is susceptible to compaction by external forces increases (Aksakal, Angin, & Oztas, 2013). Application of 6% (*w*/w) WBC decreased the soil liquid limit (θ_{ll}), i.e., the water content at which the behavior of soil changes

Table 1 Salient impacts of biochar on soil physical properties.

Biochar source	Pyrolysis temperature	Soil type	Physical properties	References
Biochar derived from lignocellulosic biomass e.g., cacao shell, rice husk, pecan shell, maple woodchips	500–600°C	Silt clay soil, fine loamy soil, silt loamy soil Sandy soil Clayey soil	Decrease in density, increase in surface area, increase in porosity, decrease in soil penetration resistance, increase in water holding capacity Increase in hydraulic conductivity Decrease in hydraulic conductivity	Abel et al. (2013); Busscher et al. (2010); Eastman (2011); Ibrahim et al. (2013); Kameyama et al. (2014); Lei and Zhang (2013); Liu et al. (2012); Reddy et al. (2015); Ulyett et al. (2013)
Animal waste-derived biochar e.g., dairy manure	300–450°C		Decrease in density, increase in surface area, increase in porosity, decrease in soil penetration resistance, increased hydrophobicity, increase in hydraulic conductivity	Kinney et al. (2012); Yi et al. (2015)
Biochar derived from plant biomass e.g., eucalypt green waste Blady grass Olive tree pruning, Mallee	650–750°C 450°C 720°C	Black clay loam, red loam, brown sandy loam	Decrease in density, increase in surface area, increase in porosity, decrease in soil penetration resistance, higher water holding capacity, moderate increase in hydraulic conductivity	

from plastic to liquid, and consequently decreased the plasticity index (*PI*) of a clayey soil (Zong, Xiao, & Lu, 2016). Further, the WBC-amended clayey soil became more plastic, even though WBC itself is nonplastic (Brewer et al., 2014). Therefore, the increase in θ_{pl} and θ_{ll} of clayey soil amended with WBC could be attributed to the high absorptive capacity of the WBC for water (Guo, Tang, Li, & Xie, 2013). This increased affinity of WBC for water could result in WBC-amended soil requiring more water to behave in a plastic or liquid manner.

Biochar amendment improves soil structure and creates pores (Hardie et al., 2014). Biochar alters soil porosity and increases soil surface area. Accordingly, a soil/biochar mix tends to display a higher water holding capacity than the unamended soil (Basso, Miguez, Laird, Horton, & Westgate, 2012). Therefore, the soil hydrological characteristics such as available water capacity (AWC), soil moisture retention, and hydraulic conductivity are improved by biochar amendment (Abel et al., 2013; Eastman, 2011; Hardie et al., 2014; Liu, Han, & Zhang, 2012). Nevertheless, this additional water may not be readily available to a plant, since water in very small saturated pores is not suitable for uptake (Sohi, Krull, Lopez-Capel, & Bol, 2010). Biochar exhibits higher infiltration and water retention at saturation if produced at higher temperatures. This might be attributable to the influence of high temperature on biochar pore volume and pore tortuosity (Kameyama, Miyamoto, & Shiono, 2014). However, the influence of biochar on soil AWC is not always positive (Hardie et al., 2014). Mukherjee and Lal (2013) observed a decrease in AWC when biochar was applied in powdered form to fine-textured soils. Applying biochar to sandy soil at a dosage of 60 t ha^{-1} resulted in a significant increase in soil water retention, which was attributed to the porous structure of biochar (Ulyett, Sakrabani, Kibblewhite, & Hann, 2013). In another study (Kameyama et al., 2014), biochar amendment at a rate of 20 g kg^{-1} increased water retention in sandy loam soil, resulting in a 12% reduction in cumulative evaporation. Furthermore, the soil water retention at saturation and at field capacity increased by 30% and 16%, respectively. The authors attributed this effect to the lower bulk density and higher surface area of woodchip biochar. On the other hand, Ibrahim, Al-Wabel, Usman, and Al-Omran (2013) observed that soil water content increased with an increase in the quantity of biochar added to the soil. However, this effect depends on the soil properties. For instance, adding up to 6% (*w*/w) biochar to clayey soil decreased saturated hydraulic conductivity (k_{sat}) by up to 78% (Liu et al., 2012), whereas adding 4% of *Conocarpus* biochar increased sandy soil k_{sat} by 25% (Ibrahim et al., 2013). Biochar produced at temperatures below 450°C can retain organic compounds that repel water (Kinney et al., 2012; Yi, Witt, Chiu, Guo, & Imhoff, 2015). This hydrophobicity may also cause erosion due to increased overflow (Conte et al., 2012). The hydrophobicity of biochar can be controlled by selecting appropriate feedstock and pyrolysis conditions. If necessary, biochar treatment with water can decrease biochar hydrophobicity (Yi et al., 2015).

Despite the growing number of studies on the influence of biochar on the agricultural soil hydraulic characteristics, the influence of biochar particle size on soil pore size distribution and hydraulic conductivity remains unclear, especially for compacted agricultural soils. Moreover, existing soil failure models should be employed to determine tillage power requirements relative to changes in the parameters of soil/biochar mix shear.

4.2.2 Impacts on soil chemical properties

Previous studies reported that biochar amendment directly and indirectly improves soil chemical properties. These effects are influenced by the physicochemical properties of biochar and its interaction with microbial and enzymatic function in soil (Tan, Lin, Ji, & Rainey, 2017). Biochar can affect soil pH, cation exchange capacity (CEC), C content, and the availability of nitrogen (N) and phosphorus (P) for plant use. Salient impacts of biochar on soil chemical properties are presented in Table 2.

4.2.2.1 pH

The application of biochar increases soil pH, but its effect depends on the biochar pH and its liming value (Lehmann et al., 2011; Lévesque, Rochette, Ziadi, Dorais, & Antoun, 2018; Tian et al., 2018), which is associated with their high base cations content (e.g., potassium, calcium, and magnesium) and alkalinity. It can also be attributed to the decreased exchangeable aluminum (Al) and hydrogen (H) ions in soil due to their adsorption on the biochar surface (Syuhada, Shamshuddin, Fauziah, Rosenani, & Arifin, 2016; Tan et al., 2017). A reduction of aluminum concentration in soil through chelation of this element with organic compounds from the biochar could also occur and explain the increased pH in acidic soil (Butnan, Deenik, Toomsan, Antal, & Vityakon, 2015). The increase in soil pH by biochar amendment seems to be less evident for calcareous soils (van Zwieten et al., 2010). However, the addition of 1% and 2% (*w*/w) of maple woodchip biochars produced in the range 500–650°C (pH=9.7) in two alkaline chernozemic soils has shown a significant increase in soil pH after 70 days of incubation compared to the nonamended soils (Chathurika et al., 2016). For alkaline soils, the combination of biochar and mineral fertilizer might be a more efficient strategy than the addition of biochar alone to avoid a significant increase of soil pH limiting the nutrient availability for plants (Chathurika et al., 2016).

4.2.2.2 Cation exchange capacity

In general, biochar amendments cause improvement of CEC in soils with a low fertility (Agegnehu, Bass, Nelson, & Bird, 2016). In addition, the results of Chathurika et al. (2016) have shown an increase in CEC in a clay loam amended with biochar. The availability of cationic nutrients in soil can thereby be increased due to the carboxylic and phenolic functional groups in biochar, promoting an electrostatic attraction force of positive charges on the biochar surface (Chathurika et al., 2016; Tian et al., 2018). The interaction between soil organic matter (SOM) and biochar amendment might explain the different effects of biochar on soil CEC (Gul, Whalen, Thomas, Sachdeva, & Deng, 2015).

Table 2 Salient impacts of biochar on soil chemical properties.

Biochar source	Pyrolysis temperature	Soil type	Chemical properties	References
Biochar derived from lignocellulosic biomass e.g., cacao shell, rice husk, pecan shell, maple woodchips	500–600°C	Silt clay soil, fine loamy soil, silt loamy soil Sandy soil Clayey soil	Increase in pH, increase in cations (K^+, Ca^{2+}, and Mg^{2+}), increase in cation exchange capacity, increase in C contents, increase in C bioavailability	Agegnehu et al. (2016); Butnan et al. (2015); Chathurika et al. (2016); Kameyama et al. (2016); Smebye et al. (2016)
	700–800°C		Higher nitrate adsorption capacity	
Animal waste-derived biochar e.g., dairy manure	300–450°C		Increase in pH, increase in cations (K^+, Ca^{2+}, and Mg^{2+}), increase in cation exchange capacity, increase in C, N, and P contents, increase in C and N bioavailability	Chathurika et al. (2016); Efthymiou et al. (2018); Gul and Whalen (2016)
Biochar derived from plant biomass e.g., eucalypt green waste	650–750°C	Black clay loam, red loam, brown sandy loam	Increase in pH, increase in cations (K^+, Ca^{2+}, and Mg^{2+}), increase in cation exchange capacity, increase in C and N contents, increase in C and N bioavailability	Abujabhah, Doyle, et al. (2016); Clough et al. (2013); Tan et al. (2017); Zhang et al. (2016)
Blady grass	450°C		Increased P content in soil	
Olive tree pruning, Mallee	720°C		Decreased P content in soil	

4.2.2.3 Soil carbon content

Depending on the soil type, a metaanalysis on the biochar stability in soil reported that the biochar amendment slightly retarded the mineralization of SOM compared to the nonamended soil (Wang, Xiong, & Kuzyakov, 2015). Biochar amendment increases the soil C content and soil water and air contents stimulating soil microbial biomass (Jiang, Denef, Stewart, & Cotrufo, 2016) that influences the native soil organic C (SOC) decomposition (Abujabhah, Bound, Doyle, & Bowman, 2016; Tan et al., 2017). The presence of biochar has slowed down the native SOC mineralization with a high availability of N and low molecular weight C compounds (Jiang et al., 2016). The addition of N can influence the SOC decomposition because of the increase in microbial biomass growth and priming effect (Jiang, Cao, & Zhang, 2014). On the other hand, biochars produced at low pyrolysis temperature have more labile C than those produced at higher temperature, increasing the dissolved organic carbon (DOC) in a horticultural growing medium (Lévesque et al., 2018). It was reported that the increase in DOC solubility and DOC desorption from mineral soil amended with biochar was caused by an increase in soil pH (Jiang et al., 2016). For instance, the increase in soil pH from 4.9 to 8.7 in acidic acrisol amended with 10% (*w*/w) cacao shell biochar made at 600°C or rice husk biochar made at 500°C increased soil DOC from 4.5 to 69 $mg\,L^{-1}$ (Smebye et al., 2016). The increase in soil pH by biochar can also influence the deprotonation of weak acid functional groups in the dissolved organic matter molecules in soil, which increases the solubility of organic carbon (Smebye et al., 2016).

4.2.2.4 Soil nitrogen content

Soil N availability is affected by the type of biochar amendment. Its physicochemical properties can influence abiotic mechanisms (adsorption/desorption) and biotic mechanisms such as mineralization, immobilization, nitrification, denitrification, and biological N_2 fixation by promoting nitrogen-fixing microorganisms in soil, such as rhizobia (Lévesque et al., 2018; Nguyen et al., 2017; Tan et al., 2017). A metaanalysis of 56 short-term studies ($\leq$5 years) with 1080 experimental cases, published between 2010 and 2015, evaluated the effects of biochar on soil available inorganic nitrogen (SIN) such as ammonium and nitrate (Nguyen et al., 2017). According to this metaanalysis, biochar generally reduced SIN regardless of experimental conditions, but 95% of cases occurred within the first year after biochar amendment (Nguyen et al., 2017). This metaanalysis identified different factors influencing SIN after biochar amendment such as residence time of biochar in soil, source of fertilization, soil properties such as texture and pH, application rate of biochar, feedstock use and pyrolysis conditions, surface area, CEC, and labile organic carbon content of biochar. Moreover, the electrical conductivity (EC) and particle sizes of biochar, C/N ratio of organic fertilizer combined with biochar, and SOM content should be considered (Nguyen et al., 2017).

In the case where there is an increase in N immobilization and decrease in nitrification, the availability of ammonium and nitrate in soils amended with biochar is reduced (Lévesque et al., 2018). For instance, Abujabhah, Doyle, Bound, and Bowman (2016) observed that the increase in N consumption by microorganisms in a red loam was associated with the increase in the application rate of eucalypt green waste biochar in this soil. The authors suggested that the presence of biochar had stimulated the decomposition of organic C, thereby reducing the N availability in the red loam (Abujabhah, Doyle, et al., 2016). Furthermore, positively charged ammonium ions can be adsorbed on biochar surfaces via oxygen active groups, such as carboxyl (—COOH) or hydroxyl (—OH) ions (Clough, Condron, Kammann, & Müller, 2013; Tan et al., 2017), and nitrate can also be chemically adsorbed on the biochar surface in soils. Kameyama, Miyamoto, Iwata, and Shiono (2016) observed that biochars produced at 700°C and 800°C were effective for nitrate adsorption due to their high surface basic functional groups. In addition, a strong correlation ($r = 0.81$) between nitrate adsorption and carbon content in biochar produced at 800°C was observed (Kameyama et al., 2016). Because wood-based biomass has higher C content than nonwood-based biomass, Kameyama et al. (2016) also suggested that biochars produced from wood may have a better efficiency to adsorb nitrate in soil. The limitation of N in soil by biochar for a short term (Nguyen et al., 2017) could be alleviated by the addition of organic compounds such as compost with biochar (Agegnehu et al., 2016). Nguyen et al. (2017) suggest applying biochar 1 month prior to the sowing date to avoid N deficiency. Furthermore, Nguyen et al. (2017) also recommended applying woody biochar instead of plant-derived biochars to limit the reduction of N in soil. In addition, plant-derived biochars with higher carbohydrate content should be combined with organic fertilizer to offset the depletion of SIN. Finally, the addition of high application rates of biochar should be combined with an organic fertilizer to alleviate the effect of N immobilization.

4.2.2.5 Soil phosphorus content

The availability of P in soil is also affected directly and indirectly by biochar amendments through different mechanisms such as adsorption versus desorption, precipitation versus solubilization, and mineralization versus immobilization (Chathurika et al., 2016; Gul & Whalen, 2016; Zhang et al., 2016). The P content in biochar can directly supply P in soil (Chathurika et al., 2016), but varies according to the feedstock used, and the biochar effect differs with initial P concentration in the soil solution (Zhang et al., 2016). The P availability content in biochar increases with increasing pyrolysis temperature (Tan et al., 2017). However, a significant loss of P can occur when the pyrolysis temperature is higher than 700°C, thus limiting the P input in soil (Zhang et al., 2016). For instance, Zhang et al. (2016) tested nine biochars, and the biochar produced from the pyrolysis of blady grass (*Imperata cylindrica*) at 450°C showed the highest contribution to bring available P in soil. On the other hand, biochar from Mallee (*Eucalyptus polybractea*) pyrolyzed at 720°C showed the highest retention of P from fertilizer, because of its high P adsorption capacity.

The P content in biochar can be oxidized by the aluminum and iron content of soil, allowing greater availability of P in soils (Tan et al., 2017). Because biochar can increase soil pH, it can also indirectly increase P mineralization in soil through the stimulation of extracellular enzymatic activity produced by plant roots such as alkaline phosphatase and by P solubilizing microorganisms living in soil (Chathurika et al., 2016; Efthymiou, Grønlund, Müller-Stöver, & Jakobsen, 2018; Gul & Whalen, 2016). However, due to alkalinity and exchangeable calcium (Ca), magnesium (Mg), and iron (Fe) on its surface, biochar can reduce P availability in the soil environment by promoting P precipitation or sorption on its surface (Tian et al., 2018). Zhang et al. (2016) proposed two stages in P adsorption on a biochar surface in soil. The first step is a rapid chemisorption on surface $CaCO_3$ and the second step is to control surface diffusion when the surface adsorption sites of biochar are saturated (Zhang et al., 2016). Schneider and Haderlein (2016) suggested that the application of alkaline biochars and those with most P in Ca-associated forms might be of benefit by increasing P availability in acidic soils, because of the increase in soil pH by biochar and rendering P already present in the soil. In contrast, the addition of biochar with most of the P in Fe-/Al-associated forms might be beneficial by improving P availability in alkaline soils (Schneider & Haderlein, 2016).

4.2.3 Impact on soil biological properties

Biochar amendments can cause significant changes in soil biological properties, which can modify soil microbial and faunal diversity and activities (Gul et al., 2015; Kolton, Graber, Tsehansky, Elad, & Cytryn, 2016; Zhang et al., 2017). A metaanalysis revealed that biochars produced from the slow pyrolysis of various feedstocks at temperatures from 300°C to 600°C consistently increased some microbial parameters (i.e., abundance and community structure of microorganisms) in a vast number of soils during short (<90 days) laboratory incubations and longer (1–3 years) field studies (Gul et al., 2015). The biochar-mediated changes in soil biological properties appeared to be a function of soil texture and biochar characteristics such as surface area, porosity, and pH (Gul et al., 2015). Consequently, biochar properties as well as the induced changes in soil properties have different impacts on ecosystem functioning in soil and the rhizosphere (Hussain et al., 2017; Kolton et al., 2016). Salient impacts of biochar on soil biological properties are presented in Table 3.

Biochar can act as a hotspot for microbial activity, which promotes abundance of soil biota (Lehmann et al., 2011). In fact, the physical properties of biochar such as specific surface area (SSA) and porosity, as well as the sorption capacity of nutrients and organic matter around biochar particles, can improve the use of resources by microorganisms. Biochar morphology can also provide a refuge for soil microorganisms (0.3–3.0 μm for bacteria; 2–80 μm for fungi; and 7–30 μm for protozoa), protecting them from larger predators in soil such as microarthropods (Gul et al., 2015; Lehmann et al., 2011; Warnock, Lehmann, Kuyper, & Rillig, 2007).

4.2.3.1 Impact on metabolic quotient of carbon

According to a metaanalysis of Zhou et al. (2017), biochars produced at temperatures over 500°C or having a pH >9 could decrease the metabolic quotient of carbon (qCO_2) value, which is used as an indicator of environmental stresses on soil

Table 3 Salient impacts of biochar on soil biological properties.

Biochar source	Pyrolysis temperature	Soil type	Biological properties	References
Biochar derived from lignocellulosic biomass e.g., cacao shell, rice husk, pecan shell, maple woodchips	500–800°C	Silt clay soil, fine loamy soil, silt loamy soil	Increased activity of alkaline phosphatases, reduced activity of carbon cycling enzymes, e.g., α-glucosidase, β-glucosidase, β-xylosidase, peroxidase, and dehydrogenase, decrease in fungi/bacteria ratio, meso- and macrofauna such as earthworms, enchytraeids, nematodes, collembolans, mites, and macroarthropods, increase in beneficial microbes and suppression of pathogens	Chen et al. (2013, 2017); Elad et al. (2010); Gul et al. (2015); Jaiswal et al. (2014, 2017); Kameyama et al. (2014); Kolton et al. (2016); Warnock et al. (2007); Zhang et al. (2017)
Maple bark biochar	400°C 550°C		Decrease in metabolic quotient of carbon (qCO_2) Increase in qCO_2	
Animal waste-derived biochar e.g., dairy manure	300–450°C		Decrease in fungi/bacteria ratio, meso- and macrofauna such as earthworms, enchytraeids, nematodes, collembolans, mites, and macroarthropods, increase in beneficial microbes and suppression of pathogens	Khadem and Raiesi (2017)
Biochar derived from plant biomass e.g., eucalypt green waste	650–750°C	Black clay loam, red loam, brown sandy loam	Decrease in fungi/bacteria ratio, increase in bacterial population of *Pseudomonas stutzeri* and *Pedobacter saltans*, meso- and macrofauna such as earthworms, enchytraeids, nematodes, collembolans, mites, and macroarthropods, increase in beneficial microbes and suppression of pathogens	Anderson et al. (2011); Harter et al. (2016); Kolton et al. (2016); Warnock et al. (2010)
Olive tree pruning, Mallee	720°C		Decrease in arbuscular mycorrhizal fungi root colonization	

Table 4 Effect of different biochars on microbial metabolic quotient (qCO_2) in a peat-based growing medium after 58 days of incubation.

Treatment	qCO_2 (mg CO_2-C/mg C microbial biomass)
Control	1.45 ± 0.16 b
M400	0.95 ± 0.06 cd
M550	0.78 ± 0.08 d
M700	1.28 ± 0.08 bc
P700	1.93 ± 0.15 a
W400	1.58 ± 0.23 ab

Means ± standard errors in a column followed by different letters are significantly different at $P < .05$ (Tukey's test). Control, without biochar; biochar treatments received 15% (*v*/v) of maple bark produced at 400°C, M400; 550°C, M550; and 700°C, M700, pine chips produced at 700°C, P700, or willow chips produced at 400°C, W400 (Lévesque et al., 2018).

microbial communities. However, the feedstock used to produce the biochar could be a factor. For example, the addition of 15% (*v*/v) biochar produced from the pyrolysis of maple bark at 400°C or 550°C in a horticultural growing medium decreased the qCO_2 value as compared to the maple bark biochar produced at 700°C (Table 4) (Lévesque et al., 2018). Thus, biochar could prevent carbon exhaustion by microorganisms producing their energy by promoting their growth due to colocation of various nutrients in and around biochar (Lévesque et al., 2018; Zhou et al., 2017).

4.2.3.2 Impact on the abundance of bacteria and fungi

The aromatic C content of biochar can also increase the growth of Gram (+) bacteria compared to Gram (−) bacteria, because Gram (+) bacteria have a better ability to consume recalcitrant organic compounds (Gul et al., 2015). The low availability of carbon in biochar because of its high stability can also reduce the activity of several enzymes involved in carbon cycling such as α-glucosidase, β-glucosidase, β-xylosidase, peroxidase, and dehydrogenase (Chen et al., 2013, 2017).

The fungi-to-bacteria ratio decreases with the increase in pyrolysis temperature and the application rates of biochar because of the increase in recalcitrant C by biochar in soil (Khadem & Raiesi, 2017). For example, in a rice paddy, the addition of 20 and 40 t ha^{-1} of wheat straw biochar produced between 350°C and 550°C in sandy loam soil increased bacterial gene abundance by 28% and 64% and reduced fungal gene abundance by 35% and 46%, respectively, compared to the nonamended soil (Chen et al., 2013). The higher abundance of bacteria than fungi in soil amended with biochar could be associated with an improvement of carbon stabilization limiting its mineralization when there is an increase in application rates of biochar (Chen et al., 2013).

4.2.3.3 Impact on microbial communities

Biochar affects the soil food web and the growth of microorganisms involved in N, P, and C cycling (Anderson et al., 2011; Harter et al., 2014, 2016; Kolton et al., 2016). For instance, an incubation study has shown that 10% *w*/w of green waste biochar pyrolyzed at 700°C added in a sandy loam increased the abundance of bacteria *Pseudomonas stutzeri* and *Pedobacter saltans*. These bacterial species might contribute to the nosZ gene activity and thereby be involved in N_2O reduction (Harter et al., 2016). Addition of 0.5%, 1%, 2%, and 4% (*w*/w) peanut biochar made at 360°C increased the available P by 101%, while arbuscular mycorrhizal fungi root colonization decreased by 74% (Warnock et al., 2010). Kolton et al. (2016) suggested that the release of nonlabile biochar-associated organic compounds in soil and the porous carbon structure of biochar increase the diversity of beneficial bacteria associated with phenolic compounds and carbohydrate metabolism and reduce grazing in soil. The change in carbon-source utilization in soil amended with biochar can cause a significant increase in microbial taxonomic and functional diversity and activity in the rhizosphere (Jaiswal et al., 2017).

Biochar amendment seems to have a higher impact on microbial diversity in nutrient-poor soil than in nutrient-rich soil (Wang, Xiong, & Kuzyakov, 2016). A random matrix theory-based molecular ecological network analysis revealed that soil microbial community assemblages were more complex in a low fertility Oxisol amended with biochar composed of 80% woodchips and 20% anaerobic digester residues and produced at 600°C when compared to a high fertility Mollisol (Yu, Deem, Crow, Deenik, & Penton, 2018).

The presence of negative charges on the biochar surface can increase soil pH buffering capacity in acidic soils and can positively increase microbial and enzymatic activities. For example, addition of 20 and 40 t ha^{-1} of wheat straw biochar with a pH (H_2O) of 10.4 in an acid sandy loam soil increased by 67% and 73% the activity of alkaline phosphatases (Chen et al., 2013).

4.2.3.4 Impact on soil-borne pathogens

Several studies also reported that biochar can promote beneficial bacteria and fungi for plant growth and suppress those pathogens due to the improvement of soil properties by biochar (Elad et al., 2010; Jaiswal et al., 2017; Jaiswal, Elad, Graber, & Frenkel, 2014; Kolton et al., 2016; Warnock et al., 2007). These changes in soil amended with biochar may together explain the reduction in soil-borne pathogens. The increase in the application rates of eucalyptus and greenhouse waste biochars at 0.5%, 1%, and 3% *v*/v in potting mixture has shown an attenuation of the disease severity on tomato plants by the pathogen FORL (isolate 8112) compared to the control (Jaiswal et al., 2017). A better plant protection against the infection by tobacco bacterial wilt was also obtained in soil amended with straw biochar (Zhang et al., 2017). These authors reported that increase in bacterial richness and abundance of bacteria in the rhizosphere induced by rice straw biochar compared to the nonamended soil could explain this beneficial effect (Zhang et al., 2017).

4.2.3.5 Impact on soil fauna

Biochar can (1) directly affect soil fauna by its chemical properties such as available organic matter contents, salts, and toxic and volatile organic compounds and (2) indirectly affect soil fauna by the modification of soil environmental conditions such as pH, water retention capacity, porosity, and aeration, as well as shifts in soil microbial communities interacting with fauna (Domene, 2016).

Few studies on the effect of biochar on soil fauna are available, except on earthworms (Domene, 2016; Lehmann et al., 2011; McCormack, Ostle, Bardgett, Hopkins, & Vanbergen, 2013). In the conditions where biochar creates a favorable environment for soil microorganisms, it is also probable to attract meso- and macrofauna such as earthworms, enchytraeids, nematodes, collembolans, mites, and macroarthropods toward these enriched zones of foods (Domene, 2016). For example, Abujabhah, Bound, et al. (2016) observed a higher population of nematodes in sandy loam soil amended with tree green waste biochar produced at 550°C compared to a nonamended soil. The authors suggested that shifts in the composition of eukaryote community might be related to an improvement in macroporosity and bioturbation in soils amended with biochar (Abujabhah, Bound, et al., 2016). However, the larger particle sizes of biochars can have negative impacts on their ingestion by earthworms and enchytraeids, particularly with those from woodchip biochars (McCormack et al., 2013). These particles are less likely to be consumed and transported or aggregated by bioturbators (McCormack et al., 2013).

4.2.4 Impacts on plant growth

Biochar amendment has shown a positive effect on plant growth. A metaanalysis of 371 independent studies from 114 published articles found that biochar amendment increased plant productivity and crop yield in different soils and climate environments (Biederman & Harpole, 2012). For example, the addition of 5%, 10%, and 15% (*v*/v) of maple biochar in a horticultural growing medium fertilized at 100% of the recommended level increased sweet pepper plant productivity compared to the control without biochar after 63 days of transplanting (Fig. 2).

Biochar application can increase carbon input and reduce nutrient leaching in soil. Hussain et al. (2017) reported that these effects are among the most important on the improvement of crop yield.

The release of nonlabile biochar-associated organic compounds to the soil might interact with plant roots and stimulate plant growth and defense mechanisms (Kolton et al., 2016). Previously, it was reported that biochar can influence the abundance of soil microorganisms by its interference in signal transmission between plants and microorganisms. Thus, it can impact the bacteria involved in plant growth and those in plant resistance and the root infections by arbuscular mycorrhizal fungi (Elad et al., 2010; Warnock et al., 2007).

Although some biochars such as those from chicken manure and coffee husk can contain a significant concentration of mineral elements, other biochars do not have sufficient fertilizing elements to allow an improvement in plant productivity (Domingues et al., 2017). For instance, Syuhada et al. (2016) observed that the addition of oil palm empty fruit bunch biochar produced between 300°C and 350°C in a sandy Podzol soil was not suitable to supply enough nutrients to gain healthy growth of corn, despite the increase in CEC by biochar in the soil.

Fig. 2 Effect of biochar produced from the pyrolysis of maple bark at 700°C (M700) amended at different rates (5%, 10%, and 15% *v*/v) on sweet pepper plant productivity compared to the control without biochar. Picture was taken 63 days after transplanting (Lévesque, V., unpublished).

The authors recommended applying an inorganic or organic fertilizer with biochar to maximize soil productivity and plant growth. A synergistic effect between fertilizers and biochar amendment on plant productivity was also recently reported (Hussain et al., 2017). However, other authors observed that this effect was low or undetectable on plant growth when biochar was combined with a fertilizer (Biederman & Harpole, 2012; Seehausen et al., 2017).

The pyrolysis conditions used to produce biochar such as temperature can influence P availability for plant growth (Wang, Camps-Arbestain, & Hedley, 2014). The inaccessibility of P in biochar for plant growth could, however, be remedied by the inoculation of phosphate-solubilizing microorganisms in soil amended with biochar containing a high P concentration. For example, Efthymiou et al. (2018) tested different strains of P-solubilizing *Penicillium* fungi. They have showed that the inoculation of *Penicillium aculeatum* increased the P fertilizer value of P-rich sewage sludge biochar pyrolyzed at 600°C in semisterile sandy loam. This association between biochar and *P. aculeatum* improved growth of wheat and induced a 6% higher shoot P content in plant biomass. The production of citrate by *P. aculeatum* seems to explain the beneficial association for plant growth because of chelation of P in biochar (Efthymiou et al., 2018).

Biochar amendment may have a better impact on the yield of cereal crops when used in an intercropping system than in a monoculture. For instance, the application of wheat straw biochar pyrolyzed at 450°C in maize (*Zea mays* L.) intercropping with either soybean (*Glycine max* L.) or peanut (*Arachis hypogaea* L.) systems showed a higher increase in yield than in a monoculture of maize (Liu et al., 2017). The authors observed a higher nodule weight and nitrogenase activity of the legume crops under intercropping systems amended with biochar than without amendment, especially in the intercropping maize with peanut system. Using the ^{15}N isotope dilution method it was shown that the addition of 10 and 20 $g\,kg^{-1}$ biochar in soil increased peanut N fixation by 15.5% and 14.1% and increased N transfer from peanut to maize by

32.7% and 36.1%, respectively (Liu et al., 2017). The authors suggested that biochar amendment in a legume-based intercropping system promotes a better N uptake by cereal crops, allowing better yields than cereal monoculture.

4.3 Biochar as a remediation tool for organic and inorganic pollutants in agricultural soils

4.3.1 Adsorption of heavy metals

Contamination of agricultural soils by heavy metals and metalloids such as arsenic (As), cadmium (Cd), lead (Pb), and zinc (Zn), frequently identified as heavy metal(loid)s, can cause health risks through soil food chain transfer (Bian et al., 2014). Heavy metal(loid)s are highly resistant to either biologically or chemically induced degradation, and thus can persist in soil for a long time after being introduced (Khalid et al., 2017). Biochar has desirable properties for soil contamination remediation (Inyang, Gao, Pullammanappallil, Ding, & Zimmerman, 2010) and has received significant attention to immobilize heavy metal(loid)s in soil (Khalid et al., 2017). The immobilization of these elements on biochar reduces the risk of pollution by leachate and runoff, which can provide environmental benefits. Biochar was identified as an immobilizing agent to decrease metal mobility, bioavailability, and bioaccessibility. In fact, biochars have the capacity to adsorb pollutants, and particularly heavy metals (Uchimiya, Klasson, Wartelle, & Lima, 2011), due to their high pH, surface area, CEC, anion exchange capacity (AEC), hydrophobicity, and negative surface charge.

Biochar selection must be made based on its characteristics, soil properties, and target application. Biochars produced at a relatively high pyrolysis temperature are generally more effective to adsorb contaminants (Ahmad et al., 2014; Tang, Zhu, Kookana, & Katayama, 2013). Complexation with organic oxygen-containing functional groups was found to be involved in heavy-metal sorption by biochars (Lu et al., 2012). According to Uchimiya, Wartelle, Klasson, Fortier, and Lima (2011), surface functional groups of biochar, which govern pH, volatile matter, and oxygen contents, control their ability to retain heavy metals. The authors found that biochars with high oxygen-containing functional groups are effective for heavy metal stabilization in acidic soils with low organic carbon content. In fact, biochar with high pH is associated with more effective metal stabilization. Houben, Evrard, and Sonnet (2013) demonstrated that biochar helped to reduce the extractability and bioavailability of Cd, Zn, and Pb, particularly because it raised the soil pH.

Similarly, Bian et al. (2014) performed a field study over a 3-year period in which a rice paddy was amended with biochar made from the slow pyrolysis of wheat straw at temperatures between 350°C and 550°C. They found that biochar soil amendment significantly decreased extractable Cd and Pb in soil, as well as Cd concentration in rice plant tissues and Pb concentration in root tissue. The effect was associated with an increase in soil pH and organic carbon.

Despite the possible remediation of heavy metal(loid)s-contaminated soil with biochar amendment, special care has to be taken regarding the composition of heavy metal(loid)s in biochar, which will depend on biomass feedstocks. Lucchini, Quilliam, DeLuca, Vamerali, and Jones (2014) demonstrated that biochar produced from the pyrolysis of wood is of low risk due to inherently low concentration of metals. However, biochar produced from anthropogenic wastes as municipal sludge can increase metal loading in soil (Lucchini et al., 2014).

4.3.2 Adsorption of pesticides

The extensive use of pesticides in agricultural soils raises questions of their potential effects on public health and the environment. A few methods were developed to remediate pesticide contaminated soils, including containment-immobilization technologies, separation technologies, and destruction technologies, which can be chemical or biological remediation (Morillo & Villaverde, 2017). Biochar amendment has been proposed as an immobilization technique to remediate soils contaminated by pesticides, and thus to control diffuse herbicide pollution (Ouyang, Zhao, Tysklind, & Hao, 2016). In fact, biochar amendment can induce a rapid and strong sorption of hydrophobic organic compounds such as pesticides (Khorram et al., 2016). The adsorption and desorption of pesticides by biochar will therefore affect their leaching and bioavailability. For example, Delwiche, Lehmann, and Walter (2014) reported that application of pine chip biochar reduced cumulative atrazine leaching by 52% in homogenized soil columns.

The effect of biochar on herbicide behavior is dependent on its properties, which are influenced by pyrolysis conditions, biomass, and soil types (Sun et al., 2012). High organic carbon content, high SSA, and more porous structures will increase the adsorption capacity (Khorram et al., 2016), providing sorption sites for pesticides via hydrophobic partition and pore filling (Zhang, Sun, Ren, Min, & Zhang, 2018). These properties are generally obtained at a higher pyrolysis temperature and can vary based on biomass feedstock. However, Zhang et al. (2018) reported that biochar produced at higher temperature will contain more ash, which lowers the accessibility of inner sorption sites of organic moiety.

A few studies were carried out to evaluate the effect of different biochars on pesticide fate. Ouyang et al. (2016) tested four biochars produced from slow pyrolysis of rice straw, corn stalk, corn cob, and soybean at 450°C. They found that biochars made from rice straw and corn stalk had the biggest removal rate for initial atrazine concentrations of 1 and 20 $mg\,L^{-1}$ in a clay loam soil. Cabrera et al. (2014) reported that the source and amount of organic matter on the biochar can alter pesticide sorption and these effects can be different for different types of pesticides. The authors found that two biochars made from the slow pyrolysis of woodchip pellets at >500°C, with SSA (measured by nitrogen surface sorption) between 17.8 and 28.8 $m^2\,g^{-1}$, increased the retention of aminocyclopyrachlor by 95%–240% and of bentazone by 20%–35% as compared to unamended soil. However, the sorption of pesticides in soil amended with biochars produced from macadamia nut shells and hardwood (oak/hickory), which had a lower SSA of 3.3 and 8.0 $m^2\,g^{-1}$, was lower than unamended soil. Zhang et al. (2018) studied the affinity for neonicotinoid pesticides of 24 biochars produced from the pyrolysis of maize straw and pig manure and found that biochars produced at 700°C have a higher sorption affinity than biochars produced at 200–500°C.

From a remediation point of view for contaminated soil, the advantage of biochar can readily become a disadvantage as biochar can compromise the efficiency of pesticides (Kookana, Sarmah, Van Zwieten, Krull, & Singh, 2011; Tang et al., 2013) by decreasing their bioavailability to plants, microbial communities, earthworms, and other organisms in soil. Therefore, biochar application in soil will require special attention. Graber, Tsechansky, Khanukov, and Oka (2011) suggested that biochars with low SSA (up to $10 m^2 g^{-1}$), and hence a lower sorption capability, would be a better option for pest and weed control. Aging of biochar was found to reduce its sorption capacity (Martin, Kookana, Van Zwieten, & Krull, 2012). Thus, large-scale long-term studies are needed to evaluate the impact of biochar on pesticide efficiency in soil.

4.4 Biochar as a tool to mitigate climate change

4.4.1 Biochar recalcitrance in soil

During pyrolysis, 20%–50% of feedstock C is transformed into condensed aromatic C (Fidel, Laird, & Parkin, 2017), which is responsible for the chemical and biological recalcitrance of biochar in soils (Schimmelpfennig & Glaser, 2012). In contrast to organic matter that is mostly decomposed after 5 years in soil, the C content of biochar could remain stable for more than 100 years (Lehmann, Gaunt, & Rondon, 2006), and up to 4000 years (Kuzyakov, Bogomolova, & Glaser, 2014) due to its condensed aromatic structure. In fact, Smith (2016) reported that biochar production has a useful negative emission potential of $0.7 Gt C_{eq} year^{-1}$.

Based on a metaanalysis study, Wang et al. (2016) estimated that 97% of biochar C is recalcitrant and could be stable in soil for 556 years. However, there is a small proportion of labile C in biochar that can be rapidly mineralized when applied to soil, on average in the first 108 days (Wang et al., 2016). Biochar labile C is the CO_2 fraction that is mineralized over a short period of time through biotic and abiotic processes (Enders, Hanley, Whitman, Joseph, & Lehmann, 2012). There are mainly two fractions of labile carbon: (1) organic C, which is incomplete pyrolyzed biopolymers and low molecular weight volatile organic compounds, and (2) inorganic C (water soluble), which consists of carbonates (HCO_3^- and CO_3^{2-}) associated with base cations in the ash (Fidel et al., 2017). Hydrolysis of inorganic C and mineralization of labile organic C may increase CO_2 emissions from biochar-amended soil during short-term incubation (<1–6 months) (Fidel et al., 2017). After a 140-day incubation of wood and corn stover biochars in soil, the authors reported that the inorganic C pool impacts on CO_2 emission was mostly restricted to the first 48 h of incubation, whereas labile organic C pool impacts were evident throughout the incubation period. Once the labile C was exhausted, CO_2 emissions generally decrease in a biochar-amended soil (Yoo & Kang, 2012).

Priming effect, the alteration of SOM due to substrate addition has been observed with biochar addition (Cely, Tarquis, Paz-Ferreiro, Méndez, & Gascó, 2014). Priming can be either negative or positive, depending on biochar properties, soil type, and duration of experiments. For example, Luo, Durenkamp, De Nobili, Lin, and Brookes (2011) found that biochar produced at 350°C caused a larger positive priming effect than biochar produced at 700°C, the latter having a very limited amount of available C. In fact, soil biochar amendment generally contributes to a negative priming effect by retarding the mineralization of SOM (Melas, Ortiz, & Alacañiz, 2017; Wang et al., 2016). As proposed by Keith, Singh, and Dijkstra (2015), a decreased priming effect with biochar may be due to the sorption or interaction of the SOC with the biochar surface. Moreover, a reduced mineralization of the native SOM can be caused by a significant soil N immobilization due to the high C/N ratio of biochar (Melas et al., 2017).

Long-term studies are needed to evaluate the long-term stability of biochar in soil. However, simple indicators can also be used to evaluate the approximate C sequestration capacity of biochar. Biochars with an O/C_{org} ratio <0.2, H/C_{org} ratio <0.7, and volatile matter below 80% generally have high C sequestration potential (Brassard, Godbout, & Raghavan, 2016). Budai et al. (2013) reported that a biochar H/C_{org} ratio <0.4 would have a BC_{+100} of 70% (i.e., at least 70% of the C in biochar must remain in soil for more than 100 years) and a biochar H/C_{org} ratio between 0.4 and 0.7 would indicate a BC_{+100} of 50%. Harvey et al. (2012) proposed a recalcitrance index, called the R_{50} (Eq. 1), using the energy required for thermal oxidation of the biochar as a measure of recalcitrance:

$$R_{50,x} = T_{50,\text{biochar}}/T_{50,\text{graphite}}, \tag{1}$$

where $T_{50,\text{biochar}}$ and $T_{50,\text{graphite}}$ are the temperatures corresponding to 50% oxidation/volatilization of biochar and graphite, respectively. Biochars with $R_{50} \geq 0.70$ are highly recalcitrant (comparable to graphite), while biochars with $R_{50} \leq 0.50$ are the least recalcitrant, having a sequestration potential comparable to plants.

4.4.2 Mitigation of greenhouse gas emissions from agricultural soils

4.4.2.1 Nitrous oxide (N_2O)

Due to its impact on soil physical, chemical, and biological properties, biochar will influence soil non-CO_2 greenhouse gas (GHG) emissions (N_2O and CH_4). N_2O is a highly potent GHG with a global warming potential 265 times higher than CO_2 over a 100-year time horizon (IPCC, 2015). Agricultural soils appear as the main source of anthropogenic N_2O emissions. Three main mechanisms of the nitrogen cycle contribute to N_2O emissions from soil: microbial nitrification, denitrification, and nitrate ammonification (Baggs, 2011). Nitrification is an aerobic process in which NH_4^+ is oxidized to NO_3^-, producing N_2O as a by-product. Denitrification is an anaerobic process reducing NO_3^- to N_2 with N_2O as an intermediary (Verhoeven & Six, 2014). These two mechanisms, contributing approximately 70% of global N_2O emissions from soils (Butterbach-Bahl, Baggs, Dannenmann, Kiese, & Zechmeister-Boltenstern, 2013), can be modified with biochar. Bruun, Müller-Stöver, Ambus, and Hauggaard-Nielsen (2011) reported that around 8- to 23-fold more N_2O is generated under conditions of high moisture content (80%) than at lower moisture content (40%) because denitrification is favored (Qambrani, Rahman, Won, Shim, & Ra, 2017).

According to the literature, N_2O emissions seem to decline following biochar amendments (Bruun et al., 2011; Cayuela et al., 2014; Kettunen & Saarnio, 2013). Brassard et al. (2016) carried out a systematic literature review in which the results from 40 research studies were compiled. The authors found that 53 biochars contributed to reduce soil N_2O emissions, 41 biochars had no effect, and only 6 biochars increased N_2O emissions. Biochars that caused significant increase in N_2O emissions following their addition to soil had a higher N content and a lower C content, resulting in a low C/N ratio. In fact, Cayuela et al. (2014) reported that biochars with a C/N ratio >30 could be better designed for the mitigation of soil N_2O emissions. These biochars are generally produced from agricultural biomass and have a high ash content. Biochars with a low H/C_{org} ratio (<0.3) have increased aromaticity and are also associated with the reducing effect of biochar on N_2O emissions (Cayuela, Jeffery, & van Zwieten, 2015). In addition to biochar properties, the effect on soil N_2O emissions will also depend on the application rate of biochar. For example, Harter et al. (2014) reported that N_2O emission was reduced by up to 96% in the presence of 10% (*w*/w) biochar, but the reduction decreased to 47% in the presence of 2% (*w*/w) biochar. Moreover, the impact can be different depending on the type of soil. Brassard et al. (2018) found a significant decrease in N_2O emission from a silt loam amended with biochar made from the pyrolysis of switchgrass at 591°C as compared to the control soil without biochar, but the difference was not significant in a loamy sand.

Most of the mechanisms that have been identified so far as being responsible for a decrease in N_2O emissions after soil biochar amendment are biotic, involving a change in soil microbial abundance (Cayuela et al., 2014; Clough et al., 2013; Kettunen & Saarnio, 2013; Rogovska et al., 2011; Sohi et al., 2010; Sun, Li, Chen, Wang, & Xiong, 2014; van Zwieten et al., 2010), but abiotic mechanisms have also been proposed by Cayuela et al. (2014) (Table 5).

Table 5 Mechanisms responsible for the decrease in soil N_2O emissions.

Mechanisms
Biotic
The liming effect of biochar that creates an optimal environment for N_2O reductase activity
The reduction of the availability of inorganic nitrogen (NO_3^- and NH_4^+) to the microorganisms involved in nitrification and/or denitrification and producing N_2O
An enhanced soil aeration that inhibits the denitrification process
The labile C content of the biochar promoting total N denitrification
The release of toxic/inhibitory compounds like ethylene inhibiting soil biological activity
Abiotic
Induced redox reaction by biochar that could act as a catalyst
The adsorption of N_2O on the biochar surface
Improved transfer of electrons through the soil due to biochar acting as an electron shuttle

4.4.2.2 Methane (CH_4)

Methane is an important GHG, with a global warming potential 28 times higher than CO_2 over a 100-year time horizon (IPCC, 2015). The agricultural sector is responsible for about half of the global anthropogenic emissions of CH_4, which is mainly due to rice production (Brevik, 2012). In aerobic conditions, soils are CH_4 sinks; however, CH_4 is emitted by soil microorganisms under anaerobic conditions through methanogenesis (Jeffery, Verheijen, Kammann, & Abalos, 2016). In the literature, it is observed that the effect of biochar on soil CH_4 emissions varies widely, depending on the physical and chemical properties of biochar, the type of soil, and water and fertilizer management (Brassard et al., 2016; Qambrani et al., 2017). Wang et al. (2012) reported that biochar can promote CH_4 emission in an anoxic environment because the labile C added from biochar can become a source of substrate for methanogenesis.

Jeffery et al. (2016) found that biochar could mitigate soil CH_4 emissions, particularly from paddy fields and acidic soils that use periods of flooding. The main mechanism that was found to be responsible for an increased CH_4 uptake in soil is improved soil aeration and CH_4 diffusion through the soil (Yaghoubi, Yargicoglu, & Reddy, 2014). Moreover, the authors concluded that biochar may affect soil CH_4 fluxes by the sorption of CH_4 to biochar's surface. Biochar with high pH and high surface area could be better designed for mitigating soil CH_4 emissions (Jeffery et al., 2016).

4.4.3 Global perspective

Biochar soil amendment was identified as a negative emission technology due to its capacity to sequester carbon in soil (IPCC, 2015; Smith, 2016; UNEP, 2016) and to reduce soil GHG emissions (Brassard et al., 2016). Moreover, GHG emissions associated with the decay of biomass are avoided when it is used to produce biochar (Smith, 2016). However, biochars need to be produced from materials that would otherwise decompose and that do not compete with food production.

In fact, the effect of biochar on a global GHG balance will depend on many factors. To study the full life-cycle GHG emission balance of biochar systems, a life-cycle analysis (LCA) is needed (Dutta & Raghavan, 2014). LCA is defined as a "cradle-to-grave" approach used to study the environmental aspects and potential impacts of a product or a system throughout its life, from raw material acquisition through production, use, and disposal (Hunkeler, 2016). Thus, in the LCA of biochar systems, the following activities must be considered: biomass cultivation or collection, handling and conditioning (grinding, drying), pyrolysis process, amendment of biochar to soil (spreading, C sequestration, mitigation on soil GHG emissions, substitution of synthetic fertilizers), and valorization pyrolysis coproducts (bio-oil and syngas) as energy sources (Brassard, Godbout, Pelletier, Raghavan, & Palacios, 2018).

The GHG abatement potential for biochar remains mostly sensitive to the absolute stability of the C stored in biochar (Smith, 2016). The BC_{+100} value (i.e., the proportion of C (%) in biochar that will remain in soil after 100 years) is used as an indicator of biochar C stability. This value depends on biochar properties. LCA studies used BC_{+100} values of 50% (Brassard, Godbout, Pelletier, et al., 2018), 68% (Ibarrola, Shackley, & Hammond, 2012), 70% (Brassard, Godbout, Pelletier, et al., 2018), 80% (Roberts, Gloy, Joseph, Scott, & Lehmann, 2010), and 90% (Peters, Iribarren, & Dufour, 2015).

Woolf, Amonette, Street-Perrott, Lehmann, and Joseph (2010) estimated that annual net GHG emissions could be reduced by a maximum of 1.8 Gt CO_2 eq. $year^{-1}$ (12% of current anthropogenic CO_2 eq. emissions). Smith (2016) proposed that the maximum technical potential of biochar is 1.3 Gt CO_2 eq. $year^{-1}$, because the global implementation levels ranged from 0.4 to 0.7 Gt CO_2 eq. $year^{-1}$ at carbon prices from \$70 to 370 t^{-1}C eq. This estimation did not consider the impact of biochar on non-CO_2 soil GHG emissions. However, it was considered that some biochars such as manure biochars can provide N-P-K nutrients and thus substitute synthetic fertilizers.

4.5 Key parameters of biochar production for agricultural use

4.5.1 Key biochar properties

Biochars used for environmental or agronomic applications should be produced with specific characteristics (Sun et al., 2014). The physicochemical properties of biochar will influence strongly its capacity to improve soil chemical and biological properties, to supply nutrients to plants, to increase crop yields, to adsorb organic and inorganic pollutants such as pesticides and heavy metals, to sequester carbon in soil, and to reduce soil GHG emissions.

Biochar with a high surface area and high porosity is needed to improve soil physical properties. For example, biochars with high pore volume and pore tortuosity exhibit higher infiltration and water retention at saturation (Kameyama et al., 2014). High absorptive capacity of biochar for water is due to its void content and large surface area (Ulyett et al., 2013). Moreover, Reddy et al. (2015) reported that soil compressibility decreases with a decrease in biochar particle size.

The requested biochar properties for the improvement of soil chemical and biological properties usually include high pH, high C content, high available N and P contents, high SSA, and high porosity. Biochar is generally alkaline, which is desired for agricultural soil amendment, specifically in acidic soils. Brassard et al. (2016) compiled the properties of 76 biochars produced from 40 research studies and found an average biochar pH of 8.6. The use of biochar to increase soil alkalinity instead of calcium oxide, which can react with $2H^+$ to produce H_2O and CO_2, can be advantageous (Allaire & Lange, 2013). It was reported that biochar pH as well as readily metabolizable C source from biochar appear to be the most important elements in the functioning of biodiversity and shifts in soil microbial communities (Abujabhah, Bound, et al., 2016; Abujabhah, Doyle, et al., 2016; Gul et al., 2015; Kolton et al., 2016). In addition, biochars with a high SSA can improve the soil CEC (Gul et al., 2015). Furthermore, a larger surface area and porous microstructure in fine-sized biochar (<0.05 mm) than in medium (0.05–1.0 mm) and coarse (1.0–2.0 mm) could increase its interaction with the soil matrix (pH, EC, and nutrients availability) allowing a better habitat for microorganisms and increasing the microbial abundance (Chen et al., 2017). Microbial community structure and enzymatic activities are also affected by the particle size and application rate of biochar (Chen et al., 2017).

Some biochars have the capacity to adsorb organic and inorganic pollutants in agricultural soils due to their high pH, surface area, CEC, AEC, hydrophobicity, and negative surface charge (Uchimiya, Klasson, et al., 2011). High organic carbon content and more porous structures will increase the adsorption capacity (Khorram et al., 2016). For this application, biochars with high ash content must be avoided because they lower the accessibility of inner sorption sites of organic moiety (Zhang et al., 2018).

Biochar chemical properties (C, N, and H contents, pH) can be used as indicators of its potential to mitigate climate change. Biochars with a high C content and low N, H, and O contents could be more suitable for the mitigation of soil N_2O emissions and could be more recalcitrant in soil (Brassard et al., 2016; Cayuela et al., 2014).

The International Biochar Initiative (IBI, 2015) defined three test categories (A, B, and C) for biochar used for soil amendment. In category A, basic biochar characteristics should be analyzed, including physical properties (particle size and moisture) and chemical properties (H, C, and N contents), ash content, EC, and pH/liming ability. However, criteria are given only for organic carbon content (class 1: $\geq$60%; class 2: $\geq$30% and <60%; and class 3: $\geq$10% and <30%) and H/C_{org} ratio (0.7 maximum). Category B involves toxicant assessment, which is divided into two categories: toxicants that may be present in the feedstock, including metals and polychlorinated biphenyls, and toxicants that may be produced by the thermochemical conversion process (polycyclic aromatic hydrocarbon and dioxins/furans). IBI (2015) provides a test method and the ranges of maximum allowed thresholds suggested for each parameter. Parameters in both test categories A and B are required for all biochars. The third optional category C is related to advanced analysis and soil enhancement properties. It includes available N, total and available P, K, Ca, Mg, S, Ca, volatile matter, total surface area, and external surface area.

4.5.2 Biomass properties

Not all residual materials are suitable for producing biochar (Lehmann et al., 2006), in part because significant differences have been observed in the biochar and pyrolysis products among different feedstocks (He et al., 2018). The feedstock utilized for biochar production influences biochar yield and characteristics (Spokas et al., 2012). In some cases, the type of biomass used for pyrolysis is even more important than production parameters. For example, Lehmann et al. (2006) showed that within 350–500°C, a common temperature range for pyrolysis, the efficiency of C conversion of biomass to biochar was highly dependent on the type of feedstock, but was not significantly affected by the pyrolysis temperature. Higher ash content as well as greater concentrations of lignin increase C recovery because lignin preferentially forms char during pyrolysis (Lehmann et al., 2006).

Soil amendment with biochar derived from different feedstocks affects soil hydrology differently. For example, biochars produced from dairy manure or woodchips mixed at 5% (*w*/w) with soil led to an increase in k_{sat} of the amended soils, but the soil k_{sat} and its water content were greater with woodchip biochar than dairy manure biochar amendment (Lei & Zhang, 2013).

Yao, Gao, Zhang, Inyang, and Zimmerman (2012) reported that biochars made from different biomass feedstocks show significant differences in adsorbing phosphate from aqueous solutions. The combination of chemical composition and physical properties of a feedstock could define the potential of the biochar to decrease soil N_2O emissions (Cayuela et al., 2014). The authors reported that plant-derived materials are the most promising feedstocks to produce biochar for the mitigation of soil N_2O emissions.

4.5.3 Production parameters

As seen in Chapter 2, pyrolysis operating parameters have to be selected carefully to obtain the desired products (Brassard, Godbout, & Raghavan, 2017). For example, with increasing temperature, biochars showed decreased contents of volatile matter, H, O, and (O+N)/C ratios, while also showing increased ash, fixed C, and total P and K contents (He et al., 2018). Sun, Gao, et al. (2014) affirmed that pyrolysis temperature is a key factor in determining application for biochars. For example, studies showed that biochars produced at high temperature could be more effective at mitigating N_2O (Cayuela et al., 2015) and CH_4 (Jeffery et al., 2016) emissions, and at sequestering C in soil (Al-Wabel, Al-Omran, El-Naggar, Nadeem, & Usman, 2013). The study carried out by Sun, Gao, et al. (2014), where biochar was produced from various feedstocks, also showed that high pyrolysis temperature is needed to produce biochars suitable to mitigate soil GHG emissions and to sequester C in soil.

Biochar characteristics and yields are also highly influenced by the type of pyrolysis process as reported in Chapters 2 and 3. The study carried out by Bruun, Ambus, Egsgaard, and Hauggaard-Nielsen (2012) showed that the pyrolysis method had a considerable influence on biochar and on the mineralization/immobilization of N in biochar-amended soil. However, soil application of the biochar materials had the potential to sequester C independently of the pyrolysis method. Kong, Liaw, Gao, Yu, and Wu (2014) highlighted that fast pyrolysis biochars (from Mallee biomass) contain fewer water-soluble inorganic nutrients than slow pyrolysis biochars. Thus, fast pyrolysis biochars provide more plant available inorganic nutrients than slow pyrolysis biochars.

4.6 Challenges for sustainable biochar production

4.6.1 Regulations

Biochar cannot be removed from soil; thus, special attention must be paid before using it as an amendment. Each type of biochar is unique, having different physicochemical characteristics. Some biochars could have adverse effects on the environment. For example, biochar can contain heavy metals if it is produced from a contaminated feedstock. Therefore, a policy was developed and implemented to control use of biochar in agricultural fields. A multitude of policies and legislative measures influence the development of the biochar industry (Van Laer et al., 2014).

According to some regulations, biochar can be considered as a waste or a hazardous material (Van Laer et al., 2014), and its application to the field can be prohibited. Biochar can also be seen as a "by-product," which is defined by the European Directive 2008/98/EC on waste as a substance or object, resulting from a production process, the primary aim of which is not the production of that item. It can also have the status of "end of waste," which is a waste that has undergone a recovery, including recycling, operations, and complies with specific criteria. If biochar is classified as "by-product" or "end of waste," it does not have to comply with all the legal regulations regarding waste (Van Laer et al., 2014).

Certification programs can help legislators to modify regulations to permit soil amendments with certified biochars. For example, in Canada, biochar can be registered as a soil supplement at the Canadian Food Inspection Agency. Results of chemical analysis must respect the IBI guidelines (IBI, 2015) and prove that biochar is safe for the environment, with toxicant concentrations below the maximum allowable values.

4.6.2 Large-scale production and application

The optimal application rate of biochar in soil is not always known and can be high to provide the desired benefits. For example, Harter et al. (2014) found that N_2O emission was reduced by up to 96% in the presence of 10% (*w*/w) biochar and by 47% in the presence of 2% (*w*/w) biochar. A similar conclusion was observed in a study by Wu, Jia, Wang, Chang, and Startsev (2013), in which biochar added at $25\,t\,ha^{-1}$ suppressed N_2O emissions by 96%, while application at $10\,t\,ha^{-1}$ resulted in a reduction of 55%. Therefore, important quantities of biochar may be needed to cover large surfaces.

Biochar can be produced at different scales, making it applicable to a variety of socioeconomic situations (Woolf et al., 2010). Large scale production of biochar can provide considerable economies for capital cost and operating cost because they can operate continually and be automated with a homogeneous biochar product between each batch. Sustainability concerns in the manufacture of biochar include the potential release of air pollutants, GHG emissions, and conversion efficiency of biomass resources to biochar and energy products.

Finally, because biochar has a low density it is difficult to spread on the field because it can be lost through wind (Fig. 3; Husk & Major, 2010) and water erosion. One proposed solution is to produce densified biochar pellets, which would facilitate its spreading and incorporation in the field. The benefits of pelletized versus non-pelletized biochar in soil should be analyzed.

4.6.3 Economic considerations

The high cost associated with the production of biochar and high application rates to soil can be a significant challenge to its widespread use (Saifullah, Dahlawi, Naeem, Rengel, & Naidu, 2018). Because there is no major industrial biochar market, it is difficult to give a realistic cost estimation of biochar (Dickinson et al., 2014), which will

Fig. 3 Clockwise from top left: Biochar losses during handling, transportation to the field, application, and incorporation (Husk & Major, 2010).

depend on a range of factors (Ahmed, Zhou, Ngo, & Guo, 2016). Based on a literature review carried out by Ahmed et al. (2016), the mean price for biochar was US$2.65 kg^{-1}, ranging from US$0.09 kg^{-1} in the Philippines to US$8.85 kg^{-1} in the United Kingdom. In fact, the price of biochar will depend on the feedstock purchased and transport cost, production cost related to mainly pyrolysis equipment and maintenance cost, and on labor and storage costs (Ahmed et al., 2016).

Many external factors have to be considered to assess the economic impact of biochar production for soil amendments. Biochar can lead to agronomic benefits, including an increase in crop yield and a reduced need for synthetic fertilizers, which will increase revenue and decrease expenses. Moreover, if the exact volume of carbon credits generated by biochar application to soil is known, the revenue generated by the registration and sale of carbon offsets would encourage farmers to produce biochar. However, if they purchase biochar, the value of the carbon offset alone could not cover the current biochar price (De Gryze and Cullen, 2010). Finally, coproducts, such as syngas and bio-oil, whose yields and properties vary with production technology, can be sold for heat production.

4.7 Conclusion

Biochar amendment is expected to reduce the environmental impact of agriculture. It can provide many benefits, such as the improvement of soil physicochemical and biological properties resulting in an increase in crop yields, the sequestration of C in soil, the reduction of soil GHG emissions, and the adsorption of organic and inorganic pollutants. However, not all biochars will have the desired effect because its properties will depend on production technology and biomass feedstock. Therefore, special care has to be taken when producing biochar for agricultural soil amendments. Biochar with beneficial characteristics from an environmental perspective can be produced when production parameters and biomass feedstock are suitably selected; these types of biochars are defined as engineered biochars (Brassard et al., 2017).

Biochar should be produced in a sustainable way, in efficient reactors, and from agricultural and forest residues that would otherwise decompose and emit GHG. The global environmental impact of biochar should be evaluated through an LCA considering every step of biochar production, from the collection of biomass to the spreading of biochar on the field and considering the valorization of coproducts. In addition to climate change impact category that was evaluated in a few studies, impact categories such as acidification, ozone depletion (stratospheric), resource demand, ozone formation (near surface), land use, eutrophication (terrestrial and aquatic), ecotoxicity, and human toxicity should be studied.

In recent years, several tests have been carried out to study the impact of biochar in soil in laboratory or incubation conditions. However, the physicochemical characteristics of biochar can be altered with time (Gul et al., 2015; Harter et al., 2016). Therefore, long-term studies in greenhouses and in fields, in the presence of crops and considering different soil conditions and climates, are recommended to validate the results obtained on a small scale.

References

Abel, S., Peters, A., Trinks, S., Schonsky, H., Facklam, M., & Wessolek, G., (2013). Impact of biochar and hydrochar addition on water retention and water repellency of sandy soil. *Geoderma 202–203*, 183–191. https://doi.org/10.1016/j.geoderma.2013.03.003.

Abujabhah, I. S., Bound, S. A., Doyle, R., & Bowman, J. P., (2016). Effects of biochar and compost amendments on soil physico-chemical properties and the total community within a temperate agricultural soil. *Applied Soil Ecology 98*, 243–253. https://doi.org/10.1016/j.apsoil.2015.10.021.

Abujabhah, I. S., Doyle, R., Bound, S. A., & Bowman, J. P., (2016). The effect of biochar loading rates on soil fertility, soil biomass, potential nitrification, and soil community metabolic profiles in three different soils. *Journal of Soils and Sediments 16*, 2211–2222. https://doi.org/10.1007/s11368-016-1411-8.

Agegnehu, G., Bass, A. M., Nelson, P. N., & Bird, M. I., (2016). Benefits of biochar, compost and biochar–compost for soil quality, maize yield and greenhouse gas emissions in a tropical agricultural soil. *Science of The Total Environment 543*, 295–306. https://doi.org/10.1016/j.scitotenv.2015.11.054.

Ahmad, M., Rajapaksha, A. U., Lim, J. E., Zhang, M., Bolan, N., Mohan, D., et al. (2014). Biochar as a sorbent for contaminant management in soil and water: a review. *Chemosphere 99* 19–33. https://doi.org/10.1016/j.chemosphere.2013.10.071.

Ahmed, M. B., Zhou, J. L., Ngo, H. H., & Guo, W., (2016). Insight into biochar properties and its cost analysis. *Biomass and Bioenergy 84*, 76–86. https://doi.org/10.1016/j.biombioe.2015.11.002.

Aksakal, E. L., Angin, I., & Oztas, T., (2013). Effects of diatomite on soil consistency limits and soil compactibility. *CATENA 101*, 157–163. https://doi.org/10.1016/j.catena.2012.09.001.

Allaire, S. E., & Lange, S., (2013). Le biochar dans les milieux poreux: une solution miracle en environnement? *Vecteur Environnement* 58–67.

Al-Wabel, M. I., Al-Omran, A., El-Naggar, A. H., Nadeem, M., & Usman, A. R. A., (2013). Pyrolysis temperature induced changes in characteristics and chemical composition of biochar produced from conocarpus wastes. *Bioresource Technology 131*, 374–379. https://doi.org/10.1016/j.biortech.2012.12.165.

Anderson, C. R., Condron, L. M., Clough, T. J., Fiers, M., Stewart, A., Hill, R. A., et al. (2011). Biochar induced soil microbial community change: implications for biogeochemical cycling of carbon, nitrogen and phosphorus. *Pedobiologia 54*, 309–320. https://doi.org/10.1016/j.pedobi.2011.07.005.

Asai, H., Samson, B. K., Stephan, H. M., Songyikhangsuthor, K., Homma, K., Kiyono, Y., et al. (2009). Biochar amendment techniques for upland rice production in Northern Laos: 1. Soil physical properties, leaf SPAD and grain yield. *Field Crops Research 111*, 81–84. https://doi.org/10.1016/j.fcr.2008.10.008.

Baggs, E. M., (2011). Soil microbial sources of nitrous oxide: recent advances in knowledge, emerging challenges and future direction. *Current Opinion in Environmental Sustainability 3*, 321–327. https://doi.org/10.1016/j.cosust.2011.08.011.

Basso, A. S., Miguez, F. E., Laird, D. A., Horton, R., & Westgate, M., (2012). Assessing potential of biochar for increasing water-holding capacity of sandy soils. *GCB Bioenergy 5*, 132–143. https://doi.org/10.1111/gcbb.12026.

Bekele, A., Roy, J. L., & Young, M. A., (2015). Use of biochar and oxidized lignite for reconstructing functioning agronomic topsoil: effects on soil properties in a greenhouse study. *Canadian Journal of Soil Science 95* 269–285. https://doi.org/10.4141/cjss-2014-008.

Bian, R., Joseph, S., Cui, L., Pan, G., Li, L., Liu, X., et al. (2014). A three-year experiment confirms continuous immobilization of cadmium and lead in contaminated paddy field with biochar amendment. *Journal of Hazardous Materials 272*, 121–128. https://doi.org/10.1016/j.jhazmat.2014.03.017.

Biederman, L. A., & Harpole, W. S., (2012). Biochar and its effects on plant productivity and nutrient cycling: a meta-analysis. *GCB Bioenergy 5* 202–214. https://doi.org/10.1111/gcbb.12037.

Boone, F. R., & Veen, B. W., (1994). Mechanisms of crop responses to soil compaction. In B. D. Soane & C. van Ouwerkerk (Eds.), *Soil compaction in crop production* (pp. 237–264). Elsevier. https://doi.org/10.1016/B978-0-444-88286-8.50019-2 (chapter 11).

Brassard, P., Godbout, S., Palacios, J. H., Jeanne, T., Hogue, R., Dubé, P., et al. (2018). Effect of six engineered biochars on GHG emissions from two agricultural soils: a short-term incubation study. *Geoderma 327*, 73–84.

Brassard, P., Godbout, S., Pelletier, F., Raghavan, V., & Palacios, J. H., (2018). Pyrolysis of switchgrass in an auger reactor for biochar production: a greenhouse gas and energy impacts assessment. *Biomass and Bioenergy 116*, 99–105. https://doi.org/10.1016/j.biombioe.2018.06.007.

Brassard, P., Godbout, S., & Raghavan, V., (2016). Soil biochar amendment as a climate change mitigation tool: key parameters and mechanisms involved. *Journal of Environmental Management. 181*, https://doi.org/10.1016/j.jenvman.2016.06.063.

Brassard, P., Godbout, S., & Raghavan, V., (2017). Pyrolysis in auger reactors for biochar and bio-oil production: a review. *Biosystems Engineering. 161*, https://doi.org/10.1016/j.biosystemseng.2017.06.020.

Brassard, P., Godbout, S., Raghavan, V., Palacios, H. J., Grenier, M., & Zegan, D., (2017). The production of engineered biochars in a vertical auger pyrolysis reactor for carbon sequestration. *Energies. 10*, https://doi.org/10.3390/en10030288.

Brevik, E. C., (2012). Soils and climate change: gas fluxes and soil processes. *Soil Horizons. 53*, https://doi.org/10.2136/sh12-04-0012.

Brewer, C. E., Chuang, V. J., Masiello, C. A., Gonnermann, H., Gao, X., Dugan, B., et al. (2014). New approaches to measuring biochar density and porosity. *Biomass and Bioenergy 66*, 176–185. https://doi.org/10.1016/j.biombioe.2014.03.059.

Bruun, E. W., Ambus, P., Egsgaard, H., & Hauggaard-Nielsen, H., (2012). Effects of slow and fast pyrolysis biochar on soil C and N turnover dynamics. *Soil Biology and Biochemistry 46*, 73–79. https://doi.org/10.1016/j.soilbio.2011.11.019.

Bruun, E. W., Müller-Stöver, D., Ambus, P., & Hauggaard-Nielsen, H., (2011). Application of biochar to soil and N2O emissions: potential effects of blending fast-pyrolysis biochar with anaerobically digested slurry. *European Journal of Soil Science 62*, 581–589. https://doi.org/10.1111/j.1365-2389.2011.01377.x.

Budai, A., Zimmerman, A. R., Cowie, A. L., Webber, J. B. W., Singh, B. P., Glaser, B., et al. (2013). *Biochar Carbon Stability Test Method: An assessment of methods to determine biochar carbon stability*: (pp. 1–10). International Biochar Initiative.

Busscher, W. J., Novak, J. M., Evans, D. E., Watts, D. W., Niandou, M. A. S., & Ahmedna, M., (2010). Influence of pecan biochar on physical properties of a Norfolk loamy sand. *Soil Science 175*, 10–14.

Butnan, S., Deenik, J. L., Toomsan, B., Antal, M. J., & Vityakon, P., (2015). Biochar characteristics and application rates affecting corn growth and properties of soils contrasting in texture and mineralogy. *Geoderma 237–238*, 105–116. https://doi.org/10.1016/j.geoderma.2014.08.010.

Butterbach-Bahl, K., Baggs, E. M., Dannenmann, M., Kiese, R., & Zechmeister-Boltenstern, S., (2013). Nitrous oxide emissions from soils: how well do we understand the processes and their controls? *Philosophical Transactions of the Royal Society of London Series B, Biological Sciences 368*, 20130122. https://doi.org/10.1098/rstb.2013.0122.

Cabrera, A., Cox, L., Spokas, K., Hermosín, M. C., Cornejo, J., & Koskinen, W. C., (2014). Science of the Total Environment In fl uence of biochar amendments on the sorption – desorption of aminocyclopyrachlor, bentazone and pyraclostrobin pesticides to an agricultural soil. *Science of the Total Environment, The 470–471*, 438–443. https://doi.org/10.1016/j.scitotenv.2013.09.080.

Carter, M. R., (1990). Relative measures of soil bulk density to characterize compaction in tillage studies on fine sandy loams. *Canadian Journal of Soil Science 70*, 425–433. https://doi.org/10.4141/cjss90-042.

Cayuela, M. L., Jeffery, S., & van Zwieten, L., (2015). The molar H:Corg ratio of biochar is a key factor in mitigating N2O emissions from soil. *Agriculture, Ecosystems & Environment 202*, 135–138. https://doi.org/10.1016/j.agee.2014.12.015.

Cayuela, M. L., van Zwieten, L., Singh, B. P., Jeffery, S., Roig, A., & Sánchez-Monedero, M. A., (2014). Biochar's role in mitigating soil nitrous oxide emissions: a review and meta-analysis. *Agriculture, Ecosystems & Environment 191*, 5–16. https://doi.org/10.1016/j.agee.2013.10.009.

Cely, P., Tarquis, A. M., Paz-Ferreiro, J., Méndez, A., & Gascó, G., (2014). Factors driving the carbon mineralization priming effect in a sandy loam soil amended with different types of biochar. *Solid Earth 5*, 585–594. https://doi.org/10.5194/se-5-585-2014.

Chathurika, J. A. S., Kumaragamage, D., Zvomuya, F., Akinremi, O. O., Flaten, D. N., Indraratne, S. P., et al. (2016). Woodchip biochar with or without synthetic fertilizers affects soil properties and available phosphorus in two alkaline, chernozemic soils. *Canadian Journal of Soil Science 96*, 472–484. https://doi.org/10.1139/cjss-2015-0094.

Chen, G., & Weil, R. R., (2011). Root growth and yield of maize as affected by soil compaction and cover crops. *Soil and Tillage Research 117*, 17–27.

Chen, J., Li, S., Liang, C., Xu, Q., Li, Y., Qin, H., et al. (2017). Response of microbial community structure and function to short-term biochar amendment in an intensively managed bamboo (*Phyllostachys praecox*) plantation soil: effect of particle size and addition rate. *Science of The Total Environment 574*, 24–33. https://doi.org/10.1016/j.scitotenv.2016.08.190.

Chen, J., Liu, X., Zheng, J., Zhang, B., Lu, H., Chi, Z., et al. (2013). Biochar soil amendment increased bacterial but decreased fungal gene abundance with shifts in community structure in a slightly acid rice paddy from Southwest China. *Applied Soil Ecology 71*, 33–44. https://doi.org/10.1016/j.apsoil.2013.05.003.

Clough, J. T., Condron, M. L., Kammann, C., & Müller, C., (2013). A review of biochar and soil nitrogen dynamics. *Agronomy*. https://doi.org/10.3390/agronomy3020275.

Conte, P., Valentina, M., Claudio, D. P., Salvatore, B., Massimo, V., Alessandro, P., et al. (2012). Nature of water-biochar interface interactions. *GCB Bioenergy 5*, 116–121. https://doi.org/10.1111/gcbb.12009.

De Gryze, S., & Cullen, M., (2010). Evaluation of the opportunities for generating carbon offsets from soil sequestration of biochar. *Climate Action Reserve* 1–99.

Delwiche, K. B., Lehmann, J., & Walter, M. T., (2014). Atrazine leaching from biochar-amended soils. *Chemosphere 95*, 346–352. https://doi.org/10.1016/j.chemosphere.2013.09.043.

Dickinson, D., Balduccio, L., Buysse, J., Ronsse, F., Huylenbroeck, G., & Prins, W., (2014). Cost-benefit analysis of using biochar to improve cereals agriculture. *GCB Bioenergy 7*, 850–864. https://doi.org/10.1111/gcbb.12180.

Domene, X., (2016). A critical analysis of meso- and macrofauna effects following biochar supplementation. In *Biochar application* (pp. 268–292). Elsevier. https://doi.org/10.1016/B978-0-12-803433-0.00011-4 (chapter 11).

Domingues, R. R., Trugilho, P. F., Silva, C. A., de Melo, I. C. N. A., Melo, L. C. A., Magriotis, Z. M., et al. (2017). Properties of biochar derived from wood and high-nutrient biomasses with the aim of agronomic and environmental benefits. *PLOS One 12*(5), e0176884. https://doi.org/10.1371/journal.pone.0176884.

Dutta, B., & Raghavan, V., (2014). A life cycle assessment of environmental and economic balance of biochar systems in Quebec. *International Journal of Energy and Environmental Engineering 5*, 106. https://doi.org/10.1007/s40095-014-0106-4.

Eastman, C. M., (2011). *Soil physical characteristics of an Aeric Ochraqualf amended with biochar*. The Ohio State University.

Efthymiou, A., Grønlund, M., Müller-Stöver, D. S., & Jakobsen, I., (2018). Augmentation of the phosphorus fertilizer value of biochar by inoculation of wheat with selected Penicillium strains. *Soil Biology and Biochemistry 116*, 139–147. https://doi.org/10.1016/j.soilbio.2017.10.006.

Elad, Y., David, D. R., Harel, Y. M., Borenshtein, M., Kalifa, H. B., Silber, A., et al. (2010). Induction of systemic resistance in plants by biochar, a soil-applied carbon sequestering agent. *Phytopathology 100*, 913–921. https://doi.org/10.1094/PHYTO-100-9-0913.

Enders, A., Hanley, K., Whitman, T., Joseph, S., & Lehmann, J., (2012). Characterization of biochars to evaluate recalcitrance and agronomic performance. *Bioresource Technology 114*, 644–653. https://doi.org/10.1016/j.biortech.2012.03.022.

Fidel, R. B., Laird, D. A., & Parkin, T. B., (2017). Impact of biochar organic and inorganic carbon on soil CO2 and N2O emissions. *Journal of Environmental Quality 46*, 505–513. https://doi.org/10.2134/jeq2016.09.0369.

Glaser, B., & Birk, J. J., (2012). State of the scientific knowledge on properties and genesis of Anthropogenic Dark Earths in Central Amazonia (terra preta de Índio). *Geochimica et Cosmochimica Acta 82*, 39–51. https://doi.org/10.1016/j.gca.2010.11.029.

Glaser, B., Haumaier, L., Guggenberger, G., & Zech, W., (2001). The "Terra Preta" phenomenon: a model for sustainable agriculture in the humid tropics. *Naturwissenschaften 88*, 37–41. https://doi.org/10.1007/s001140000193.

Graber, E. R., Tsechansky, L., Khanukov, J., & Oka, Y., (2011). Sorption, volatilization, and efficacy of the fumigant 1,3-dichloropropene in a biochar-amended soilfile. *Soil Science Society of America Journal 75*, 1365–1373. https://doi.org/10.2136/sssaj2010.0435.

Gul, S., & Whalen, J., (2016). Biochemical cycling of nitrogen and phosphorus in biochar-amended soils. *Soil Biology and Biochemistry 103*, 1–15. https://doi.org/10.1016/j.soilbio.2016.08.001.

Gul, S., Whalen, J. K., Thomas, B. W., Sachdeva, V., & Deng, H., (2015). Physico-chemical properties and microbial responses in biochar-amended soils: mechanisms and future directions. *Agriculture, Ecosystems & Environment 206*, 46–59. https://doi.org/10.1016/j.agee.2015.03.015.

Guo, Y., Tang, H., Li, G., & Xie, D., (2013). Effects of cow dung biochar amendment on adsorption and leaching of nutrient from an acid yellow soil irrigated with biogas slurry. *Water, Air, & Soil Pollution 225*, 1820. https://doi.org/10.1007/s11270-013-1820-x.

Hardie, M., Clothier, B., Bound, S., Oliver, G., & Close, D., (2014). Does biochar influence soil physical properties and soil water availability? *Plant and Soil 376*, 347–361. https://doi.org/10.1007/s11104-013-1980-x.

Harris, W., (1971). The soil compaction process. In K. K. Barnes (Ed.), *Compaction of agricultural soils* (pp. 9–44). St-Joseph, MI: American Society of Agricultural Engineers.

Harter, J., Krause, H.-M., Schuettler, S., Ruser, R., Fromme, M., Scholten, T., et al. (2014). Linking N2O emissions from biochar-amended soil to the structure and function of the N-cycling microbial community. *ISME J 8*, 660–674.

Harter, J., Weigold, P., El-Hadidi, M., Huson, D. H., Kappler, A., & Behrens, S., (2016). Soil biochar amendment shapes the composition of N2O-reducing microbial communities. *Science of The Total Environment 562*, 379–390. https://doi.org/10.1016/j.scitotenv.2016.03.220.

Harvey, O. R., Kuo, L.-J., Zimmerman, A. R., Louchouarn, P., Amonette, J. E., & Herbert, B. E., (2012). An index-based approach to assessing recalcitrance and soil carbon sequestration potential of engineered black carbons (biochars). *Environmental Science & Technology 46*, 1415–1421. https://doi.org/10.1021/es2040398.

He, X., Liu, Z., Niu, W., Yang, L., Zhou, T., Qin, D., et al. (2018). Effects of pyrolysis temperature on the physicochemical properties of gas and biochar obtained from pyrolysis of crop residues. *Energy 143*, 746–756. https://doi.org/10.1016/j.energy.2017.11.062.

Hossain, M. K., Strezov, V., Yin Chan, K., & Nelson, P. F., (2010). Agronomic properties of wastewater sludge biochar and bioavailability of metals in production of cherry tomato (Lycopersicon esculentum). *Chemosphere 78*, 1167–1171. https://doi.org/10.1016/j.chemosphere.2010.01.009.

Houben, D., Evrard, L., & Sonnet, P., (2013). Mobility, bioavailability and pH-dependent leaching of cadmium, zinc and lead in a contaminated soil amended with biochar. *Chemosphere 92*, 1450–1457. https://doi.org/10.1016/j.chemosphere.2013.03.055.

Hu, Q., Yang, H., Yao, D., Zhu, D., Wang, X., Shao, J., et al. (2016). The densification of bio-char: effect of pyrolysis temperature on the qualities of pellets. *Bioresource Technology 200*, 521–527. https://doi.org/10.1016/j.biortech.2015.10.077.

Hunkeler, D., (2016). Life cycle assessment (LCA): a guide to best practice. *The International Journal of Life Cycle Assessment 21*, 1063–1066. https://doi.org/10.1007/s11367-016-1083-z.

Husk, B., & Major, J., (2010). *Commercial scale agricultural biochar field trial in Québec, Canada over two years: Effects of biochar on soil fertility, biology and crop productivity and quality*. Drummondville, Québec, Canada.

Hussain, M., Farooq, M., Nawaz, A., Al-Sadi, A. M., Solaiman, Z. M., Alghamdi, S. S., et al. (2017). Biochar for crop production: potential benefits and risks. *Journal of Soils and Sediments 17*, 685–716. https://doi.org/10.1007/s11368-016-1360-2.

Ibarrola, R., Shackley, S., & Hammond, J., (2012). Pyrolysis biochar systems for recovering biodegradable materials: a life cycle carbon assessment. *Waste Management 32*, 859–868. https://doi.org/10.1016/j.wasman.2011.10.005.

IBI, (2015). *Standardized product definition and product testing guidelines for biochar that is used in soil. International Biochar Initiative 23. http://www.biochar-international.org/characterizationstandard 22.*

Ibrahim, H. M., Al-Wabel, M. I., Usman, A. R. A., & Al-Omran, A., (2013). Effect of conocarpus biochar application on the hydraulic properties of a sandy loam soil. *Soil Science 178*.

Inyang, M., Gao, B., Pullammanappallil, P., Ding, W., & Zimmerman, A. R., (2010). Biochar from anaerobically digested sugarcane bagasse. *Bioresource Technology 101*, 8868–8872. https://doi.org/10.1016/j.biortech.2010.06.088.

IPCC, (2015). Agriculture, forestry and other land use (AFOLU). In *Climate change 2014: Mitigation of climate change: Working group III contribution to the IPCC fifth assessment report* (pp. 811–922). Cambridge: Cambridge University Press. https://doi.org/10.1017/CBO9781107415416.017.

Jaiswal, A. K., Elad, Y., Graber, E. R., & Frenkel, O., (2014). Rhizoctonia solani suppression and plant growth promotion in cucumber as affected by biochar pyrolysis temperature, feedstock and concentration. *Soil Biology and Biochemistry 69*, 110–118. https://doi.org/10.1016/j.soilbio.2013.10.051.

Jaiswal, A. K., Elad, Y., Paudel, I., Graber, E. R., Cytryn, E., & Frenkel, O., (2017). Linking the belowground microbial composition, diversity and activity to soilborne disease suppression and growth promotion of tomato amended with biochar. *Scientific Reports. 7*, 44382https://doi.org/10.1038/srep44382.

Jeffery, S., Verheijen, F. G. A., Kammann, C., & Abalos, D., (2016). Biochar effects on methane emissions from soils: a meta-analysis. *Soil Biology and Biochemistry 101*, 251–258. https://doi.org/10.1016/j.soilbio.2016.07.021.

Jiang, X., Cao, L., & Zhang, R., (2014). Changes of labile and recalcitrant carbon pools under nitrogen addition in a city lawn soil. *Journal of Soils and Sediments 14*, 515–524. https://doi.org/10.1007/s11368-013-0822-z.

Jiang, X., Denef, K., Stewart, C. E., & Cotrufo, M. F., (2016). Controls and dynamics of biochar decomposition and soil microbial abundance, composition, and carbon use efficiency during long-term biochar-amended soil incubations. *Biology and Fertility of Soils 52*, 1–14. https://doi.org/10.1007/s00374-015-1047-7.

Kameyama, K., Miyamoto, T., Iwata, Y., & Shiono, T., (2016). Influences of feedstock and pyrolysis temperature on the nitrate adsorption of biochar. *Soil Science and Plant Nutrition 62*, 180–184. https://doi.org/10.1080/00380768.2015.1136553.

Kameyama, K., Miyamoto, T., & Shiono, T., (2014). Influence of biochar incorporation on TDR-based soil water content measurements. *European Journal of Soil Science 65*, 105–112. https://doi.org/10.1111/ejss.12083.

Keith, A., Singh, B., & Dijkstra, F. A., (2015). Biochar reduces the rhizosphere priming effect on soil organic carbon. *Soil Biology and Biochemistry 88*, 372–379. https://doi.org/10.1016/j.soilbio.2015.06.007.

Kettunen, R., & Saarnio, S., (2013). Biochar can restrict N2O emissions and the risk of nitrogen leaching from an agricultural soil during the freeze-thaw period. *Agricultural and Food Science 22*(4), 373–379.

Khadem, A., & Raiesi, F., (2017). Responses of microbial performance and community to corn biochar in calcareous sandy and clayey soils. *Applied Soil Ecology 114*, 16–27. https://doi.org/10.1016/j.apsoil.2017.02.018.

Khalid, S., Shahid, M., Niazi, N. K., Murtaza, B., Bibi, I., & Dumat, C., (2017). A comparison of technologies for remediation of heavy metal contaminated soils. *Journal of Geochemical Exploration 182*, 247–268. https://doi.org/10.1016/j.gexplo.2016.11.021.

Khorram, M. S., Zhang, Q., Lin, D., Zheng, Y., Fang, H., & Yu, Y., (2016). Biochar: a review of its impact on pesticide behavior in soil environments and its potential applications. *Journal of Environmental Sciences 44*, 269–279. https://doi.org/10.1016/j.jes.2015.12.027.

Kinney, T. J., Masiello, C. A., Dugan, B., Hockaday, W. C., Dean, M. R., Zygourakis, K., et al. (2012). Hydrologic properties of biochars produced at different temperatures. *Biomass and Bioenergy 41*, 34–43. https://doi.org/10.1016/j.biombioe.2012.01.033.

Kolton, M., Graber, E. R., Tsehansky, L., Elad, Y., & Cytryn, E., (2016). Biochar-stimulated plant performance is strongly linked to microbial diversity and metabolic potential in the rhizosphere. *New Phytologist 213*, 1393–1404. https://doi.org/10.1111/nph.14253.

Kong, Z., Liaw, S. B., Gao, X., Yu, Y., & Wu, H., (2014). Leaching characteristics of inherent inorganic nutrients in biochars from the slow and fast pyrolysis of mallee biomass. *Fuel 128*, 433–441. https://doi.org/10.1016/j.fuel.2014.03.025.

Kookana, R. S., Sarmah, A. K., Van Zwieten, L., Krull, E., & Singh, B., (2011). *Biochar application to soil*: (pp. 103–143). . https://doi.org/10.1016/B978-0-12-385538-1.00003-2.

Kuzyakov, Y., Bogomolova, I., & Glaser, B., (2014). Biochar stability in soil: decomposition during eight years and transformation as assessed by compound-specific 14C analysis. *Soil Biology and Biochemistry 70*, 229–236. https://doi.org/10.1016/j.soilbio.2013.12.021.

Lehmann, J., Gaunt, J., & Rondon, M., (2006). Bio-char sequestration in terrestrial ecosystems—a review. *Mitigation and Adaptation Strategies for Global Change 11*, 403–427. https://doi.org/10.1007/s11027-005-9006-5.

Lehmann, J., Rillig, M. C., Thies, J., Masiello, C. A., Hockaday, W. C., & Crowley, D., (2011). Biochar effects on soil biota—a review. *Soil Biology and Biochemistry 43*, 1812–1836. https://doi.org/10.1016/j.soilbio.2011.04.022.

Lei, O., & Zhang, R., (2013). Effects of biochars derived from different feedstocks and pyrolysis temperatures on soil physical and hydraulic properties. *Journal of Soils and Sediments 13*, 1561–1572. https://doi.org/10.1007/s11368-013-0738-7.

Lévesque, V., Rochette, P., Ziadi, N., Dorais, M., & Antoun, H., (2018). Mitigation of CO2, CH4 and N2O from a fertigated horticultural growing medium amended with biochars and a compost. *Applied Soil Ecology 126*, 129–139. https://doi.org/10.1016/j.apsoil.2018.02.021.

Liu, X. H., Han, F. P., & Zhang, X. C., (2012). Effect of biochar on soil aggregates in the Loess Plateau: results from incubation experiments. *International Journal of Agriculture and Biology 14*, 975–979.

Liu, L., Wang, Y., Yan, X., Li, J., Jiao, N., & Hu, S., (2017). Biochar amendments increase the yield advantage of legume-based intercropping systems over monoculture. *Agriculture, Ecosystems & Environment 237*, 16–23. https://doi.org/10.1016/j.agee.2016.12.026.

Lu, H., Zhang, Y. Y., Huang, X., Wang, S., & Qiu, R., (2012). Relative distribution of Pb^{2+} sorption mechanisms by sludge-derived biochar. *Water Research 46*, 854–862.

Lucchini, P., Quilliam, R. S., DeLuca, T. H., Vamerali, T., & Jones, D. L., (2014). Does biochar application alter heavy metal dynamics in agricultural soil? *Agriculture, Ecosystems & Environment 184*, 149–157. https://doi.org/10.1016/j.agee.2013.11.018.

Luo, Y., Durenkamp, M., De Nobili, M., Lin, Q., & Brookes, P. C., (2011). Short term soil priming effects and the mineralisation of biochar following its incorporation to soils of different pH. *Soil Biology and Biochemistry 43*, 2304–2314. https://doi.org/10.1016/j.soilbio.2011.07.020.

Martin, S. M., Kookana, R. S., Van Zwieten, L., & Krull, E., (2012). Marked changes in herbicide sorption–desorption upon ageing of biochars in soil. *Journal of Hazardous Materials 231–232*, 70–78. https://doi.org/10.1016/j.jhazmat.2012.06.040.

McCormack, S. A., Ostle, N., Bardgett, R. D., Hopkins, D. W., & Vanbergen, A. J., (2013). Biochar in bioenergy cropping systems: impacts on soil faunal communities and linked ecosystem processes. *GCB Bioenergy 5*, 81–95. https://doi.org/10.1111/gcbb.12046.

Melas, G. B., Ortiz, O., & Alacañiz, J. M., (2017). Can biochar protect labile organic matter against mineralization in soil? *Pedosphere 27*, 822–831. https://doi.org/10.1016/S1002-0160(17)60421-1.

Morillo, E., & Villaverde, J., (2017). Advanced technologies for the remediation of pesticide-contaminated soils. *Science of The Total Environment 586*, 576–597. https://doi.org/10.1016/j.scitotenv.2017.02.020.

Mukherjee, A., & Lal, R., (2013). Biochar impacts on soil physical properties and greenhouse gas emissions. *Agronomy*. https://doi.org/10.3390/agronomy3020313.

Nguyen, T. T. N., Xu, C.-Y., Tahmasbian, I., Che, R., Xu, Z., Zhou, X., et al. (2017). Effects of biochar on soil available inorganic nitrogen: a review and meta-analysis. *Geoderma 288*, 79–96. https://doi.org/10.1016/j.geoderma.2016.11.004.

Ouyang, W., Zhao, X., Tysklind, M., & Hao, F., (2016). Typical agricultural diffuse herbicide sorption with agricultural waste-derived biochars amended soil of high organic matter content. *Water Research 92*, 156–163. https://doi.org/10.1016/j.watres.2016.01.055.

Peters, J. F., Iribarren, D., & Dufour, J., (2015). Biomass pyrolysis for biochar or energy applications? A life cycle assessment. *Environmental Science & Technology 49*, 5195–5202. https://doi.org/10.1021/es5060786.

Qambrani, N. A., Rahman, M. M., Won, S., Shim, S., & Ra, C., (2017). Biochar properties and eco-friendly applications for climate change mitigation, waste management, and wastewater treatment: a review. *Renewable and Sustainable Energy Reviews 79*, 255–273. https://doi.org/10.1016/j.rser.2017.05.057.

Reddy, K. R., Yaghoubi, P., & Yukselen-Aksoy, Y., (2015). Effects of biochar amendment on geotechnical properties of landfill cover soil. *Waste Management & Research 33*, 524–532. https://doi.org/10.1177/0734242X15580192.

Roberts, K. G., Gloy, B. A., Joseph, S., Scott, N. R., & Lehmann, J., (2010). Life cycle assessment of biochar systems: estimating the energetic, economic, and climate change potential. *Environmental Science & Technology 44*, 827–833. https://doi.org/10.1021/es902266r.

Rogovska, N., Laird, D., Cruse, R., Fleming, P., Parkin, T., & Meek, D., (2011). Impact of biochar on manure carbon stabilization and greenhouse gas emissions. *Soil Science Society of America Journal 75*, 871–879. https://doi.org/10.2136/sssaj2010.0270.

Saifullah, Dahlawi, S., Naeem, A., Rengel, Z., & Naidu, R., (2018). Biochar application for the remediation of salt-affected soils: challenges and opportunities. *Science of The Total Environment 625*, 320–335. https://doi.org/10.1016/j.scitotenv.2017.12.257.

Schimmelpfennig, S., & Glaser, B., (2012). One step forward toward characterization: some important material properties to distinguish biochars. *Journal of Environmental Quality 41*, 1001–1013. https://doi.org/10.2134/jeq2011.0146.

Schneider, F., & Haderlein, S. B., (2016). Potential effects of biochar on the availability of phosphorus—mechanistic insights. *Geoderma 277*, 83–90. https://doi.org/10.1016/j.geoderma.2016.05.007.

Seehausen, L. M., Gale, V. N., Dranga, S., Hudson, V., Liu, N., Michener, J., et al. (2017). Is there a positive synergistic effect of biochar and compost soil amendments on plant growth and physiological performance? *Agronomy*. https://doi.org/10.3390/agronomy7010013.

Sharma, A., Wang, S., Pareek, V., Yang, H., & Zhang, D., (2014). CFD modeling of mixing/segregation behavior of biomass and biochar particles in a bubbling fluidized bed. *Chemical Engineering Science 106*, 264–274. https://doi.org/10.1016/j.ces.2013.11.019.

Smebye, A., Alling, V., Vogt, R. D., Gadmar, T. C., Mulder, J., Cornelissen, G., et al. (2016). Biochar amendment to soil changes dissolved organic matter content and composition. *Chemosphere 142*, 100–105. https://doi.org/10.1016/j.chemosphere.2015.04.087.

Smith, P., (2016). Soil carbon sequestration and biochar as negative emission technologies. *Global Change Biology 22*, 1315–1324. https://doi.org/10.1111/gcb.13178.

Sohi, S. P., Krull, E., Lopez-Capel, E., & Bol, R., (2010). A review of biochar and its use and function in soil. *Advances in agronomy* (pp. 47–82). Vol. 105(pp. 47–82). Academic Press. https://doi.org/10.1016/S0065-2113(10)05002-9 (chapter 2).

Spokas, K. A., Cantrell, K. B., Novak, J. M., Archer, D. W., Ippolito, J. A., Collins, H. P., et al. (2012). Biochar: a synthesis of its agronomic impact beyond carbon sequestration. *Journal of Environmental Quality 41*, 973–989. https://doi.org/10.2134/jeq2011.0069.

Sun, K., Gao, B., Ro, K. S., Novak, J. M., Wang, Z., Herbert, S., et al. (2012). Assessment of herbicide sorption by biochars and organic matter associated with soil and sediment. *Environmental Pollution 163*, 167–173. https://doi.org/10.1016/j.envpol.2011.12.015.

Sun, Y., Gao, B., Yao, Y., Fang, J., Zhang, M., Zhou, Y., et al. (2014). Effects of feedstock type, production method, and pyrolysis temperature on biochar and hydrochar properties. *Chemical Engineering Journal 240*, 574–578. https://doi.org/10.1016/j.cej.2013.10.081.

Sun, L., Li, L., Chen, Z., Wang, J., & Xiong, Z., (2014). Combined effects of nitrogen deposition and biochar application on emissions of N2O, CO2 and NH3 from agricultural and forest soils. *Soil Science and Plant Nutrition 60*, 254–265. https://doi.org/10.1080/00380768.2014.885386.

Syuhada, A. B., Shamshuddin, J., Fauziah, C. I., Rosenani, A. B., & Arifin, A., (2016). Biochar as soil amendment: impact on chemical properties and corn nutrient uptake in a Podzol. *Canadian Journal of Soil Science 96*, 400–412. https://doi.org/10.1139/cjss-2015-0044.

Tan, Z., Lin, C. S. K., Ji, X., & Rainey, T. J., (2017). Returning biochar to fields: a review. *Applied Soil Ecology 116*, 1–11. https://doi.org/10.1016/j.apsoil.2017.03.017.

Tang, J., Zhu, W., Kookana, R., & Katayama, A., (2013). Characteristics of biochar and its application in remediation of contaminated soil. *Journal of Bioscience and Bioengineering 116*, 653–659. https://doi.org/10.1016/j.jbiosc.2013.05.035.

Tian, X., Li, C., Zhang, M., Wan, Y., Xie, Z., Chen, B., et al. (2018). Biochar derived from corn straw affected availability and distribution of soil nutrients and cotton yield. *PLoS ONE 13*, e0189924.

Uchimiya, M., Klasson, K. T., Wartelle, L. H., & Lima, I. M., (2011). Influence of soil properties on heavy metal sequestration by biochar amendment: 1. Copper sorption isotherms and the release of cations. *Chemosphere 82*, 1431–1437. https://doi.org/10.1016/j.chemosphere.2010.11.050.

Uchimiya, M., Wartelle, L. H., Klasson, K. T., Fortier, C. A., & Lima, I. M., (2011). Influence of pyrolysis temperature on biochar property and function as a heavy metal sorbent in soil. *Journal of Agricultural and Food Chemistry 59*, 2501–2510. https://doi.org/10.1021/jf104206c.

Ulyett, J., Sakrabani, R., Kibblewhite, M., & Hann, M., (2013). Impact of biochar addition on water retention, nitrification and carbon dioxide evolution from two sandy loam soils. *European Journal of Soil Science 65*, 96–104. https://doi.org/10.1111/ejss.12081.

UNEP, (2016). *The emissions gap report 2016*. . ISBN 978-92-807-3617-5.

Van Laer, T., Smedt, P., Ronsse, F., Ruysschaert, G., Boeckx, P., Verstraete, W., et al. (2014). Legal constraints and opportunities for biochar: a case analysis of EU law. *GCB Bioenergy 7*, 14–24. https://doi.org/10.1111/gcbb.12114.

van Zwieten, L., Kimber, S., Morris, S., Chan, K. Y., Downie, A., Rust, J., et al. (2010). Effects of biochar from slow pyrolysis of papermill waste on agronomic performance and soil fertility. *Plant and Soil 327*, 235–246. https://doi.org/10.1007/s11104-009-0050-x.

van Zwieten, L., Kimber, S., Morris, S., Downie, A., Berger, E., Rust, J., et al. (2010). Influence of biochars on flux of N2O and CO2 from Ferrosol. *Soil Research 48*, 555–568.

Verhoeven, E., & Six, J., (2014). Biochar does not mitigate field-scale N2O emissions in a Northern California vineyard: an assessment across two years. *Agriculture, Ecosystems & Environment 191*, 27–38. https://doi.org/10.1016/j.agee.2014.03.008.

Wang, T., Camps-Arbestain, M., & Hedley, M., (2014). The fate of phosphorus of ash-rich biochars in a soil-plant system. *Plant and Soil 375*, 61–74. https://doi.org/10.1007/s11104-013-1938-z.

Wang, J., Pan, X., Liu, Y., Zhang, X., & Xiong, Z., (2012). Effects of biochar amendment in two soils on greenhouse gas emissions and crop production. *Plant and Soil 360*, 287–298. https://doi.org/10.1007/s11104-012-1250-3.

Wang, J., Xiong, Z., & Kuzyakov, Y., (2015). Biochar stability in soil: meta-analysis of decomposition and priming effects. *GCB Bioenergy 8*, 512–523. https://doi.org/10.1111/gcbb.12266.

Wang, J., Xiong, Z., & Kuzyakov, Y., (2016). Biochar stability in soil: meta-analysis of decomposition and priming effects. *GCB Bioenergy 8*, 512–523. https://doi.org/10.1111/gcbb.12266.

Warnock, D. D., Lehmann, J., Kuyper, T. W., & Rillig, M. C., (2007). Mycorrhizal responses to biochar in soil—concepts and mechanisms. *Plant and Soil 300*, 9–20. https://doi.org/10.1007/s11104-007-9391-5.

Warnock, D. D., Mummey, D. L., McBride, B., Major, J., Lehmann, J., & Rillig, M. C., (2010). Influences of non-herbaceous biochar on arbuscular mycorrhizal fungal abundances in roots and soils: Results from growth-chamber and field experiments. *Applied Soil Ecology 46*(3), 450–456. https://doi.org/10.1016/j.apsoil.2010.09.002.

Woolf, D., Amonette, J. E., Street-Perrott, F. A., Lehmann, J., & Joseph, S., (2010). Sustainable biochar to mitigate global climate change. *Nature Communications 1*, 56.

Wu, F., Jia, Z., Wang, S., Chang, S. X., & Startsev, A., (2013). Contrasting effects of wheat straw and its biochar on greenhouse gas emissions and enzyme activities in a Chernozemic soil. *Biology and Fertility of Soils 49*, 555–565. https://doi.org/10.1007/s00374-012-0745-7.

Yaghoubi, P., Yargicoglu, E. N., & Reddy, K. R., (2014). Effects of biochar-amendment to landfill cover soil on microbial methane oxidation: initial results. In: *Geotechnical Special Publication, Proceedings*https://doi.org/10.1061/9780784413272.181.

Yao, Y., Gao, B., Zhang, M., Inyang, M., & Zimmerman, A. R., (2012). Effect of biochar amendment on sorption and leaching of nitrate, ammonium, and phosphate in a sandy soil. *Chemosphere 89*, 1467–1471. https://doi.org/10.1016/j.chemosphere.2012.06.002.

Yi, S., Witt, B., Chiu, P., Guo, M., & Imhoff, P., (2015). The origin and reversible nature of poultry litter biochar hydrophobicity. *Journal of Environmental Quality 44*, 963–971. https://doi.org/10.2134/jeq2014.09.0385.

Yoo, G., & Kang, H., (2012). Effects of biochar addition on greenhouse gas emissions and microbial responses in a short-term laboratory experiment. *Journal of Environmental Quality 41*, 1193–1202. https://doi.org/10.2134/jeq2011.0157.

Yu, J., Deem, L. M., Crow, S. E., Deenik, J. L., & Penton, C. R., (2018). Biochar application influences microbial assemblage complexity and composition due to soil and bioenergy crop type interactions. *Soil Biology and Biochemistry 117*, 97–107. https://doi.org/10.1016/j.soilbio.2017.11.017.

Zhang, H., Chen, C., Gray, E. M., Boyd, S. E., Yang, H., & Zhang, D., (2016). Roles of biochar in improving phosphorus availability in soils: a phosphate adsorbent and a source of available phosphorus. *Geoderma 276*, 1–6. https://doi.org/10.1016/j.geoderma.2016.04.020.

Zhang, C., Lin, Y., Tian, X., Xu, Q., Chen, Z., & Lin, W., (2017). Tobacco bacterial wilt suppression with biochar soil addition associates to improved soil physiochemical properties and increased rhizosphere bacteria abundance. *Applied Soil Ecology 112*, 90–96. https://doi.org/10.1016/j.apsoil.2016.12.005.

Zhang, P., Sun, H., Ren, C., Min, L., & Zhang, H., (2018). Sorption mechanisms of neonicotinoids on biochars and the impact of deashing treatments on biochar structure and neonicotinoids sorption. *Environmental Pollution 234*, 812–820. https://doi.org/10.1016/j.envpol.2017.12.013.

Zhou, H., Zhang, D., Wang, P., Liu, X., Cheng, K., Li, L., et al. (2017). Changes in microbial biomass and the metabolic quotient with biochar addition to agricultural soils: a meta-analysis. *Agriculture, Ecosystems & Environment 239*, 80–89. https://doi.org/10.1016/j.agee.2017.01.006.

Zong, Y., Xiao, Q., & Lu, S., (2016). Acidity, water retention, and mechanical physical quality of a strongly acidic Ultisol amended with biochars derived from different feedstocks. *Journal of Soils and Sediments 16*, 177–190. https://doi.org/10.1007/s11368-015-1187-2.

Further reading

Augustenborg, C. A., Hepp, S., Kammann, C., Hagan, D., Schmidt, O., & Muller, C., (2012). Biochar and earthworm effects on soil nitrous oxide and carbon dioxide emissions. *Journal of Environmental Quality 41*, 1203–1209. https://doi.org/10.2134/jeq2011.0119.

Case, S. D. C., McNamara, N. P., Reay, D. S., & Whitaker, J., (2012). The effect of biochar addition on N2O and CO2 emissions from a sandy loam soil – The role of soil aeration. *Soil Biology and Biochemistry 51*, 125–134. https://doi.org/10.1016/j.soilbio.2012.03.017.

Downie, A., Crosky, A., & Munroe, P., (2009). Physical properties of biochar. In J. Lehmann & S. Joseph (Eds.), *Biochar for environmental management: Science and technology* (pp. 13–32). London: Earthscan.

DuLeca, T. H., Mackenzie, M. D., & Gundale, M. J., (2009). Biochar effects on soil nutrient transformations. In J. Lehmann & S. Joseph (Eds.), *Biochar for environmental management: Science and technology* (pp. 251–270). London: Earthscan.

Herath, H. M. S. K., Camps-Arbestain, M., & Hedley, M., (2013). Effect of biochar on soil physical properties in two contrasting soils: an Alfisol and an Andisol. *Geoderma 209–210*, 188–197. https://doi.org/10.1016/j.geoderma.2013.06.016.

Imparato, V., Hansen, V., Santos, S. S., Nielsen, T. K., Giagnoni, L., Hauggaard-Nielsen, H., et al. (2016). Gasification biochar has limited effects on functional and structural diversity of soil microbial communities in a temperate agroecosystem. *Soil Biology and Biochemistry 99*, 128–136. https://doi.org/10.1016/j.soilbio.2016.05.004.

Jeffery, S., Verheijen, F. G. A., van der Velde, M., & Bastos, A. C., (2011). A quantitative review of the effects of biochar application to soils on crop productivity using meta-analysis. *Agriculture, Ecosystems & Environment 144*, 175–187. https://doi.org/10.1016/j.agee.2011.08.015.

Jenkins, J. R., Viger, M., Arnold, E. C., Harris, Z. M., Maurizio, V., Franco, M., et al. (2016). Biochar alters the soil microbiome and soil function: results of next-generation amplicon sequencing across Europe. *GCB Bioenergy 9*, 591–612. https://doi.org/10.1111/gcbb.12371.

Jien, S.-H., & Wang, C.-S., (2013). Effects of biochar on soil properties and erosion potential in a highly weathered soil. *CATENA 110*, 225–233. https://doi.org/10.1016/j.catena.2013.06.021.

Kameyama, K., Miyamoto, T., Shiono, T., & Shinogi, Y., (2012). Influence of sugarcane bagasse-derived biochar application on nitrate leaching in calcaric dark red soil. *Journal of Environmental Quality 41*, 1131–1137. https://doi.org/10.2134/jeq2010.0453.

Wu, H., Zeng, G., Liang, J., Chen, J., Xu, J., Dai, J., et al. (2016). Responses of bacterial community and functional marker genes of nitrogen cycling to biochar, compost and combined amendments in soil. *Applied Microbiology and Biotechnology 100*, 8583–8591. https://doi.org/10.1007/s00253-016-7614-5.

Char combustion

5

Besma Khiari, Mejdi Jeguirim†, Simona Bennici†, Lionel Limousy†*
*National School of Engineers of Carthage, Tunis, Tunisia, †University of Strasbourg, University of Upper Alsace, Institute of Materials Science of Mulhouse (IS2M—UMR CNRS 7361), Mulhouse, France

Chapter Outline

5.1 Introduction

With increasing energy demand and continuously deteriorating environmental issues, the use of renewable energies has become an urgent requirement. Among the various energy resources, biomass feedstocks as well as their derived chars represent an inexpensive and highly effective way to replace conventional fuels. While biomass resources have been used from ancient times, it is only recently that chars have retained stakeholders' attention as combustion fuels. Compared to their corresponding original biomass, it has been demonstrated that chars have better combustion characteristics, including lower content of moisture and volatiles, lower H/C and O/C ratios, higher heating values and energy density, increased steady-state burning, and shorter ignition times.

Char and Carbon Materials Derived from Biomass. https://doi.org/10.1016/B978-0-12-814893-8.00005-5

Thermogravimetry analysis (TGA) is a valuable tool to understand the combustion behavior of biomass-derived chars for energy production. Thermographs may also help to assess the complexities of the thermal decomposition of the material and by way of consequence deduce the combustion kinetics. These latter are in use in the design and control of combustion reactors.

To attain good combustion efficiencies and to reduce costs of storage, handling, and transportation, densification is often implemented to obtain biomass char pellets, logs, and briquettes. Specifications and combustion characteristics of these may vary according to their final usage in household activities or in industrial applications.

This chapter is divided in several sections. While the first section is dedicated to char combustion characteristics in terms of thermogravimetry (TG)/differential thermogravimetry (DTG) and kinetics, the second section examines the densification operations and how far they ameliorate the combustibility and ignition of the obtained pellets and briquettes. Finally, the use of the compacted biomass-derived chars is assessed by exposing some of their possible applications in domestic or industrial fields.

It is worth mentioning that in this chapter distinction is made between charcoal and char: charcoal is relative to wood being processed thermally, usually by carbonization, whereas char is the by-product of the thermal treatment (pyrolysis, hydrothermal carbonization, gasification, etc.) of any type of biomass (lignocellulosic, sludge, organic waste, etc.).

5.2 Char combustion in TGA/DTG

TG or TGA is one of the most common thermal analysis techniques for materials characterization. Applied to biomass-derived chars, it is also a powerful tool to assess their suitability for processing by thermal conversion in general and to better approach their oxidation mechanisms during combustion in particular.

Char combustion is indeed a complex chemical process involving many steps, depending on the properties of the combustible material. It is initiated by external factors such as heat to attain the ignition temperature. Significant physical phenomena also occur during char combustion, including mass and energy transfers by diffusion and/or convection. Investigations have been carried out using TGA/DTG to make clearer all these aspects.

According to the charring conditions, combustion thermographs of biomass chars are generally made in one or two stages. For example, if the char was prepared at low temperature, then the two phases correspond respectively to devolatilization (of the remaining in-char polymers) and oxidation of the char. If the char was yielded at high temperatures, one single phase is observed, indicating the decomposition by oxidation of the solid residue to burnout. The char process also makes distinction in the thermograms for the same biomass and the same temperatures. For instance, prior hydrothermal carbonization (Islam, Asif, & Hameed, 2015) and pyrolysis (Islam, Auta, Kabir, & Hameed, 2016) of Karanj fruit hulls gave very different DTG profiles (Fig. 1). The dual stages of the char's DTG at the respective characteristic temperature ranges and peaks revealed the complex nature of the hull's pyrolytic char combustion.

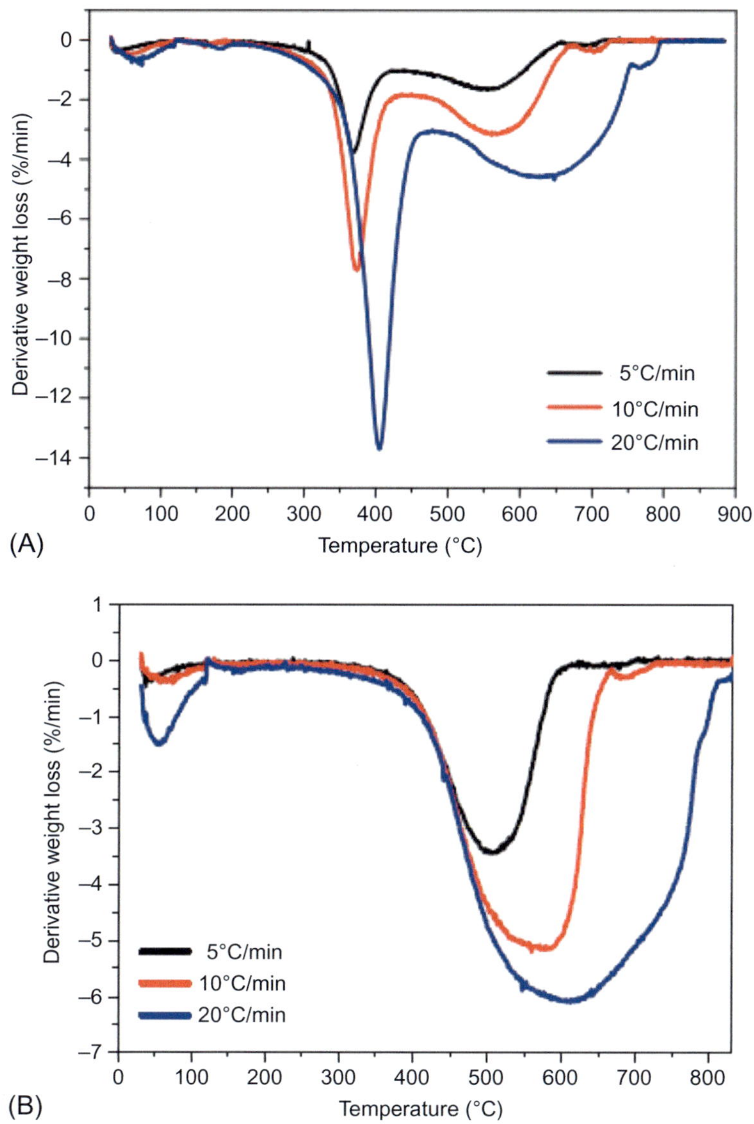

Fig. 1 Differential thermogravimetry of Karanji fruit hulls: (A) hydrochar and (B) pyrolytic char during combustion at different heating rates.
Reprinted with permission from (A) Islam, M. A., Asif, M., Hameed, B. H. (2015). Pyrolysis kinetics of raw and hydrothermally carbonized Karanj (Pongamia pinnata) fruit hulls via thermogravimetric analysis. *Bioresource Technology*, *179*, 227–233. (B) Islam, M. A., Auta, M., Kabir, G., Hameed, B. H. (2016). A thermogravimetric analysis of the combustion kinetics of karanja (Pongamia pinnata) fruit hulls char. *Bioresource Technology*, *200*, 335–341. Copyright 2018, Elsevier.

In the second stage, peaks are broad, meaning that char oxidation is more rapid because of an increasing faculty to overcome the mass transfer resistance layer easily, because oxygen disperses inside and outside of the char particles, which helps maximum char conversion at peak temperature. In fact, this is due to the dominance of chemical reactions that accelerate the heat release and ameliorate the heat permeability to degrade the char, compared to slow pyrolysis (Senneca, Chirone, & Salatino, 2002).

Away from the peak combustion rate, and when char is not yet reduced to its mineral and ash fraction, the mass transfer resistance continuously builds up and limits the penetration of the oxygen to promote oxidation. The diffusion layer becomes stringent to oxygen permeability and the final stagnated mass loss indicates the end of the char life-cycle.

A thorough examination of the literature may give an accurate explanation of the char multistep combustion process. Mass loss during the first phase took place at the same time where volumes of gas, mainly carbon oxides and small quantities of aerosols, were recorded. Moreover, it was reported that chars obtained under mild conditions (low temperatures) undergo, in particular, two major degradation reactions: pyrolysis of intact and partially decomposed glycosyl units and decomposition of paraffinic and carbonyl groups, whereas chars at severe conditions (higher temperatures) show only the last stage (Antal & Grønli, 2003).

In the same vein, the so-obtained low-temperature biomass chars, being more stable with relatively ordered carbon structure, are interesting for aromatic compound production in low-temperature processes, such as smoldering combustion and slow pyrolysis (Di Blasi, 2009).

The usually reported mechanisms during char combustion are based on mass transfer. At the beginning of the process, oxygen diffuses to the active site within the char. This diffusion is facilitated by the numerous pathways left unoccupied after the volatile's departure. Chemisorption of oxygen on the carbon surface follows according to the following schemes:

$$C + O_2 \rightarrow CO_2$$
$$2\,C + O_2 \rightarrow 2\,CO$$
$$2\,CO + O_2 \rightarrow 2\,CO_2$$
$$C + CO_2 \rightarrow 2\,CO$$
$$2\,C_f + O_2 \rightarrow 2\,C(O)$$
$$C(O) \leftrightarrow C{-}O$$
$$2\,C(O) \rightarrow CO_2\ (+m\ C_f)$$
$$C(O) \rightarrow CO\ (+n\ C_f)$$

Then, the chemisorbed oxygen reacts with the reactive components that constitute the char (Zhou et al., 2013). Subsequently, the produced gas is desorbed from the carbon surface and expelled across the active sites and pathways (Liu, Zhang, & Wu, 2010) by several phenomena but dominated by diffusion mass transfer, which facilitates the combustion reactions and leads to mass loss (de la Puente, Fuente, & Pis, 2000). In the absence of mass transport limitations, char reactivity is mainly controlled by char active surface area (Islam, Kabir, Asif, & Hameed, 2015).

It is widely admitted in the literature that the reactivity of chars from lignocellulosic materials is high. It is, for example, twice the reactivity of pyrolytic chars from tire-derived fuel, ebonite, plastic materials (Senneca et al., 2002), or charcoal (Sima-Ella, Yuan, & Mays, 2005). As a consequence, their combustion starts at lower temperatures. The explanation comes from the char morphology (porosity and disordered structure), which is favorable to short ignition times, thanks to the high volatile contents in the biomass that have left available sites in the material during the prior treatment. However, some authors reached such high reactivity for only a low heating rate, while for fast heating rates the importance of the devolatilization step is reduced with respect to the subsequent oxidation phase (Haykiri-Acma, Yaman, & Kucukbayrak, 2012).

The heating rates also affect the maximum burning rate and its corresponding temperature, both of which were taken into account as indicators of combustibility and reactivity in many investigations (He et al., 2017). For example, Haykiri-Acma et al. (2012) found that the high reactivity of pyrolytic char from hazelnut shells is correlated with low maximum burning temperature and that increasing heating rate shifted this temperature as well as the maximum burning rates to higher levels.

It is worth pointing out here that, in addition to the aforementioned factors, char behavior during combustion depends also on the oxygen diffusion rate, on particle size, and on the final char production temperature (de la Puente et al., 2000; Román, Nabais, Laginhas, Ledesma, & González, 2012). As an illustration of the impact of this latter point, chars from palm empty fruit bunches were obtained at different final temperatures of the hydrothermal carbonization process (Fig. 2). With hydrochars obtained at 150°C and 250°C, combustion took place in two stages, whereas chars yielded at 350°C DTG were characterized by only one major combustion stage from 310°C to 590°C. Between 280°C and 400°C, the profiles exhibited the highest weight loss rate, correlated to hemicellulose and cellulose degradation during

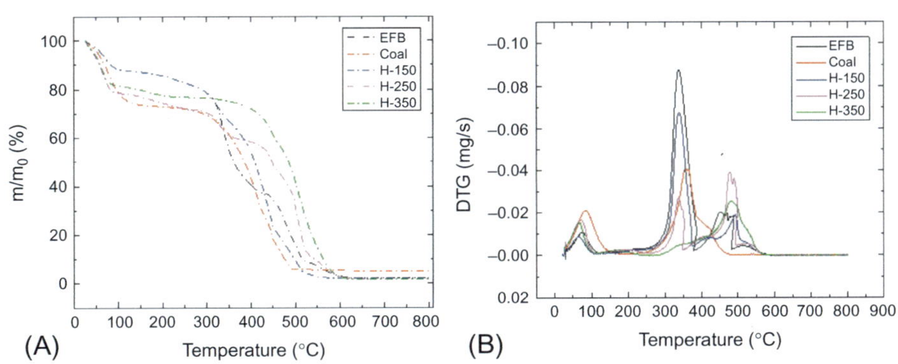

Fig. 2 Thermogravimetry (A) and differential thermogravimetry (DTG) (B) of palm empty fruit bunches hydrochar.
Reprinted with permission from Parshetti, G. K., Kent Hoekman, S., Balasubramanian, R. (2013). Chemical, structural and combustion characteristics of carbonaceous products obtained by hydrothermal carbonization of palm empty fruit bunches. *Bioresource Technology, 135*, 683–689. Copyright 2018, Elsevier.

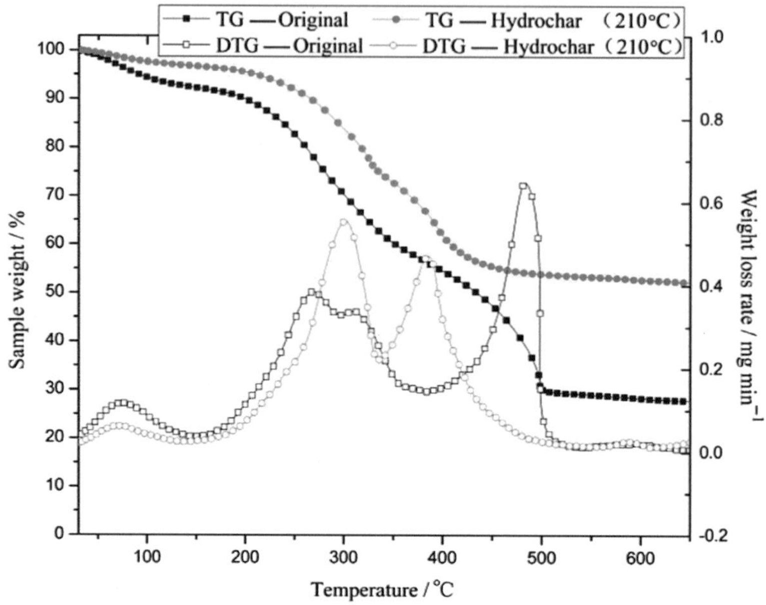

Fig. 3 Thermogravimetry (TG) and differential thermogravimetry (DTG) curves for combustion profiles of original sludge and corresponding hydrochar obtained at 210°C. Reprinted with permission from Wang, L., Li, A., Chang, Y. (2016). Hydrothermal treatment coupled with mechanical expression at increased temperature for excess sludge dewatering: heavy metals, volatile organic compounds and combustion characteristics of hydrochar. *Chemical Engineering Journal, 297*, 1–10. Copyright 2018, Elsevier.

hydrothermal carbonization. Above 400°C (the zone where lignin decomposes), the DTG poorly shaped peaks became smaller. For chars obtained at 350°C, these peaks are fused into a single broad peak (Parshetti, Kent Hoekman, & Balasubramanian, 2013).

It is well known that the TG and DTG profiles of a given biochar are different from its original biomass, indicating important changes in behavior during combustion.

An example is given in Fig. 3, where significantly modified profiles are revealed, although the sludge hydrochar was obtained at low temperature (210°C). As mentioned above, the main disparities are attributed to the decrease in volatile matter and the increase in ash content. The peak below 160°C related to moisture in the original sludge is still observed with chars, which is explained by the removal of inherent moisture (Han et al., 2016).

5.3 Kinetics of char combustion

To confirm the mass loss behavior of biomass chars during combustion, kinetic studies are usually carried out and kinetic parameters are determined based on Arrhenius' equation.

Early kinetic models of wood char combustion, using integral thermogravimetric dynamic data for parameter estimation, consist of a one-step global reaction characterized by relatively low values of the activation energy (less than 100 kJ/mol) (Di Blasi, 2009). Although these provide very poor predictions of differential thermogravimetric results, some researchers adopted them assuming that the mechanisms of combustion for mineral coal chars are also appropriate for lignocellulosic chars (Hurt & Calo, 2001).

The most widely used simple global reaction is:

$C + O_2 \rightarrow CO/CO_2$

The corresponding conversion rate is given by:

$$r_c = k_0 e^{\left(-\frac{Ea}{RT}\right)} P_{O2}^n \tag{1}$$

where P_{O2}^n is the partial pressure of O_2.

Some examples are given in Table 1. Nevertheless, and as stated earlier, since all TG/DTG profiles display evidently separated stages of devolatilization and combustion, the kinetic parameters of each stage may be determined independently by means of the linear regression method following the corresponding data of different stages. This treatment produces more accurate predictions and gives rise to higher activation energies of the combustion reaction (140–230 kJ/mol).

An assumption of a first-order reaction mechanism is most frequently used for char combustion. Therefore, the preexponential and activation energies associated with the combustion process were calculated according to the first-order combustion reaction under nonisothermal conditions. Such results were reached by Han et al. (2016) in their investigations on sludge hydrochar combustion. These authors found that the devolatilization and combustion reactions were best fitted to the first-order mechanism with high correlation coefficients ($R^2 > 0.97$). Isoconversional analysis revealed that the release of volatiles and the oxidation of hydrochar structural components were all multistage reaction processes. The generated hydrochar as solid fuel proved to have good combustion performance.

Table 1 Examples of conversion rate for different biomass-derived chars

Charring conditions	Conversion rate	References
Pinewood pyrolytic char; 300°C/s until 600°C	$r_c = 5.3 \times 10^5 \exp\left(\frac{-125}{RT}\right) P_{O_2}^{0.53} (1-X)^{0.49}$	Janse, de Jonge, Prins, and van Swaaij (1998)
Sugarcane bagasse pyrolytic char; 15°C/min until 700°C	$r_c = A \exp\left(\frac{-180}{RT}\right) P_{O_2}^{0.65} (1-X)$	Luo and Stanmore (1992)
Corncob char; flash carbonization	$r_c = 8.12 \times 10^8 \exp\left(\frac{-151}{RT}\right) P_{O_2}^{0.53} f(X)$	Várhegyi, Mészáros, Antal, Bourke, and Jakab (2006)

Table 2 Combustion activation energies (E_a) calculated by the Kissinger-Akahira-Sunose (KAS) and Flynn-Wall-Ozawa (FWO) models

	KAS model	FWO model	References
Pine sawdust char	212.4 (kJ/mol)	212.7 (kJ/mol)	Chen, Cai, and Liu (2015)
Karanj hulls char	130.5 (kJ/mol)	135.9 (kJ/mol)	Islam, Asif, and Hameed (2015)

More precisely, as temperature of the hydrothermal carbonization is increased from 120°C to 210°C, the devolatilization activation energy was increased from 52 to 56 kJ/mol, while in the second stage (volatiles and char combustion phase) the activation energy was significantly reduced from 132.6 to 64 kJ/mol. Since the activation energy of all hydrochars in the first reaction was much lower than that in the second reaction, the kinetic pathway of char combustion was volatile matter release followed by the oxidation of the volatiles and char. The higher activation energy during the devolatilization stage means that more energy is required to release volatiles leading to an improvement in thermal stability of hydrochar. More importantly, the lower activation energy during the volatiles and char combustion stage improved the flammability of hydrochar, finally reducing the ignition temperature.

Preexponential factor (k_0) values, on the other hand, indicated that the hydrochar produced at increased temperature tended to be a good solid fuel in terms of easy and stable combustion.

The same conclusions have been reported by other authors adopting the Kissinger-Akahira-Sunose, Flynn-Wall-Ozawa, and Friedman, Ozawa, and Coats-Redfern methods to describe the kinetic parameters of biomass char combustion. Examples are found in Table 2.

The degree of char conversion increases with increasing temperature since the activation energies diminish to promote rapid char degradation. The reaction order and diffusion models of the Coats-Redfern method disclosed the firm dependence of the char combustion on a complex multistep mechanism.

5.4 Char densification

Densification is often implemented to improve storage, handling, and transportation costs but also and particularly to reach good combustion efficiencies (Jeguirim, Limousy, & Dutournie, 2014; Lajili, Limousy, & Jeguirim, 2014). Indeed, densification achieves a uniformly shaped and sized solid porous fuel with increased density, optimum hardness, and high carbon concentration, which will help increase the heating value.

In addition, materials can be used either exclusively or blended with other substrates to ameliorate the solid fuel quality and to optimize the densification process (Kraiem et al., 2014; Kraiem, Lajili, Limousy, Said, & Jeguirim, 2016). Chars obtained from biomass pyrolysis are one of the most interesting candidates for densification because they have low bulk on the one hand and interesting heating values on the other hand. Pyrolysis, torrefaction, or carbonization was carried out prior to densification for several types of biomass (woody and herbaceous residues, crops, agricultural goods, animal wastes, municipal wastes) and numerous activities such as food processing and sewage sludge treatment. Results gave a superior quality of fuels compared to densification of raw materials (Hirunpraditkoon, Intharit, Srisumran, & Wongvirojtana, 2014).

However, in many cases, the densification of pyrolyzed biomass was difficult to carry out, because the treated materials underwent structural modifications in fiber, cohesiveness, flowability, and moisture content (Pimchuai, Dutta, & Basu, 2010). Reversing the chain order was then a clever solution to enhance the quality of the biomass char (Wang, Han, Gao, Li, & Lu, 2017).

Densified chars might be in the form of pellets, briquettes, or logs. The two first forms are discussed in the following paragraphs in terms of manufacturing and properties.

5.4.1 Pellet production

A solid by-product from biomass pyrolysis can be densified into pellets (Fig. 4) with favorable properties, not only inherited from char (the quality of which itself depends on the pyrolysis operating parameters) but also enhanced by the densification process. Indeed, compacting conditions such as compressive strength, proportion of binders, or water addition content influence the resultant char pellets.

Throughout the literature, different systems were applied to various chars, with the goal to produce pellets. For instance, a 200kN universal material testing machine (CMT5205, MTS, China) was used by Hu et al. (2016) to pelletize chars derived from woody shavings pyrolysis at 550–650°C. A hydraulic press powder pelletizer

Fig. 4 Biomass-derived char pellets.

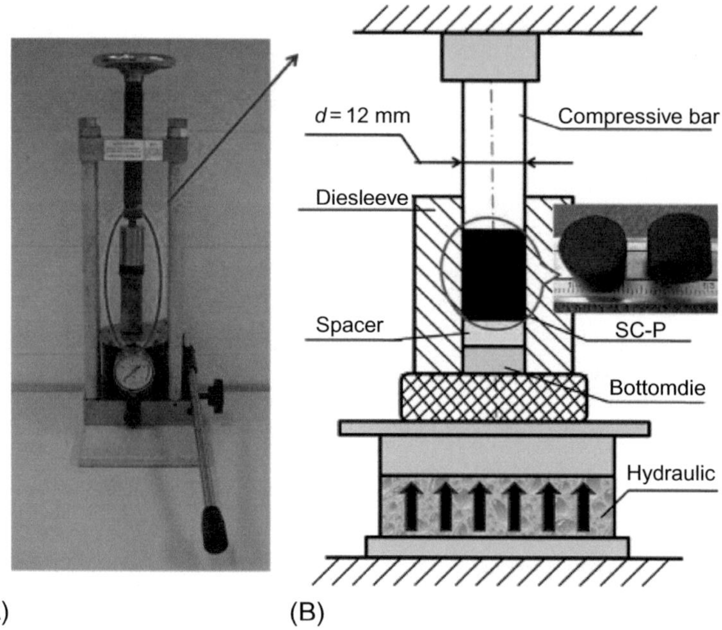

Fig. 5 Hydraulic press powder pelletizer (A) with the corresponding mold (B).
Reprinted with permission from Kong, L., Tian, S., Li, Z., Luo, R., Chen, D., Tu, Y., et al. (2013). Conversion of recycled sawdust into high HHV and low NOx emission bio-char pellets using lignin and calcium hydroxide blended binders. *Renewable Energy, 60*, 559–565. Copyright 2018, Elsevier.

(HuaiYu Tianjin, China) was employed by Kong et al. (2013) to densify spruce sawdust char blended with 10% of lignin (Fig. 5).

Bazargan, Rough, and McKay (2014) processed chars issued from palm kernel shell pyrolysis using a modified 50kN fully instrumented strain frame (Zwick/Roell, Germany).

The properties of the obtained pellets and the optimum conditions for pelletizing are summarized in Table 3.

Before densification, chars from biomass present very low densities ranging between 270 for woody char and $288 \pm 35\,kg/m^3$ for shell chars. After pelletizing, these values more than double and exceed even $700\,kg/m^3$, which will enable and facilitate their routing to classical combustion plants and installations. Aside logistic benefits, Rhén, Öhman, Gref, and Wästerlund (2007) mentioned that automation and optimization were possible for pellets, because of their interesting combustion yields in terms of residual amounts and process efficiency. Indeed, heating values were increased (by 95%, for example, for spruce wood sawdust), and are actually compared to coal energy content (29MJ/kg).

Compressive pressure is a parameter that determines pellet quality (together with its shape and resistance), as it increases molecular connections and promotes the

Table 3 Properties and optimum parameters of some char pellets

Chars derived from	Woody shavings	Spruce wood sawdust	Palm kernel shell
Apparent density (kg/m^3)	693.26	540	701 ± 36
Compressive strength (MPa)	128	2.5–8	60
Moisture uptake content (wt%)	12	17.11–25.79	40
Diameter (mm)	20	12	24.9
Height (mm)	12–20	–	10–14
Water content addition (%)	35	10	–
Impact resistance index	–	200	50
High heating value (MJ/kg)	–	29.49	28.18
References	Hu et al. (2016)	Kong et al. (2013)	Bazargan et al. (2014)

coherence of the adjacent particles, but it also comes into play to calculate the energy consumption. For woody shavings pyrolytic chars, this latter is estimated between 19.9 and 21.02 J/g. Beyond this value, it is a waste of energy and quality, since fractures and dilatations occur without any gain in the volume density.

A minimum amount of water is needed to supply enough contact area and increase cohesive van der Waals forces within the pellets. Humidification ratios ranging from 1:10 to 1:3 are met in the literature (Hu et al., 2016; Kong et al., 2013). Since chars derived from biomass pyrolysis often lack plasticity, a sticking or an agglomerating material, preferably combustible, is added to enable the formation of solid pellets (Rousset, Caldeira-Pires, Sablowski, & Rodrigues, 2011). Indeed, the net heat output depends not only on the amount of charcoal in the pellet, but also on the amounts and combustibility of starch, lignin, soil, animal dung, waste paper, gum arabic, etc. used in binding sawdust and chars (Sotannde, Oluyege, & Abah, 2010).

5.4.2 Briquette production

Char briquettes issued from biomass pyrolysis are usually used in cooking, barbecuing, camping, or heating. A wide range of shapes and sizes is available on the market: oblong, oblate, hexagonal, octagonal, honeycomb, beehive, pillow shaped, etc. (Fig. 6)

Briquetting technologies of biomass-derived chars are different from those of raw biomasses, since chars are highly abrasive and are not suitable for modern large-scale equipment, even those intended for torrefied biomass.

In fact, briquetting requires a binder to be mixed with minute char particles and a press to transform the mixture into a briquette, which is then dried so that it becomes strong enough either to be handled or to be transported. Another way to acquire char briquettes from direct pyrolysis of biomass briquettes is done for pine sawdust

Fig. 6 Different shapes and sizes of chars.

biomass briquettes (Liu et al., 2018) and for maize straw briquettes (Wang, Han, Gao, Wang, & Lu, 2017).

Many technologies are available in a wide capacity range, from very small to very large and with varying degrees of mechanization and automation. For instance, hand presses and screw extruders (manual or electrical) are disseminated across rural areas in developing countries lacking electricity supply (Fig. 7). They require low investment but are very labor intensive with low capacities. Also, briquette quality is relatively low and distribution is limited to neighborhood or own community.

Another small-scale briquetting process applied in several developing countries is agglomeration technology. The charcoal is milled to powder, mixed with binders, and

Fig. 7 Manual and electrical extruder for char briquetting.

then agglomerated, usually with a motor-driven agglomerator that typically delivers spherical and very strong briquettes, intended for domestic and industrial purposes as well (Fig. 8).

In medium- to large-scale briquetting technologies, systems used may be a press unit (roller, heating, hydraulic) or a screw extruder (conical, endless, with die unit), and up to a few tons per day of medium- to high-quality char briquettes may be produced and distributed to regular customers and in a well-organized retail network.

Successful briquette production is found mostly in developed countries (Fig. 9). For example, about 50 tons of fine charcoals derived from waste sawdust and bark are briquetted daily in an industry based in the south of the United States (Division, 1987). On the other side of the planet, and in spite of obvious advantages of char briquettes such as price, burn time, and environmental sustainability, their use instead of solid fuels in sub-Saharan Africa is still very limited, compared to Asian countries such as India or China. Around the world, policies (through stricter regulation and targeted tax exemptions), technical interventions (R&D), and marketing approaches (consumer preferences in fuel and stove attributes) are made (Mwampamba, Owen, & Pigaht, 2013).

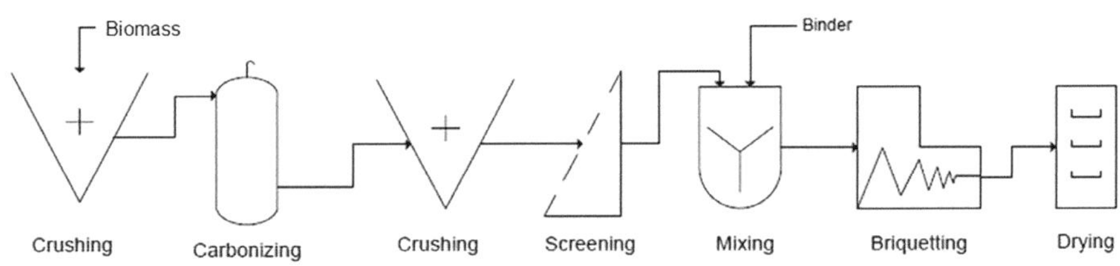

Fig. 8 Flow chart diagram of biomass-derived char briquetting process.
Adapted from Abdu Zubairu, S. A. G. (2014). Production and characterization of briquette charcoal by carbonization of agro-waste. *Energy and Power, 4*(2), 41–47.

Fig. 9 Charcoal briquette plant (China).

Systems used for briquetting may be a press unit (hand, roller, heating, hydraulic) or a screw extruder (conical, endless, with die unit). Numerous biomass chars were briquetted in the literature. Table 4 depicts the properties of some char briquettes.

In all these studies, it was stated that the energy content was increased when compared to briquettes of raw biomass. For some biomass feedstock, depending on the ratio of blend or on the final pyrolysis temperature, heating values exceeded that of charcoal. As an example, Liu et al. (2018) found that the high heating value (HHV) of briquettes of pine sawdust chars, when the final pyrolysis temperature was 400°C, was 30.75 MJ/kg and increased to 35.67 MJ/kg when the final temperature was 800°C. An optimum at 600°C is obtained as a compromise between energy consumption and briquette properties. In their study, during pyrolysis of the biomass briquettes, a CO_2-containing atmosphere was applied with different percentages (0%, 10%, 30%, 60%, and 100%). The increase in CO_2 ratio developed char microporosity, but did not enhance the HHV, the volume density, or the durability of char briquettes.

Char briquetting may also take into account mechanisms of flue-gas desulfurization and denitrification as pointed out by Kim and Li (2005). Indeed, they found that pulp black liquor issued from paper production industries used as a binder reduces approximately 35% of SO_x and 40% of NO_x during the char combustion process.

Investment costs (mill, briquetting equipment, drying, and packaging) and operational costs such as procurement and transport of raw materials (biomass char, sticker, binder), energy, and human resources are the main factors that fix biomass char briquette costs. In Africa, production costs are calculated at 0.15–0.25 Euro/kg of char briquettes, without counting purchase, transport, and labor costs. Indeed, location of the production facility near the raw material and/or the customer base may contribute more than twice to the briquette price.

Wu, Zhang, Wang, Mu, and Huang (2018) tried to prepare barbecue charcoal without binders. To achieve this goal, cotton stalk and wood sawdust were dried, torrefied, and hydrothermally treated at 200°C, 230°C, and 260°C, respectively. The obtained materials were densified into briquettes then carbonized at 400°C to prepare charcoal briquettes. Hydrothermal charcoal briquette physical properties were more interesting than those of the commercial barbecue-bonded charcoal. Mass densities and compressive strengths on the one hand and fixed carbon and ash yield on the other hand were low. The combustion character index (S) (defined in Section 5.5.2 by Eq. 6), which reflects the combustion reactivity of the char during the whole combustion process, was similar for the two biomasses to that of commercial barbecue charcoals and both were the best materials for charcoal briquette making (mass density: 887.2 and 866.7 kg/m^3, HHV: 27.5 and 29.8 MJ/kg, fixed carbon: 68.4 and 64.7 wt%, respectively), when preprocessed at 230°C (Wu et al., 2018).

Table 4 Properties of some biomass char briquettes

Chars derived from	Wood sourced	Pine sawdust	Sewage sludge/fine charcoal	Vegetable market waste	Durian peel	Sugarcane bagasse
Apparent density (kg/m^3)	700	700–750	750–910	509–747	250–290	1120
Compressive strength (MPa)	24.5	15	21–42	–	–	>7
Dimensions (mm)	60 × 50 × 30	$L=20$ $D=10$	$L=200$ $D=70$	$D=50$	$L=60$ $D=12$	$D=30$
High heating value (MJ/kg)	26.5	31.22–35.67	13.21–23.10	10.26–13.70	22.6–24.8	25.551
References	Borowski, Stępniewski, and Wójcik-Oliveira (2017)	Liu et al. (2018)	de Oliveira, Palacio, da Silva, Mariani, and Reinehr (2017)	Srivastava, Narnaware, Makwana, Singh, and Vahora (2014)	Hirunpraditkoon et al. (2014)	Teixeira, Pena, and Miguel (2010)

5.5 Combustion characteristics

5.5.1 Combustion of biomass-derived chars

To assess performance, combustion indices such as fuel ratio (FR), combustibility index (CI), and volatile ignitability (VI) may be determined. These indices are calculated using the following equations:

$$FR = \frac{\text{Fixed carbon}^{db}}{\text{Volatile matter}^{db}} \tag{2}$$

$$CI = \frac{HHV^{db}}{FR} \times \left(115 - \text{Ash}^{db}\right) \times \frac{1}{105}\,(\text{MJ/kg}) \tag{3}$$

$$VI = \left[\frac{HHV^{db} - 0.338\,\text{fixed carbon}^{db}}{\text{Total volatiles}}\right] \times 100\,(\text{MJ/kg}) \tag{4}$$

It is usually assumed that when FR values are greater than 2, problems of ignition and flammability may occur. Indeed, a low fuel ratio suggests that the solid fuel easily ignites; degradation is thus rapid and difficult to control, resulting in incomplete combustion of the fuel and in smoke emissions.

A suitable value of CI, which is used to evaluate the compatibility of a material for mixed combustion with coal, is 23 MJ/kg. Finally, the minimum for VI should be 14 MJ/kg, which is the available energy provided by the total volatiles, assuming that the fixed carbon constituting the fuel sample is solely made up of pure carbon (Conag, Villahermosa, Cabatingan, & Go, 2017).

Conag et al. (2017) concluded that torrefaction of sugarcane bagasse at 250°C would require a longer time (>60 min) to lower the volatile matter content to meet the desired FR. Meanwhile, torrefaction at 300°C and 350°C only required about 30 min for the volatile matter to decrease to an amount to achieve the desired FR. In view of the CI, the initial value of the untreated sugarcane bagasse was reduced from 82 to 16 MJ/kg. Torrefaction at 350°C for at least 30 min resulted in the reduction of the CI to the desirable value, but also in the increase in the VI from 11 to 16 MJ/kg. All three torrefaction temperatures were able to improve the VI to the desired value but would require a longer torrefaction time of 45 min at 250°C and as short as 15 min at 350°C.

In addition, bulk densities were found to be suitable for use as either briquettes (de Oliveira et al., 2017; Rousset et al., 2011; Sotannde et al., 2010) or as direct feed in fluidized combustion chambers (Vamvuka, Zografos, & Alevizos, 2008) and fixed beds (Mitchell et al., 2016). Sugarcane bagasse char kept under 6% moisture content even after 15 days of sun exposure was found to be less hygroscopic. It was concluded that sugarcane bagasse could be well accepted in charcoal production in areas where these are produced in large volumes (Conag et al., 2017).

To help attain cleaner and more efficient combustion of biomass chars, Qi, Han, Wang, and Gao (2017) loaded additives to the samples and found that calcium oxide and ammonium phosphate dibasic have reduced alkali content to different degrees, while calcium carbonate promoted the retention of alkali metals. It is well known that prior pyrolysis reduces SO_2 emissions during char combustion, but adding the three earlier-mentioned additives during pyrolysis progress brings them to almost zero emissions. These additives also enhanced the NO emission to 1.2–3.5 mg/g, which is still under the limit. Yet, the three additives were judged economic and suitable for industrial application (Qi et al., 2017).

Laboratory-scale drop-tube furnaces were used to study char as fuel in many works (Fig. 10). Mallee bark raw biomasses and their six derived char samples were combusted in such equipment at 1300°C. Particle and gaseous emissions were analyzed and the particle size distributions of particulate matter $(PM)_{10}$ were found to be bimodal and unimodal, respectively, for raw biomass combustion and for char combustion. It is widely admitted that during pyrolysis, the majority of inorganic species is retained in the biochar; yet, the combustion of this latter significantly reduces PM_1 (and the mass of Na, K, and Cl in PM_1) when compared to biomass combustion, as a consequence of volatile and Cl removal from the raw biomass during pyrolysis. At the same time, a considerable increase in the emission of PM_{1-10} as well as of Mg and Ca masses in PM_{1-10} is declared during the char combustion process, due probably to its more porous structure and increased ash loadings (Gao & Wu, 2011).

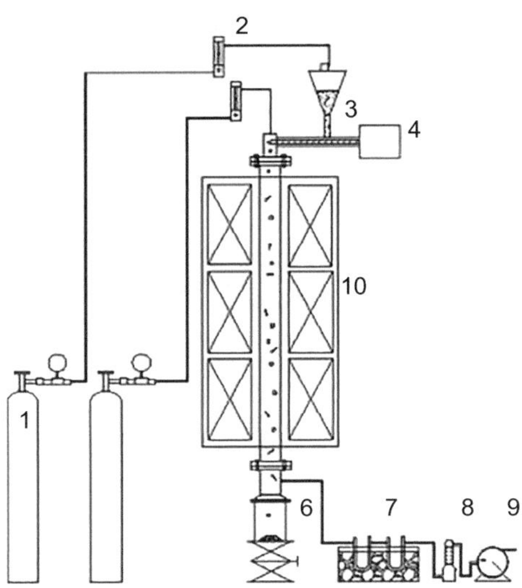

(1) N_2 cylinder; (2) Rotameter; (3) Hopper; (4) Screw feeder; (5) Drop tube reactor; (6) Char collector; (7) Condensing unit; (8) Filter unit; (9) Cumulative flow meter; (10) Furnace.

Fig. 10 Schematic diagram of a drop-tube furnace.
Reprinted with permission from Li, S., Chen, X., Wang, L., Liu, A., Yu, G. (2013). Co-pyrolysis behaviors of saw dust and Shenfu coal in drop tube furnace and fixed bed reactor. *Bioresource Technology, 148*, 24–29. Copyright 2018, Elsevier.

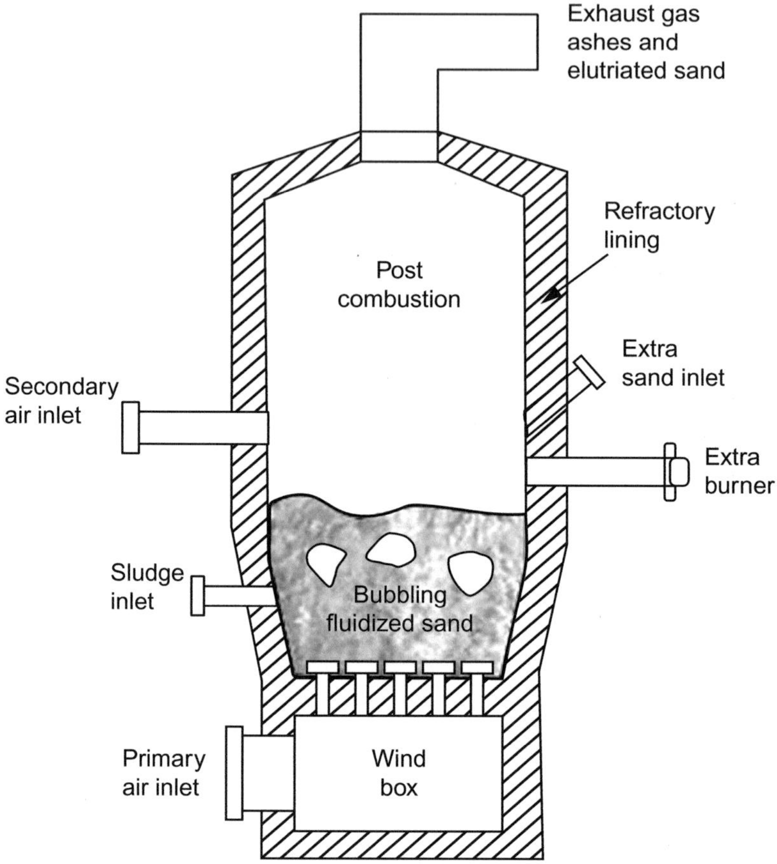

Fig. 11 Experimental setup of a fluidized-bed reactor.
Reprinted with permission from Khiari, B., Marias, F., Zagrouba, F., Vaxelaire, J. (2008). Transient mathematical modelling of a fluidized bed incinerator for sewage sludge. *Journal of Cleaner Production, 16*, 178–191. Copyright 2018, Elsevier.

Combustion of char was also investigated at bench scales. In a fluidized-bed combustor, for example, fixed carbon of *Robinia pseudoacacia* conversion to CO or CO_2 was almost always complete, half via direct coarse char combustion and half via elutriable carbon fines generation of coarse char and their subsequent postcombustion during their residence time in the bed (Scala, Salatino, & Chirone, 2000). This efficiency was reached thanks to the high reactivity of the char and the long fines residence time in the reactor. Similar conclusions were reached by Chouchene, Jeguirim, Khiari, Trouvé, and Zagrouba (2010) during the oxidation of char derived from olive wastes in a fixed-bed reactor (Chouchene et al., 2010).

Scala, Chirone, and Salatino (2006) also investigated fluidized-bed combustion (Fig. 11) of three other biomass chars issued from pine seed shells, olive husk, and woodchips in a bench-scale combustor, with a focus on fixed carbon fate and on relevance of attrition phenomena to char particle size distribution in the bed and their relative combustion rate (Scala et al., 2006). The same conclusions as Scala et al. (2000) were reached concerning extensive primary and secondary fragmentation of char particles, except that this time 25%–45% of fixed carbon consumption was due to percolative fragmentation. The biogenous character of the char fuels played a significant role in the afterburning of the attrited fines. Also, particle shrinkage was significant for all three samples after devolatilization, and the materials developed a highly porous structure characterized by macropore dominance (Scala et al., 2006).

Two other biomass chars issued from pine and switch grass were subjected to combustion at 1600 K in a laminar flow reactor (Wornat, Hurt, Yang, & Headley, 1995). The two solid residues underwent numerous structural and compositional modifications as combustion proceeded. First, devolatilization led to the removal of amorphous material and the release of oxygen- and hydrogen-rich gases. Then, several inorganic transformations occurred: vaporization (namely potassium and sodium), surface migration and coalescence, embodiment of metals, such as calcium, into silicate, etc. By the end of the combustion, the amorphous structure of inorganic compounds shifted to crystalline forms.

Biomass chars also fueled chemical looping combustors (CLC), the goal of which is to capture carbon at minimum costs. Linderholm, Schmitz, Knutsson, Källén, and Lyngfelt (2014) published findings from a 100 kW CLC fed by Swedish wood char and Mexican petcoke. The carbon capture rate was high with wood char (93%–97%) but lower with petcoke. Gas conversion was also declared high for the former (90%–95.3%), especially when compared to the 84% yielded by bituminous coal in the same conditions (Linderholm et al., 2014).

5.5.2 Combustion of densified biomass-derived chars

The combustion performance of densified biomass chars can be evaluated by a number of parameters such as the ignition and burnout temperatures (T_i and T_f), and the maximum and average combustion rates (V_{max} and V_{mean}).

Ignition index (D_i) and combustion characteristic index (S) are also commonly used to assess the general performance of biochar briquettes and are defined as (Han et al., 2016; Niu, Chen, Li, & Xue, 2016):

$$D_i = \frac{V_{max}}{T_i T_{max}} \left(\%/\left(\min \cdot {}^\circ C^2\right)\right) \tag{5}$$

$$S = \frac{V_{max} \times V_{mean}}{T_i^2 T_f} \left(\%^2/\left(\min^2 \cdot {}^\circ C^3\right)\right) \tag{6}$$

D_i indicates how fast the fuel is ignited, whereas S expresses the fuel comprehensive characteristics. The higher are D_i, V_{mean}, and S, the better is the fuel combustion activity.

Mass and energy yield ratios (η_M and η_E) of a given char briquette are calculated on a dry ash basis (Bridgeman, Jones, Shield, & Williams, 2008; Park & Jang, 2012) by:

Table 5 Combustion properties of biomass char briquettes

Sample	HHV (MJ/kg)	η_M (%)	η_E (%)	D_i (%/(S °C^2))	S (10^{-9} %^{2}/(S^2°C^3)
Maize straw briquette 250	17.79	79.2	83.28	1.13	0.97
Maize straw briquette 250 +$NH_4H_2PO_4$	–	–	–	1.00	0.98
Maize straw briquette 250+Ca$(H_2PO_4)_2$	–	–	–	1.19	1.38
Maize straw briquette 550	21.05	33.32	41.46	0.75	0.13
Maize straw briquette 550 +$NH_4H_2PO_4$	–	–	–	0.75	0.05
Maize straw briquette 550+Ca$(H_2PO_4)_2$	–	–	–	1.01	0.14
Torrefied rice husk briquette 250	15.89	77.5	86	–	–

$$\eta_M = 100\frac{M_t}{M_0} \tag{7}$$

$$\eta_E = 100\eta_M\frac{HHV_t}{HHV_0} \tag{8}$$

With the help of these parameters, Wang, Aparu, Kurogawa, and Sugiyama (2014) (Wang, Han, Gao, Li, & Lu, 2017) investigated the combustion properties of briquettes of maize straw char obtained at different final pyrolysis temperature. Some of the results are summarized in Table 5 where rice husk biomass char pellets are also presented. In another study, these same parameters helped authors to state that blending maize straw char with phosphorous additives in briquettes did not ameliorate combustion characteristics, although it was thought first that the mixing of mineral-based additives may reduce problems with respect to alkali compounds present in green biomass such as bed agglomeration, fouling, and corrosion in combustion reactors (Boström et al., 2012; Vamvuka et al., 2008; Werkelin, Skrifvars, Zevenhoven, Holmbom, & Hupa, 2010).

Manures of cow and poultry are other biomasses that have been proved to efficiently displace solid fossil fuels when briquetted and charred (Wang et al., 2014).

In a blast furnace, Kong et al. (2013) comparatively evaluated the combustion characteristics of sawdust and pellets of char derived from sawdust to investigate the conversion from raw mass to densified char. They found that pyrolysis plays a positive role in influencing the combustion properties of the char sawdust pellets, especially because both weight loss and NO_x emissions of torrefied pellets decreased in comparison with raw sawdust. This last fact was explained by the release of volatiles or N-containing volatiles such as HCN and NH_3 during pyrolysis of char, whereas these former volatiles are quickly combusted within the raw cane (Kong et al., 2013).

Hirunpraditkoon et al. (2014) compared combustion of commercial coconut shell briquettes with durian peel charcoal briquettes (Hirunpraditkoon et al., 2014). Ignition time of the former was twice longer than the latter at all different mass ratios. For both, briquette sample burning starts from the edges to the center. Approximately 30 min later, both briquettes were covered with ash on their surface. Total combustion time of durian peel charcoal briquette was around 140 min, which was 30 min shorter than the commercial coconut shell charcoal briquette.

Combustion emissions of different charcoals issued from pyrolysis of different biomass residues are reported in Table 6. These factors are issued from briquettes, pellets, and logs of charcoals that have been burned in a tube furnace under different temperature and airflow conditions to simulate charcoal combustion at high temperature (Huang, Lee, & Wu, 2016).

As for modeling, the combustion process of char briquettes happens in two consecutive steps: volatile then char combustion. The first stage happens over the whole pellet, whereas the second proceeds in a shrinking-core mode, with a flame sheet moving from the surface toward the center of the biomass char briquette (Kim, Lu, Naruse, Yuan, & Ohtake, 2000).

Household market requirements are different from those in industrial applications as far as packing styles, marketing habits, technical specifications, etc. in biomass char production are concerned. Thus, the two biomass char usages (domestic and industrial) are exposed separately in dedicated sections.

5.6 Domestic-scale applications

Nearly half of the world's population uses solid fuels and more than a third relies on solid biomass energy (wood, grass, crop straw, dung, etc.) for household activities, mainly in developing countries where meals are cooked and homes are heated with homemade traditional stoves or open fires.

The direct firing of raw solid fuel in traditional stoves emits hazardous particles and pollutant gases (particle matter, black carbon, CO, NO_x, SO_2), causing health risks (indoor air pollution) and environment problems (greenhouse gases, air pollution, deforestation). In addition, with inefficient stoves, more time is required for cooking and fuel gathering, a burden usually endured by women and children, raising social problems such as diverting time away from education, equity, employment, and gender issues (Lewis & Pattanayak, 2012).

Yet, household stoves are not likely to be replaced soon in many areas due essentially to limited and tough economic conditions. Attempts to convert households from these fuels to modern ones or from traditional stoves to more efficient and cleaner burning stoves are being made. According to Winijkul and Bond (2016), clean fuel may reduce emissions by 18%–25% depending on pollutant, whereas stove improvements with the best available techniques may lower emissions by 25%–82% and PM by 95% (Winijkul & Bond, 2016). The following sections try to summarize works aiming to overcome the earlier cited deficiencies, where char issued from biomass

Table 6 Emission factors of air pollutants for combustion of different types of charcoal

Charcoal	Dimensions (mm)	CO (g/kg)	CO_2 (g/kg)	NO_x (mg/kg)	HC (g/kg)	$PM_{2.5}$ (mg/kg)
China brick shape		128.2 ± 22.1	776.0 ± 45.8	834.1 ± 49.1	3.05 ± 0.74	235.3 ± 27.5
China spheroid briquette	$D=40$ mm	300.3 ± 31.3	986.8 ± 60	161.1 ± 98.6	5.27 ± 0.39	368.9 ± 47.3
Indonesia sawdust hexagon briquette with hollow core	$D=40$ mm $d=10$ mm	214.7 ± 16.6	878.7 ± 96.3	462.1 ± 68.4	1.93 ± 0.26	153.9 ± 11.4
Indonesia mangrove log	$D=30$–60 mm $L=150$–200 mm	108.5 ± 19.3	855.1 ± 92.1	277.0 ± 44.2	9.30 ± 2.99	955.7 ± 112
Indonesia spheroid briquette	$D=40$ mm	67.7 ± 56.4	723.7 ± 43.2	413.6 ± 149.3	4.05 ± 0.56	282.6 ± 91.6
Southeast Asian binshotan log	$D=20$–40 mm $L=100$–150 mm	76.6 ± 14.4	1225.2 ± 39	831.9 ± 295.9	3.92 ± 1.19	12 ± 2.4

pyrolysis is concerned. Indeed, as stated above, the use of raw or unprocessed biomass is associated with bad energy efficiencies and high emissions. However, pretreated biomass such as pyrolytic char or charcoal needs stove optimization (Bailis, Berrueta, et al., 2007; Bailis, Smith, and Edwards, 2007; Bhattacharya, Albina, & Myint Khaing, 2002).

5.6.1 Cooking stove

Inefficient combustion with unsustainably harvested biomass in cooking stoves results in the highest-potential global warming gases, such as CH_4 and CO (Sagar & Kartha, 2007) and in 22% of global emitted black carbon (Ramanathan & Carmichael, 2008), which, unlike globally distributed greenhouse gases, results in localized impacts because of its short lifetime (Smith et al., 2009). Also, because some cooking briquettes are produced using raw materials collected from dumping sites, heavy metal contamination is common in rural areas (Wang et al., 2014). Consequently, important efforts are being made worldwide to increase affordable, clean, efficient, safe, and reliable home cooking and heating practices (Sevault et al., 2017). Stove improvements include adding chimneys, retaining heat, plugging fans, etc. Prices range from $2 in Ethiopia to $15 for rocket stoves in Guatemala.

In fact, a wide variety of cooking stoves and charcoal-fired cooking stoves is available (Bhattacharya, Albina, & Myint Khaing, 2002; Gallagher, Beard, Clifford, & Watson, 2016; Lask et al., 2015; MacCarty, Still, & Ogle, 2010). It is difficult to compare the performances and emissions of stoves originating from different places and for different cooking habits unless standardized and replicable protocols are set such as the Cooking Control Test (Bailis), Kitchen Performance Test (Bailis, 2007), and Water Boiling Test (WBT) (Assistance, 1985).

This last method was adopted by many researchers (Jetter & Kariher, 2009; Lask et al., 2015; MacCarty et al., 2010), but as the WBT was originally designed for wood-burning stoves, some modifications were made to accommodate charcoal stoves (Booker et al., 2011). For instance, WBT was used by a voluntary Partnership for Clean Indoor Air with the help of the Aprovecho Research Center, Cottage Grove, Oregon, which took the challenge to design cooking stoves with improved fuel efficiency, lower pollutant emissions, and smaller-mass durable and affordable components. To achieve this goal, a number of cooking stoves and char fuel combinations were tested.

In the following text, two African charcoal stoves, namely UCODEA (Urban COmmunity DEvelopment Association) and Lakech, are described (Fig. 12) (Jetter & Kariher, 2009). The UCODEA charcoal stove, widely used in Kampala, Uganda, has a metal body with a ceramic liner and grate to hold the hot charcoal. Flowing air is controlled by two doors on the side near the bottom. The Lakech stove, used in Ethiopia, has almost the same technical characteristics except that it is shorter, with a smaller grate, and that it has only one door.

Compared to all other stoves tested in this research, the UCODEA and Lakech charcoal stoves consumed the least fuel thanks to higher energy per unit of mass. Compared to most of the wood-burning stoves under study, both charcoal stoves emitted

UCODEA stove LAKECH stove

Fig. 12 Two African cooking stoves.
UCODEA stove – Reproduced with permission from https://www.designwithoutborders.com/ugastove; LAKECH stove – From Obeng, G.Y., Mensah, E., Ashiagbor, G., Boahen, O., Sweeney, D.J. (2017). Watching the Smoke Rise Up: Thermal Efficiency, Pollutant Emissions and Global Warming Impact of Three Biomass Cookstoves in Ghana. *Energies, 10,* 641.

high CO, total hydrocarbon, and $PM_{2.5}$ and liberated larger amounts of visible smoke during the cold start compared to the hot start (Fig. 13) (Jetter & Kariher, 2009).

The additional height of UCODEA might have provided more air draft for better combustion and lower emissions. Two other improved charcoal stoves (Merchaye and Lakech) in Ethiopia showed decrease in emissions of CO, CO_2, and $PM_{2.5}$ (by 28%, 22%, and 27%) and (15%, 8%, and 13%), respectively, in comparison with a traditional charcoal stove (Mamuye, Lemma, & Woldeamanuel, 2018).

The importance of ventilation was also the conclusions of Masekameni, Makonese, and Annegarn (2016) who evaluated the performances of three other African charcoal stoves.

As seen from Table 7, Kenyan Jiko and Madagascan charcoal stoves produced comparable emissions, both higher than the Zambian one, since this last one is highly ventilated. The limitation of the Zambian stove is in terms of excess firepower, exceeding the level required for domestic cooking.

To contribute to finding a viable option for safe cooking, Njenga et al. (2013) prepared briquettes from charcoal dust bonded with paper, soil, or corn starch and from

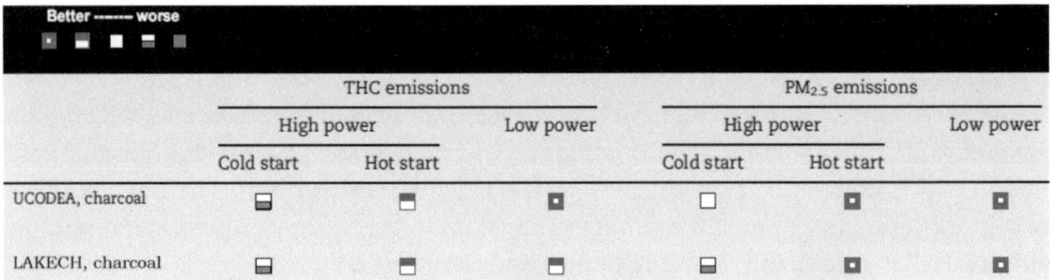

Fig. 13 Total hydrocarbon (THC) and particulate matter $(PM)_{2.5}$ emissions.
Reprinted with permission from Jetter, J. J., Kariher, P. (2009). Solid-fuel household cook stoves: characterization of performance and emissions. *Biomass and Bioenergy, 33*(2), 294–305. Copyright 2018, Elsevier.

Table 7 Emissions and firepower of different stoves

Stove	$PM_{2.5}$ (mg/MJ)	CO (g/MJ)	CO_2 (mg/MJ)	Firepower (kW)
Kenyan Jiko	6.7 ± 0.2	5.2 ± 0.2	553.1 ± 22	3.1 ± 0.2
Madagascan	7.5 ± 0.2	5.4 ± 0.1	562.5 ± 11	2.9 ± 0.2
Zambian	5.2 ± 0.1	4.3 ± 0.1	573.3 ± 25	7.3 ± 0.2

sawdust bonded with gum arabic (Njenga et al., 2013). The quality of briquettes produced from charcoal dust and sawdust presents them as a good source of cooking fuel. The calorific values of the briquettes were acceptable at 23.6 and 21.4 MJ/kg, respectively. Soil contamination was found to increase ash content in charcoal dust briquettes. Contaminants like Hg, Pb, and Cr were high in briquettes made from dumpsites. Charcoal dust briquette bonded with soil combustion was the safest in terms of indoor CO, CO_2, and fine PM concentrations. Burning sawdust briquettes bonded with gum arabic also released high fine $PM_{2.5}$.

Different traditional and improved designs of cooking stoves from different Asian countries were tested (Fig. 14). Their efficiencies (Table 8) varied widely (12%–27%, the highest recorded for wood fired was 19.7%) (Bhattacharya, Albina, & Abdul Salam, 2002).

The efficiency can be expressed as follows:

- during high-power phase:

$$\text{efficiency}\,(\%) = \frac{M_w C_w (T_f - T_i)}{H_c (M_{ci} - M_{cf})} \tag{9}$$

- during low-power phase:

$$\text{efficiency}\,(\%) = \frac{H_w (M_{wi} - M_{wf})}{H_c (M_{ci} - M_{cf})} \tag{10}$$

where:

M_w: average water mass in the cook pot from prestart to first boiling
C_w: heat capacity of water (4.184 kJ/kg °C)
T_f: temperature of first boiling (°C)
T_i: initial temperature of water in the pot (°C)
H_w: heat of vaporization of water (2.26 MJ/kg)
M_{ci}: initial mass of char in the cooking stove (g)
M_{cf}: final mass of charcoal in the cooking stove (g)
H_c: energy content of charcoal (29 MJ/kg)

As for emissions per kg of fuel, lesser rates of CO_2, CO, and NO_x and comparable quantities of CH_4 and total nonmethane organic compounds are measured for wood stoves in front of charcoal stoves (Table 9).

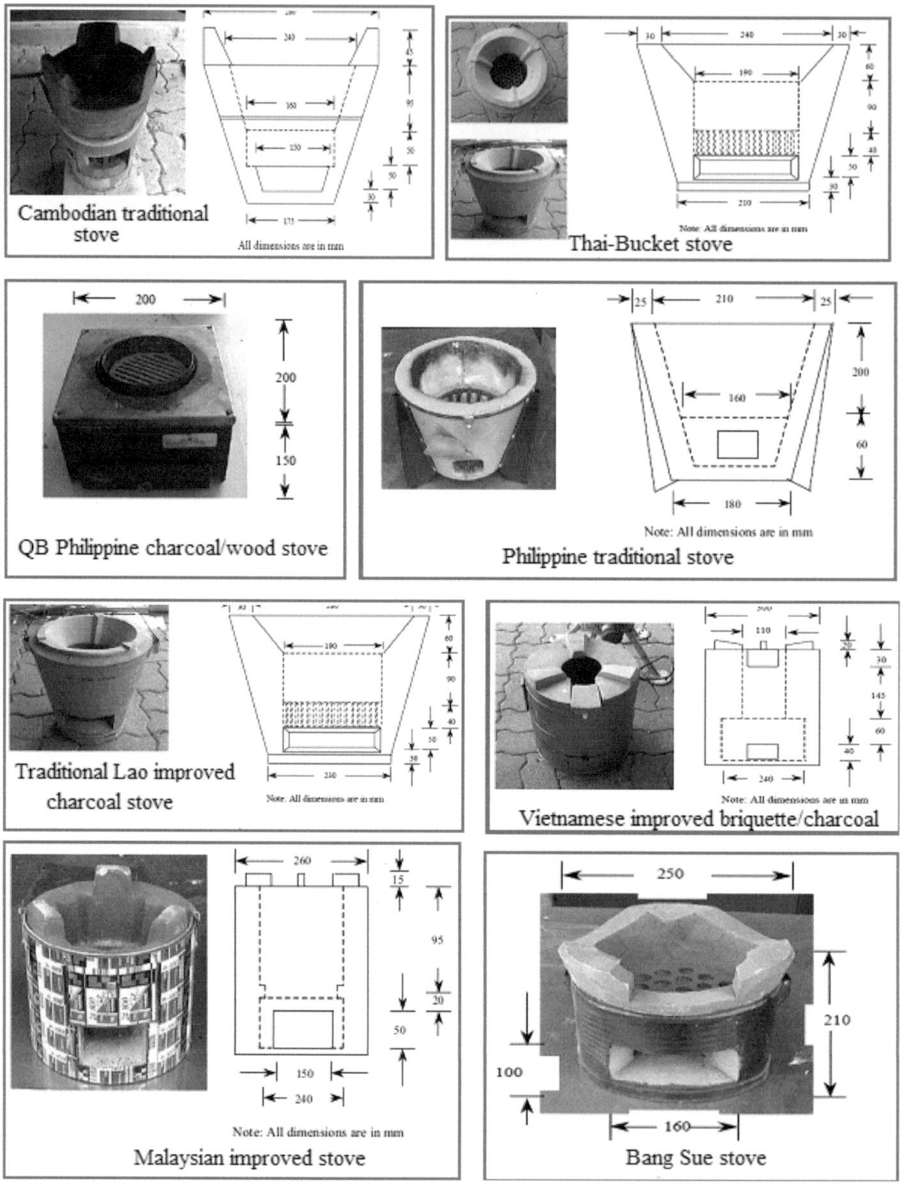

Fig. 14 Some Asian cooking stoves.
Reprinted with permission from Bhattacharya, S.C., Albina, D.O., Abdul Salam, P. 2002. Emission factors of wood and charcoal-fired cookstoves. Biomass and Bioenergy, 23(6), 453–469. Copyright 2018, Elsevier.

Table 8 Efficiencies of stoves using charcoal and wood (Bhattacharya, Albina, & Abdul Salam, 2002)

Name of cooking stove	Charcoal-fired stove efficiency (%)	Wood-fired stove efficiency (%)
Cambodian traditional	14.5	11
Thai-bucket stove	16.2	14
Chinese traditional	12.5	12.2
QB Philippine charcoal/ firewood	27	23
Philippine charcoal/firewood	21.5	12
Lao improved	16.5	18.4
Vietnamese improved briquette/charcoal	25	17.5
Malaysian improved	18	19.7
Bang Sue stove	18.2	18.2

5.6.2 Space heating stove

Very few studies have been performed with space-heating char stoves. An example is the one studied by Koyuncu and Pinar (2007) (Koyuncu & Pinar, 2007) and presented in Fig. 15. One section of their paper was dedicated to the flue-gas emissions and the thermal efficiency of an improved space-heating charcoal stove. The charcoal used had the characteristics mentioned in Table 10.

The authors compared different biomass fuels with charcoal and found that charcoal was the most appropriate for use in space-heating biomass stoves since it had the least total flue-gas emissions: more precisely less smoke, less NO_x, and less SO_2 (Fig. 16). In addition, the thermal efficiency of the stove, when fed with charcoal, was approximately 46%, which classified it in the category of improved stoves (Koyuncu & Pinar, 2007).

As far as polycyclic aromatic hydrocarbon compounds are concerned, charcoal burnt in an insulated Thai double-skinned stove also emitted much less in terms of factors, rates, and concentrations in the smoke than other biomass fuels in the same operating conditions (Kim Oanh, Bætz Reutergårdh, & Dung, 1999).

5.7 Industrial applications

Char products from biomass pyrolysis can be utilized in all industrial applications where other fuels are used, whether they are solid, liquid, or gas, namely in drying, gasification, steam, and electricity generation. Chars can be burnt alone in dedicated burners or in combination with other fuels. They can be used as a solid for cofiring directly with pulverized coal as in electric power plants or metallurgical units and as an upgraded feedstock for densified and pyrolyzed biomass fuels such as in brick making, lime production, and pottery.

Table 9 Emission factors ($g\,kg^{-1}$ fuel used) (Bhattacharya, Albina, & Abdul Salam, 2002)

	CO		CO_2		CH_4		TNMOC		NO_x	
Name of cooking stove	**Charcoal**	**Wood**	**Charcoal**	**Wood**	**Charcoal**	**Wood**	**Charcoal**	**Wood**	**Charcoal**	**Wood**
Cambodian traditional	34.2	25.3	2352	1580	7.7	2.7	6.5	22.4	0.07	0.01
Thai-bucket stove	35.7	26.4	2155	1596	6.8	10	5.8	8.9	0.03	0.12
Chinese traditional	175	24.4	2436	1570	7.8	5.4	8.5	8.4	0.3	0.1
QB Philippine charcoal/ firewood	198	45.1	2276	1568	8	9.7	9.7	9.5	0.22	0.22
Phillipine charcoal/firewood	155	28.6	2567	1603	7.8	8.9	8.5	7.5	0.14	0.17
Lao improved	134	51.9	2451	1565	9.8	6	6.3	10	0.19	0.19
Vietnamese improved	87.2	47.1	2233	1577	10.8	5.3	4.8	7.6	0.3	0.15
Malaysian improved	155	18.7	2576	1603	8.2	13.7	6.2	9.3	0.43	0.11
Bang Sue stove	178	22.1	2555	1585	8.7	12.4	7.8	9.5	0.42	0.11

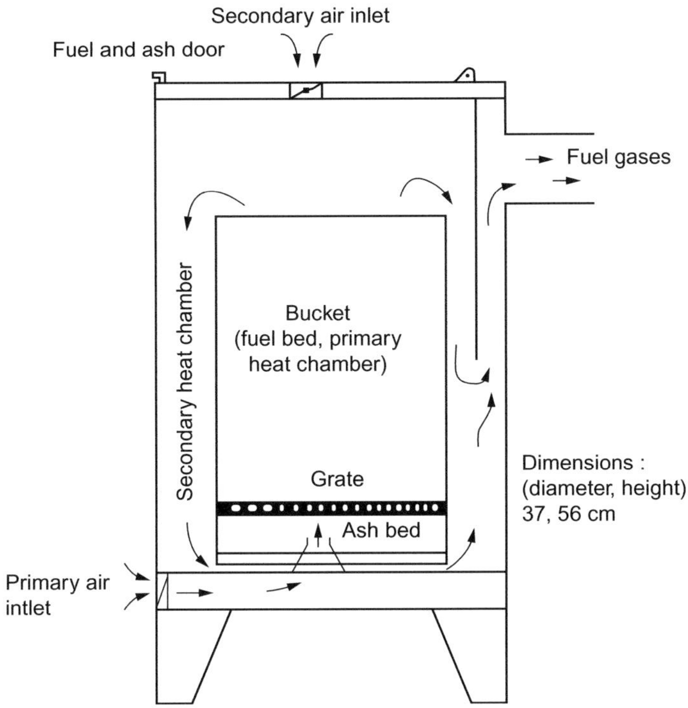

Fig. 15 Schematic representation of a charcoal-heating stove.
Reprinted with permission from Koyuncu, T., Pinar, Y. 2007. The emissions from a space-heating biomass stove. *Biomass and Bioenergy, 31*(1), 73–79. Copyright 2018, Elsevier.

Table 10 Characteristics of charcoal used in Koyuncu and Pinar (2007)

Moisture (%)	Ash (%)	Density (kg/m^3)	Calorific value (MJ/kg)	Burning rate (kg/h)
4.14	5.86	291	19.25	1.13

Char viability in industries is due to their high ratio of carbon to ash, high surface area, stable pore structure, and good reduction ability. Also, chars are almost smokeless and have relatively few and unreactive inorganic impurities (Emrich, 1985). Moisture, little ash, and low sulfur and nitrogen of chars make them ideal contenders for commercial coals, especially because no expensive posttreatment equipment is needed. Another advantage is their short plant start-up time in comparison with uncarbonized biomass fuels.

Char quality depends on the particle dimensions of the initial feedstock, the pyrolytic kiln used, the process considered (Guo & Lua, 2000), and the original plant species. As far as this last aspect is concerned, it has been found, for example, that the structure of chars was rather amorphous in the case of silver fir and stone fir

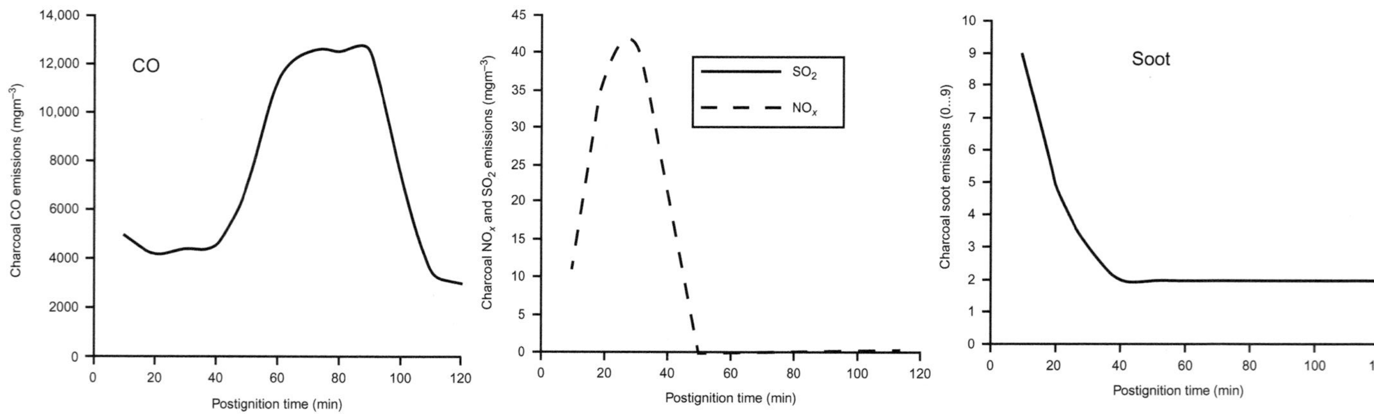

Fig. 16 Emissions of a charcoal-heating stove.
Reprinted with permission from Koyuncu, T., Pinar, Y. (2007). The emissions from a space-heating biomass stove. *Biomass and Bioenergy, 31*(1), 73–79. Copyright 2018, Elsevier.

(López, Centeno, García-Díaz, & Alguacil, 2013), and crystalline in holm oak (Valenzuela-Calahorro, Bernalte-Garcia, Gómez-Serrano, & Bernalte-García, 1987) and Pyrenean oak (López et al., 2013).

5.7.1 Application in iron and steel making

Since ancient times, charcoal was extensively used by blacksmiths, steel makers, and physicians. Charcoal is still in demand in today's iron-making and steel-making industries, Brazil is the leader in charcoal use within the metallurgical sector, with injection rates in small blast furnaces up to 190 kg/t of hot metal (Suopajärvi, Pongrácz, & Fabritius, 2013) thanks to short-rotation forests used as feedstock (Suopajärvi, Kemppainen, Haapakangas, & Fabritius, 2017).

In Russia, however, charcoal is the most expensive type of fuel for combustion in furnaces, so only high-quality materials mixed with other alloys are used in this country. In fact, char systems do have similar energy and enthalpy inputs to conventional iron- and steel-making systems, with lower ash content, higher volatiles, lower mechanical stability, and substantially reduced greenhouse gases. In the condition that feedstock is sustainably supplied, goals like energy saving, emission cutting, and cost reduction could be achieved to different extents. As a consequence, charcoal and torrefied biomass feedstock are used in many stages of the process.

Indeed, chars can be blended with coal for coke making. It can be used simultaneously for iron ore agglomeration and sintering processes to replace coke fines, which lowers SO_2 and NO_x emissions. In the iron-making process, ground char can be injected into blast furnaces from the same tuyères as pulverized coals. Moreover, a certain quality of biomass chars may reduce efficiently the coke rate if loaded into the blast furnace. Another usage of chars can be found in coal-based direct and smelting reduction. The resultant pure and high-quality hot metal is used in the steel-making process.

Finally, char and charcoals can successfully take part in other annexed activities such as prereduction and magnetization of iron ore, baking pellets, and preheating of furnaces.

Further details can be found in Chapter 11, which is wholly dedicated to the use of chars in the steel- and iron-making industries.

5.7.2 Application in electricity generation

On a life-cycle basis, electricity produced from char (or char mixed with coal) emits significantly less CO_2, SO_x, and Hg than coal. Moreover, new rules around the world on the disposal of coal combustion residuals from electricity generation place a certain technical requirement for safe disposal of coal ash. Char ash generally has high concentrations of important plant mineral nutrients, which have marketable agroeconomic benefits. Extending generation life assets and avoiding stranded and early remediation costs are among other financial benefits of biomass char use in power plants. This is truer in rural areas, where small-scale char-based thermal power plants may resolve supply and transmission issues, and where small industries like dairy milk production or cereals milling can be initiated (Kumar & Gupta, 1998).

In this vein, pellets of corncob char are cofired with coal in the 17-MW Willmar municipal power plant by the Chippewa Valley Ethanol Company (CVEC) in Benson with the help of the Agricultural Utilization Research Institute in Minnesota, United States, which decreases power plant emissions and aids compliance with the new mandates.

According to Cunningham Michael and Linda (2016), fecal sludge char could potentially be used as a solid fuel in industries with large and consistent fuel demands such as coal power plants and cement kilns (Cunningham Michael and Linda, 2016). One of the main potential benefits is increased revenue; in Kampala, Uganda, dried fecal sludge is sold for $10 per ton as a soil conditioner, while coconut husks and sawdust are being sold at $150 per ton when transformed into solid fuels (Gold et al., 2017).

For assessing alternative pyrolysis char use options, several scenarios of cycle assessment analysis were adopted by different authors. For example, part of fossil coal in power plants is displaced by biomass chars in a way that the energy content of the biochar is equal to that of the coal being substituted. Fossil coal displacement by cofiring is the scenario with the most important and significant environmental impact reduction due to the initial hazardous effects associated with mining and combustion of mineral coal and the negative impacts of agricultural activity and electricity consumption. The scenario of coal substitution in a power plant together with a more efficient gas cleaning system would mitigate the positive effect on the environment (Peters, Iribarren, & Dufour, 2015).

Another scenario tested the effect of using poplar char to replace natural gas in heat and power generation. Using char as a substitute for natural gas instead of coal increases the minimum selling price from $1.35 to $1.36 per normal liter and decreases greenhouse gases from 38.9 to 32.8 gCO_2eq/MJ (Kulas, Winjobi, Zhou, & Shonnard, 2018).

5.7.3 Application in DCFC

An innovative application of char as a fuel was recently tested within a direct carbon fuel cell (DCFC). It consists of its integration in molten carbonate fuel cells to produce electricity through direct conversion from the chemical energy contained in the char to electrical output. DCFC thermodynamic advantages at high temperature are profitable to transform the solid phase of the feedstock into the gaseous phase within the products. The potential of coupling fuel cell stack with biomass char was proved for power generation, thanks to the high-power density levels achieved on DCFC, greater by 60%–70% than that of mineral coal under the same conditions (Ahn et al., 2013).

Lim et al. (2014) also found that palm shell pyrolyzed at 750°C, with a heating rate of 10°C/min and residence time of 1 h, yielded a highly potential char for DCFC applications (Lim et al., 2014).

5.8 Conclusions, perspectives, and challenges

It has been demonstrated that biomass-derived char is a high-quality fuel that could totally or partially displace fossil-based fuels thanks to high carbon content, interesting HHV, and low particle and gas emissions. These advantages are because of several physical and chemical characteristics, such as sustainability, carbon neutrality, low sulfur, and nitrogen contents, interesting porous medium qualities, as well as good combustibility and reactivity.

As a consequence and to date there is a real capacity for biomass char to be a market-driven uptake of the technology, unless the relevant technoeconomic processing challenges are overcome.

Research is mostly concentrated on woody charcoal, and little work is done on other biomass residues, which may be less expensive and more widely apportioned. However, adequate supplies to feed medium to large industrial units are a real concern for the system operators.

Also, when it comes to pricing, it is clear that chars are not currently able to compete with coal in terms of economics in industrial applications (Norgate, 2009). From an economic point of view, it was suggested that starting a new venture has a higher chance of success when it is developed as an add-on to an existing business so that production facilities (land, building, back office staff, etc.) can be shared. An alternative is that donors contribute funding for a start-up period (Frederik, Sven, Dane, & Wolter, 2013).

At another level, biomass chars may cause some technical inconveniences in combustors. For instance, irregular morphology can complicate operations such as handling or feeding. Also, high combustion reactivity may rise because of the disordered nature of the carbon structure on the one hand and the presence of alkali and alkaline earth metals on the other hand (Wornat et al., 1995). Reliability and maintenance issues are also of consideration, especially because of high concentrations of alkalis and silicon in the feedstock (Ellem Gary & Mulligan Cara, 2011). In fact, the vaporization and recondensation of alkali metals during biomass char combustion may lead to the release of submicron aerosols and to fouling and corrosion of burner surfaces. The turning of calcium and/or alkali metals into silicates mitigates the ash softening temperature, which would increase ash deposits (Wornat et al., 1995). Leaching and mechanical separation processes are key technologies in reducing the alkali and mineral content of biomass for fuel quality production.

Finally and as stated by Wang, Li, Zeng, and Tong (2012), if a significant decrease in emissions could be combined with low costs, revolutionary low-emission steel could be produced, especially if char-based reduction agents are introduced in the industries worldwide (Wang et al., 2012).

References

Ahn, S. Y., Eom, S. Y., Rhie, Y. H., Sung, Y. M., Moon, C. E., Choi, G. M., et al. (2013). Utilization of wood biomass char in a direct carbon fuel cell (DCFC) system. *Applied Energy*, *105*, 207–216.

Antal, M. J., & Grønli, M. (2003). The art, science, and technology of charcoal production. *Industrial & Engineering Chemistry Research*, *42*(8), 1619–1640.

Assistance, V. I. T. (1985). *Testing the efficiency of wood-burning cookstoves: international standards*. Virginia: Arlington.

Bailis, R. (2007). *Controlled Cooking Test (CCT), version 2.0*. University of California-Berkeley.

Bailis, R., Berrueta, V., Chengappa, C., Dutta, K., Edwards, R., Masera, O., et al. (2007). Performance testing for monitoring improved biomass stove interventions: experiences of the Household Energy and Health. *Energy for Sustainable Development*, *11*(2), 57–70.

Bailis, R., Smith, K. R., & Edwards, R. (2007). Kitchen Performance Test (KPT), version 3.0. In *University of California-Berkeley*.

Bazargan, A., Rough, S. L., & McKay, G. (2014). Compaction of palm kernel shell biochars for application as solid fuel. *Biomass and Bioenergy*, *70*, 489–497.

Bhattacharya, S. C., Albina, D. O., & Abdul Salam, P. (2002). Emission factors of wood and charcoal-fired cookstoves. *Biomass and Bioenergy*, *23*(6), 453–469.

Bhattacharya, S. C., Albina, D. O., & Myint Khaing, A. (2002). Effects of selected parameters on performance and emission of biomass-fired cookstoves. *Biomass and Bioenergy*, *23*(5), 387–395.

Booker, K., Han, T. W., Granderson, J., Jones, J., Lask, K., Yang, N., et al. (2011). *Performance of Charcoal Cookstoves for Haiti, Part 1: Results from the Water Boiling Test*. Berkley Lab.

Borowski, G., Stępniewski, W., & Wójcik-Oliveira, K. (2017). Effect of starch binder on charcoal briquette properties. *International Agrophysics*, *31*, 571.

Boström, D., Skoglund, N., Grimm, A., Boman, C., Öhman, M., Broström, M., et al. (2012). Ash transformation chemistry during combustion of biomass. *Energy & Fuels*, *26*(1), 85–93.

Bridgeman, T. G., Jones, J. M., Shield, I., & Williams, P. T. (2008). Torrefaction of reed canary grass, wheat straw and willow to enhance solid fuel qualities and combustion properties. *Fuel*, *87*(6), 844–856.

Chen, T., Cai, J., & Liu, R. (2015). Combustion kinetics of biochar from fast pyrolysis of pine sawdust: isoconversional analysis. *Energy Sources, Part A: Recovery, Utilization, and Environmental Effects*, *37*(20), 2208–2217.

Chouchene, A., Jeguirim, M., Khiari, B., Trouvé, G., & Zagrouba, F. (2010). Study on the emission mechanism during devolatilization/char oxidation and direct oxidation of olive solid waste in a fixed bed reactor. *Journal of Analytical and Applied Pyrolysis*, *87*(1), 168–174.

Conag, A. T., Villahermosa, J. E. R., Cabatingan, L. K., & Go, A. W. (2017). Energy densification of sugarcane bagasse through torrefaction under minimized oxidative atmosphere. *Journal of Environmental Chemical Engineering*, *5*(6), 5411–5419.

Cunningham Michael, G. M., & Linda, S. (2016). Literature review: Slow pyrolysis of feacal sludge. In *Eawag aquatic research*. E. Zurich.

de la Puente, G., Fuente, E., & Pis, J. J. (2000). Reactivity of pyrolysis chars related to precursor coal chemistry. *Journal of Analytical and Applied Pyrolysis*, *53*(1), 81–93.

de Oliveira, R. S., Palacio, S. M., da Silva, E. A., Mariani, F. Q., & Reinehr, T. O. (2017). Briquettes production for use as power source for combustion using charcoal thin waste and sanitary sewage sludge. *Environmental Science and Pollution Research International*, *24*(11), 10778–10785.

Di Blasi, C. (2009). Combustion and gasification rates of lignocellulosic chars. *Progress in Energy and Combustion Science*, *35*(2), 121–140.

Division, F. I. (1987). Simple technologies for charcoal making. *FAO forestry paper 41*. In *Vol. 41*. Rome: F.A.A.O.O.T.U. NATIONS.

Ellem Gary, K., & Mulligan Cara, J. (2011). Biomass char as a fuel for internal combustion engines. *Asia-Pacific Journal of Chemical Engineering*, *7*(5), 769–776.

Emrich, W. (1985). End-Use Markets for Charcoal and Charcoal by-Products. In W. Emrich (Ed.), *Handbook of charcoal making: The traditional and industrial methods* (pp. 178–207). Dordrecht: Springer Netherlands.

Frederik, R., Sven, v. H., Dane, D., & Wolter, P. (2013). Production and characterization of slow pyrolysis biochar: influence of feedstock type and pyrolysis conditions. *GCB Bioenergy*, *5* (2), 104–115.

Gallagher, M., Beard, M., Clifford, M. J., & Watson, M. C. (2016). An evaluation of a biomass stove safety protocol used for testing household cookstoves, in low and middle-income countries. *Energy for Sustainable Development*, *33*, 14–25.

Gao, X., & Wu, H. (2011). Biochar as a Fuel: 4. emission behavior and characteristics of PM1 and PM10 from the combustion of pulverized biochar in a drop-tube furnace. *Energy & Fuels*, *25*(6), 2702–2710.

Gold, M., Ddiba, D. I. W., Seck, A., Sekigongo, P., Diene, A., Diaw, S., et al. (2017). Faecal sludge as a solid industrial fuel: A pilot-scale study. *Journal of Water Sanitation and Hygiene for Development*, *7*(2), 243–251.

Guo, J., & Lua, A. C. (2000). Effect of heating temperature on the properties of chars and activated carbons prepared from oil palm stones. *Journal of Thermal Analysis and Calorimetry*, *60*(2), 417–425.

Han, K., Wang, Q., Zhao, J., Luo, K. H., Li, H., Chen, Y., et al. (2016). Combustion pattern, characteristics, and kinetics of biomass and chars from segmented heating carbonization. *Asia-Pacific Journal of Chemical Engineering*, *11*(5), 812–822.

Haykiri-Acma, H., Yaman, S., & Kucukbayrak, S. (2012). Combustion characteristics of sodium-free pyrolytic char from hazelnut shell. *Fuel Processing Technology*, *96*, 169–174.

He, X.-M., Yi, S., Fu, P.-R., Zeng, X.-C., Zhang, D., & Cheng, X.-H. (2017). Combustion reactivity of biochar and char generated from Co-pyrolysis of coal and four additives: application in blast furnace. *Journal of Energy Engineering*, *143*(1), 04016023.

Hirunpraditkoon, S., Intharit, S., Srisumran, S., & Wongvirojtana, P. (2014). Combustion properties of briquette charcoal from Durian peel. *Applied Mechanics and Materials*, *666*, 41–45.

Hu, Q., Yang, H., Yao, D., Zhu, D., Wang, X., Shao, J., et al. (2016). The densification of biochar: effect of pyrolysis temperature on the qualities of pellets. *Bioresource Technology*, *200*, 521–527.

Huang, H.-L., Lee, W.-M. G., & Wu, F.-S. (2016). Emissions of air pollutants from indoor charcoal barbecue. *Journal of Hazardous Materials*, *302*, 198–207.

Hurt, R. H., & Calo, J. M. (2001). Semi-global intrinsic kinetics for char combustion modeling. *Combustion and Flame*, *125*(3), 1138–1149.

Islam, M. A., Asif, M., & Hameed, B. H. (2015). Pyrolysis kinetics of raw and hydrothermally carbonized Karanj (Pongamia pinnata) fruit hulls via thermogravimetric analysis. *Bioresource Technology*, *179*, 227–233.

Islam, M. A., Auta, M., Kabir, G., & Hameed, B. H. (2016). A thermogravimetric analysis of the combustion kinetics of karanja (Pongamia pinnata) fruit hulls char. *Bioresource Technology*, *200*, 335–341.

Islam, M. A., Kabir, G., Asif, M., & Hameed, B. H. (2015). Combustion kinetics of hydrochar produced from hydrothermal carbonisation of Karanj (Pongamia pinnata) fruit hulls via thermogravimetric analysis. *Bioresource Technology*, *194*, 14–20.

Janse, A. M. C., de Jonge, H. G., Prins, W., & van Swaaij, W. P. M. (1998). Combustion kinetics of char obtained by flash pyrolysis of pine wood. *Industrial & Engineering Chemistry Research*, *37*(10), 3909–3918.

Jeguirim, M., Limousy, L., & Dutournie, P. (2014). Pyrolysis kinetics and physicochemical properties of agropellets produced from spent ground coffee blended with conventional biomass. *Chemical Engineering Research and Design*, *92*(10), 1876–1882.

Jetter, J. J., & Kariher, P. (2009). Solid-fuel household cook stoves: characterization of performance and emissions. *Biomass and Bioenergy*, *33*(2), 294–305.

Kim, H.-J., Lu, G.-Q., Naruse, I., Yuan, J., & Ohtake, K. (2000). Modeling on combustion characteristics of biocoalbriquettes. *Journal of Energy Resources Technology*, *123*(1), 27–31.

Kim, H., & Li, T. (2005). Denitrification mechanism in combustion of biocoal briquettes. *Environmental Science & Technology*, *39*(4), 1180–1183.

Kim Oanh, N. T., Bætz Reutergårdh, L., & Dung, N. T. (1999). Emission of polycyclic aromatic hydrocarbons and particulate matter from domestic combustion of selected fuels. *Environmental Science & Technology*, *33*(16), 2703–2709.

Kong, L., Tian, S., Li, Z., Luo, R., Chen, D., Tu, Y., et al. (2013). Conversion of recycled sawdust into high HHV and low NOx emission bio-char pellets using lignin and calcium hydroxide blended binders. *Renewable Energy*, *60*, 559–565.

Koyuncu, T., & Pinar, Y. (2007). The emissions from a space-heating biomass stove. *Biomass and Bioenergy*, *31*(1), 73–79.

Kraiem, N., Jeguirim, M., Limousy, L., Lajili, M., Dorge, S., Michelin, L., et al. (2014). Impregnation of olive mill wastewater on dry biomasses: impact on chemical properties and combustion performances. *Energy*, *78*, 479–489.

Kraiem, N., Lajili, M., Limousy, L., Said, R., & Jeguirim, M. (2016). Energy recovery from Tunisian agri-food wastes: evaluation of combustion performance and emissions characteristics of green pellets prepared from tomato residues and grape marc. *Energy*, *107*, 409–418.

Kulas, D., Winjobi, O., Zhou, W., & Shonnard, D. (2018). Effects of coproduct uses on environmental and economic sustainability of hydrocarbon biofuel from one- and two-step pyrolysis of poplar. *ACS Sustainable Chemistry & Engineering*, *6*(5), 5969–5980.

Kumar, M., & Gupta, R. C. (1998). Industrial uses of wood char. *Energy Sources*, *20*(7), 575–589.

Lajili, M., Limousy, L., & Jeguirim, M. (2014). Physico-chemical properties and thermal degradation characteristics of agropellets from olive mill by-products/sawdust blends. *Fuel Processing Technology*, *126*, 215–221.

Lask, K., Booker, K., Han, T., Granderson, J., Yang, N., Ceballos, C., et al. (2015). Performance comparison of charcoal cookstoves for Haiti: laboratory testing with water boiling and controlled cooking tests. *Energy for Sustainable Development*, *26*, 79–86.

Lewis, J. J., & Pattanayak, S. K. (2012). Who adopts improved fuels and cookstoves? A systematic review. *Environmental Health Perspectives*, *120*(5), 637–645.

Lim, S.-H., Yong, S.-T., Ooi, C.-W., Chai, S.-P., Doshi, V., & Daud, W. R. W. (2014). Pyrolysis of palm waste for the application of direct carbon fuel cell. *Energy Procedia*, *61*, 878–881.

Linderholm, C., Schmitz, M., Knutsson, P., Källén, M., & Lyngfelt, A. (2014). Use of low-volatile solid fuels in a 100 kW chemical-looping combustor. *Energy & Fuels*, *28*(9), 5942–5952.

Liu, Z., Zhang, F.-S., & Wu, J. (2010). Characterization and application of chars produced from pinewood pyrolysis and hydrothermal treatment. *Fuel*, *89*(2), 510–514.

Liu, Z., Zhang, F., Liu, H., Ba, F., Yan, S., & Hu, J. (2018). Pyrolysis/gasification of pine sawdust biomass briquettes under carbon dioxide atmosphere: Study on carbon dioxide reduction (utilization) and biochar briquettes physicochemical properties. *Bioresource Technology*, *249*, 983–991.

López, F. A., Centeno, T. A., García-Díaz, I., & Alguacil, F. J. (2013). Textural and fuel characteristics of the chars produced by the pyrolysis of waste wood, and the properties of activated carbons prepared from them. *Journal of Analytical and Applied Pyrolysis*, *104*, 551–558.

Luo, M., & Stanmore, B. (1992). The combustion characteristics of char from pulverized bagasse. *Fuel*, *71*(9), 1074–1076.

MacCarty, N., Still, D., & Ogle, D. (2010). Fuel use and emissions performance of fifty cooking stoves in the laboratory and related benchmarks of performance. *Energy for Sustainable Development*, *14*(3), 161–171.

Mamuye, F., Lemma, B., & Woldeamanuel, T. (2018). Emissions and fuel use performance of two improved stoves and determinants of their adoption in Dodola, southeastern Ethiopia. *Sustainable Environment Research*, *28*(1), 32–38.

Masekameni, D., Makonese, T., & Annegarn, H. J. (2016). Performance evaluation of three charcoal stoves. In *2016 International Conference on the Domestic Use of Energy (DUE), 30-31 March 2016* (pp. 1–7).

Mitchell, E. J. S., Lea-Langton, A. R., Jones, J. M., Williams, A., Layden, P., & Johnson, R. (2016). The impact of fuel properties on the emissions from the combustion of biomass and other solid fuels in a fixed bed domestic stove. *Fuel Processing Technology*, *142*, 115–123.

Mwampamba, T. H., Owen, M., & Pigaht, M. (2013). Opportunities, challenges and way forward for the charcoal briquette industry in Sub-Saharan Africa. *Energy for Sustainable Development*, *17*(2), 158–170.

Niu, S., Chen, M., Li, Y., & Xue, F. (2016). Evaluation on the oxy-fuel combustion behavior of dried sewage sludge. *Fuel*, *178*, 129–138.

Njenga, M., Karanja, N., Jamnadass, R., Kithinji, J., Sundberg, C., & Jirjis, R. (2013). Quality of cooking fuel briquettes produced locally from charcoal dust and sawdust in Kenya. *Journal of Biobased Materials and Bioenergy*, *7*(3), 315–322.

Norgate, T., & Langberg, D. (2009). Environmental and economic aspects of charcoal use in steelmaking. *ISIJ International*, *49*, 587–595.

Park, S.-W., & Jang, C.-H. (2012). Effects of pyrolysis temperature on changes in fuel characteristics of biomass char. *Energy*, *39*(1), 187–195.

Parshetti, G. K., Kent Hoekman, S., & Balasubramanian, R. (2013). Chemical, structural and combustion characteristics of carbonaceous products obtained by hydrothermal carbonization of palm empty fruit bunches. *Bioresource Technology*, *135*, 683–689.

Peters, J. F., Iribarren, D., & Dufour, J. (2015). Biomass pyrolysis for biochar or energy applications? A life cycle assessment. *Environmental Science & Technology*, *49*(8), 5195–5202.

Pimchuai, A., Dutta, A., & Basu, P. (2010). Torrefaction of agriculture residue to enhance combustible properties. *Energy & Fuels*, *24*(9), 4638–4645.

Qi, J., Han, K., Wang, Q., & Gao, J. (2017). Carbonization of biomass: Effect of additives on alkali metals residue, SO_2 and NO emission of chars during combustion. *Energy*, *130*, 560–569.

Ramanathan, V., & Carmichael, G. (2008). Global and regional climate changes due to black carbon. *Nature Geoscience*, *1*, 221.

Rhén, C., Öhman, M., Gref, R., & Wästerlund, I. (2007). Effect of raw material composition in woody biomass pellets on combustion characteristics. *Biomass and Bioenergy*, *31*(1), 66–72.

Román, S., Nabais, J. M. V., Laginhas, C., Ledesma, B., & González, J. F. (2012). Hydrothermal carbonization as an effective way of densifying the energy content of biomass. *Fuel Processing Technology*, *103*, 78–83.

Rousset, P., Caldeira-Pires, A., Sablowski, A., & Rodrigues, T. (2011). LCA of eucalyptus wood charcoal briquettes. *Journal of Cleaner Production*, *19*(14), 1647–1653.

Sagar, A. D., & Kartha, S. (2007). Bioenergy and sustainable development? *Annual Review of Environment and Resources*, *32*(1), 131–167.

Scala, F., Chirone, R., & Salatino, P. (2006). Combustion and attrition of biomass chars in a fluidized bed. *Energy & Fuels*, *20*(1), 91–102.

Scala, F., Salatino, P., & Chirone, R. (2000). Fluidized bed combustion of a biomass char (Robinia pseudoacacia). *Energy & Fuels*, *14*(4), 781–790.

Senneca, O., Chirone, R., & Salatino, P. (2002). A thermogravimetric study of nonfossil solid fuels. 2. Oxidative pyrolysis and char combustion. *Energy & Fuels*, *16*(3), 661–668.

Sevault, A., Khalil, R. A., Enger, B. C., Skreiberg, Ø., Goile, F., Wang, L., et al. (2017). Performance evaluation of a modern wood stove using charcoal. *Energy Procedia*, *142*, 192–197.

Sima-Ella, E., Yuan, G., & Mays, T. (2005). A simple kinetic analysis to determine the intrinsic reactivity of coal chars. *Fuel*, *84*(14), 1920–1925.

Smith, K. R., Jerrett, M., Anderson, H. R., Burnett, R. T., Stone, V., Derwent, R., et al. (2009). Public health benefits of strategies to reduce greenhouse-gas emissions: Health implications of short-lived greenhouse pollutants. *Lancet*, *374*(9707), 2091–2103.

Sotannde, O. A., Oluyege, A. O., & Abah, G. B. (2010). Physical and combustion properties of briquettes from sawdust of Azadirachta indica. *Journal of Forestry Research*, *21*(1), 63–67.

Srivastava, N. S. L., Narnaware, S. L., Makwana, J. P., Singh, S. N., & Vahora, S. (2014). Investigating the energy use of vegetable market waste by briquetting. *Renewable Energy*, *68*, 270–275.

Suopajärvi, H., Kemppainen, A., Haapakangas, J., & Fabritius, T. (2017). Extensive review of the opportunities to use biomass-based fuels in iron and steelmaking processes. *Journal of Cleaner Production*, *148*, 709–734.

Suopajärvi, H., Pongrácz, E., & Fabritius, T. (2013). The potential of using biomass-based reducing agents in the blast furnace: a review of thermochemical conversion technologies and assessments related to sustainability. *Renewable and Sustainable Energy Reviews*, *25*, 511–528.

Teixeira, S. R., Pena, A. F. V., & Miguel, A. G. (2010). Briquetting of charcoal from sugar-cane bagasse fly ash (SCBFA) as an alternative fuel. *Waste Management*, *30*(5), 804–807.

Valenzuela-Calahorro, C., Bernalte-Garcia, A., Gómez-Serrano, V., & Bernalte-García, M. J. (1987). Influence of particle size and pyrolysis conditions on yield, density and some textural parameters of chars prepared from holm-oak wood. *Journal of Analytical and Applied Pyrolysis*, *12*(1), 61–70.

Vamvuka, D., Zografos, D., & Alevizos, G. (2008). Control methods for mitigating biomass ash-related problems in fluidized beds. *Bioresource Technology*, *99*(9), 3534–3544.

Várhegyi, G., Mészáros, E., Antal, M. J., Bourke, J., & Jakab, E. (2006). Combustion kinetics of corncob charcoal and partially demineralized corncob charcoal in the kinetic regime. *Industrial & Engineering Chemistry Research*, *45*(14), 4962–4970.

Wang, Q., Aparu, P., Kurogawa, H., & Sugiyama, K. (2014). Recovery briquetting technologies of waste biomass and pyrolyzed waste char produced from solid industrial and agricultural organic wastes. *International Journal of Sustainable Development and Planning*, *9*(5), 705–716.

Wang, Q., Han, K., Gao, J., Li, H., & Lu, C. (2017). The pyrolysis of biomass briquettes: effect of pyrolysis temperature and phosphorus additives on the quality and combustion of biochar briquettes. *Fuel*, *199*, 488–496.

Wang, Q., Han, K., Gao, J., Wang, J., & Lu, C. (2017). Investigation of maize straw char briquette ash fusion characteristics and the influence of phosphorus additives. *Energy & Fuels*, *31*(3), 2822–2830.

Wang, W. Q., Li, P. F., Zeng, C. S., & Tong, C. (2012). Evaluation of silicate iron slag as a potential methane mitigating method. *Advanced Materials Research*, *468–471*, 1626–1630.

Werkelin, J., Skrifvars, B.-J., Zevenhoven, M., Holmbom, B., & Hupa, M. (2010). Chemical forms of ash-forming elements in woody biomass fuels. *Fuel*, *89*(2), 481–493.

Winijkul, E., & Bond, T. C. (2016). Emissions from residential combustion considering end-uses and spatial constraints: part II, emission reduction scenarios. *Atmospheric Environment*, *124*, 1–11.

Wornat, M. J., Hurt, R. H., Yang, N. Y. C., & Headley, T. J. (1995). Structural and compositional transformations of biomass chars during combustion. *Combustion and Flame*, *100* (1), 131–143.

Wu, S., Zhang, S., Wang, C., Mu, C., & Huang, X. (2018). High-strength charcoal briquette preparation from hydrothermal pretreated biomass wastes. *Fuel Processing Technology*, *171*, 293–300.

Zhou, Z., Hu, X., You, Z., Wang, Z., Zhou, J., & Cen, K. (2013). Oxy-fuel combustion characteristics and kinetic parameters of lignite coal from thermo-gravimetric data. *Thermochimica Acta*, *553*, 54–59.

Further reading

Abdu Zubairu, S. A. G. (2014). Production and characterization of briquette charcoal by carbonization of agro-waste. *Energy and Power*, *4*(2), 41–47.

Li, S., Chen, X., Wang, L., Liu, A., & Yu, G. (2013). Co-pyrolysis behaviors of saw dust and Shenfu coal in drop tube furnace and fixed bed reactor. *Bioresource Technology*, *148*, 24–29.

Char gasification

6

Andrius Tamošiūnas, Mejdi Jeguirim†*
*Lithuanian Energy Institute, Plasma Processing Laboratory, Kaunas, Lithuania,
†University of Strasbourg, University of Upper Alsace, Institute of Materials Science of Mulhouse (IS2M—UMR CNRS 7361), Mulhouse, France

Chapter Outline

Char and Carbon Materials Derived from Biomass. https://doi.org/10.1016/B978-0-12-814893-8.00006-7

6.1 Introduction

Today, the main source of global energy relies on fossil fuels derived from coal, petroleum, and natural gas. However, prior to the use of fossil fuels, biomass, especially wood, was the primary source for energy and heat production, whose importance has diminished over the years due its lower energy density. In addition, the increasing use of fossil fuels causes serious environmental and anthropogenic problems such as global warming and weather change (floods, hurricanes, glaciers melting). Therefore, taking into account these circumstances, searching for other renewable sources of energy is crucial. Furthermore, fossil fuel resources are not distributed uniformly around the globe making many countries heavily dependent on the import and/or search for alternative solutions.

The use of biomass for energy production is considered as a carbon neutral process. There are three main ways of converting biomass to biofuels by different methods, which are broadly classified as thermochemical and biochemical (Kumar, Jones, & Hanna, 2009; McKendry, 2002a). Gasification is a thermochemical process that converts carbonaceous materials into useful products such as gaseous fuels or chemical feedstock. This chapter introduces char/biomass gasification for producing biofuels, power, and chemicals. It summarizes the basic steps involved in the gasification process, biomass/char reactivity, and chemical kinetics, as well as some gasification technologies and methodologies that help to estimate gasification performance.

6.2 Bioenergy from gasification

Thermochemical conversion technologies rely mostly on combustion, pyrolysis, and gasification (Basu, 2013; Parascanu, Sandoval-Salas, Soreanu, Valverde, & Sanchez-Silva, 2017). The combustion process is the most popular, well-established, and technically easiest process, whereas pyrolysis and gasification are still under development. Gasification is a thermochemical process that converts carbonaceous materials such as biomass to gaseous fuels constituted mainly of hydrogen and carbon monoxide, also known as syngas. Usually, this process is based on several steps (Fig. 1) (Seo, Lee, Kang, Hwang, & Yu, 2010):

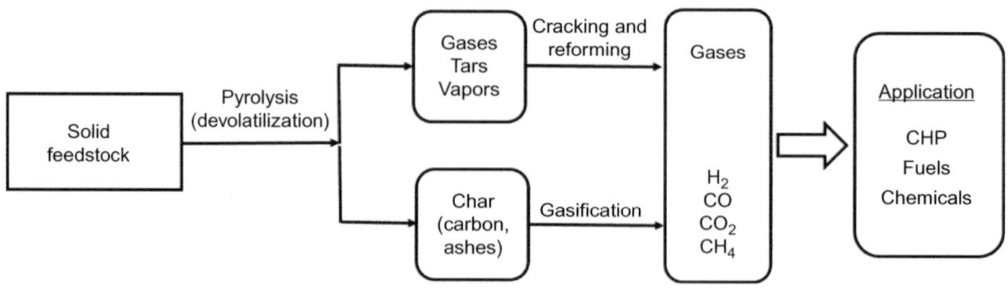

Fig. 1 Flow of the gasification process. *CHP*, combined heat and power.

(1) Pyrolysis is the decomposition process of biomass into a range of useful products (chars, gases: CH_4, CO_2, CO, and tars) at elevated temperatures (300–650°C) in the absence of oxidizing agent or with a limited amount.

(2) Char (carbon and ashes) left after the pyrolysis of biomass undergoes gasification with an external gasifying agent (CO_2, H_2O, O_2, air), thus leading to the formation of H_2 and CO (syngas). This process typically proceeds at the temperature range of 700–1000°C and higher.

The produced syngas can be used in different applications such as the production of chemicals (methanol, hydrogen) or liquid fuels via Fischer-Tropsch synthesis. The syngas can be also combined with advanced technologies such as gas turbines and/or fuel cells (solid oxide fuel cell, proton exchange membrane).

6.3 Gasification steps

Basically, the gasification process can be divided into several steps (Fig. 2).

6.3.1 Upstream processing

Size reduction: Upstream processing includes biomass feedstock pretreatment to adapt for gasification operations and to increase the overall efficiency. In particular, size reduction (grinding, milling) is needed to obtain appropriate particle sizes, thus increasing the surface area per unit mass and accelerating the heat transfer mechanism as well as reaction rates during gasification. Smaller size biomass can significantly increase the conversion efficiency and dry gas yield, whereas the contents of char and tar decrease (Susastriawan, Saptoadi, & Purnomo, 2017). However, gasification of very tiny biomasses has a high-pressure drop problem and high dust content in producer gas. On the contrary, larger particle size reduces gasification reactivity and causes starting and bridging problems as well as reduced yield of producer gas (Patel, Upadhyay, & Patel, 2014). Therefore, the optimal feed particle sizes are defined in the range of 20–80 mm.

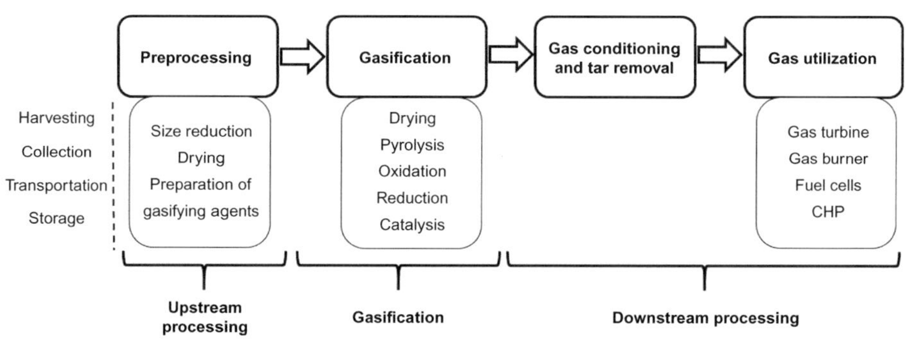

Fig. 2 Processes involved in biomass gasification (Kumar et al., 2009). *CHP*, combined heat and power.

Drying: Usually, biomass harvested and collected from farm and forest lands contains high moisture. Therefore, drying is needed to attain a desired range of water content in biomass to sustain the overall thermal efficiency of the gasification system. Drying is an energy-intensive process because first biomass feedstock is heated from room temperature to 100°C, then vaporized, and finally water vapor is heated to the gasification temperature (Sansaniwal, Rosen, & Tyagi, 2017). Therefore, biomass should contain a limited amount of moisture ($\leq$40%) to achieve good-quality producer gas with a higher content of H_2 and CO yielding a higher lower heating value (LHV). McKendry (2002b) reported that biomass with more than 30% moisture content reduces the LHV of synthesis gas due to incomplete pyrolysis during gasification.

Densification: Many types of biomass feedstock such as rice husk, straw, grass, etc. are bulky and fibrous, and cannot be used directly in the gasification process due to their stickiness (Chevanan et al., 2010). Such a constraint could be solved by utilizing densification, during which the bulk density of biomass is increased (Jeguirim, Limousy, & Dutournie, 2014; Lajili, Guizani, Escudero Sanz, & Jeguirim, 2018).

Torrefaction: Prior to densification, biomass feedstock can be thermally treated in a temperature range of 230–300°C. During this process, called torrefaction or mild pyrolysis, water and some volatiles are removed (Guizani, Haddad, Jeguirim, Colin, & Limousy, 2016). The obtained torrefied biomass is dry, has hydrophobic character, and can be densified to increase its mass and energy density.

6.3.2 Gasification

Gasification occurs at high temperatures in the presence of a gasifying agent (air, oxygen, steam, carbon dioxide, or their mixture) (Guizani, Escudero Sanz, Jeguirim, Gadiou, & Salvador, 2015). Heat is supplied to the gasifier directly or indirectly to reach a desired gasification temperature.

6.3.2.1 Basic chemistry

The major chemical reactions taking place during gasification can be summarized as follows.

Pyrolysis

As indicated in Section 6.2, during pyrolysis, biomass components are decomposed into lighter molecules and eventually into permanent noncondensable gases (CO, CO_2, H_2, CH_4, and lighter hydrocarbons), liquid (tars), ash, and char (Jeguirim, Elmay, Limousy, Lajili, & Said, 2014). The decomposition takes place between the gas-gas phase (homogeneous reactions) and gas-solid phase (heterogeneous reactions). Basically, biomass pyrolysis can be represented by the reaction in Eq. (1):

$$C_nH_mO_p(\text{Biomass}) \xrightarrow{\text{Heat}} \sum\nolimits_{\text{liq}} C_xH_yO_z + \sum\nolimits_{\text{gas}} C_aH_bO_c + H_2O + \text{Char} \tag{1}$$

Char gasification reactions

Chars produced after biomass pyrolysis contain C, H, O, and some amount of N, S, and minerals (Guizani, Jeguirim, Valin, Limousy, & Salvador, 2017). Generally, chars derived from biomass are more porous and reactive compared to coal char (Basu, 2013). Al-Qayim, Nimmo, Hughes, and Pourkashanian (2017) showed that biomass char reactivity was two to three times higher than coal chars at 1273 K. Such behavior was attributed to the highly porous char structure produced at higher temperatures (Guizani, Jeguirim, Esceduro Sainz, Gadiou, & Salvador, 2016). Moreover, it was shown that biomass released around 90% of the total volatile matter content at 773 K, while mineral coal released less than 66%, meaning that higher temperatures are requested for complete devolatilization. Gasification of biomass char (C) includes several exothermic and endothermic reactions (oxidation, reduction) between the char and the gasifying agents.

Heterogeneous reactions:

Partial oxidation (PO)

$$C + 0.5O_2 \rightarrow CO, \quad -111\,\text{kJ/mol} \tag{2}$$

Complete oxidation (CO)

$$C + O_2 \rightarrow CO_2, \quad -394\,\text{kJ/mol} \tag{3}$$

PO of char produces carbon monoxide (Eq. 2), while total oxidation of char yields carbon dioxide and three times more heat (Eq. 3).

Boudouard reaction

$$C + CO_2 \rightarrow 2CO, \quad +172\,\text{kJ/mol} \tag{4}$$

Water-gas (or steam gasification) reaction

$$C + H_2O \rightarrow CO + H_2, \quad +131\,\text{kJ/mol} \tag{5}$$

Methanation (or hydrogasification)

$$C + 2H_2 \rightarrow CH_4, \quad -74.8\,\text{kJ/mol} \tag{6}$$

Eqs. (2) through (6) show how char (C) directly reacts with gasifying agents such as O_2, CO_2, H_2, and steam to convert it into gases such as CO, H_2, and CH_4.

Homogeneous reactions:

An important gas-gas phase reaction helps to increase the hydrogen content of the gasification product at the expense of CO. This reaction is the so-called "water-gas shift (WGS) reaction":

$$CO + H_2O \leftrightarrow CO_2 + H_2, \quad -41.2\,\text{kJ/mol}. \tag{7}$$

Char gasification includes both endothermic (Eqs. 4, 5) and exothermic (Eqs. 5, 6) reactions. However, the following exothermic gas phase reactions such as methanation and steam-methane reforming may influence the gasification process (Kumar et al., 2009):

$$CO + 3H_2 \leftrightarrow CH_4 + H_2O, \quad -206\,kJ/mol \tag{8}$$

$$CH_4 + H_2O \leftrightarrow CO + 3H_2, \quad +206\,kJ/mol. \tag{9}$$

6.3.2.2 Gasification operating conditions

Gasification process parameters such as temperature, pressure, equivalence ratio (ER), consumption rate, and residence time have to be considered for gasifier design.

Temperature

Gasification temperature is an important parameter influencing the producer gas composition and properties. A higher temperature in the gasifier promotes tar cracking and reforming, and enhances char gasification rate yielding higher producer gas with increased calorific value. Liu, Hu, Wang, Wang, and Li (2012) showed that tar conversion increased from 17% to 54% when the temperature increased from 750°C to 950°C in an updraft gasifier combined with a copper slag reformer. Above 750–800°C, water-gas and steam-methane reforming reactions (Eqs. 5, 9) are in favor resulting in increasing H_2 and decreasing CH_4 contents. Above 850–900°C, both the Boudouard and steam reforming reactions (Eqs. 4, 9) prevail, resulting in CO content increase.

Gasifying agents

Air, O_2, steam, CO_2, and their mixtures can be used as gasification agents. Generally, air is used due to its abundance and low cost. Gasification medium is defined differently depending on the gasifying agent. For instance, for air the parameter is ER; for steam, steam-to-biomass ratio (S/B ratio); for carbon dioxide, CO_2-to-biomass ratio (CO_2/biomass), and for a steam-oxygen mixture, (steam + O_2)-to-biomass ratio. Syngas quality depends on the gasifying medium. Usually, three typical product gas qualities are distinguished as shown in Table 1.

The desired heating value of the producer gas determines the choice of gasification medium. In some cases, the LHV of product gas may exceed 40 MJ/Nm^3 if hydrogen is used as a gasifying agent. However, often economic factors are a primary choice and therefore air is the most popular gasifying medium. The major disadvantage is the

Table 1 Net calorific value of product gas after gasification in different gasifying media.

Caloric value	Value range (MJ/Nm^3)	Gasifying agent
Low	4–6	Air
Medium	10–18	Steam
High	12–28	Oxygen

presence of nitrogen with air diluting the syngas and giving a lower calorific value of 4–6 MJ/Nm^3. The use of pure oxygen is too expensive and a steam–air mixture allows calorific values ranging from 10 to 20 MJ/Nm^3.

Equivalence ratio

The ER is an important process and gasifier design parameter. ER is defined as the ratio of actual air used in gasification to stoichiometric air (excess air, EA, in combustion):

$$ER(<1.0)_{\text{gasification}} = \frac{\text{Actual air}}{\text{Stoichiometric air}} = EA(>1.0)_{\text{combustion}} \quad (10)$$

The stoichiometric amount of air is calculated based on the fuel ultimate analysis. Basically, ER determines the performance of the gasifier. For effective gasification, a typical ER is in the range of 0.2–0.4. With a lower ER, charcoal production can be maximized since the char is not fully converted into gases. However, a lower ER leads to higher tar production in the producer gas. Liu et al. (2012) found that by increasing ER from 0.1 to 0.5, the H_2, CO, CH_4, and C_2H_x contents decreased, while the CO_2 content increased. Furthermore, the producer gas yield increased from 1.30 to 2.84 Nm^3/kg, while the LHV decreased from 7.26 to 4.15 MJ/Nm^3 and the tar yield decreased from 80.4 to 31.9 g/kg. Guo, Dong, Dong, and Guo (2014) showed that when the ER increased from 0.18 to 0.27, the cold gas efficiency (CGE) increased sharply reaching 65%. However, the CGE decreased at an ER $\geq$0.3, indicating a decrease of total energy conversion. Moreover, when the ER ranged between 0.18 and 0.37, the higher temperature led to tar cracking and thus the tar yield decreased from 2.5 to 0.52 g/Nm^3. Generally, ER can be adjusted by changing the air flow rate. Higher air flow leads to higher temperature, higher biomass/char conversion, and higher producer gas. On the contrary, the EA may decrease energy content of the producer gas since biomass is partially combusted.

Steam-to-biomass ratio

During steam gasification, the H_2O partial pressure inside the gasifier increases and favors water-gas, methane reforming, and WGS reactions (Eqs. 5–7, 6.9) leading to an increase in H_2 and CH_4 production. Therefore, gasification temperature needs to be high enough (above 800°C). Moreover, a higher S/B ratio leads to higher char conversion. Zhai et al. (2017) demonstrated that sawdust char conversion increased with temperature and steam flow rate. In addition, H_2 content increased from 53% to 60% when temperature increased from 800°C to 1200°C. Waheed, Wu, and Williams (2016) investigated steam gasification of char derived from sugarcane bagasse. The authors found that the gasification temperature increased from 750°C to 950°C, which increased the H_2 yield from 45.3 to 187.3 mmol/g. A similar tendency was observed when the steam flow rate increased from 6 to 15 mL/h, and the H_2 yield increased from 187.3 to 208.4 mmol/g.

The reduction of tars may also occur at higher S/B ratios through their steam reforming with an increased steam partial pressure. Meng, de Jong, Fu, and

Verkooijen (2011) examined the effect of S/B ratio on producer gas distribution and tar formation during woody biomass gasification. It was revealed that with increased S/B ratio from 1.13 to 1.45 at 770°C, the tar content was decreased from 10.2 to 7.2 g/Nm^3. Further tar reduction is also possible combining steam reforming with catalysts such as dolomite, alkali metals, and nickel.

CO_2-to-biomass ratio

Recently, CO_2 as a gasifying agent has received particular attention since it could be an issue for the captured CO_2 from the power plant. Several investigations have examined CO_2 gasification (Bouraoui et al., 2015; Sadhwani, Adhikari, & Eden, 2016). However, Sadhwani et al. (2016) during wood gasification in a bubbling fluidized-bed (BFB) gasifier did not observe a significant effect on the output depending on CO_2-to-biomass ratio. Alternatively, a mixture of CO_2 and H_2O or CO_2 and O_2/air is used with special catalysts to achieve the highest reaction rates and producer gas quality.

Pressure

Pressure has a minor effect on the char and biomass conversion process. However, increasing reactor pressure may influence tar formation and conversion. Knight (2000) reported that during woodchip gasification the oxygenated compounds, especially phenols, were completely eliminated when the gasifier operated at a pressure of 21.4 bar. However, the fraction of polyaromatic hydrocarbons with 4+ rings increased. Generally, gasifiers can be operated at elevated pressures especially with the goal of installing a gas turbine to generate power.

Residence time

The residence time of biomass feedstock and gas products in the gasifier is also an important parameter. The flow rate of gasifying agent may affect not only process temperature but also the residence time, which may influence the extent of biomass (char) conversion and product gas quality. Longer residence time (from seconds to minutes) ensures a complete conversion of biomass into gas products. Therefore, fewer tars and solid by-products are generated. Otherwise, particles must be finely grounded to reach higher biomass/char conversion at shorter residence time (Hernandez, Aranda-Almansa, & Bula, 2010). James, Yuan, Boyette, Wang, and Kumar (2014) observed an increase of 36% in the syngas calorific value using a Ni-based catalyst of 5%–10% loading in an updraft biomass gasifier as the residence time changed from 0.2 to 0.3 s. However, high residence time (several minutes to an hour) may cause serious problems with ash melting and slag formation.

6.3.3 Downstream processing

The producer gas needs to be processed further for effective utilization, e.g., combined heat and power, and production of biofuels and/or chemicals. The involved processes are designed as downstream processing and include basic steps such as syngas cleaning, biopower generation, reforming, and conversion to biofuels. Product gas cleaning is a complicated and costly process. In fact, syngas contains particles, tar, alkali, nitrogen, and sulfur-containing compounds that need to be eliminated before final use. The required purity of syngas depends on the technological process and its final application. For instance, high hydrogen content in syngas is desirable for fuel cell applications, whereas the specific range of H_2/CO (usually 2:1) is desirable for producing biofuels and chemicals. On the other hand, combustion systems do not require very "clean" fuels and can operate with relatively high amounts of tar and other contaminants.

6.4 Char gasification reactivity

Generally, reactivity is a property of a solid fuel defining the value of the reaction rate under well-defined conditions of gasifying medium, temperature, and pressure. Appropriate values of char reactivity are necessary for all types of gasifiers. Biomass char gasification differs from fossil coal in terms of reactivity and mechanisms. Biomass-derived chars are more reactive, have higher volatiles content, and require lower temperature for gasification than coal.

6.4.1 Effect of char production conditions on char reactivity

The conditions under which the char is produced affect its reactivity (Guizani et al., 2017). Septien, Escudero Sanz, Salvador, and Valin (2018) investigated the effect of pyrolysis heating rate on wood char production and gasification. The authors found that the increase in heating rate (from 1°C/s to 1000°C/s) led to faster gasification rate and to a more reactive solid and higher ash catalytic activity. Lu, Huang, (Thanos) Bourtsalas, Chi, and Yan (2017) performed experimental research on basic properties and reactivity of two types of waste-derived chars (biomass-derived fuel (BDF) produced from rice straw and refuse-derived fuel (RDF) composed of municipal solid waste (MSW) and sewage sludge) pyrolized at 600°C. It was reported that the BDF char was more porous and active than the RDF char. The BDF char was characterized by a much larger total surface area (around 209.7 m^2/g) and less ash content (10 wt.% for the BDF char and 50 wt.% for the RDF char). Also, the activation energies under gasification conditions were 119.5 kJ/mol for the BDF char and 137.9 kJ/mol for the RDF char. It was also reported that due to high temperatures the number of active sites of reaction as well as the number of edge atoms of the char was reduced. Moreover, longer residence time at high temperatures during char production may also reduce its reactivity.

6.4.2 Catalytic effect of AAEM on char gasification and reactivity

Biomass contains a number of alkali and alkaline earth metals (AAEMs) such as potassium, sodium, calcium, and magnesium (Haddad et al., 2017). These elements can act as catalysts and accelerate the char gasification process (Bouraoui, Dupont,

Jeguirim, Limousy, & Gadiou, 2016; Eom et al., 2012; Jiang et al., 2015). The presence of AAEMs can favor heterogeneous char-steam, char-O_2, and char-CO_2 reactions (Eqs. 2–5) as well as homogeneous reactions (Eqs. 7–9), thus enhancing the production of synthesis gas. Lahijani, Alimuddin, Mohamed, and Mohammadi (2013) studied the influence of alkali (Na, K), alkaline earth (Ca, Mg), and transition (Fe) metal nitrates on CO_2 gasification reactivity of pistachio nut shell (PNS) char in a bench-scale reactor. It was highlighted that char gasification reactivity enhanced in the order of Na > Ca > Fe > K > Mg > raw char. Among various catalysts, 5 wt.% $Na(NO_3)_3$ had the best impact on enhancing gasification reactivity of PNS char. The activation energy was 53 kJ/mol lower than that of char without catalytic activation. Jiang et al. (2015) explored the catalytic effects of AAEMs on synthesis gas composition, concentration, and tar cracking during steam gasification of lignocellulosic biomass. It was found that the presence of AAEMs had a significant catalytic effect on thermal cracking/reforming of tar and production of H_2 and CO_2. Also, the presence of AAEMs inhibited the production of CO, CH_4, C_2H_4, and C_2H_6. The basic mechanism was the part of AAEMs volatized into the gaseous phase, thus involving homogeneous gas-gas reactions (hydrocarbon reforming, WGS reactions), whereas the other part of AAEMs remained in biochar, thus involving heterogeneous reactions (char-gasifying agent, char-volatile reactions), which altered the reaction pathway of biomass gasification. Tilghman and Mitchell (2015) noted that ash impurities within the carbon matrix may influence the reactivity of the char. Such impurities like Mg and Ca tend to increase char reactivity to both gasifying agents O_2 and CO_2, whereas the presence of Si and Al tends to reduce char reactivity to H_2O.

6.4.3 Intrinsic reaction rate and mass transfer control

Usually, char gasification reactions take place on its particle surface, i.e., on the outer surface area of the particle. Since the char particles are highly porous, the surface area of their inner and outer pore walls differs significantly. For instance, the actual surface area (Brunauer-Emmett-Teller) of an internal pore of a 1 mm diameter beech wood is 660 cm^2, while its outer surface is only 3.14 cm^2 (Basu, 2013). Therefore, if there is no physical limitation, the reacting gas can easily enter the pores of the char particle and react on their walls. As a result, a higher overall char conversion rate can be achieved. For this reason, two char particles of the same size may have different reaction rates because of their different internal surface areas and structures.

From a scientific point of view, it is reasonable to express the surface reaction rate in terms of the actual pore surface (intrinsic) area on which the reaction occurs rather than the external surface area. The reaction rate based on the actual pore wall surface area is known as the *intrinsic reaction rate*, whereas the rate based on the external surface area of the char particle is the *apparent reaction rate*. Basically, the intrinsic reaction rates represent the kinetic data measured under chemical controlled processes with minimized influence of heat and mass transfer (Gao, Zhang, Li, Zhao, & Jiang, 2016). Intrinsic kinetics is important in the modeling of gasification processes, hereby combining the intrinsic reaction rates with underlying heat and mass transfer mechanisms.

Table 2 Gasification reaction control mechanisms.

Kinetic control reaction	**Diffusion control reaction**
Diffusion rate $\gg$ kinetic rate	Diffusion rate $\ll$ kinetic rate

For the gasification reaction to proceed on the inner surface area of the char's particle pore, the reacting gas (gasifying agent) must enter the pores. If the entrance of the reacting gas is restricted, then it is entirely consumed on the outer surface of the char particle. In such a case, gasification is limited to the external surface area. This can occur due to the limitation of the mass transfer of the reacting gas to the surface of the char. How fast the reacting gas can diffuse to the char surface to react with the active carbon sites and form reaction products can be characterized by kinetic rate or diffusion rate, which controls the reaction. If the kinetic rate of the reaction is much faster than the diffusion of the reacting gas (H_2O, O, CO_2) to the surface of the char, the reaction is called a *diffusion control reaction*. On the contrary, if the kinetic rate of the reaction is slower compared to the diffusion rate of the reacting gas, then the reacting gas will diffuse into the pores of the char and react on their walls. The reaction in this situation is called a *kinetic control reaction*. An illustration is shown in Table 2.

The intrinsic reaction rate model, taking into account both diffusion and kinetic effects, is mostly applied for simulation of the biomass gasification process and comparison with experimental data.

6.4.4 Char reactivity under different gasifying agents

Char gasification reactivity depends on the used gasification agent. In particular, the rate of char gasification reaction depends on the rate of chemical reaction and diffusion rate of the gasifying agent into the pores of the char. Therefore, among the gasifying media used such as O_2/air, H_2O, CO, and H_2, the reaction involving oxygen is the most active, followed by steam, carbon monoxide, and hydrogen:

$$R_{C+O_2} \gg R_{C+H_2O} \gg R_{C+CO_2} \gg R_{C+H_2}, \tag{11}$$

where R represents the relative reaction rate.

According to Di Blasi (2009) the reaction rate of steam gasification of wood char is about 2–5 orders of magnitude faster than that of CO_2 and gives a higher calorific value of the produced syngas. Generally, char gasification rates are always lower than char oxidation rates. Basu (2013) reports that char gasification with steam is 3–5 orders of magnitude slower than char oxidation reaction. In addition, char gasification reaction with CO_2 (Boudouard reaction) is 6–7 orders of magnitude slower than char oxidation reaction. Finally, char-hydrogen reaction (hydrogasification) is the slowest of all the above-mentioned reactions.

6.4.4.1 *Char reactivity to H_2O (waster-gas reaction)*

Char gasification in steam, known as the *waster-gas reaction*, is considered as the most important heterogeneous gasification reaction (Eq. 5). It is reported that at a fixed steam concentration, char reactivity increases with increase in temperature, and in contrast, char reactivity to steam increases with increase in steam flow rate at a fixed temperature (Tilghman & Mitchell, 2015). However, ash impurities such as silicon and aluminum reduce char reactivity. It was also found that the addition of 2.7% of hydrogen to the reaction decreased the reactivity. Char reactivity also increases with increase in steam partial pressure. Le and Kolaczkowski (2015) found that if partial pressure of steam increased from 0.333 to 0.667 bar, at 850°C, then RDF-derived char reactivity increased. However, at higher conversion rates (>60%), its apparent reactivity decreased and was similar to chars derived from coal.

6.4.4.2 *Char reactivity to CO_2 (Boudouard reaction)*

CO_2 char gasification is known as the Boudouard reaction (Eq. 4). Tilghman and Mitchell (2015) reported that the reactivity of pure corn stover char to CO_2 was affected by temperature and content of CO_2. Char conversion rates increased with increasing temperature and CO_2 concentration. Moreover, upon the addition of CO to the gaseous environment, the mass loss rate was lowered, and CO inhibits heterogeneous reactions. Di Blasi (2009) also indicated that char gasification in the presence of CO_2 does not occur significantly for temperatures below 1000 K.

6.4.4.3 *Char reactivity to O_2*

Since the main gasification reactions are endothermic, a certain amount of external heat is required for heating (reaction start-up), drying, and pyrolysis. Therefore, this heat can be obtained by exothermic reactions (Eqs. 2, 3) occurring to some extent in a gasifier. CO reaction (3) gives the highest amount of heat (394 kJ/mol) and its rate is much faster, whereas PO reaction (2) provides only 111 kJ/mol of heat and is relatively slow. However, during PO reaction, carbon monoxide is produced, which constitutes a portion in syngas. Char reactivity in the ambient air (or oxygen) increases with increasing temperature as well as air (oxygen) content. Air flow rate, which is directly proportional to both ER and superficial velocity, influences the gasification products and affects the gasification temperature. Higher air flow rate results in higher temperature, thus leading to higher char conversion rate and a higher quality of syngas. However, the excess rate of oxidant may decrease energy content of the produced gas by diluting it with carbon dioxide and nitrogen, and will increase the temperature in the gasifier unnecessarily.

6.5 Chemical kinetics of char gasification

Since char gasification reaction is a heterogeneous reaction proceeding in various gasifying media, it is therefore important to know the reaction kinetics to provide information for a gasifier's design and models simulating the gasification process. Char reactivity models are mostly referred to as a structural rate or volumetric rate. Structural reaction rate models are used to describe the internal solid matrix or the internal pore structure of the particle during char conversion. In contrast, for the volumetric reaction rate model, the changes in the porous structure of the char particle are described by empirical correlations where porosity or pore properties do not appear explicitly (Di Blasi, 2009).

6.5.1 Kinetics of solid-gas reactions

6.5.1.1 Carbon-CO_2 reaction mechanism

A char gasification kinetic mechanism in a carbon dioxide medium can be described through several steps followed by an oxygen exchange model (Eqs. 12–14). In the first step, CO_2 dissociates at an active carbon site (C_f), releasing CO and forming a carbon-oxygen complex C(O). This reaction can proceed in the opposite direction, thus forming an active site and carbon dioxide (the second step). In the third step, the C(O) complex forms a molecule of CO:

$$C_f + CO_2 \xrightarrow{k_1} C(O) + CO, \tag{12}$$

$$C(O) + CO \xrightarrow{k_2} C_f + CO_2, \tag{13}$$

$$C(O) \xrightarrow{k_3} CO, \tag{14}$$

where k_1, k_2, and k_3 are the Arrhenius rate constants, C_f is an active carbon site, and C(O) a carbon-oxygen complex.

When the CO concentrations are rather small and/or the inhibiting effect lowering the steady-state concentration of C(O), which is caused by the presence of CO, is not taken into account, a simple global model can be applied:

$$C + CO_2 \rightarrow 2CO, \quad r_c = A \exp(-E/RT) P^n_{CO_2}, \tag{15}$$

where r_c is the chemical kinetic term, A is the preexponential factor, E is the activation energy, R is the universal gas constant, and n is an empirical parameter.

6.5.1.2 Carbon-H_2O reaction mechanism

Char gasification under steam is more complex than CO_2 gasification since the involvement of H_2, CO_2, and CO depends on the equilibrium of the WGS reaction (Eq. 7). Basically, the carbon-steam reaction mechanism involves two models: the oxygen exchange model and the hydrogen inhibition model. Therefore, the carbon-H_2O reaction has several steps:

$$C_f + H_2O \xrightarrow{k_1} C(O) + H_2, \tag{16}$$

$$C(O) + H_2 \xrightarrow{k_2} C_f + H_2O, \tag{17}$$

$$C(O)\xrightarrow{k_3}CO, \tag{18}$$

$$C_f + H_2\xrightarrow{k_4}C(H)_2, \tag{19}$$

$$C(H)_2\xrightarrow{k_5}C_f + H_2, \tag{20}$$

$$C_f + 0.5H_2\xrightarrow{k_6}C(H), \tag{21}$$

$$C(H)\xrightarrow{k_7}C_f + 0.5H_2, \tag{22}$$

where k_1–k_7 are Arrhenius rate constants.

The oxygen exchange mechanism is based on reversible equations (16)–(18), whereas the hydrogen inhibition model is based on irreversible equations (16), (18)–(20). A different version of the hydrogen inhibition reaction mechanism substitutes Eqs. (19), (20) with Eqs. (21), (22). In that case, the hydrogen inhibition mechanism involves the following steps: (16), (18), (21), and (22). Each model has a different explanation of the inhibition effect of hydrogen. According to the oxygen exchange model, it is due to the equilibrium of the dissociation reactions (16), (17), releasing hydrogen and forming a surface oxide complex C(O). For the traditional hydrogen inhibition model, the formation of the $C(H)_2$ complex is the reason for inhibition. Finally, the second version of the hydrogen inhibition model involves a dissociative chemisorption of hydrogen on the active sites, blocking them for the oxygen transfer reaction with steam. Basu (2013) showed that hydrogen can have an inhibiting effect on the char gasification rate in steam. A proposed solution for the effective accelerating of the char-H_2O reaction and therefore minimizing hydrogen inhibition is continuous removal of hydrogen from the reaction site. Similar to CO_2 gasification, the kinetic mechanism considers a global model for char gasification with steam:

$$C + H_2O \rightarrow CO + H_2, \quad r_c = A\exp(-E/RT)P^n_{H_2O}. \tag{23}$$

Di Blasi (2009) detailed the expressions of the char conversion rates with parameter values.

6.5.1.3 Carbon-O_2 reaction mechanism

The char combustion mechanism for coal chars (Hurt & Calo, 2001) can also, in principle, be applicable for the explanation of the biomass-derived char combustion mechanism. The most widely used mechanism is based on a simple global reaction (Di Blasi, 2009):

$$C + O_2 \rightarrow CO/CO_2, \quad r_c = k_0\exp(-E/RT)P^n_{O_2}, \tag{24}$$

where P_{O_2} is the partial pressure of O_2. The expressions of the char oxidation rates with parameter values are very well defined by Di Blasi (2009). Generally, the carbon-O_2 reaction mechanism includes series adsorption and desorption processes:

$$2C_f + O_2\xrightarrow{k_1}2C(O), \tag{25}$$

$$C(O) \xrightarrow{k_2} CO, \tag{26}$$

$$2C(O) \xrightarrow{k_3} CO_2 + C_f, \tag{27}$$

$$C(O) + O_2 \xrightarrow{k_4} CO_2/CO + C(O), \tag{28}$$

$$C_f + CO_2 \xrightarrow{k_5} CO + C(O), \tag{29}$$

where k_1–k_5 are Arrhenius rate constants, C_f is an active carbon site, and C(O) is a carbon-oxygen complex. Eq. (25) represents the chemisorption of O_2 on active carbon sites producing a C(O) complex and Eq. (26) the formation of CO through a desorption mechanism.

6.5.1.4 Hydrogasification reaction mechanism

Since char gasification with hydrogen leads to the formation of methane, the reaction is known as a hydrogasification reaction (Eq. 6). However, its rate is much slower than other gasification reactions and does not have any significant impact on the process. Therefore, a deeper discussion regarding this reaction is not considered:

$$C + 2H_2 \rightarrow CH_4.$$

6.5.2 Kinetics of gas-phase reactions

The most important gas-phase reaction is the WGS (Eq. 7) and to some extent the intermediate methanation reaction (Eqs. 8, 9).

6.5.2.1 Water-gas shift reaction mechanism

The WGS reaction is used to increase H_2 content in the produced synthesis gas at the expense of CO, or in the other words to control the ratio of H_2/CO present in the synthesis gas and its further utilization for energy/fuels or chemicals production. It is a reverse slightly exothermic reaction mostly depending on temperature:

$$CO + H_2O \leftrightarrow CO_2 + H_2$$

The WGS reaction reaches equilibrium at higher temperatures (above 1000°C). However, increasing temperature decreases the yield of H_2 but the reaction rate increases. Therefore, at an industrial scale to balance the reaction rate (at higher temperatures) and achieve higher yields of H_2 (at lower temperatures), a two-stage WGS reactor with catalysts is used (Gradisher, Dutcher, & Fan, 2015). The first stage, or high-temperature stage, uses iron-based catalysts with chromium oxides and proceeds between 350°C and 450°C, giving a high reaction rate, but a low conversion. The second stage, or low-temperature stage, proceeds at 200–260°C and involves copper, zinc oxide, and alumina-based catalysts. Thus, the conversion is increased at the expense of reaction rate. Usually, catalytic materials used to produce heterogeneous catalysts include iron, copper, cobalt/molybdenum, zinc, gold, platinum, ruthenium, and rare earth metals such as cerium, samarium, gadolinium, and lanthanum (Lang, Secordel, Kiennemann, & Courson, 2017; LeValley, Richard, & Fan, 2014).

6.6 Gasification technologies and their performance evaluation

Generally, gasifiers are classified into three basic types: fixed or moving bed, fluidized bed, and entrained flow (Pereira & Martins, 2017). This classification is based on how solid feedstock interacts with gasifying medium (gas-solid contact) as well as type of gasifying agent. Molino, Chianese, and Musmarra (2016) and Arena (2012) distinguish two more types of gasifiers: rotary-kiln and plasma gasifiers. Each type can be subdivided into a specific commercial-type gasifier. For instance, the moving-/fixed-bed gasifiers refer to the Lurgi Mark IV (dry bottom) and the British Gas Lurgi slagging gasifier. Numerous entrained-flow commercial gasifiers are produced by Conoco-Philips E-gas (slurry fed), General Electric (slurry fed), Shell (dry fed), Prenflo, Siemens, and Mitsubishi Heavy Industries. Finally, the fluidized-bed gasifiers include the U-Gas process, Winkler, and KBR transport gasifier (Breault, 2010; Watson, Zhang, Si, Chen, & de Souza, 2018). A scheme illustrating the types of gasifiers is shown in Fig. 3.

Each type of gasifier has its specific range of operational capacity to operate properly and efficiently (Fig. 4). As an example, the downdraft gasifiers are characterized by a smaller capacity, while the entrained-flow gasifiers are the largest capacity gasifiers. Operational capacity range and size of rotary-kiln and plasma gasifiers are close to the entrained-flow gasifiers.

Comparison between different gasifiers' parameters is summarized in Table 3.

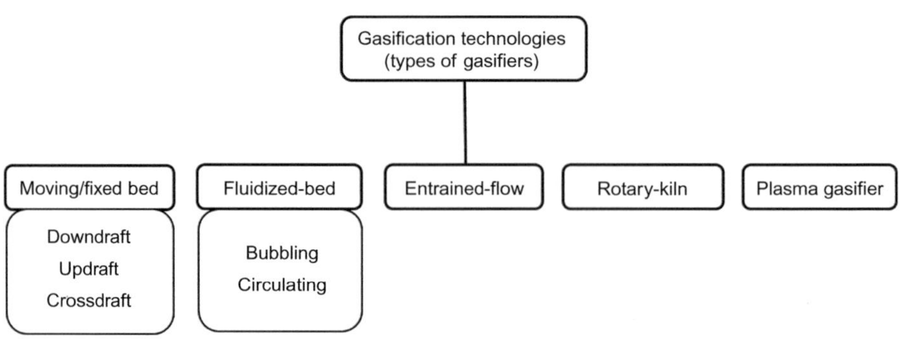

Fig. 3 Types of gasification technologies.

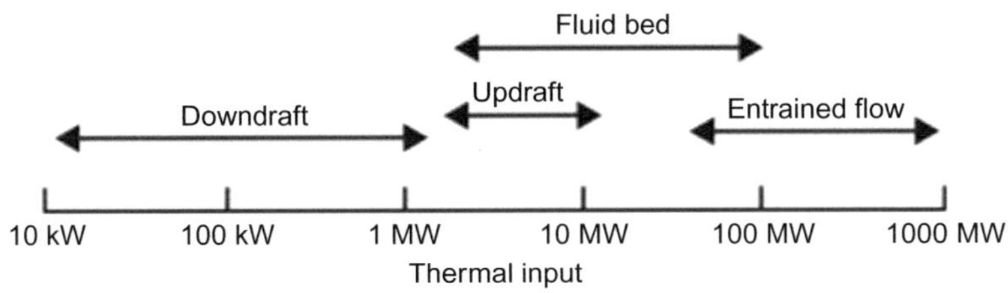

Fig. 4 Range of operational capacity of various types of gasifiers. Copyright 2013 Elsevier.

Table 3 Comparison of characteristics of different gasifiers.

	Fixed bed/moving bed	**Fluidized bed**	**Entrained flow**	**Rotary kiln**	**Plasma**
Capacity range	$<10\,MW_{th}$	5–100 MW_{th}	$>50\,MW_{th}$	$>50\,MW_{th}$	$>50\,MW_{th}$
Feedstock flexibility	Poor	Good	Good	Very good	Very good
Particle size	No larger than 100 mm	<6 mm	0.1–1 mm	Fines to large (any size)	Fines to large (any size)
Morphology	Uniform	Uniform	Uniform	No problem	No problem
Moisture content	Up to 60%—updraft gasifier Up to 25%—downdraft gasifier	<55%	<15%	No problem	<60%
Ash content (dry basis)	<5%–15%	<25%	<20%	<40%	<3%–10%
Ash melting point	>1000–1250°C	>1000°C	<1250°C	No problem	No problem
Bottom ash status	Dry	Dry	Slagging	Slagging	Slagging
Operational temperature	700–900°C	800–900°C	1300–1500°C	850–1200°C	Up to 10,000°C 2000–5000°C (in contact area with waste)
Tar content	Very low to very high <0.01–$0.1\,g/Nm^3$ crossdraft gasifier $3\,g/Nm^3$ downdraft gasifier 30–150 g/Nm^3 updraft gasifier	Low $10\,ng/m^3$	Very low	High	Very low $<1\,g/Nm^3$
Oxidant requirements	Low	Moderate	High	High	Low

Continued

Table 3 Continued

	Fixed bed/moving bed	Fluidized bed	Entrained flow	Rotary kiln	Plasma
Start-up/ shut-down/ control	Moderate 20–30 min to start downdraft gasifier More than 30 min to start updraft gasifier	Easy—bubbling fluidized bed Difficult to control—circulating fluidized bed	Good	Difficult with limited process flexibility	Very fast
Cold gas efficiency	80%	89%	80%	–	40%–90%
Typical product gas composition					
Air gasification	H_2 (vol%): 11–16 CO (vol%): 13–18 CO_2 (vol%): 12–16 CH_4 (vol%): 2–6 N_2 (vol%): 45–60				
Oxygen gasification	H_2 (vol%): 23–28 CO (vol%): 45–55 CO_2 (vol%): 10–15 CH_4 (vol%): 0–1 N_2 (vol%): 0–1				
Steam gasification	H_2 (vol%): 35–40 CO (vol%): 25–30 CO_2 (vol%): 20–25 CH_4 (vol%): 9–11 N_2 (vol%): 0–5				
Gas LHV	4–6 MJ/Nm^3 (fuel: wood, charcoal + air) 6–11 MJ/Nm^3 with steam	4–11 MJ/Nm^3 (fuel: coal + oxygen) Use of steam leads to a higher LHV rate	4–11 MJ/Nm^3 (fuel: coal + oxygen) Use of steam leads to a higher LHV rate	10–12 MJ/ Nm^3 (fuel: biomass + steam) Use of steam leads to a higher LHV rate	6–12 MJ/Nm^3 (fuel: coal, char + steam) Use of steam leads to a higher LHV rate
Carbon conversion	High (downdraft)	High	Very high	High	High
Residence time	Minutes to hours	Minutes to hours	Very short (few seconds)	Very long (1 – 2h)	Extremely short
Scale-up	Good to limited No problems—updraft gasifier Limited—downdraft gasifier	Good	No problems	No problems	No problems
Costs (investment, maintenance)	Relatively law	High investments Maintenance relatively cheap	High (plant and maintenance costs)	High maintenance costs Low investment costs	High

LHV, lower heating value.

6.6.1 Fixed-bed/moving-bed gasifiers

In a fixed-bed (or moving-bed) gasifier, the feedstock (biomass, etc.) moves down the reactor and settles on a grate. Fixed-bed gasifiers are simple, reliable, and robust small-sized technology. Due to the small size of the reactor they are inexpensive to build and are an attractive solution in small-scale applications. Furthermore, due to limited flexibility to load, requirement of feedstock homogeneity and size uniformity, and relatively poor heat transfer and mixing, it is difficult to achieve uniform distribution of feedstock, temperature, and producer gas composition across the cross-section of the gasifier (Arena, 2012; Belgiorno, De Feo, Della Rocca, & Napoli, 2003; Molino et al., 2016). Three types of fixed-bed/moving-bed gasifiers are distinguished according to the technological solutions of how treated material and gasifying agent are supplied to the gasifier: updraft, downdraft, and crossdraft gasifiers (similar to downdraft).

6.6.1.1 Updraft gasifier

In the updraft gasifier the feedstock is loaded from the top (moves downward), while the gasifying agent (air, oxygen, or steam) travels upward (at the bottom). Therefore, the feedstock and the gasifying agent (as well as the producer gas) are in countercurrent mode (Fig. 5).

In the updraft gasifier, feedstock with a high content of moisture (up to 60%) can be used, since the gases pass through the different zones (drying, pyrolysis, reduction, and oxidation) successfully and thus the feedstock is dried in the top of the gasifier. The methane and tar-rich syngas leave at the top of the gasifier, whereas the ash falls from the grate for collection at the bottom (Arena, 2012). Tar production is very high ranging from 30 to 150 g/Nm^3, which makes it unsuitable for high-volatility

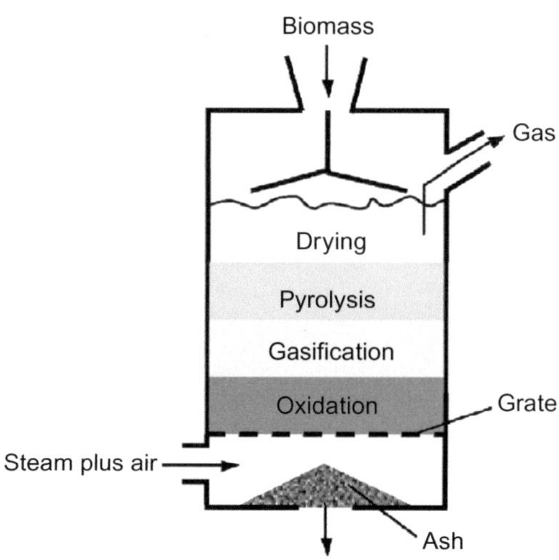

Fig. 5 Schematic of an updraft gasifier (Basu, 2013).
Copyright 2013 Elsevier.

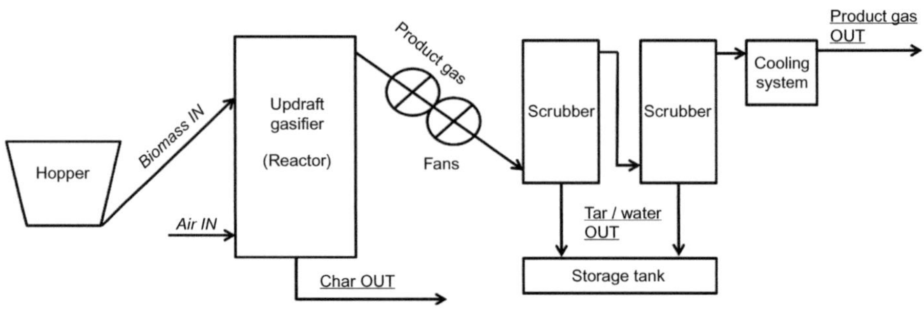

Fig. 6 The scheme of the updraft gasification unit used for poultry litter gasification (Taupe et al., 2016).

feedstocks. However, the updraft gasifier utilizes combustion heat very efficiently, thus achieving high CGE (Basu, 2013).

There are a number of investigations using updraft gasifiers for various biomass gasifications. Taupe et al. (2016) gasified poultry litter in the updraft gasifier using air as a gasifying agent. The moisture and ash content in the feedstock were around 30% (wt.% as received) and 20% (wt.% dry basis), respectively. The scheme of the updraft gasification unit is presented in Fig. 6.

Due to the low ash melting temperature of the poultry litter, which was determined to be 639°C, the reactor temperature was kept around this value. As a consequence of lower reactor temperature the process performance parameters were low. The CGE was 26% and the carbon conversion efficiency (CCE) was 44%. The mass balance showed that poultry litter was converted to raw product gas and char. The product gas yield represented 57.8 wt.% with a LHV of 3.39 MJ/Nm^3, the uncovered char accounted for 15.6 wt.%, and the tar-water emulsion accounted for 20.3 wt.% of the total input stream. The energy balance showed that 55.7% of the LHV was stored in the products, whereas the remaining 44.3% of the energy was utilized to evaporate moisture and dry feedstock, heat the feedstock, and manage thermal losses to the environment. The carbon balance revealed that a significant amount of carbon remained unconverted in the char (27.2 wt.%), while 44.7 wt.% was turned to lower molecular weight gaseous products. The gasification char was heterogeneous in nature due to uneven temperature distribution inside the gasifier. It was also found that the char exceeded the concentration threshold for Zn and Cu to achieve European biochar certification.

6.6.1.2 Downdraft gasifier

In the downdraft gasifier, the feedstock is supplied at the top of the gasifier, while the oxidizing agent is supplied at a certain height below the top or the sides, i.e., feedstock and gases move in the same co-current direction. The producer gas also moves downdraft in the same zones of the updraft gasifier but in a different order and leaves from the lower section of the gasifier through a bed of hot ashes (Fig. 7). Due to feedstock

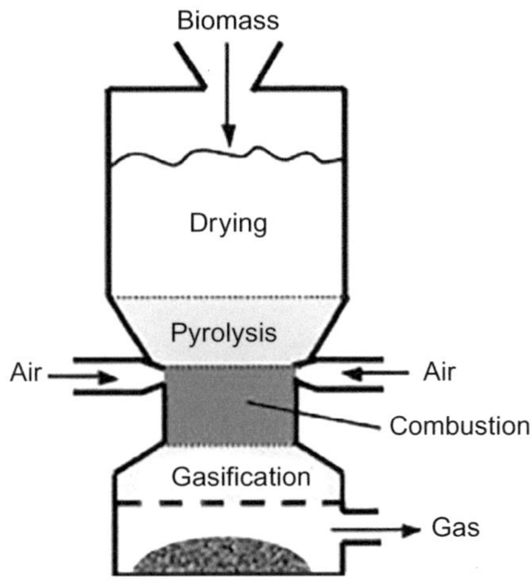

Fig. 7 Schematic of a downdraft gasifier (Basu, 2013).
Copyright 2013 Elsevier.

combustion in the gasifier's oxidizing zone, a bed of hot char/ashes forms. Formed producer gases have to pass through this bed. Therefore, a high-temperature zone ensures effective cracking of tars existing in the producer gas. As a result, the downdraft gasifier has the lowest tar production rate (up to 0.1–3 g/Nm^3) (Belgiorno et al., 2003; Breault, 2010; Molino et al., 2016; Striūgas, Zakarauskas, Džiugys, Navakas, & Paulauskas, 2014; Watson et al., 2018).

A downdraft gasifier also has the same drawbacks as an updraft gasifier, especially regarding its limited flexibility to load, requirement of homogeneous and uniform size (no larger than 100 mm) of feedstock, and low coefficient of heat transfer. Another major drawback is the limited ability of scale-up due to the problems of temperature control. However, the updraft gasifier has no problems with scale-up. Striūgas, Valinčius, Pedišius, Poškas, and Zakarauskas (2017) performed an investigation on sewage sludge gasification combining a downdraft gasifier with and without plasma-assisted cleaning of the producer gas. The comparative results are shown in Table 4.

Table 4 shows that the composition of the constituents of the producer gas decreased except N_2. The latter concentration slightly increased from 48.80 to 52.75 vol.% due to the extra air used for plasma stream generation. However, slightly reduced concentrations of H_2, CO, and CH_4 affected the producer gas calorific value, but increased the yield. In fact, the LHV decreased from 5.28 to 4.82 MJ/Nm^3, but the producer gas yield increased from 2.63 to 3.68 nm^3/kg. After the use of plasma, the tar content and chlorine compounds in the form of hydrochloric acid (HCl) were reduced from 2.03 and 2.34 mg/Nm^3 to 0.09 and 1.89 mg/Nm^3, respectively. The solid by-product obtained after gasification was found in the form of char (3.4 kg/h). The char was 17 wt.% from the initial feedstock, in which 46.6 wt.% of carbon was obtained. The performance of the sewage sludge gasification with plasma cleaning

Table 4 Comparison of gasification performance parameters with and without plasma cleaning (Striūgas et al., 2017).

Parameter	After gasifier	After plasma reactor
Fuel load, kg/h	19	19
Plasma power, kW	–	52
Air flow, Nm^3/h:	32.6	52
Primary	20	20
Secondary	1	1
Tertiary	11.6	11.6
Air to plasma reactor	–	19.4 (7 g/s)
Air equivalence ratio	0.36	0.58
Producer gas, Nm^3/h	50	70
Producer gas yield, Nm^3/kg	2.63	3.68
Average producer gas composition, vol.% (dry):		
H_2	14.62	13.19
CO	22.20	23.42
CH_4	2.37	1.17
C_2H_2	0.11	0.01
C_2H_6	0.13	0.01
C_3H_8	0.03	n.d.
CO_2	10.30	8.72
N_2	48.80	52.75
Producer gas LHV, MJ/Nm^3	5.28	4.82
Tar, g/Nm^3	2.03	0.09
Chlorine (HCl), mg/Nm^3	2.34	1.89
Hexachlorobenzene, $\mu g/Nm^3$	n.d.	n.d.
Solid particulates (PM) before ESP, mg/Nm^3		270
Solid particulates (PM) after ESP, mg/Nm^3		60
Char yield, kg/h		3.4
Carbon content in char, wt.%		46.6
Cold gas efficiency, %	74.3	65.0
Hot gas efficiency, %	83.2	81.7

ESP, *electrostatic precipitator; LHV*, lower heating value; *n.d.*, no data; *PM*, particulate matter.

decreased from 74.3% to 65%. This was affected by the plasma torch running on electrical energy. However, despite the reduced calorific value of the syngas and decrease in CGE, the total produced energy per 1 kg of feedstock with plasma-assisted gas cleaning increased from 15.2 to 22.6 MJ/kg.

6.6.2 Fluidized-bed gasifiers

Fluidized-bed gasification was developed to solve operational problems, such as gas-solid mixing and contact, temperature uniformity and control, efficiency, etc., inherent in fixed-/moving-bed gasifiers. The fluidized-bed reactor includes a bed of inert granular material, typically silica or sand (optionally olivine or dolomite). The bed material is transformed into a liquid-like state (fluidized state) by contact with an upward flowing gasification agent through it at appropriate velocities (2–3 or 5–10m/s) (Belgiorno et al., 2003). The feedstock is generally fed into a bed of hot materials and the ash, containing organic/inorganic matter, is drained from the bottom of the bed. The fluidized-bed gasifiers operate without different reaction zones under nearly isothermal conditions and are suitable for medium-size gasification units with capacity <25 MW_{th}. They are also characterized by low tar content in the produced syngas, which is between updraft and downdraft gasifiers and is around 10 g/Nm^3. There are two basic types of fluidized-bed gasifiers: bubbling and circulating.

6.6.2.1 Bubbling fluidized-bed gasifier

In the BFB the bed of inert material behaves like a liquid and is continuously stirred by the presence of gas bubbles (gasifying agent) whose mobility ensures uniform gasification conditions (Fig. 8). The fluidized state ensures an intense mixing and exchange of heat and mass transfer between gas and solid contact. Also, depending on operating conditions, BFB gasifiers can operate at low and high temperatures, and at atmospheric or elevated pressures.

Most gasification reactions in the BFB gasifier take place inside the bed and the region just above the top of the bed, where there is maximum turbulence induced by bubble eruption (Arena, Zaccariallo, & Mastellone, 2010; Watson et al., 2018). The produced syngas flows upward along the freeboard (upper part of the gasifier where only the gas phase is present and is not reached by the bed material). The bed temperature is kept below 900°C to avoid ash melting and consequent agglomeration. In the BFB gasifiers the velocity of the upward flowing gasifying agent is between 1 and 3 m/s and the bed material and particles of feedstock do not emerge from the reactor because of the low velocity (Belgiorno et al., 2003). Since there

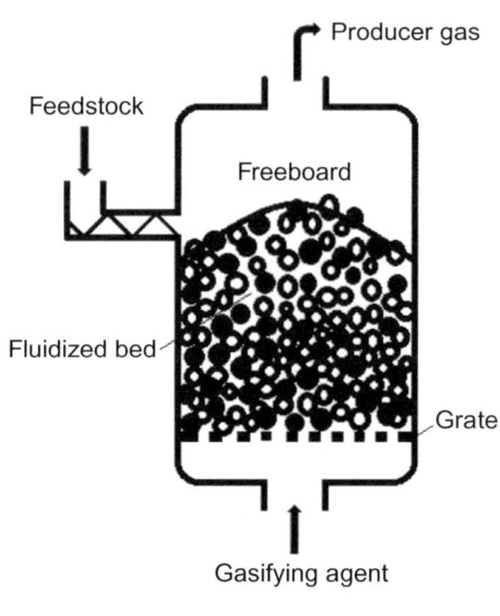

Fig. 8 Schematic of the bubbling fluidized-bed gasifier.

are no moving parts inside the gasifier, maintenance is simple and relatively cheap. However, one of the biggest disadvantages of this type of gasifier is the need to have a relatively low process temperature to avoid fluidization of the bed. Such conditions appear when the temperature is lower than the softening point of the solid residues. Moreover, some residual carbon can be lost in the ashes during the gasification process (Higman & van der Burgt, 2008).

Pio, Tarelho, and Matos (2017) demonstrated forest biomass (pine wood, eucalyptus, wood pellets) gasification in an 80 kW_{th} pilot-scale BFB gasifier operating on air. The operating conditions were, namely, ER between 0.17 and 0.36 and bed temperature in the range of 700–850°C. The process was autothermal and operated under steady-state conditions. The composition of the producer gas was in the range of 14.0%–21.4% CO, 14.2%–17.5% CO_2, 3.6%–5.8% CH_4, 1.3%–2.4% C_2H_4, 2.0%–12.7% H_2, and 48.9%–61.1% N_2. Therefore, the LHV of syngas was between 4.4 and 6.9 MJ/Nm^3. It was determined that process performance parameters such as the gas yield, CGE, and CCE were influenced by the fuel type, operating conditions, and reactor type. As a result, gas production was between 1.2 and 2.2 Nm^3/kg biomass, CGE was between 41.1% and 62.6%, and CCE was between 60% and 87.5%.

6.6.2.2 Circulating fluidized-bed gasifier

In a circulating fluidized-bed (CFB) gasifier the superficial gas velocity increases significantly and is much higher than the terminal solids velocity. The increase in superficial velocity is influenced by the upward flowing gasifying agent, with velocity around 5–10 m/s. Therefore, the expended bed occupies the entire reactor (there is no freeboard any more) and a fraction of the bed material and treated biomass particles (char) are carried out of the gasifier together with the gas stream through a riser. Further, this fraction is captured, recycled, and returned to the bottom of the reactor using a cyclone (Cahyadi et al., 2017). Due to the recirculation, the gasifier is called a "circulating fluidized bed" (Fig. 9).

A CFB gasifier is especially suitable for fuels with a high content of volatiles. The temperature in the reactor varies from 800°C to 900°C to avoid ash melting and sintering. Despite the advantageous characteristics, such as load and feedstock flexibility, high heat and mass transfer rates, uniformity of high temperatures, improved CCE, etc. these types of gasifiers suffer from high tar and dust-related problems as well as complex technology, difficult control, and high investment costs. There are also some developed versions of CFB gasifiers varying only in engineering details: transport gasifier, twin reactor system, and chemical looping gasifier. These gasifiers are described in more detail in the literature (Basu, 2013).

Gomez-Barea, Ollero, and Leckner (2013) studied the fundamentals and methods of how to improve char and tar conversion in fluidized-bed reactors utilized for biomass and waste gasification. The defined problem relies on relative low gasification temperature inside the gasifier's reactor to prevent ash agglomeration. As a consequence, carbon conversion slows down and generated gas is rich in high concentrations of tars, thus resulting in reduced overall gasification efficiency. It was stated that char conversion can be increased by solids recirculation in a directly heated

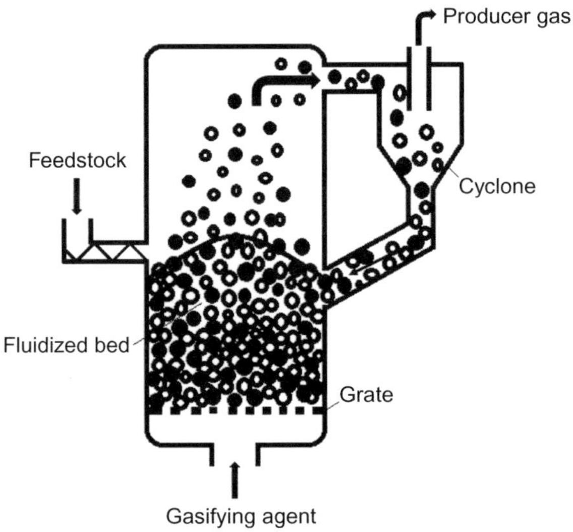

Fig. 9 Schematic of a circulating fluidized-bed gasifier.

fluidized-bed reactor (stand-alone units) or by burning the char in a separate chamber in an indirectly heated reactor. In other words, the use of staged gasification may help to reach a higher char conversion rate, yielding a gas with a low-level concentration of tars.

6.6.3 Entrained-flow gasifiers

An entrained-flow gasifier is used for large-scale (>50 MW_{th}) gasification. This technology is very flexible to the feedstock composition (coal, petroleum coke, refinery residues, plastic waste, etc.). However, due to a very short residence time (a few seconds) in the reactor, the fed material needs to be very fine (0.1–1 mm), and therefore a grinding process is required. Optionally, slurry feed with added water can be used with a concentration of solids >60%. In such a case, water serves as a transport medium, temperature moderator, and even a reactant promoting hydrogen formation (Arena, 2012). This type of gasifier operates at high pressures (about 25 bar) and temperatures higher than 1400°C (Fig. 10).

Such extreme conditions lead to very little tar content in the producer gas and high quality of syngas. Entrained-flow gasifiers can be divided into slagging and non-slagging. In the slagging gasifier the ash is turned into vitrified inert slag, whereas in the nonslagging gasifier slag is not produced (1% is the maximum allowable ash content).

Entrained-flow gasifiers have a number of advantages over other types of gasifiers, such as fuel flexibility, uniform and high temperatures, high-pressure operation, very low tar content in the produced syngas, very high carbon conversion, short residence time, ash slagging and vitrification, and no problems in scale-up. Despite the mentioned advantages there are a number of disadvantages, for instance, large oxidant requirements, high level of sensible heat in producer gas, requirement of heat recovery

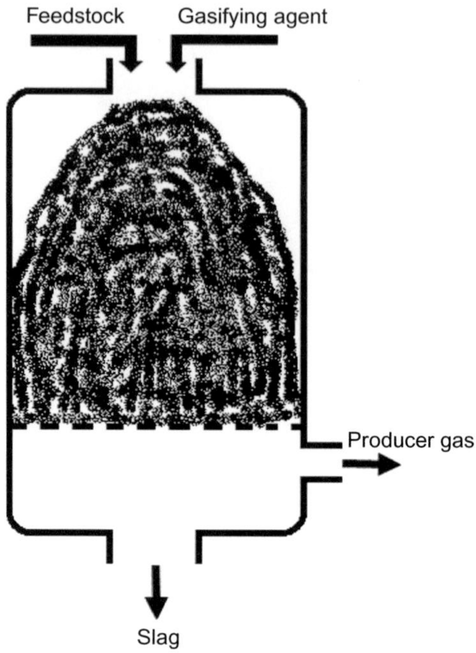

Fig. 10 Schematic of an entrained-flow gasifier.

to improve efficiency, low CGE, requirement of size reduction and preparation supply of feedstock, high plant and maintenance costs, and short life of system components (Molino et al., 2016).

Schneider, Grube, Herrmann, and Ronsch (2016) developed a small-scale entrained-flow gasifier for decentral use of woody biomass. The experiments were performed at temperatures between 950°C and 1100°C, ER 0.38–0.43, and using various fractions of biomass particles with diameters ranging from 0.04 to 1 mm. The gas concentrations of H_2, CO, CO_2, and N_2 varied between 2% and 10%, 4% and 16%, 6% and 11%, and 67% and 83%, respectively. These depended on gasification temperature and particle diameter. The H_2/CO ratio was in the range of 0.55 and 0.81, indicating that such values are typical for gasifiers that operate on dry biomass and oxygen/air gasification media. The best process performance was obtained at a temperature of 1100°C and particle sizes less than 0.2 mm. The values of carbon conversion, CGE, and specific syngas volume varied between 63% and 100%, 14% and 61%, and 0.6 and 1.4 Nm^3/kg, respectively.

6.6.4 Rotary-kiln gasifiers

This type of reactor is usually used for industrial waste incineration and cement production. A rotary-kiln reactor is typically comprised of a slightly inclined (1%–3%) steel cylindrical chamber (drum) that slowly rotates around its own axis (Fig. 11).

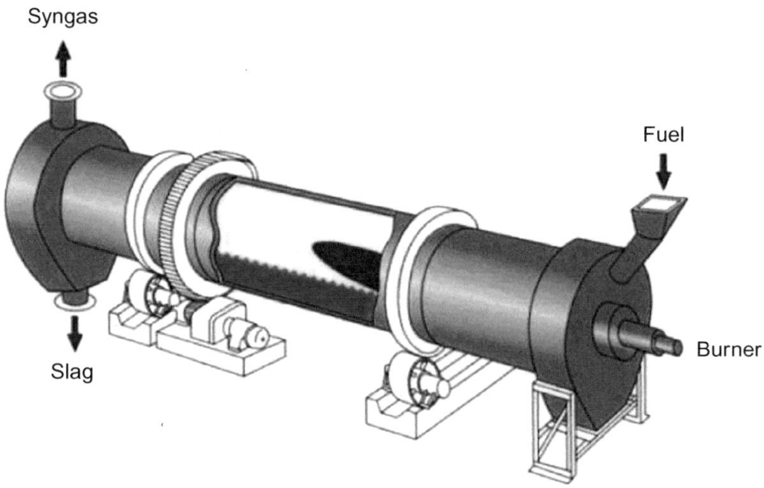

Fig. 11 Schematic of a rotary-kiln gasifier (Molino et al., 2016).
Copyright 2016 Elsevier.

The gas-solid contact is ensured by periodic rotation of the drum that exposes the new solid surfaces to the gasification agent. However, heat and mass transfer between the treated waste and the gasifying medium is not very effective, but the residence time is much higher (1–2 h) than other already mentioned gasification technologies. These types of reactors do not have shortcomings regarding the sensitivity of waste (composition, humidity), scale-up, conversion rate, reliability in operation, and simplicity of construction. However, the major drawbacks are low capacity of heat and mass exchange, high content of tar and dust, limitation in temperature controlling and start-up, high maintenance costs, etc. (Schneider et al., 2016).

Freda, Cornacchia, Romanelli, Valerio, and Grieco (2018) investigated the potential of energy recovery from sewage sludge in a rotary-kiln gasifier. The experiments were carried out under the conditions described in Table 5.

The main results on producer gas composition and energy contents are described in Table 6.

The CGE was estimated equal to 67% at 850°C, ER 0.16. The solid by-product was char composed mainly of ash (70 wt.%).

6.6.5 Plasma gasification

Plasma gasification is a new and promising thermochemical treatment method. The heart of the technology lies in the gasification setup, called plasma torch, which heats the feedstock at high temperatures (from several thousands to tens of thousands). Such high temperatures are reached due to the electric arc burning between two electrodes in the discharge chamber of the plasma torch. A gasifying medium is fed into the discharge chamber by passing through the region of the burning arc and is heated

Table 5 Experimental conditions, gas and char yield of sewage sludge treatment in the rotary-kiln reactor.

Test	Kiln rotation speed (rpm)	Kiln slope (degrees)	Sludge mass rate (g/h)	Air vol rate (NL/min)	ER	Gasification temperature (°C)	Char yield (wt.%)	Gas yield (Nm^3/kg)	Char residence time (min)
I	2	+1.5	261	1	0.05	750	60.8	0.45	55
II	2	+1.5	171	2	0.15	800	54.0	0.97	54
III	2	+1.5	237	3	0.16	850	43.7	1.09	36
IV	2	+1.5	244	4.5	0.24	850	36.0	1.13	34

ER, equivalence ratio.

Table 6 Producer gas composition and heating values (Freda et al., 2018).

Test	ER	Temperature (°C)	Gas composition (vol.%)							HHV (MJ/Nm^3)	LHV (MJ/Nm^3)
			CO_2	C_2H_6	C_3H_8	H_2	N_2	CH_4	CO		
I	0.05	750	17	0.9	0.3	23	26	13	20	11.00	9.73
II	0.15	800	15	0.1	0.0	16	47	5	17	6.40	5.64
III	0.16	850	14	0.1	0.0	23	32	7	24	8.64	7.90
IV	0.24	850	15	0.0	0.0	20	38	4	23	7.15	6.47

ER, equivalence ratio; *HHV*, higher heating value; *LHV*, lower heating value.

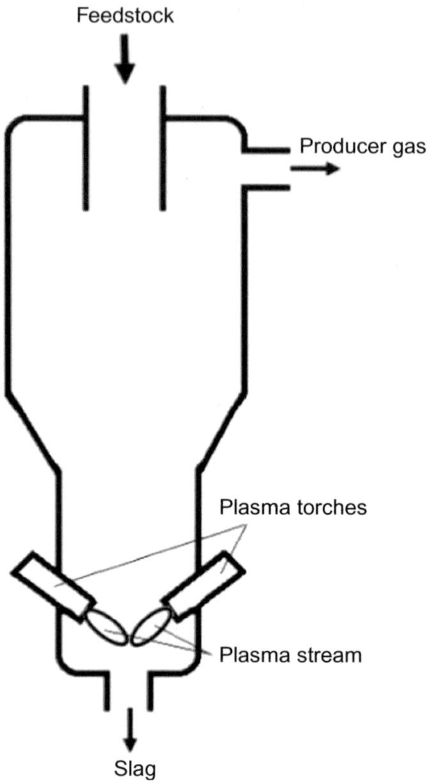

Fig. 12 Schematic of the plasma gasifier.

by joule heating. Where treated material is in contact with the plasma stream, the temperature exceeds 2000–5000°C. Therefore, one of the main advantages of plasma gasification is its ability to treat any hazardous or nonhazardous organic and inorganic feedstock by simply breaking it down into elemental molecules. Organic matter is converted into a high-quality syngas with a very small amount of tar, whereas inorganic fraction is vitrified into inert slag and exits the reactor at the bottom (Fig. 12).

There are several pilot-scale and near-to-market plasma or plasma-assisted gasification systems developed by several well-known companies throughout the world, such as Westinghouse Plasma (AlterNRG), Plasco Energy Group, Europlasma, and InENTec. Plasma gasification can be considered as the reference technologies. Europlasma and Plasco combine plasma gasification with traditional gasification, where plasma acts as a medium for raw synthesis gas cleaning from impurities (tars, inorganics) coming from conventional gasifiers. A plasma torch is also used to melt and vitrify the slag to inert material, which can then be disposed of safety or used as an admixture in construction applications. AlterNRG uses plasma directly for MSW treatment.

The use of plasma in waste/biomass gasification has a number of advantages such as high enthalpy to significantly enhance the reaction rate and kinetics, high chemical reactivity due to production of active radicals and particles, rapid quenching rate, ability to easily control the process parameters, extremely short residence time due to very high temperatures, ash melting to vitrified slag, etc. (Petitpas et al., 2007; Tamošiūnas, Grigaitienė, Valatkevičius, & Valinčius, 2012). It also has some drawbacks. Currently, the major bottleneck limiting plasma technology for a wider commercial application is electrical energy, which is used to run plasma torches. There are other disadvantages too, such as erosion of plasma torch electrodes and their frequent change, high costs of power sources and instability to sustain plasma stream, especially for microwave and radio frequency plasmas, high investment costs, etc. Plasma technology has enormous prospects once the aforementioned limitations have been resolved.

So far, there has been very little research on plasma charcoal gasification. Tamošiūnas et al. (2017) applied a direct current (DC) arc plasma torch operating on water vapor for olive pomace charcoal gasification. A scheme of such a plasma gasification unit is shown in Fig. 13.

The experimental setup consists of a 52.2 kW power DC arc plasma torch with water vapor supply (1), a plasma-chemical reactor (PCHR) (2), a charcoal supply system (3), and syngas sampling and analysis unit (4). The impact of the water vapor-to-charcoal ratio and the plasma torch power on producer gas concentration and quality was investigated. The authors determined that with an increase in water vapor-to-charcoal ratio from 1.85 to 3.15, the volumetric concentrations of H_2 and CO_2 increased from 20.3% and 12.3% to 41.17% and 18.68%, respectively. At the 3.15 to 3.57 ratio, the concentrations of H_2 and CO_2 decreased to 36.19% and 13.48%, respectively, whereas the content of CO increased continually from 8.35% to 14.15%. Such variation was explained by the Boudouard (Eq. 4), water-gas (Eq. 5), and WGS (Eq. 7) reactions. The H_2/CO ratio was in the range of 2.43–3.15, indicating that produced syngas required the adjustment to produce biodiesel via Fischer-Tropsch synthesis. However, a high ratio of H_2/CO indicates produced syngas suitable for hydrogen or biomethane production.

The LHV of synthesis gas changed depending on the water vapor-to-olive pomace charcoal ratio. At the ratio of 1.85 to 3.15, the LHV increased from 3.24 to 6.09 MJ/nm^3, but at the ratio of 3.15 to 3.57, the LHV of the produced syngas decreased from 6.09 to 5.67 MJ/nm^3. This change was explained with regard to the decrease in H_2 concentration determined by a reversed WGS reaction (Eq. 7).

6.6.6 Performance evaluation of gasification system

To evaluate the performance and efficiency of any gasification system, it is convenient to use a commonly accepted methodology, including some quantifying measures. Usually, for the gasification process the following most important performance parameters are used (Petitpas et al., 2007; Tamošiūnas et al., 2016, Tamošiūnas, Valatkevičius, et al., 2017):

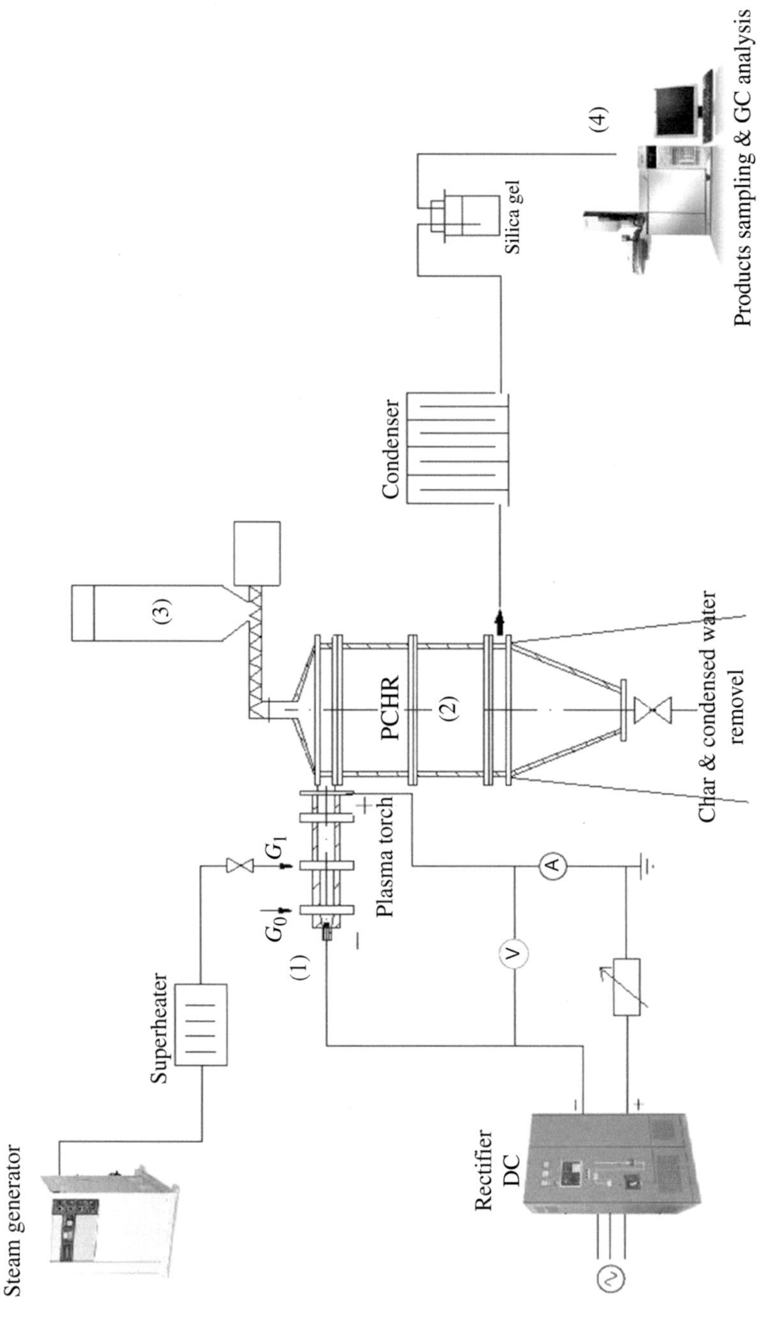

Fig. 13 Scheme of a plasma gasification unit (Tamošiūnas, Chouchene, et al., 2017).

Cold gas efficiency (CGE), defined as the ratio between the chemical energy of the produced syngas and chemical energy of the feedstock fed to the gasifier:

$$\mathrm{CGE} = \frac{m_{\mathrm{syngas}} \times \mathrm{LHV}_{\mathrm{syngas}}}{m_{\mathrm{fuel}} \times \mathrm{LHV}_{\mathrm{fuel}}} \times 100\%, \tag{30}$$

where m_{syngas} and m_{fuel} are the mass flow rates of product gas and feedstock (kg or Nm^3), respectively. $\mathrm{LHV}_{\mathrm{syngas}}$ and $\mathrm{LHV}_{\mathrm{fuel}}$ are net calorific values of product gas and feedstock (MJ/kg or $\mathrm{MJ/Nm}^3$), respectively.

Hot gas efficiency (HGE) is the ratio between the sum of chemical energy and sensible heat of the produced syngas and the sum of chemical energy and sensible heat of the feedstock fed to the gasifier:

$$\mathrm{HGE} = \frac{m_{\mathrm{syngas}} \times \mathrm{LHV}_{\mathrm{syngas}} + m_{\mathrm{syngas}} c_{\mathrm{p}} (T_{\mathrm{f}} - T_0)}{m_{\mathrm{fuel}} \times \mathrm{LHV}_{\mathrm{fuel}}} \times 100\%, \tag{31}$$

where c_{p} is the specific heat capacity (J/kg K), T_{f} is the temperature at the gasifier exit (K), and T_0 is the temperature of the fuel (feedstock) entering the gasifier (K).

Carbon conversion efficiency (CCE) is the ratio between the carbon converted to gaseous products and the carbon existing in the solid feedstock fed to the gasifier. It gives an indication of the amount of unconverted material, thus providing a measure of chemical efficiency of the process:

$$\mathrm{CCE}(\%) = 12 \times Y_{\mathrm{dry\,gas}} \times \left\{ \frac{[\mathrm{CO} + \mathrm{CO_2} + \mathrm{CH_4}] + 2 \times [\mathrm{C_2H_2} + \mathrm{C_2H_4} + \mathrm{C_2H_6}]}{22.4 \times \mathrm{C}} \right\} \times 100\%, \tag{32}$$

where Y_{drygas} is a dry gas yield in Nm^3 per kg of dry feedstock (Nm^3/kg), CO, CO_2, CH_4, C_2H_2, C_2H_4, and C_2H_6 are in % (*v*/v), and C is in % of carbon in dry feedstock.

Energy conversion efficiency (ECE): this parameter is very similar to the CGE, but includes external energy input (allothermal process) such as in the case of plasma used:

$$\mathrm{ECE} = \frac{m_{\mathrm{syngas}} \times \mathrm{LHV}_{\mathrm{syngas}}}{P_{\mathrm{plasma}} + m_{\mathrm{fuel}} \times \mathrm{LHV}_{\mathrm{fuel}}} \times 100\%, \tag{33}$$

where P is a plasma torch power (W).

Specific energy requirements (SER) is the ratio between the input electrical power used by plasma and the amount of the produced syngas:

$$\mathrm{SER} = \frac{P_{\mathrm{plasma}}}{m_{\mathrm{syngas}} \times M_{\mathrm{syngas}}}, (\mathrm{kJ/mol\ or\ kJ/kg\ or\ kWh/kg}), \tag{34}$$

where m_{syngas} is a mass flow rate of syngas (mol/s) and M_{syngas} is the molar mass of syngas (kg/mol).

Higher heating value (HHV) is defined as the amount of heat released when fuel is completely combusted and the products have returned to a temperature of 25°C. The latent heat of water vaporization is included in the total measured heat:

$$\begin{aligned}\mathrm{HHV(syngas)} &= 12.74\mathrm{H_2}(\%) + 12.63\mathrm{CO}(\%) \\ &\quad + 39.82\mathrm{CH_4}(\%) + \cdots + X\mathrm{C}_x\mathrm{H}_y(\%), \left(\mathrm{MJ/Nm^3}\right),\end{aligned} \tag{35}$$

where $H_2(\%)$, $CO(\%)$, $CH_4(\%)$, and $C_xH_y(\%)$ are the contents of gaseous products in producer gas after gasification.

Lower heating value (LHV), also known as a net calorific value, is the amount of heat released when fuel is fully combusted but the latent heat of vaporization of water in the fuel and the reaction products is not recovered, i.e., the latent heat of vaporization of water vapor formed by the combustion is subtracted from the HHV:

$$\mathrm{LHV} = \mathrm{HHV} - L_{\mathrm{steam}}\left(\frac{9H}{100} + \frac{M}{100}\right), \ (\mathrm{kJ/kg}), \tag{36}$$

where L_{steam} is the steam latent heat (2260 kJ/kg), and H and M are hydrogen and moisture percentages.

For calculation of a net calorific value of the produced syngas, it is convenient to use the following equation (Eq. 37):

$$\begin{aligned}\mathrm{LHV(syngas)} &= 10.78\mathrm{H_2}(\%) + 12.63\mathrm{CO}(\%) \\ &\quad + 35.88\mathrm{CH_4}(\%) + \ldots + X\mathrm{C}_x\mathrm{H}_y(\%), \left(\mathrm{MJ/Nm^3}\right),\end{aligned} \tag{37}$$

where $H_2(\%)$, $CO(\%)$, $CH_4(\%)$, and $C_xH_y(\%)$ are the contents of gaseous products in producer gas after gasification.

6.7 Challenges and development

In spite of the growing interest in renewable technologies and the numerous advantages provided by waste/biomass gasification, this conversion process has not yet been able to fully compete with traditional energy production methods, such as combustion and anaerobic digestion. In fact, a wider exploitation and dissemination of gasification technologies is not capable of achieving successful commercialization due to a number of challenges and barriers that have to be properly solved. These challenges include institutional, informational, financial, policy, market, social and environmental, and technological barriers (Dasappa, Sridhar, Sridhar, Paul, & Mukunda, 2013; Ravindranath, 1993; Silveira, 2005).

6.7.1 Institutional challenges

Lack of communication and action coordination between stakeholders (local authorities, communities, NGOs, industry, and science and research institutions) lead to hacked and incomplete policy formation regarding the technological development and promotion of bioenergy production. Instead of developing and disseminating new "greener" technologies, institutions usually have plans with inaccessible targets, a lack of planning, implementation, and management, and are unable to empower local communities. An example illustrating poor supervision and coordination by local authorities is in the United Kingdom where eight biomass gasifiers failed to run. However, starting from the Second World War, Europe has had a long history of development of gasification technologies. However, nothing more tangible has been developed yet. Therefore, more effective management and monitoring by local authorities is required to strengthen the bioenergy sector.

6.7.2 Informational challenges

Unequal awareness and lack of basic knowledge and education of the renewable technologies, especially gasification, among various groups of society are other challenges for a wider spread and dissemination of these technologies. During the past few years, solar, wind, and geothermal/aerothermal energy have been widely applied to domestic households, whereas the energy (heat, electricity) produced from biomass/or waste, regardless of firewood used in domestic stoves in rural areas, is usually used at a higher scale such as combined heat and power plants operating for district heating needs and/or electricity production. Therefore, the scope and potential benefits of gasification technology, including its relation to climate change, global warming, and carbon footprint, do not reach the stakeholders at the proper level.

6.7.3 Financial challenges

Another barrier to the successful spread of gasification technology is financial challenges due to high capital and operational costs of gasification plants. The cost of gasification facilities widely varies and depends on the size and complexity of the plant, the processing of the feedstock, and the syngas application. For instance, the capital costs of a gasification facility include costs of site construction, the biomass pretreatment unit (drying, chips, or briquetting), the gasifier reactor, the producer gas conditioning system (tar and ash removal, filtration, purification, etc.), the engine or gas/steam turbine, and electricity distribution network (Buragohain, Mahanta, & Moholkar, 2010). Moreover, often new technologies require technological upgrade based on R&D for technology advancements and sustainability, which is also costly and not always successfully achieved.

Generally, to promote "green" energy production, a governmental mechanism for subsidies is required since these technologies cannot compete with fossil fuel-based ones. Therefore, a joint financial agreement is needed on financing technology, research, and innovation challenges worldwide, keeping the planet cleaner for future generations. The first commitment was made in 1992 through the Kyoto Protocol with further continuity in 2015 under the Paris Agreement on greenhouse gas emissions mitigation, adaption, and finance starting in the year 2020. A financial agreement was approved by 174 countries to raise an amount of US$100 billion each year starting from 2020 to project completion to reduce greenhouse gas emissions.

6.7.4 Policy barriers

For large-scale gasification plants, one of the revenue streams, despite the other two streams such as gate fee and produced syngas utilization for electricity/fuel production, depends on policy aspects. Renewable energy credits for electricity production and carbon credits for every ton of CO_2 prevented from entering the atmosphere by country differ significantly. For instance, according to the RES LEGAL Europe database, statistics covering all 28 Member States, the EFTA countries, and Members of the Energy Community, as of 2017 the highest feed-in tariff for electricity production from biomass/waste in Lithuania was €ct 6.6 per kWh, Latvia—€ct 0.14 per kWh, Germany—€ct 13.32 per kWh, Italy—€ct 24.6 per kWh, Austria—€ct 22 per kWh, and Turkey—€ct 11.3 per kWh. Striūgas et al. (2016) state that investing in a gasification plant of 500 kWh_{el} would be economically feasible if the energy tariff for sold electricity were around €ct 14 per kWh. Carbon credits can also provide economic benefits by saving a substantial amount of money if nonfossil fuel feedstock or wastes for energy production are being used.

6.7.5 Market barriers

During the last few decades, the worldwide market of gasification technologies has increased significantly. There are more than 272 operating gasification plants worldwide with 686 gasifiers and a total gasification capacity of over 100 GW_{th} of syngas. However, the majority (over 400) are fed by coal as a primary feedstock, followed by petroleum coke (around 100) and natural gas (around 30), which is not in line with the targets of climate change mitigation and carbon footprint. The number of biomass and waste feed gasifiers is currently very small (around 15%), but this feedstock category is expected to grow in the future.

The International Energy Agency (IEA) Bioenergy Task with 33 working groups of international experts from 10 countries (Austria, Denmark, Finland, Germany, Italy, Sweden, Switzerland, the Netherlands, Norway, and the United States) aims to promote the commercialization of efficient, economical, and environmentally preferable thermal biomass gasification processes. Their database of biomass gasification projects indicates around 86 commercial, demonstration, and pilot facilities in the IEA Bioenergy member states, with 62 in operation, 5 under construction, 2 planned, 16 on hold, and 1 where the status is not clear.

The overall gasification market trends are becoming both larger and smaller. The large industrial gasification projects with primary feedstock of coal and petroleum coke (for chemical, H_2, power) are increasing in the number of gasifiers used per project instead of increasing in size, and are concentrated in Asia (China, India) and the Middle East. At the same time, demand for smaller-size, modular gasifiers for biomass and waste gasification is increasing, especially in Europe and the United States. In such a case, the best policy is to use a gasifier that is just big enough to handle the waste of a particular city, rural areas of developing countries, or temporary military bases. Worldwide gasification capacity is expected to grow with the primary growth in

Asia (China, India, South Korea, and Mongolia), which will be based on primary gasification feedstock coming from fossil fuels.

6.7.6 Social and environmental impact

Life-cycle assessment (LCA) of conventional (incineration with energy recovery and landfill with electricity recovery) and advanced energy-from-waste technologies (gasification-plasma (G-PI), fast pyrolysis with combustor (FP-C), air blown gasification (G-SC)) showed better environmental performances over conventional MSW treatment technologies in terms of carbon footprint, global warming potential, acidification potential, eutrophication potential, photochemical ozone creation potential, and human toxicity potential (Evangelisti et al., 2015). Moreover, it was reported that net electrical efficiency was higher for the case of the combination of fluidized-bed gasification with plasma processes over the other two advanced methods used. The net electrical efficiency of the G-PI process was 28%, whereas that of the G-SC was only 8%. The net electrical efficiency of the FP-C process was 26% and was close to that for G-PI. However, this value does not take into account the energy content of the oil used in the producer gas cleaning part. If it is included in the LCA, the net electrical efficiency will be lower.

Another study of gasification and incineration technologies for the disposal of MSW showed quite similar results. It was found that gasification led to four times fewer emissions of SO_2 than incineration processes, leading to less acidification of the air. Further, 33% fewer NO_x emissions were emitted for gasification than incineration leading to less smog production (Maya, Sarmiento, de Sales Oliveira, Lora, & Andrade, 2016). Watson et al. (2018) in a summarized review reported that particulate matter ($PM_{2.5}$) emissions were three times lower for the gasification process than for the combustion process due to the lower NO_x emissions. Such a difference was equal to -0.05 and -0.15 kg $PM_{2.5}$ eq. for the moving-grate combustor and the gasifier, respectively. To sum up, the gasification process leads to net environmental and social benefits over the combustion process.

6.7.7 Technological challenges

Gasification technology has a number of technological challenges, which also prevent its wider prevalence. The presence of tar in syngas is a bottleneck if high-value chemicals are planned to be produced from syngas. In such a case, the tar content should be less than 1 mg/m^3. For electricity production in gas engines or turbines, the syngas should have a tar level lower than 30 mg/m^3. The tar content can be reduced by using catalysts during the gasification process. De Diego et al. (2016) reported that the use of catalytic fibers made of MgO and NiO in a dual fluidized-bed reactor reduced the tar content up to 95%, which was less than 0.2 g/m^3 of the total gas. However, the use of catalysts has another issue, such as poisoning and deactivation. Gasification products of waste usually contain sulfur and other compounds, which are responsible for poisoning of catalysts even at concentrations <10 ppm.

Another major challenge is the separation of gaseous products. A pure stream of syngas is required if high-value chemical products or liquid fuels are planned to be produced. Generally, there are some constructional and design problems such as scale-up, periodic replacement of refractory materials, or even more unique issues if plasma is applied. As a result, conventional incineration combined with the usage of fossil fuels for energy/heat or chemicals production is more flexible, simpler, and cheaper to use.

6.8 Conclusion

In this chapter, char/biomass gasification for producing biofuels, power, and chemicals was reviewed. In particular, basic steps involved in the gasification process, reactivity, and chemical kinetics for different gasification agents were summarized. Particular attention was paid to available gasification technologies and methodologies enabling the evaluation of gasification process performance. In addition, current status, challenges, and development perspectives of gasification technologies were also discussed.

Generally, biomass/char gasification is a promising technique to mitigate climate change. The use of gasification technologies may help to reduce the consumption of fossil fuels and CO_2 emission. Among other renewable energy conversion methods, gasification has great prospects due to the flexibility of the feedstock and for its ability to produce not only energy, but also a wide range of valuable fuels and chemicals. Despite the emerging challenges in commercialization and development issues, gasification has a capability to bridge a gap between cleaner and efficient energy production technologies.

References

Al-Qayim, K., Nimmo, W., Hughes, K., & Pourkashanian, M. (2017). Kinetic parameters of the intrinsic reactivity of woody biomass and coal chars via thermogravimetric analysis. *Fuel*, *210*, 811–825.

Arena, U. (2012). Process and technological aspects of municipal solid waste gasification. A review. *Waste Management*, *32*, 625–639.

Arena, U., Zaccariallo, L., & Mastellone, M. L. (2010). Fluidized bed gasification of waste-derived fuels. *Waste Management*, *30*, 1212–1219.

Basu, P. (2013). *Biomass gasification, pyrolysis and torrefaction: Practical design and theory* (2nd ed.). Elsevier.

Belgiorno, V., De Feo, G., Della Rocca, C., & Napoli, R. M. A. (2003). Energy from gasification of solid wastes. *Waste Management*, *23*(1), 1–15. https://doi.org/10.1016/S0956-053X(02)00149-6.

Bouraoui, Z., Dupont, C., Jeguirim, M., Limousy, L., & Gadiou, R. (2016). CO_2 gasification of woody biomass chars: the influence of K and Si on char reactivity. *Comptes Rendus Chimie*, *19*, 457–465.

Bouraoui, Z., Jeguirim, M., Guizani, C., Limousy, L., Dupont, C., & Gadiou, R. (2015). Thermogravimetric study on the influence of structural, textural and chemical properties of biomass chars on CO_2 gasification reactivity. *Energy*, *88*, 703–710.

Breault, R. W. (2010). Gasification processes old and new: a basic review of the major technologies. *Energies*, *3*, 216–240.

Buragohain, B., Mahanta, P., & Moholkar, V. S. (2010). Biomass gasification for decentralized power generation: the Indian perspective. *Renewable and Sustainable Energy Reviews*, *14*, 73–92.

Cahyadi, A., Anantharaman, A., Yang, S., Karri, S. B. R., Findlay, J. G., Cocco, R. A., et al. (2017). Review of cluster characteristics in circulating fluidized bed (CFB) risers. *Chemical Engineering Science*, *158*, 70–95.

Chevanan, N., Womac, A. R., Bitra, V. S. P., Igathinathane, C., Yang, Y. T., Miu, P. I., et al. (2010). Bulk density and compaction behavior of knife mill chopped switchgrass, wheat straw, and corn stover. *Bioresource Technology*, *101*, 207–214.

Dasappa, S., Sridhar, H. V., Sridhar, G., Paul, P. J., & Mukunda, H. S. (2013). Biomass gasification—a substitute to fossil fuel for heat application. *Biomass and Bioenergy*, *25*, 637–649.

De Diego, L. F., García-Labiano, F., Gayán, P., Abad, A., Mendiara, T., Adánez, J., et al. (2016). Tar abatement for clean syngas production during biomass gasification in a dual fluidized bed. *152*, 116–123.

Di Blasi, C. (2009). Combustion and gasification rates of lignocellulosic chars. *Progress in Energy and Combustion Science*, *35*, 121–140.

Eom, J. Y. K., Kim, T. S., Lee, S. M., Choi, D., Choi, I. G., & Choi, J. W. (2012). Effect of essential inorganic metals on primary thermal degradation of lignocellulosic biomass. *Bioresource Technology*, *104*, 687–694.

Evangelisti, S., Tagliaferri, C., Clift, R., Lettieri, P., Taylor, R., & Chapman, C. (2015). Life cycle assessment of conventional and two-stage advanced energy-from-waste technologies for municipal solid waste treatment. *Journal of Cleaner Production*, *100*, 212–223.

Freda, C., Cornacchia, G., Romanelli, A., Valerio, V., & Grieco, M. (2018). Sewage sludge gasification in a bench scale rotary kiln. *Fuel*, *212*, 88–94. https://doi.org/10.1016/j.fuel.2017.10.013.

Gao, X., Zhang, Y., Li, B., Zhao, Y., & Jiang, B. (2016). Determination of the intrinsic reactivities for carbon dioxide gasification of rice husk chars through using random pore model. *Bioresource Technology*, *218*, 1073–1081.

Gomez-Barea, A., Ollero, P., & Leckner, B. (2013). Optimization of char and tar conversion in fluidized bed biomass gasifiers. *Fuel*, *103*, 42–52.

Gradisher, L., Dutcher, B., & Fan, M. (2015). Catalytic hydrogen production from fossil fuels via the water gas shift reaction. *Applied Energy*, *139*, 335–349.

Guizani, C., Escudero Sanz, F. J., Jeguirim, M., Gadiou, R., & Salvador, S. (2015). The effects of textural modifications on beech wood-char gasification rate under alternate atmospheres of CO_2 and H_2O. *Fuel Processing Technology*, *138*, 687–694.

Guizani, C., Haddad, K., Jeguirim, M., Colin, B., & Limousy, L. (2016). Combustion characteristics and kinetics of torrefied olive pomace. *Energy*, *107*, 453–463.

Guizani, C., Jeguirim, M., Esceduro Sainz, J., Gadiou, R., & Salvador, S. (2016). Biomass char gasification by H_2O, CO_2 and their mixture: evolution of chemical, textural and structural properties of the chars. *Energy*, *112*, 133–145.

Guizani, C., Jeguirim, M., Valin, S., Limousy, L., & Salvador, S. (2017). Biomass chars: the effects of pyrolysis conditions on their morphology, structure, chemical properties and reactivity. *Energies*, *10*(6), 796.

Guo, F., Dong, Y., Dong, L., & Guo, C. (2014). Effect of design and operating parameters on the gasification process of biomass in a downdraft fixed bed: an experimental study. *International Journal of Hydrogen Energy*, *39*, 5625–5633.

Haddad, K., Jeguirim, M., Jellali, S., Delmotte, L., Bennici, S., & Limousy, L. (2017). Combined NMR structural characterization and thermogravimetric analyses for the assessment of the AAEM effect during lignocellulosic biomass pyrolysis. *Energy*, *134*, 10–23.

Hernandez, J. J., Aranda-Almansa, G., & Bula, A. (2010). Gasification of biomass wastes in an entrained flow gasifier: effect of the particle size and the residence time. *Fuel Processing Technology*, *91*, 681–692.

Higman, C., & van der Burgt, M. (2008). *Gasification*. Burlington: Gulf Professional Publishing.

Hurt, R. H., & Calo, J. M. (2001). Semi-global intrinsic kinetics for char combustion modeling. *Combust Flame*, *125*, 1138–1149.

James, A. M., Yuan, W., Boyette, M. D., Wang, D., & Kumar, A. (2014). In-chamber thermocatalytic tar cracking and syngas reforming using char-supported NiO catalyst in an updraft biomass gasifier. *International Journal of Agricultural and Biological Engineering*, *7*, 91–97.

Jeguirim, M., Elmay, Y., Limousy, L., Lajili, M., & Said, R. (2014). Devolatilization behavior and pyrolysis kinetics of potential Tunisian biomass fuels. *Environmental Progress and Sustainable Energy*, *33*, 1452–1458.

Jeguirim, M., Limousy, L., & Dutournie, P. (2014). Pyrolysis kinetics and physicochemical properties of agropellets produced from spent ground coffee blended with conventional biomass. *Chemical Engineering Research and Design*, *92*, 1876–1882.

Jiang, L., Hu, S., Wang, Y., Su, S., Sun, L., Xu, B., et al. (2015). Catalytic effects of inherent alkali and alkaline earth metallic species on steam gasification of biomass. *International Journal of Hydrogen Energy*, *40*, 15460–15469.

Knight, R. A. (2000). Experience with raw gas analysis from pressurized gasification of biomass. *Biomass and Bioenergy*, *18*, 67–77.

Kumar, A., Jones, D. D., & Hanna, M. A. (2009). Thermochemical biomass gasification: a review of the current status of the technology. *Energies*, *2*, 556–581.

Lahijani, P., Alimuddin, Z., Mohamed, A. R., & Mohammadi, M. (2013). CO_2 gasification reactivity of biomass char: catalytic influence of alkali, alkaline earth and transition metal salts. *Bioresource Technology*, *144*, 288–295.

Lajili, M., Guizani, C., Escudero Sanz, F. J., & Jeguirim, M. (2018). Fast pyrolysis and steam gasification of pellets prepared from olive oil mill residues. *Energy*, *150*, 61–68.

Lang, C., Secordel, X., Kiennemann, A., & Courson, C. (2017). Water gas shift catalysts for hydrogen production from biomass steam gasification. *Fuel Processing Technology*, *156*, 246–252.

Le, C. D., & Kolaczkowski, S. T. (2015). Steam gasification of a refuse derived char: reactivity and kinetics. *Chemical Engineering Research and Design*, *102*, 389–395.

LeValley, T. L., Richard, A. R., & Fan, M. (2014). The progress in water gas shift and steam reforming hydrogen production technologies—a review. *International Journal of Hydrogen Energy*, *39*, 16983–17000.

Liu, H., Hu, J., Wang, H., Wang, C., & Li, J. (2012). Experimental studies of biomass gasification with air. *Journal of Natural Gas Chemistry*, *21*, 374–380.

Lu, P., Huang, Q., (Thanos) Bourtsalas, A. C., Chi, Y., & Yan, J. (2017). Experimental research of basic properties and reactivity of waste derived chars. *Applied Thermal Engineering*, *119*, 639–649. https://doi.org/10.1016/j.applthermaleng.2017.03.099.

Maya, D. M. Y., Sarmiento, A. L. E., de Sales Oliveira, C. A. V. B., Lora, E. E. S., & Andrade, R. V. (2016). Gasification of municipal solid waste for power generation in Brazil, a review of available technologies and their environmental benefits. *Journal of Chemistry and Chemical Engineering*, *10*, 249–255. https://doi.org/10.17265/1934-7375/2016.06.001.

McKendry, P. (2002a). Energy production from biomass (part 1) overview of biomass. *Bioresource Technology*, *83*, 37–46.

McKendry, P. (2002b). Energy production from biomass (part 3): gasification technologies. *Bioresource Technology*, *83*, 55–63.

Meng, X., de Jong, W., Fu, N., & Verkooijen, A. H. M. (2011). Biomass gasification in a 100 kWth steam–oxygen blown circulating fluidized bed gasifier: effects of operational conditions on product gas distribution and tar formation. *Biomass and Bioenergy*, *35*, 2910–2924.

Molino, A., Chianese, S., & Musmarra, D. (2016). Biomass gasification technology: the state of the art overview. *Journal of Energy Chemistry*, *25*, 10–15.

Parascanu, M. M., Sandoval-Salas, F., Soreanu, G., Valverde, J. L., & Sanchez-Silva, L. (2017). Valorization of Mexican biomasses through pyrolysis, combustion and gasification processes. *Renewable and Sustainable Energy Reviews*, *71*, 509–522.

Patel, V. R., Upadhyay, D. S., & Patel, R. N. (2014). Gasification of lignite in a fixed bed reactor: influence of particle size on performance of downdraft gasifier. *Energy*, *78*, 323–332.

Pereira, E. G., & Martins, M. A. (2017). Gasification technologies. *Encyclopedia of Sustainable Technologies*, 315–325. https://doi.org/10.1016/B978-0-12-409548-9.10133-2.

Petitpas, G., Rollier, J. D., Darmon, A., Gonzalez-Aguilar, J., Metkemeijer, R., & Fulcheri, L. (2007). A comparative study of non-thermal plasma assisted reforming technologies. *International Journal of Hydrogen Energy*, *32*(14), 2848–2867. https://doi.org/10.1016/j.ijhydene.2007.03.026.

Pio, D. T., Tarelho, L. A. C., & Matos, M. A. A. (2017). Characteristics of the gas produced during biomass direct gasification in an autothermal pilot-scale bubbling fluidized bed reactor. *Energy*, *120*, 915–928.

Ravindranath, N. H. (1993). Biomass gasification: environmentally sound technology for decentralized power generation, a case study from India. *Biomass and Bioenergy*, *4*, 49–60.

Sadhwani, N., Adhikari, S., & Eden, M. (2016). Biomass gasification using carbon dioxide: effect of temperature, CO_2/C ratio, and the study of reactions influencing the process. *Industrial and Engineering Chemistry Research*, *55*, 2883–2891.

Sansaniwal, S. K., Rosen, M. A., & Tyagi, S. K. (2017). Global challenges in the sustainable development of biomass gasification: an overview. *Renewable and Sustainable Energy Reviews*, *80*, 23–43.

Schneider, J., Grube, C., Herrmann, A., & Ronsch, S. (2016). Atmospheric entrained-flow gasification of biomass and lignite for decentralized applications. *Fuel Processing Technology*, *152*, 72–82.

Seo, D. K., Lee, S. K., Kang, M. W., Hwang, J., & Yu, T. U. (2010). Gasification reactivity of biomass chars with CO_2. *Biomass and Bioenergy*, *34*, 1946–1953.

Septien, S., Escudero Sanz, F. J., Salvador, S., & Valin, S. (2018). The effect of pyrolysis heating rate on the steam gasification reactivity of char from woodchips. *Energy*, *142*, 68–78.

Silveira, S. (2005). Promoting bioenergy through the clean development mechanism. *Biomass and Bioenergy*, *28*, 107–117.

Striūgas, N., Šlančiauskas, A., Zakarauskas, K., Stravinskas, G., Paulauskas, R., & Vorotinskienė, L. (2016). Biomass and waste gasification in Lithuania: state of the art and future perspectives. *Energetika*, *62*, 219–237 (in Lithuanian).

Striūgas, N., Valinčius, V., Pedišius, N., Poškas, R., & Zakarauskas, K. (2017). Investigation of sewage sludge treatment using air plasma assisted gasification. *Waste Management*, *164*, 149–160.

Striūgas, N., Zakarauskas, K., Džiugys, A., Navakas, R., & Paulauskas, R. (2014). An evaluation of performance of automatically operated multi-fuel downdraft gasifier for energy production. *Applied Thermal Engineering*, *73*, 1151–1159.

Susastriawan, A. A. P., Saptoadi, H., & Purnomo (2017). Small-scale downdraft gasifiers for biomass gasification: a review. *Renewable and Sustainable Energy Reviews*, *76*, 989–1003.

Tamošiūnas, A., Chouchene, A., Valatkevičius, P., Gimžauskaitė, D., Aikas, M., Uscila, R., et al. (2017). The potential of thermal plasma gasification of olive pomace charcoal. *Energies*, *10*, 710.

Tamošiūnas, A., Grigaitienė, V., Valatkevičius, P., & Valinčius, V. (2012). Syngas production from hydrocarbon-containing gas in ambient of water vapor plasma. *Catalysis Today*, *196*, 81–85.

Tamošiūnas, A., Valatkevičius, P., Gimžauskaitė, D., Jeguirim, M., Mėčius, V., & Aikas, M. (2016). Energy recovery from waste glycerol by utilizing thermal water vapor plasma. *Environmental Science and Pollution Research*, *24*, 10030–10040.

Tamošiūnas, A., Valatkevičius, P., Gimžauskaitė, D., Valinčius, V., & Jeguirim, M. (2017). Glycerol steam reforming for hydrogen and synthesis gas production. *International Journal of Hydrogen Energy*, *42*, 12896–12904.

Taupe, N. C., Lynch, D., Wnetrzak, R., Kwapinska, M., Kwapinski, W., & Leahy, J. J. (2016). Updraft gasification of poultry litter at farm-scale: a case study. *Waste Management*, *50*, 324–333.

Tilghman, M. B., & Mitchell, R. E. (2015). Coal and biomass char reactivities in gasification and combustion environments. *Combustion and Flame*, *162*, 3220–3235.

Waheed, Q. M. K., Wu, C., & Williams, P. T. (2016). Hydrogen production from high temperature steam catalytic gasification of bio-char. *Journal of the Energy Institute*, *89*, 222–230.

Watson, J., Zhang, Y., Si, B., Chen, W. T., & de Souza, R. (2018). Gasification of biowaste: a critical review and outlooks. *Renewable and Sustainable Energy Reviews*, *83*, 1–17.

Zhai, M., Liu, J., Wang, Z., Guo, L., Wang, X., Zhang, Y., et al. (2017). Gasification characteristics of sawdust char at a high-temperature steam atmosphere. *Energy*, *128*, 509–518. https://doi.org/10.1016/j.energy.2017.04.083.

Biomass-derived chars used as adsorbents for liquid and gaseous effluents treatment

7

Salah Jellali, Madona Labaki†, Ahmed Amine Azzaz*, Hanene Akrout*, Lionel Limousy‡, Mejdi Jeguirim‡*

*Wastewaters and Environment Laboratory, Water Research and Technologies Centre (CERTE), Soliman, Tunisia, †Laboratory of Physical Chemistry of Materials (LCPM/PR2N), Lebanese University, Campus Fanar, Jdeidet El Metn, Lebanon, ‡University of Strasbourg, University of Upper Alsace, Institute of Materials Science of Mulhouse (IS2M—UMR CNRS 7361), Mulhouse, France

Chapter Outline

7.1 Introduction

Water resources pollution by inorganic substances (heavy metals in particular) and organic compounds (dyes, pesticides, pharmaceuticals) has been identified as a critical issue with regards to environment contamination and health risks. This topic has therefore attracted particular global attention from governmental regulatory agencies, concerned public and private organizations, civil society, and scientists (Sousa, Ribeiro, Barbosa, Pereira, & Silva, 2018).

Dissolved heavy metals and organic pollutants concentrations in discharged wastewaters from various industries often exceed the admissible contents. Even at relatively

Char and Carbon Materials Derived from Biomass. https://doi.org/10.1016/B978-0-12-814893-8.00007-9

low concentrations, these pollutants could induce severe effects through the eutrophication process, groundwater quality, plant development, animal health, and microorganism growth (Sousa et al., 2018). Indeed, it has been proven that the uncontrolled discharge of micropollutants into the environment at low concentrations (i.e., ng/L to μg/L) could contribute to their accumulation in the aquatic compartments, with important effects on aquatic ecosystems and flora (Rafatullah, Sulaiman, Hashim, & Ahmad, 2010). Moreover, human exposure to these pollutants could cause many significant health problems such as neurological or physiological impacts, especially for children (Choong, Chuah, Robiah, Gregory Koay, & Azni, 2007; Rafatullah et al., 2010). Various heavy metals and also organics (mainly pesticides) are recognized as mutagenic and carcinogenic agents (Choong et al., 2007; Sousa et al., 2018). Moreover, these pollutants could also accumulate in living organisms throughout the food chain due to their low biodegradability (Sousa et al., 2018). Although some organic compounds are, however, biodegradable they could lead to the formation of more recalcitrant and more toxic by-products (Holkar, Jadhav, Pinjari, Mahamuni, & Pandit, 2016).

Precipitation, ion exchange, and membrane filtration are the mainly used techniques for the treatment of wastewaters polluted with heavy metals (Shannon et al., 2008). However, these methods produce sludge that has to be managed, is inefficient under specific experimental conditions (weak selectivity), and is a high energy consumer, respectively (Shannon et al., 2008). For organic pollutants, microbial, coagulation-flocculation, filtration technologies as well as advanced oxidation processes are the mainly used techniques. However, these methods have the drawbacks of sludge production and/or exorbitant costs. Therefore, the development and testing of innovative but also low-cost methods for wastewater treatment has been pointed out as an urgent need to satisfy the fixed norms for wastewater discharge.

Important analytical-chemical development as well as the adoption of more stringent environmental regulations has incited the public and private institutions to develop efficient but also cost-effective and environmentally friendly technologies for wastewater treatments. During the last few decades, various natural resources have been applied for the removal of heavy metals and organic compounds from aqueous effluents. Many review articles dealing with the potential of numerous natural and synthetic adsorbents for pollutants removal from aqueous solutions have been reported (Abdolali et al., 2014; Rafatullah et al., 2010). In this context, carbonate minerals (calcite, magnesite, dolomite) have been shown to be very effective in removing heavy metals (Mlayah & Jellali, 2015). Using these materials, metals are removed through a combination of precipitation as metal oxides on carbonate surfaces due to a significant increase in aqueous pH values and also ion exchange with mainly Ca^{2+} and Mg^{2+} cations. Nowadays, there is a growing interest in the use of pristine as well as modified chars for various pollutants removal (Rosales, Meijide, Pazos, & Sanromán, 2017; Sizmur, Fresno, Akgül, Frost, & Moreno-Jiménez, 2017). This interest is justified because these materials exhibit relatively important adsorption capacities even when compared to commercially activated carbons (Kearns, Wellborn, Summers, & Knappe, 2014).

On the other hand, harmful and toxic gaseous pollutants are nowadays a major concern. Among these pollutants, methane (CH_4) is emitted in large amounts by landfills and agriculture. It is a strong greenhouse effect gas. Volatile organic compounds

(VOCs) are harmful pollutants for human health (toxicity) and the environment (photochemical smog). Many methods have been proposed for their elimination. Among them, adsorption processes exhibit many advantages in practice such as high efficiency, low energy consumption, and nonformation of toxic by-products. Carbonaceous materials are interesting adsorbents due to their high stability, high surface areas, suitable pore structure, and relative low cost. In addition, most of the VOCs adsorbed on carbonaceous materials can be recycled via desorption carried out in suitable conditions (Zhang, Gao, Creamer, Cao, & Li, 2017). Carbon dioxide (CO_2) and hydrogen sulfide (H_2S) are acidic gases and common contaminants. They are found in oil and gas production/processing, wastewater treatment plants, emissions of fossil fuel combustion, biogas, etc. (Bamdad, Hawboldt, & MacQuarrie, 2018) and can result in corrosion, problematic gaseous emissions, and represent a safety risk (Bamdad et al., 2018 and references therein). Being acidic gases, CO_2 and H_2S could be removed by absorption using monoethanolamine and diethylamine as solvents. The main disadvantages of this process are the large amount of energy required for regeneration, the corrosivity of amine solutions, and the large absorber volume (Bamdad et al., 2018; Rufford et al., 2012; Yu, Huang, & Tan, 2012). Hence, an alternative to absorption is adsorption in which the contaminants are removed from the gas mixture by porous solid adsorbents (Bamdad et al., 2018), with lower energy costs and less capital investment (Rufford et al., 2012). NO_x and SO_x gases cause acidic rain and have bad impacts on lungs and on ecosystems. Ozone (O_3) is an important atmospheric and indoor pollutant and causes respiratory problems. Mercury (Hg) vapors are also known for their severe toxic effects on human health and the environment. The elimination of all these pollutants by adsorption on low-cost materials is also advantageous.

In this chapter, a brief review regarding the use of chars for the removal of heavy metals and organic substances (dyes, pesticides, and pharmaceuticals) from aqueous solutions as well as gaseous pollutants removal is addressed. The choice of dyes, pesticides, and pharmaceutical compounds is justified by the fact that the greatest concern of organic pollutants has been focused on them (Inyang & Dickenson, 2015; Rosales et al., 2017; Sizmur et al., 2017).

Specifically, the following section will present the effect of the main parameters influencing these kinds of pollutants removal efficiencies and mechanisms, including: (1) the nature of the feedstock, (2) the pyrolysis conditions, (3) the chars' physicochemical modification, and (4) the experimental sorption conditions.

7.2 Chars used for heavy metals adsorption

The application of char-based materials for heavy metals removal from aqueous solutions has been an important scientific topic over the last decade (Cha et al., 2016; Inyang & Dickenson, 2015; Rosales et al., 2017; Tan et al., 2015). The efficiency of chars in removing metals from aqueous solutions depends mainly on the pollutant properties and the adsorbent characteristics such as surface area, porosity, pH of zero point charge (pH_{ZPC}), mineral composition, and nature of the functional groups (Sizmur et al., 2017).

7.2.1 Use of pristine chars

The efficiency of pristine chars in removing heavy metals from aqueous solutions depends mainly on the nature of the used biomass but also on the pyrolysis conditions because they affect directly the physicochemical characteristics of the produced chars such as:

(a) **The specific surface areas (SSAs) and porosities.** The SSAs are mainly linked to the distribution of micro- or mesopores. The greater the extension of the micropore network, the greater the corresponding surface area and consequently the more the number of surface sites that could adsorb heavy metals (Tan et al., 2015). For example, for the same pyrolysis conditions ($T = 300°C$ for 1 h; gradient = 17°C/min), the SSAs of chars generated from rice husk, organic fraction of municipal solid wastes, and sewage sludge were assessed to be 155, 5, and 51 m^2/g (Agrafioti, Kalderis, & Diamadopoulos, 2014). Similarly, Inyang et al. (2012), when studying lead removal by chars derived from the pyrolysis at 600°C for 2 h and for a heating rate of 10°C/min of digested dairy waste residue and digested sugar beet residue, demonstrated that their respective SSAs were 161.2 and 48.6 m^2/g. The corresponding Langmuir Pb(II) adsorption capacities were determined to be about 248 and 197 mg/g, respectively.

On the other hand, for a given feedstock, the SSAs are also very dependent on the used pyrolysis conditions, especially the pyrolysis temperature, the heating gradient, and the oxygen percentage in the reactor. In this context, Cui et al. (2016) studied Cd(II) removal from aqueous solutions by chars derived from *Canna indica* biomass (China) at temperatures of 300°C, 400°C, 500°C, and 600°C for 2 h and a heating rate of 5°C/min (slow pyrolysis). They demonstrated that the increase in the pyrolysis temperature from 300°C to 600°C increased their SSAs from 3.46 to 10.4 m^2/g and the Cd(II) removal capacity increased from 63.32 to about 140.01 mg/g. Another study reported a more important increase in the SSAs of chars derived from *Miscanthus sacchariflorus* from 0.56 to about 381 m^2/g when the pyrolysis temperature was increased from 300°C to 600°C (Kim et al., 2013). On the basis of scanning electron microscopy images, they showed a significant pore space increase with the increasing of the pyrolytic temperature that might be due to the volatiles released during cellulose and hemicellulose degradation as well as the formation of big channel structures. However, this high increase in the SSAs was not accompanied by a significant Cd(II) removal efficiency since it increased from 11.40 to only 12.96 mg/g. The authors attributed this behavior to the adsorption limitation of the other involved mechanisms, mainly complexation with functional groups and precipitation. Moreover, Kılıc, Kırbıyık, Cepeliogullar, and Pütün (2013) studied the removal of Ni(II) and Co(II) by using a char generated from the pyrolysis at 600°C of almond shell. They pointed out that the pyrolysis operation had significantly increased its overall porosity and the related size of voids (Fig. 1). Its SSA, Ni(II), and Co(II) Langmuir's adsorption capacities were determined to be 145, 20.0, and 28.1 mg/g, respectively. Finally, the presence of even low oxygen contents during the pyrolysis process of *Phyllostachys pubescens* biomass (China) at a fixed temperature and gradient of 450 and 20°C/min, respectively, induced a significant increase in the SSAs of the generated chars (Zhang, Shan, Tang, & Zhu, 2017). As an example, these SSA values

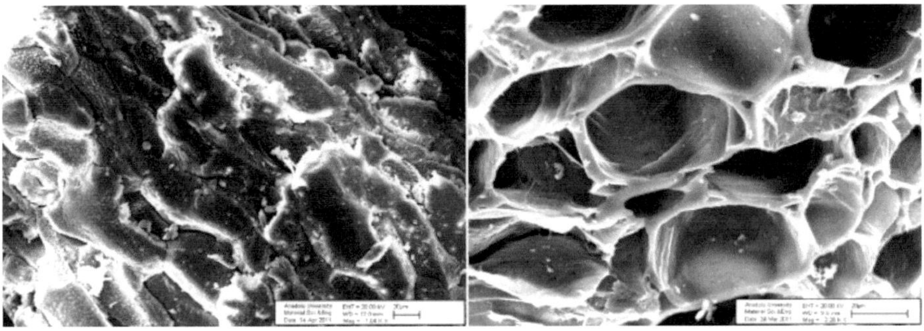

Fig. 1 Scanning electron microscopy images of raw almond shell and its derived char at a pyrolysis temperature of 600°C and a heating rate of 10°C/min.
Reprinted with permission from Kılıc, M., Kırbıyık, I., Cepeliogullar, O., Pütün, A. E. (2013). Adsorption of heavy metal ions from aqueous solutions by bio-char, a by-product of pyrolysis. *Applied Surface Science, 283*, 856–862. Copyright 2018 Elsevier.

increased from 118 (in case of total absence of O_2) to 232 m^2/g for 4% oxygen content. This rising in the SSAs has increased lead and cadmium adsorption capacities from about 27.4 and 4.4 mg/g to 40.7 and 13.1 mg/g, respectively.

It is important to underline that for some biomasses, the increase in the pyrolysis temperature induces, however, a significant decrease in the SSA of the resulting chars. For instance, Zhang, Mao, Chen, Huang, and Qiu (2013) found a decrease in the SSA of a char produced from a municipal wastewater sludge when the pyrolysis temperature increased from 400°C to 600°C. It was imputed to the blockage and collapse of the pore structures. This decrease in the SSA has induced a decrease in the Pb(II) and Cr(VI) adsorption capacity from 18.2 and 36.6 mg/g to 12.8 and 19.6 mg/g, respectively.

(b) The surface functionality. Surface functionality plays an important role in heavy metals adsorption. In this context, it has been proven that metals removal onto pristine chars occurs mainly according to the following mechanisms: cation exchange, including complexation, cation-π interaction, and precipitation. Further details regarding these mechanisms are presented in Section 7.2.2.2.

7.2.2 Use of modified chars

Various methods for char modification have been reported with the aim of improving their performance in heavy metals removal. These modifications were performed either as (Rosales et al., 2017; Sizmur et al., 2017): (1) pretreatment where the raw feedstock was treated by using slats, acidic or alkaline solutions, or (2) posttreatment through chemical and/or thermal agents. All these methods act on the physical or chemical characteristics of the chars, such as the surface area and/or surface functionality.

The most common char modifications include mainly: (1) physical activation using steam, which generally oxidizes the surface and creates oxygen-containing functional groups (Han, Boateng, Qi, Lima, & Chang, 2013), (2) chemical activation using acidic and alkaline solutions, which contributes to the creation of a network of interconnecting micropores and new "desired" functional groups (Goswami et al., 2016), and (3) the production of char-based composites through the embedding of different materials such as clay minerals, metal oxides, and carbonaceous structures (Deng et al., 2017; Jang et al., 2018).

In this context, various raw biomasses were impregnated with salts such as $MgCl_2$, $CaCl_2$, $ZnCl_2$, KCl, $MnCl_2$, $AlCl_3$, $Co(NO_3)_2$, and $KMnO_4$. These pretreatments generally contribute to the improvement of the pore's properties and also the richness in oxygen-containing functional groups (Harikishore Kumar Reddy & Lee, 2014; Jellali et al., 2016; Wang, Gao, Li, et al., 2015; Wang, Gao, Wang, et al., 2015). They have significantly increased the modified material efficiencies in removing heavy metals from aqueous solutions. For instance, Wang, Gao, Li, et al. (2015); Wang, Gao, Wang, et al. (2015) proved that the pretreatment of hickory chips (United States) by a 0.1 M $KMnO_4$ solution under ultrasonic irradiation for 2 h has significantly enhanced the adsorption characteristics of the produced char at 600°C. Indeed, the lead, cadmium, and copper adsorption capacities increased from about 71.4, 4.7, and 12.3 mg/g for the pristine char to about 153.1, 28.1, and 34.2 mg/g for the $KMnO_4$ pretreated one (Fig. 2).

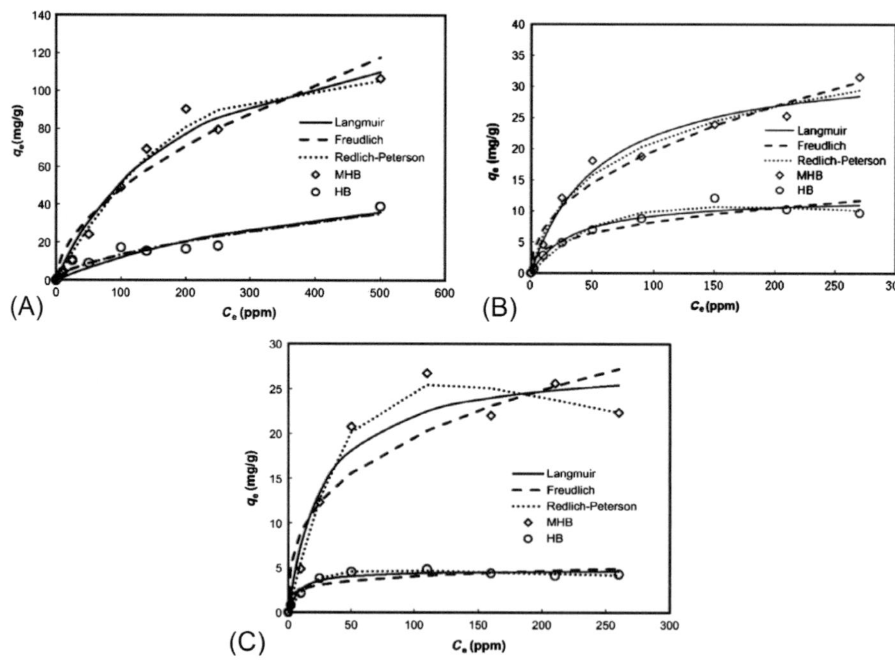

Fig. 2 Sorption isotherm data and fitted models of heavy metals onto the pristine (HB) and engineered (MHB) chars: (A) Pb(II), (B) Cu(II), and (C) Cd(II).
Reprinted with permission from Wang, H., Gao, B., Wang, S., Fang, J., Xue, Y., Yang, K. (2015). Removal of Pb(II), Cu(II), and Cd(II) from aqueous solutions by biochar derived from $KMnO_4$ treated hickory wood. *Bioresource Technolnology, 197*, 356–362. Copyright 2018 Elsevier.

For the posttreatment of chars by physical agents, Han et al. (2013) studied the adsorption of Zn(II) and Cu(II) by chars derived from switchgrass (SG), hardwood (HW), and softwood (SW) and activated by steam at 800°C for 45 min with N_2 as carrier. The results showed that the activation of chars produced at 500°C increased their surface areas from about 3.6 to 168 m^2/g for SG, from 119 to 344 m^2/g for HW, and from 95 to 384 m^2/g for SW. In addition, for HW, mesopore and micropore volumes increased from 0.046 to 15.486 cm^3/g and from 0.030 to 0.151 cm^3/g, respectively. However, the use of high activation temperature has significantly reduced the carboxylic, lactonic, and phenolic groups' contents. Indeed, for HW at an activation temperature of 800°C, their respective contents have decreased from 0.146 to 0.009 mmol/g, from 0.172 to 0.006 mmol/g, and from 0.058 to 0.008 mmol/g. The metals' adsorption tests showed a much higher binding affinity of the activated chars for Cu(II) than for Zn(II). For instance, the Cu(II) adsorption ability by the activated HW char (19.5 mg/g) was more than 51 times higher than the pristine one.

Regarding the chemical posttreatment of chars, Peng et al. (2017) studied Cu(II) and Cd(II) removal by raw chars originating from pine sawdust pyrolyzed at temperatures between 200°C and 650°C in comparison with their treated form with phosphoric acid. They showed that the SSAs of the activated chars were about two orders of magnitude higher than those of the raw ones produced at 200°C and 350°C. For those produced at 500°C and 650°C the SSAs of the modified ones were about five- and twofold larger than those of pristine chars, respectively. Moreover, they showed that at the same pyrolysis temperature, the quantities of carboxyl (—COOH) and hydroxyl (—OH) functional groups of the modified chars were larger than the pristine ones, which resulted in more effective Cu(II) and Cd(II) adsorption. Indeed, compared to the pristine chars, the H_3PO_4 modification increased the Cu(II) and Cd(II) by about 12–44 times and 4–13 times, respectively. On the other hand, the authors suggested that the newly formed phosphorus-containing groups, such as P=O and P=OOH due to the acid phosphorus modification, may interact and form complexes with the studied heavy metals.

Concerning the elaboration of biomass-derived char-based composites, Deng et al. (2017) made a novel engineered char prepared through postmodification with chitosan and pyromellitic dianhydride of a char derived from Chinese rice straw for the removal of heavy metal ions from single-metal and mixed-metal solutions (Cd, Cu, and Pb). They proved that this modification almost doubled its surface area (an increase from 32.1 to 62.6 m^2/g) and increased its pore volume from 0.038 to 0.096 cm^3/g. Furthermore, this modification introduced onto the char surface large quantities of amino groups and *O*-containing functional groups. An increase in the removal of Pb(II) and Cd(II) of about 10% has been noted when using modified char instead of the raw matrix. However, the modified char exhibited a strong selective adsorption ability to Cu(II) with an increased percentage of about 166%.

On the other hand, several studies have been performed to take advantage of the magnetic properties of iron to create magnetic chars that could efficiently remove metals from aqueous solutions (Sizmur et al., 2017; Wang, Gao, Wang, et al., 2015). For instance, Wang, Gao, Wang, et al. (2015) produced a magnetic char by pyrolyzing at 600°C for 1 h a mixture of hematite and pine wood biomass. The authors observed that the As(V) sorption capacity of this modified char was relatively high (428.7 mg/g), which is about 1.6 higher than the pristine one.

7.2.2.1 Effects of experimental adsorption conditions

Contact time—kinetic adsorption

Kinetic studies are very useful to select the suitable chars for wastewaters treatment as well as for process cost assessment. Generally, metal adsorption kinetics includes three steps. The first one is relatively fast and imputed to metals accumulation on the char surface (diffusion through the particle's boundary layer). The second step of the adsorption process is much slower and concerns the intraparticle diffusion process. The third one corresponds to an equilibrium state where the metal ions occupy the active sites (Harikishore Kumar Reddy & Lee, 2014). Various empirical models have been proposed for experimental data analysis regarding heavy metals adsorption. The mainly used models are pseudo-first-order and pseudo-second-order models (Bogusz, Oleszczuk, & Dobrowolski, 2015). The duration required for reaching the equilibrium state varies from about 1 h for adsorption of Cr(III) on a char derived from municipal solid wastes (Agrafioti et al., 2014) to about 1 day for the copper adsorption by char from corn straw (Chen et al., 2011). This equilibrium time depends on the metal and char properties and the experimental conditions (Agrafioti et al., 2014; Chen et al., 2011; Cui et al., 2016; Jang et al., 2018).

Bogusz et al. (2015) showed that Cd(II), Cu(II), and Zn(II) adsorption is faster on chars rich in hydroxyl groups (commercial char) since they are characterized by lower binding energy compared to the one containing high carboxyl groups (char from wheat straw), while Cu(II) adsorption on their commercial and laboratory chars occurs more rapidly (3–4 h) than Zn(II) (3.5–5.0 h) and Cd(II) (4.0–6.0 h). They attributed these differences to the solubility of $M(OH)_2$ and the ionic radii of these ions that varies in the following order: Cu(II) < Zn(II) < Cd(II).

The main metal kinetic adsorption studies data were fitted by the pseudo-second-order model indicating that the adsorption process is governed by chemisorption (Agrafioti et al., 2014; Cui et al., 2016; Wan et al., 2018).

Adsorption isotherm

The adsorption isotherm is essential to select the suitable chars for wastewaters treatment since it gives important information on the chars' efficiencies and the involved interactions between metal and adsorbent particles. Various empirical models have been proposed for the analysis of heavy metals adsorption on biomass-derived chars. The major models are Langmuir, Freundlich, Dubinin-Radushkevich, and Temkin equations (Jellali et al., 2016).

The Langmuir model, which suggests monolayer adsorption of the heavy metal on a homogeneous adsorbent surface, was found to be the suitable one in various studies dealing with biomass-derived chars (Bogusz et al., 2015; Chen et al., 2011; Cui et al., 2016). Other studies reported that the Freundlich model fits better the experimental data (Agrafioti et al., 2014; Wan et al., 2018; Wang and Liu, 2018). The Freundlich model is related to heterogeneous and multilayer adsorption processes (Wan et al., 2018). For instance, Agrafioti et al. (2014) found that the adsorption of As(V), Cr(III), and Cr(VI) by a char derived from rice husk at 300°C was well fitted by the Freundlich model, while Langmuir completely fails in fitting the experimental data.

Wang and Liu (2018) noticed that the Freundlich model was valid for Pb(II) removal by char derived from rice straw. When this char was chemically modified, the Pb(II) adsorption isotherm became better fitted with the Langmuir model. This finding was attributed to the changes operated on the physicochemical characteristics of the modified char.

Solution pH

pH is an important parameter during studies of heavy metals adsorption by chars. Indeed, various heavy metals could precipitate as metal oxides for basic pH (Mlayah & Jellali, 2015; Rosales et al., 2017). For this reason, the majority of adsorption tests were carried out under acidic conditions. However, raw chars have alkaline pH_{ZPC} (Zhang, Liu, & Liu, 2014) leading to an increase in pH during the shaking operation and therefore to metal precipitation.

At low pH values, the main functional groups of chars are protonated and present in positively charged form. Moreover, for aqueous pH values lower than the pH_{ZPC}, the char surface is globally positively charged favoring the removal of anions and limiting the removal of the positively charged heavy metals. Also, the H^+ ions present in the aqueous solutions may compete with the heavy metal cations for adsorption on active sites. When increasing the aqueous pH values, competition of metal ions with protons for active sites will be reduced. However, when the aqueous pH value is higher than the pH_{ZPC}, the char surfaces will be negatively charged and therefore positively charged metals adsorption will be enhanced.

Heavy metals adsorption studies showed that their removal efficiencies increased with increasing pH values. For instance, Bogusz et al. (2015) studied the effect of the initial solution pH (between 1 and 8) on the adsorption of synthetic solutions of 100 mg/L of Cd(II), Cu(II), and Zn(II) by char generated from wheat straw. They showed that at low pH ($pH < 3$), the adsorption of all tested ions was very low and did not exceed 30%. However, this percentage reached almost 100% for initial pH values of 6 for Cd(II) and Cu(II) and 7 for Zn(II). Cui et al. (2016) also studied Cd(II) removal by four chars produced during the pyrolysis of *C. indica* at different temperatures varying between 300°C and 600°C. They showed that for an initial pH of 2, no Cd(II) was removed. However, from an initial pH of 5, all the studied chars reached a quasiequilibrium (plateau of Cd(II) removal). The same trend was observed by Deng et al. (2017) when studying Pb(II), Cd(II), and Cu(II) removal by pristine and chitosan-modified char derived from rice husk at 450°C for initial pH values varying from 2 to 6.

Char dosage

During the investigations dealing with heavy metals adsorption by biomass-derived chars, the increase in the chars' dosage increased metals removal efficiency percentages. This behavior is attributed to both the presence of more active sites and also to the increase in the contact probabilities between metal ions and these sorption sites. Sekulić et al. (2018) investigated the removal of Pb(II), Cd(II), and Ni(II) at an initial concentration of 100 mg/L and a pH of 6 by varying doses of phosphoric acid-functionalized *Prunus armeniaca* stones char from 0.2 to 10 g/L. They showed that the metals removal percentages increased by increasing the used doses. Pb(II) and Cd(II) removal percentages increased, respectively, from about 12.5% to 95.1% and from 28.6% to 94.1% when the used modified char doses increased from 0.2 to 2.0 g/L.

Coexisting pollutants

The presence of coexisting pollutants, especially noted in the case of real wastewaters, could impact negatively the metals removal efficiencies of chars (Jang et al., 2018; Wang & Liu, 2018). Indeed, these pollutants compete with metals for adsorption onto the chars' active sites. In this context, Jang et al. (2018) examined the effect of competing cations (Na^+, K^+ Ca^{2+}, Mg^{2+}, and Al^{3+}) with different initial concentrations ranging from 1 to 100 mg/L on the adsorption efficiency of strontium using rice straw-based char. They showed that only high concentrations of Ca^{2+} and Na^+ significantly impacted strontium adsorption with a maximum decrease of more than 35% for an initial Na^+ concentration of 100 mg/L (Fig. 3). They attributed this finding

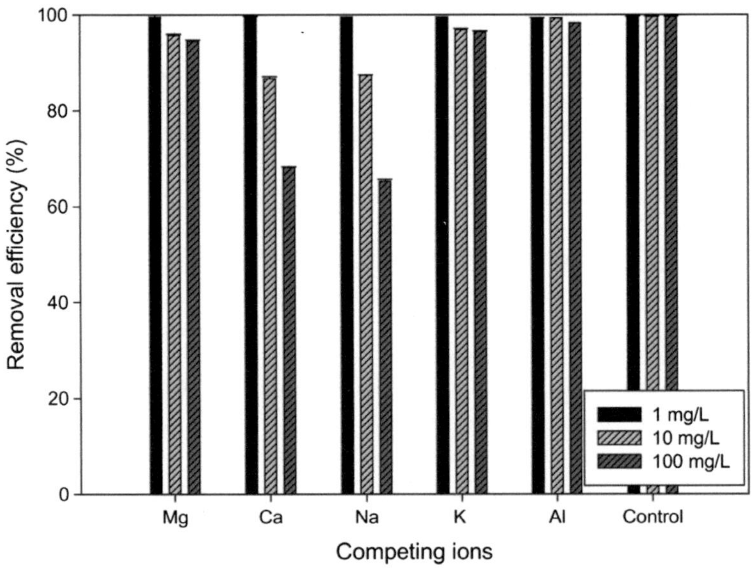

Fig. 3 Effect of competing ions on the removal of strontium ions using the rice straw-based char beads.

Reprinted with permission from Jang, J., Mirana, W., Divine, S. D., Nawaz, M., Shahzad, A., Woo, S. H., Lee, D. S. (2018). Rice straw-based biochar beads for the removal of radioactive strontium from aqueous solution. *Science of the Total Environment, 615*, 698–707. Copyright 2018 Elsevier.

to the fact that the ionic radii of these hydrated ions are relatively similar to strontium. Besides, Wang and Liu (2018) determined the single and multimetal Cd(II), Cu(II), Pb(II), and Zn(II) adsorption isotherms by pristine yak manure char and its modified form by H_2O_2. They showed that all the metals adsorption capacities by the two used chars were inhibited when their initial concentrations exceeded 50 mg/L. For instance, the Langmuir's maximum lead adsorption capacities in a multimetal system decreased from 77.1 to 35.4 mg/g for pristine char and from 155.4 to 71.7 mg/g for chemically modified char.

Adsorption temperature

Several studies have reported that the adsorption of heavy metals by chars increased with increasing temperature (Sekulić et al., 2018; Wan et al., 2018). Wan et al. (2018) studied the effect of temperature (10°C, 25°C, and 40°C) on Pb(II) and Cd(II) removal by a nanocomposite-based char. They showed that the increase in temperature led to an increase in their adsorption capacities indicating the endothermic nature of heavy metals sorption. When the used temperature increased from 10°C to 40°C, the Langmuir's adsorption capacities of lead and cadmium increased from 62.6 to 110.1 mg/g and from 23.6 to 30.9 mg/g, respectively. In another study, Sekulić et al. (2018) studied the thermodynamic adsorption of Pb(II), Cd(II), and Ni(II) onto a phosphoric acid-functionalized char derived from *P. armeniaca* stones for temperatures of 22°C, 32°C, and 42°C. The experimental results showed that for all the studied metals, their adsorption significantly increased with increasing temperature. The determination of the thermodynamic parameters, including enthalpy (ΔH), entropy (ΔS), and Gibbs free energy (ΔG), showed that metals adsorption was endothermic, spontaneous, and thermodynamically favorable. A similar trend was reported by Zhou, Liao, Lin, Qiu, and Song (2018) when studying thermodynamic aspects of the adsorption of Cd(II) and Cu(II) by a ferromanganese binary oxide char composite. The main results of recent studies regarding the effects of the above-cited parameters on char removal characteristics of heavy metals from aqueous solutions are summarized in Table 1.

7.2.2.2 Involved mechanisms

The surface functionalities of pristine or modified chars impact significantly on heavy metals removal as well as on involved mechanisms in the adsorption process. These mechanisms include (Fig. 4): (1) ion exchange through the release of mineral cations (Na^+, K^+, Ca^{2+}, or Mg^{2+}: Uchimiya, Lima, Klasson, & Wartelle, 2010) and also protons from functional groups existing on the chars' surface such as hydroxylic, carboxylic, and phenolic groups (Ahmad et al., 2014; Bogusz et al., 2015; Lu et al., 2014), (2) physical sorption that occurs between the dissolved metals in water and the delocalized cloud of electrons, which are associated with aromatic groups on the chars' surface, creating therefore cation-π interactions with the C=C aromatic bonds (Harvey, Herbert, Rhue, & Kuo, 2011), and (3) precipitation (or coprecipitation) with insoluble salts at the chars' surface that have large ash contents and also the formation of metal (hydr)oxides as a consequence of the pH rising (Kim et al., 2013; Zhang et al., 2014).

Table 1 Adsorption characteristics of various heavy metals onto raw and modified chars

Feedstock	Pretreatment	Pyrolytic conditions	Posttreatment	BET surface area (m^2/g)	Pollutant	Adsorption batch conditions	Contact time (h)	Kinetic model	Isotherm	QL (mg/g)	References
Rice husk, Greece	Drying at 103°C for 24 h	$T = 300°C$ for 1 h; gradient = 17°C/min	–	155	As(V) Cr(III) Cr(VI)	Dose = 4 g/L Ambient temperature 200 rpm $C_0 = 90–850$ μg/L	24 3 –	PSO PSO –	Freundlich	–	Agrafioti et al. (2014)
Organic fraction of municipal solid wastes, Greece			–	5			24 3 –	PSO PSO –	Freundlich		
Sewage sludge, Greece			–	51	As(V) Cr(III) Cr(VI)		24 3 –	PSO PSO –	Freundlich	13.42 94.34 64.10	
Peanut shell, China	Drying at 60°C for 24 h	$T = 400°C$ for 1 h; gradient = 15°C/min	–	2.99	Pb(II) Cd(II)	Dose = 0.2 g/L $T = 25°C$ 180 rad/min; pH = 5 C_0 (Pb(II)) = 3.0–53 mg/L C_0 (Cd(II)) = 3.0–33 mg/L	24	PSO	—	—	Wan et al. (2018)
			Treatment with $MnSO_4·4H_2O$ + binary NaClO–NaOH solution	3.57				PSO	Freundlich	110.15 24.43	
						Same conditions, $T = 10°C$	–	–		62.56 23.63	
						Same conditions, $T = 45°C$	–	–		82.11 30.90	
Pristine char	–	$T = 450°C$ for 2 h; gradient = 5°C/min	Modification with chitosan and pyromellitic dianhydride	62.6	Pb(II) Cd(II) Cu(II)	Dose = 0.4 g/L $T = 25°C$ 150 rad/min $C_0 = 5–500$ mg/L pH = 5	24	PSO PSO PSO	Langmuir	9.24 30.12 89.69	Deng et al. (2017)

Ipomoea fistulosa, India	Washing and sun drying	$T = 550°C$ for 1 h; gradient = 40°C/min		119.0	Cd(II)	Dose = 5 g/L $T = 24°C$ 200 rad/min $C_0 = 10–50$ mg/L pH = 7	6	PSO	Freundlich	41.67	Goswami et al. (2016)
			Activation with KOH	650.0						72.43	
Switchgrass	–	$T = 500°C$ for 1 h	–	3.62	Cu(II) Zn(II)	Dose = 10 g/L $T = 20°C$ 200 rad/min pH = 4.8	24	–	Both Freundlich and Langmuir	5.21 4.84	Han et al. (2013)
		$T = 500°C$ for 1 h	Steam activation	167.83						17.60 4.45	
Hardwood	–	$T = 500°C$ for 1 h; gradient = –	–	178.98						0.38 0.07	
		$T = 500°C$ for 1 h; gradient = –	Steam activation	343.91						19.51 0.20	
Softwood	–	$T = 500°C$ for 1 h; gradient = –	–	94.74						6.35 5.95	
		$T = 500°C$ for 1 h; gradient = –	Steam activation	383.66						30.25 4.45	
Rice straw, South Korea	–	$T = 450°C$ for 1 h; gradient = 1.5°C/min	Treatment with sodium alginate	71.53	Strontium	Dose = 20 g/L $T = 35°C$ 120 rpm $C_0 = 1–10$ g/L pH = 7	24	PSO	Langmuir	175.95	Jang et al. (2018)
Pine sawdust, China	Air drying				Cu(II) Cd(II)	Dose = 1 g/L Room temperature $C_0 = 5$ mg/L 100 rpm pH = 4.5	120	–	Freundlich		Peng et al. (2017)

Continued

Table 1 Continued

Pine sawdust, China	Air drying	$T = 650°C$ for 4 h	–	408.0	Cu(II) Cd(II)	Dose = 1 g/L $C_0 = 5$ mg/L 100 rpm pH = 4.5	120	–	Freundlich	1.55 4.7	Peng et al. (2017)
		$T = 650°C$ for 4 h	Treatment with H_3PO_4, then pyrolyzed again at 200°C for 2 h	900.0		Same conditions, except dose = 0.25 g/L				22.35 14.8	
Apricot stones, Serbia	Washing and air drying. Then, impregnation with H_3PO_4	$T = 180°C$ for 45 min; gradient = 10°C/min; then increase in T to 500°C for 1 h at the same gradient	–	1098.9	Pb(II) Cd(II) Ni(II)	Dose = 2 g/L $T = 22°C$ $C_0 = 5$–500 mg/L 140 rpm pH = 6.0	1	PSO	Freundlich Langmuir Freundlich	179.476 105.844 78.798	Sekulić et al. (2018)
Wasted kelp: marine macro-algae, South Korea	Washing, air drying for 24 h, exsiccation and grinding. Magnetization using M $FeCl_3$,$6H_2O$ solution	$T = 500°C$ for 2 h; gradient = 7°C/min	–	0.97	Cd(II) Cu(II) Zn(II)	Dose = 16.7 g/L $C_0 = 500$–2000 mg/L 150 rpm pH = –	24	–	Freundlich Langmuir Langmuir	34.89 46.74 95.49	Son, Poo, Chang, and Chae (2018)
Wasted hijikia: marine macro-algae, South Korea		$T = 500°C$ for 2 h; gradient = 7°C/min	–	63.33	Cd(II) Cu(II) Zn(II)	Dose = 16.7 g/L $C_0 = 500$–2000 mg/L 150 rpm pH = –	24	–	Langmuir Langmuir Freundlich	7.01 36.44 22.03	

Hickory chips, USA	Milling	$T = 600°C$ for 1 h; gradient = –	–	101	Pb(II) Cd(II) Cu(II)	Dose = 2 g/L $C_0 = 2–500$ for Pb(II) and 2–250 for Cd(II) and Cu(II) 50 rpm $T = 22°C$ pH = –	24	Elovich Richie Elovich	Freundlich Langmuir Langmuir	71.43 4.748 12.297	Wang, Gao, Li, et al. (2015)
	Milling then treatment with $KMnO_4$ solution		–	205	Pb(II) Cd(II) Cu(II)	Dose = 2 g/L $C_0 = 2–500$ for Pb(II) and 2–250 for Cd(II) and Cu(II) 50 rpm $T = 22°C$ pH = –	24	Elovich Richie Elovich	Langmuir Langmuir Langmuir	153.109 28.104 34.205	
Yak manure, China	Air drying and crushing	$T = 350°C$ for 2 h, gradient = 5°C/min	–	1.18	Cd(II) Cu(II) Pb(II) Zn(II)	Dose = 1 g/L $T = 23°C$ pH = 5.5	48	PSO	Langmuir	47.181 46.710 77.166 34.654	Wang and Liu (2018)
Yak manure, China	Air drying and crushing	$T = 350°C$ for 2 h, gradient = 5°C/min	Modification with H_2O_2	4.59						69.082 64.979 155.355 37.303	
		$T = 450°C$ for 3 h; gradient = 20°C/min; oxygen content = 4%		232.6						40.654 13.138	

PFO, pseudo-first-order kinetic model; *PSO*, pseudo-second-order kinetic model; *QL*, Langmuir's maximum adsorption capacity.

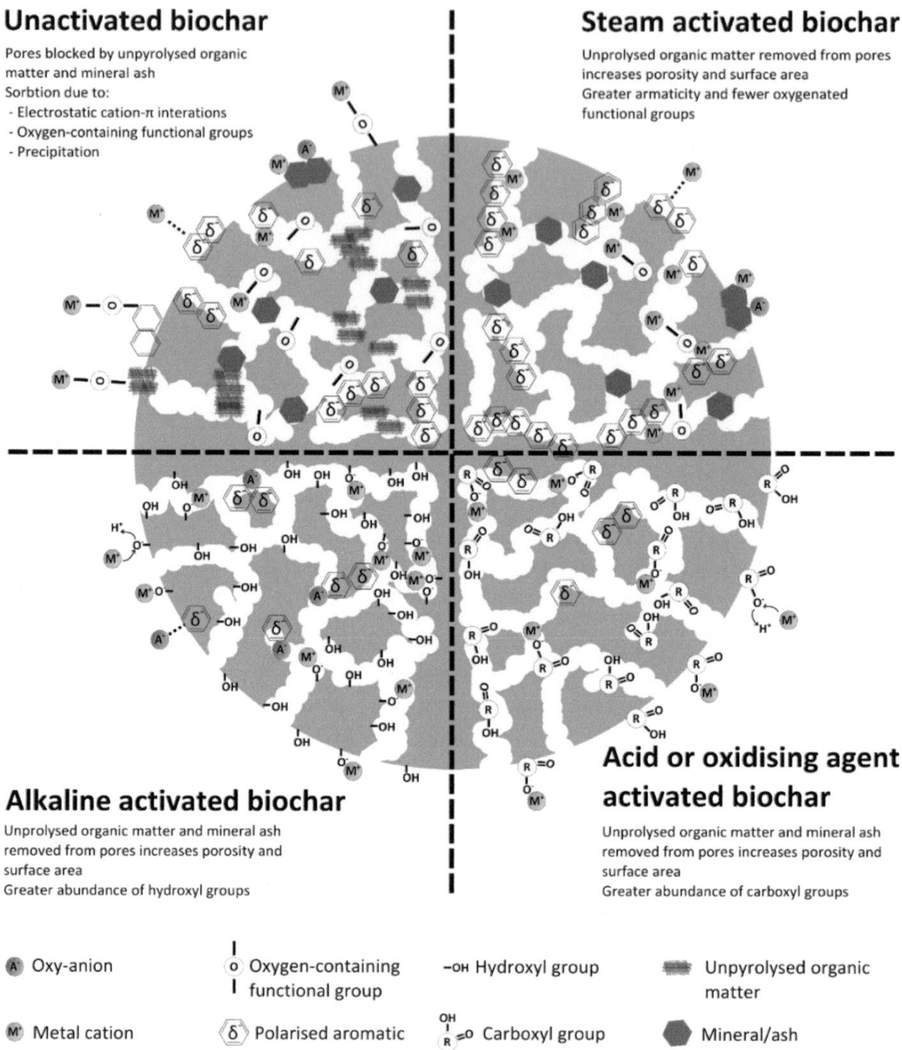

Fig. 4 Conceptual model identifying the mechanisms of metal cation (e.g., Cd^{2+}, Cu^{2+}, Hg^{2+}, Pb^{2+}, Zn^{2+}) and oxyanion (e.g., PO_4^{3-}, AsO_4^{3-}) sorption to unactivated char and char physically activated with steam or chemically activated with acids, oxidizing agents, or alkaline solutions. Reprinted with permission from Sizmur, T., Fresno, T., Akgül, G., Frost, H., Moreno-Jiménez, E. (2017). Biochar modification to enhance sorption of inorganics from water. *Bioresource Technology, 246*, 34–47. Copyright 2018 Elsevier.

In this context, Zhang et al. (2014) studied Pb(II) sorption by chars prepared during municipal wastewater sludge under different heating conditions. They pointed out that Pb(II) sorption on the chars' surface involved various mechanisms with many sorption affinities: cationic exchange, inner complex with the minerals, and organic functional groups such as —OH and —COOH, as well as coprecipitation such as $PbSiO_3$.

Furthermore, Wang, Gao, Wang, et al. (2015) determined the main mechanisms involved in the removal of lead by chars generated from peanut shells. They demonstrated that functional groups complexation, Pb^{2+}-π interactions, as well as precipitation with released minerals mutually contributed to Pb^{2+} removal. They proved also that new mineral precipitates such as $Pb_2(SO_4)O$ and $Pb_4(CO_3)_2(SO_4)(OH)_2$ were formed during Pb^{2+} sorption. They reported that for pyrolysis temperatures higher than 500°C, Pb^{2+} adsorption via Pb^{2+}-π interactions were enhanced. However, contribution of the complexation mechanism was reduced. They imputed this finding to the fact that at high temperatures, the contents of various functional groups such as aliphatic, carboxylic, and ketones were seriously decreased.

Only a few studies have tried to assess the contribution of each sorption mechanism at various initial aqueous pH values. Lu et al. (2012) demonstrated that the char derived from the pyrolysis of sewage sludge could effectively remove lead from acidic solutions (pH values between 2 and 5) with adsorption capacities of 16.11, 20.11, 24.80, and 30.88 mg/g at initial pH values of 2, 3, 4, and 5, respectively. Metal analysis of the supernatants after Pb^{2+} sorption showed that relatively important concentrations of Ca^{2+}, Mg^{2+}, K^+, and Na^+ cations were released, equivalent to 9.76, 11.60, 11.29, and 17.50 mg/g of Pb^{2+}, and accounted for about 60.6%, 57.70%, 45.5%, and 56.7% of the total removed amounts at initial pH 2, 3, 4, and 5, respectively.

On the other hand, Cui et al. (2016) demonstrated that the increase in pyrolysis temperature of *C. indica* from 300°C to 600°C increased greatly the percentage of Cd^{2+} removal by precipitation (Fig. 5) compared to the other existing mechanisms (cation exchange, complexation, and Cd^{2+}-π coordination). They attributed this result to the release of large contents of CO_3^{2-} at temperatures higher than 500°C. These carbonate ions interact with Cd^{2+} to form the precipitate $CdCO_3$.

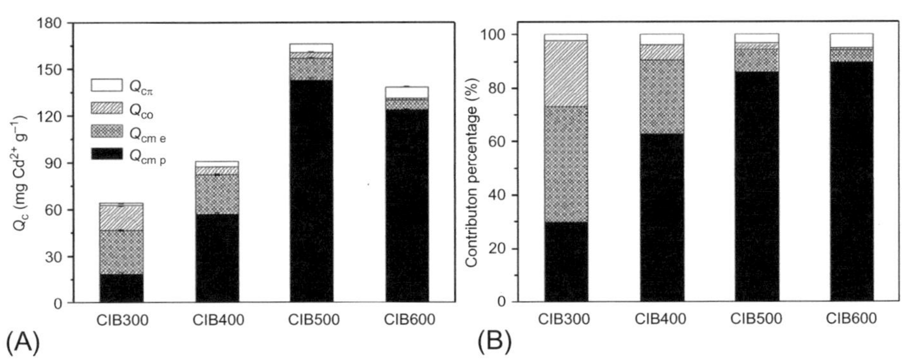

Fig. 5 (A) The contribution to Cd^{2+} sorption by metal ion exchange (Q_{cme}), precipitation with minerals (Q_{cmp}), functional groups complexation (Q_{co}), and Cd^{2+}-π coordination ($Q_{c\pi}$) on chars. (B) The contribution percentage of different mechanisms to Cd^{2+} sorption on *Canna indica*-derived chars.

Reprinted with permission from Cui, X., Fang, S., Yao, Y., Li, T., Ni, Q., Yang, X., et al. (2016). Potential mechanisms of cadmium removal from aqueous solution by Canna indica derived biochar. *Science of the Total Environment, 562*, 517–525. Copyright 2018 Elsevier.

7.3 Use of biomass-derived chars for organics removal from aqueous solutions

7.3.1 Factors affecting organics adsorption

7.3.1.1 Nature of the biomass feedstock

Several studies have been conducted to determine the impact of the biomass feedstock nature on the chars' efficiencies in removing organic compounds (Sizmur et al., 2017; Tan et al., 2015). In particular, different chars have been produced from various biomass feedstocks under the same pyrolytic conditions and then tested for organics removal from aqueous solutions (Chen, Zhou, & Lin, 2015; García-jaramillo et al., 2015; Liu, Charrua, Weng, Yuan, & Ding, 2015; Lyu et al., 2018). Chen et al. (2015) showed that three chars produced from the pyrolysis of bamboo (BB), rice husk (BRH), and wood sawdust (BWS) at 500°C removed differently the *N*-nitrosodimethylamine, which is an emerging carcinogenic disinfection by-product. Indeed, the corresponding Langmuir's adsorption capacities were determined to be 15.08, 6.35, and 6.38 mg/g for BB, BRH, and BWS, respectively.

Another study (Liu et al., 2015) looked at the use of chars derived from pyrolysis at 450°C during 2 h of six different feedstocks: soybean (SBB), corn stalk (CSB), rice stalk (RSB), poultry manure (PMB), cattle manure (CMB), and pig manure (PgMB) for atrazine (common pesticide) removal. They showed that at equilibrium, the SBB removed most of atrazine, followed by RSB and CMB. The Freundlich isotherm adsorption parameters indicated that the adsorption capacities of the used chars decreased in the order SBB > RSB > CMB > CSB > PMB > PgMB. The authors also showed that a significant correlation existed between the char adsorption efficiencies, the total pore volumes, and the pH_{ZPC} values of the different produced chars. Similar findings were reported by García-jaramillo et al. (2015) when studying tricyclazole (common fungicide) adsorption by various chars. Regarding dyes, Lyu et al. (2018) studied methylene blue (MB) removal under the same pyrolysis temperature of hickory chips (HC), sugarcane bagasse (BG), and bamboo (BB), as well as their ball-milled derived chars at different temperatures, named BMHC, BMBG, and BMBB, respectively. The related adsorption capacities were 114, 169, and 202 mg/g for BMHC300, BMBG300, and BMBB300, respectively (Fig. 6). The authors attributed this finding to higher α-cellulose contents enhancing the dyes' adsorptive capacities. In another study, the adsorption capacities of methyl violet varied with their feedstock in the following order: canola straw char > peanut straw char > soybean straw char > rice hull char (Xu, Xiao, Yuan, & Zhao, 2011).

7.3.1.2 Physicochemical properties of the produced chars

The main chars' characteristics such as the SSA, the developed porosities (macro, meso, and micro), and their richness in specific functional groups that are susceptible to react with the dissolved pollutants' molecules are key parameters for organics removal from aqueous solutions (Inyang & Dickenson, 2015; Xu et al., 2017). To

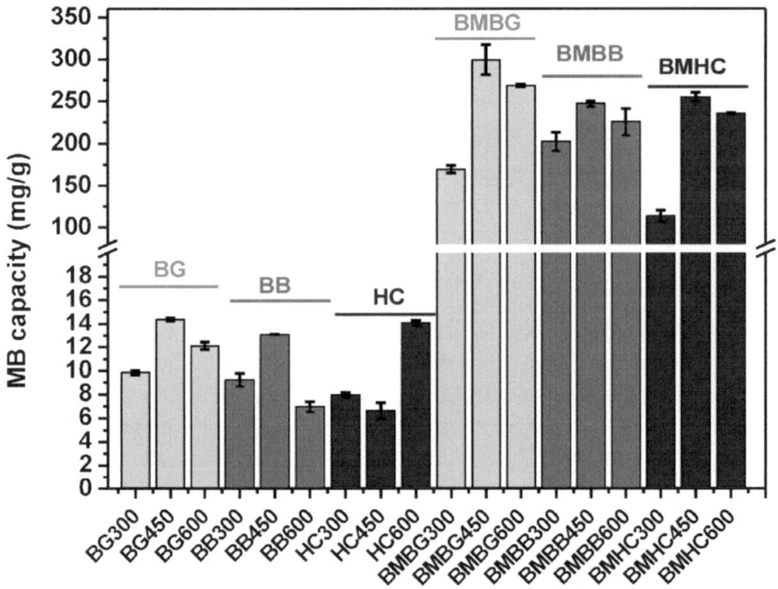

Fig. 6 Effects of feedstock type and pyrolysis temperature on methylene blue (MB) adsorption. *BB*, Bamboo; *BG*, bagasse; *BMBB*, ball-mill bamboo; *BMBG*, ball-mill bagasse; *BMHC*, ball-mill hickory chip; *HC*, hickory chip.
Reprinted with permission from Lyu, H., Gao, B., He, F., Zimmerman, A.R., Ding, C., Tang, J., et al. (2018). Experimental and modeling investigations of ball-milled biochar for the removal of aqueous methylene blue. *Chemical Engineering Journal, 335*, 110–119. Copyright 2018 Elsevier.

improve the chars' properties, several pre- or posttreatments or both have been tested (Rosales et al., 2017; Tan et al., 2016). In this context, Lian et al. (2016) studied the modification of chars derived from corn straw by using the N-doping approach for the study of dyes removal from aqueous solutions. This approach (physical activation) consists of the thermal treatment of these chars at temperatures between 600°C and 800°C by using NH_3 as a gas carrier. They demonstrated that the N-doping of a produced char at a final temperature of 800°C for 3 h increased its SSA and microporosity percentage from 61.0 to 418.7 m^2/g and from 20.5% to 71.5%, respectively. This modification has significantly increased the adsorption capacities of both anionic (acid orange) and cationic (MB) dyes up to 292 and 436 mg/g, which qualified as being among the most efficient ever reported for carbonaceous adsorbents.

A similar trend was also observed when activating chars derived from the fast pyrolysis of a mixture of sawdust and sugarcane straw (de Mendonça, da Cunha, Soares, Tristão, and Lago, 2017). Indeed, the authors pretreated this char with concentrated HCl under reflux for 12 h, then pyrolyzed it at temperatures of 400°C, 600°C, and 800°C. They reported that the SSA and microporosity increased from 3 m^2/g and 0.0001 cm^3/g (pristine char) to about 590 m^2/g and 0.231 cm^3/g for the treated one at a temperature of 800°C, respectively. On the other hand, this treatment decreased the oxygen content from about 27 to 14 wt%, which is due to decomposition of surface functionalities. This decrease significantly decreased the positively charged studied dye adsorption (MB). An opposite behavior was observed for the studied anionic dye (indigo carmine). However, the activation of char derived from safflower seed press cake at 500°C by $ZnCl_2$ with an impregnation ratio $ZnCl_2$:char = 4:1 increased MB adsorption capacity from 16.18 to 128.21 mg/g for carbonization temperatures of 600°C and 900°C, respectively (Angin, Altintig, & Köse, 2013). They imputed this increase to the high assessed SSA (801.5 m^2/g) and microporous structure (0.284 cm^3/g) as well as to the nonsignificant losses in oxygen-based functional groups.

Concerning pesticides, Mandal, Singh, and Purakayastha (2017) studied atrazine and imidacloprid removal from aqueous solution. These authors modified char prepared by rice straw pyrolysis at 600°C using phosphoric acid. They noticed that this modification reduced both the SSA (from 220.2 to 192.3 m^2/g) and their pore volume (from 0.646 to 0.161 cm^3/g), which were attributed to the increase in functional groups inside the pore's structure. However, this acid treatment has significantly increased the adsorption of both atrazine and imidacloprid capacities.

On the other hand, pyrolysis temperature and duration affect strongly the chars' properties and therefore their capacities to remove organics from aqueous solutions. Fig. 6 gives the effect of pyrolysis temperature for various chars on MB dye adsorption (Lyu et al., 2018). Zhang et al. (2011) studied simazine adsorption onto corn straw chars prepared at various pyrolysis temperatures between 100°C and 600°C with an interval of 100°C and a residence time of 6 h. The authors demonstrated that the SSA increased with the increase in the pyrolysis temperature from 2.1 to 329.0 m^2/g for BCS100 and BCS600, respectively. They proved the correlation between the adsorbed amounts and the SSA indicating the involvement of a pore-filling adsorption mechanism (see Section 7.3.2.). Indeed, the increase in pyrolysis temperature from 100 to 600°C has significantly increased the Freundlich constant (K_f) from 124 to 4054, respectively. A relatively similar trend was reported by Zhang, Sun, Yu, and Sun (2013) when exploring the adsorption of carbaryl and atrazine on pig manure-derived chars pyrolyzed at temperatures of 350°C and 700°C.

However, some studies demonstrated that pyrolysis at higher temperatures can reduce the organics removal ability. Chen et al. (2015) studied the removal of *N*-nitrosodimethylamine by various chars obtained from bamboo pyrolysis at temperatures of 400 (BC400), 500 (BC500), and 700°C (BC700). Even if the corresponding SSAs were determined to be 33.5, 38.5, and 39.3 m^2/g, their Langmuir's adsorption capacities were 9.1, 15.1, and 7.6 mg/g for BC400, BC500, and BC700, respectively. They suggested that at higher temperatures, some surface micropores would be blocked.

Finally, it has been demonstrated that the minerals in chars generally decrease some pesticides removal efficiencies (Zhang, Sun, et al., 2013). For this reason, some chars are demineralized or de-ashed before using them for organics adsorption purposes (Tan et al., 2015; Xu et al., 2017). For instance, Zhang, Sun, et al. (2013) reported that de-ashing pig manure chars resulted in a significant increase of carbaryl and atrazine adsorption efficiencies thanks to the unblocking of inaccessible organic adsorption sites in the original chars. Indeed, at a pyrolysis temperature of 700°C, the affinity Freundlich coefficient (K_f) increased from about 708 to 5011 for carbaryl and from 589 to 3548 for atrazine.

7.3.1.3 Adsorption experimental conditions

Contact time: kinetic adsorption

Organics removal kinetics from aqueous solutions by chars depends on various factors such as the chars' nature, the dye properties, and the physicochemical characteristics of the bulk solution. In the scientific literature, the required time to reach an equilibrium state varies from several minutes (Hameed & El-Khaiary, 2008) to several days (Liu et al., 2015). Indeed, Hameed and El-Khaiary (2008) found that the malachite green (MG) adsorption by char generated from rice straw, at a dosage of 1.5 g/L, a pH of 5, and initial concentrations varying between 25 and 300 mg/L, attains an equilibrium state after a contact time ranging between 40 and 70 min. The MG adsorption kinetic was mainly governed by a diffusion mechanism that is very important in the beginning and decreases versus time. The kinetic experimental data were well fitted by the pseudo-first order, confirming that the adsorption rate is initially controlled by external mass transfer. However, Mahmoud, Salleh, Karim, Idris, and Abidin (2012) showed that for MB removal by a 3 M HCl-treated kenaf char, a duration of 24 h was needed to attain equilibrium. The experimental data were well fitted with the pseudo-second order indicating that the adsorption process might be mainly governed by chemisorption. Moreover, for initial atrazine concentrations varying from 5 to 35 mg/L, Liu et al. (2015) demonstrated that the removal kinetic by chars generated from different feedstocks attained an equilibrium state after 196 h. Furthermore, the adsorption kinetic process was mainly impacted by the diffusion mechanism. Finally, Vithanage, Mayakaduwa, Herath, and Sik (2016) studied the impact of the nature of chars produced under the same pyrolysis conditions but from different feedstocks on some pesticides' kinetic adsorption. They studied carbofuran removal using chars from tea waste (TWB) and rice husk (RHB) produced at a pyrolysis temperature of 700°C for 3 h. They reported that carbofuran adsorption equilibrium was attained after 4 h for RHB. About double this time was required for TWB.

Initial pollutants concentrations: adsorption isotherm

Various empirical models were fitted to the isotherm experimental data depending on the nature of the char and the organic pollutant. In this context, the equilibrium experimental data regarding MG removal by char derived from rice straw was well fitted with the Langmuir model with a correlation coefficient higher than 0.94 (Hameed & El-Khaiary, 2008). Another study (Mahmoud et al., 2012) dealing with MB adsorption by acid-treated kenaf char showed that for temperatures of 30°C, 40°C, and 50°C, the equilibrium experimental data were well fitted with the Langmuir model with correlation coefficients higher than 0.99. Other models were found to fit better the isotherm's experimental data (Mui, Cheung, Valix, & Mckay, 2010; Yang, Lin, Wei, Zhao, & Wang, 2014). Mui et al. (2010), when studying MB, acid blue 25, and acid yellow 117 removal from aqueous solutions by a char derived from bamboo pyrolyzed at 900°C, showed that the Redlich-Peterson model is considered as the most appropriate model to describe the experimental equilibrium data. However, Yang et al. (2014), when studying char generated from the pyrolysis of the same feedstock (bamboo) at 1000°C for the removal of metal-complex dye acid black 172 from aqueous solutions, found that the Freundlich model was more suitable for fitting the equilibrium adsorption data.

Regarding pharmaceutical compounds, Boshir, Zhou, Hao, Guo, and Hasan (2017) studied the competitive sorption affinity of different antibiotics, sulfamethazine (SMT), sulfamethoxazole (SMX), sulfathiazole (STZ), and chloramphenicol (CP), toward an activated eucalyptus wood char with phosphoric acid. They showed that the Langmuir model fitted very well the experimental data for STZ and SMX, while the Freundlich isotherm model provided a better fit for SMT and CP. The Langmuir's adsorption capacities were determined to be 45.2, 28.29, 21.35, and 20.71 mg/g for STZ, SMX, CP, and SMT, respectively, in competitive mode.

For pesticides, the majority of studies showed that the Freundlich model is the best isotherm model fitting well the experimental data (Liu et al., 2015; Mandal et al., 2017; Mandal & Singh, 2017; Zhang, Sun, et al., 2013). This finding indicates that their adsorption on chars' surfaces was heterogeneous and multilayered.

Aqueous pH

Initial aqueous pH is a key factor during adsorption experiments. Indeed, pH value is linked to the surface charge of the adsorbent as well as the aqueous form of the pollutant. Hameed and El-Khaiary (2008) found that for an MG initial concentration of 50 mg/L, a dosage of 1.5 g/L, and a contact time of 2 h, its removal by a char derived from rice straw attains maximal removal for initial pH values between 3 and 7 (33 mg/g). However, when the used initial pH was fixed to 2, the adsorbed amount decreases to about 22 mg/g. They attributed this behavior to the competition between H^+ and the positively charged MG dye molecules. The same trend was observed by Mahmoud et al. (2012) where the percentage removal of MB by kenaf-derived char passes from 70% at an initial pH of 2.5 to 84% for a pH of 8.5 for an initial MB concentration of 100 mg/L, a dosage of 2 g/L, and a temperature of 30°C. Similar findings were reported by Ates and Tezcan Un (2013) when testing the use of four chars derived from hornbeam sawdust (Turkey) pyrolyzed at 500°C, 600°C, 700°C, and 800°C for disperse orange 30 (DO30) removal from synthetic solutions.

Another study dealing with ibuprofen (IB) and salicylic acid (SA) removal from aqueous solutions using pine wood fast pyrolysis char (Essandoh, Kunwar, Pittman, Mohan, & Mlsna, 2015) showed that their adsorption was higher in acidic solutions than in the basic region (Fig. 7). They explained this behavior by the fact that during the adsorption process, at a pH = 3 (the char's point of zero charge [=2.5] is lower than the solution pH), the char's surface becomes negatively charged causing an increase in the electrostatic repulsion of adsorbate carboxylate anions. Furthermore, when the p*K*a of the IB (4.91) and SA (2.98) is greater than the solution pH values, these pollutants exist primarily as their conjugate acid (R-COOH) forms. Electrostatic repulsion is minor at these solution pHs, promoting therefore the adsorption by attractive H-bonds

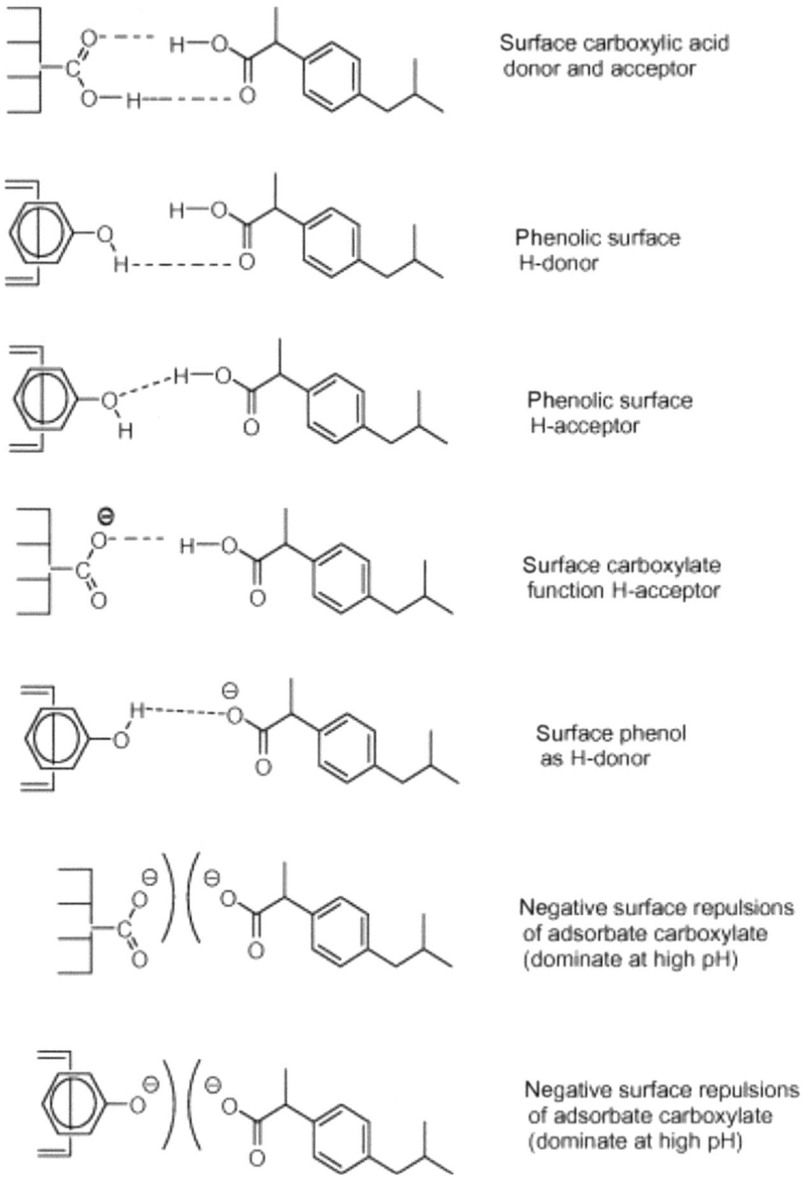

Fig. 7 Various H-bonding and charge-charge repulsive interactions. Both adsorbate and surface functions can serve as H-donors and H-acceptors.
Reprinted with permission from Essandoh, M., Kunwar, B., Pittman, C. U., Mohan, D., Mlsna, T. (2015). Sorptive removal of salicylic acid and ibuprofen from aqueous solutions using pine wood fast pyrolysis biochar. *Chemical Engineering Journal, 265*, 219–227. Copyright 2018 Elsevier.

and other interactions. Globally, similar behavior was observed when studying tetracycline (TET) removal from aqueous solutions by an iron zinc-doped sawdust char (Zhou, Liu, Xiang, et al., 2017). Indeed, both the net surface charge of the adsorbent as well as the p*K*a of TET influenced its removal efficiency.

Char dosage

Generally, when the char dose increases, organics removal percentage increases. For instance, for an initial DO30 dye concentration of 50 mg/L, a pH of 2, and a contact time of 2 h, Ates and Tezcan Un (2013) studied the impact of increasing the dosage of four chars derived from the pyrolysis of hornbeam sawdust at 500°C (SC1), 600°C (SC2), 700°C (SC3), and 800°C (SC4) from 0.2 to 8 g/L, and showed an increase in DO30 removal efficiencies from 2% to 27% for SC1, from 19% to 45% for SC2, from 15% to 65% for SC3, and from 16% to 71% for SC4. The same behavior was observed by Mondal, Aikat, Siddharth, and Sarkar (2017) when using the response surface methodology and artificial neural network approaches for the optimization of cationic pharmaceutical ranitidine hydrochloride removal by an activated char generated from *Parthenium hysterophorus* biomass.

Coexisting pollutants

It is well known that in general the coexistence of organic/mineral pollutants reduces organic pollutants removal by chars (Lin, Jiang, & Xu, 2017; Rosales et al., 2017). Lin et al. (2017) studied pharmaceuticals uptake by three commercial chars in the presence of slats and organic matter. They showed that the maximum adsorption capacity of IB in reverse osmosis concentrate, calculated using the Langmuir model, was 12.5 mg/g, which is only about 58% of the one assessed in deionized water. Zhou, Liu, Xiang, et al. (2017) explored the influence of the existence of TET and Cu(II) on their mutual adsorption onto an iron and zinc-doped sawdust char (Fe/Zn char). They observed that in comparison with the single TET assays, the presence of low concentrations of Cu(II), i.e., 5 and 10 mg/L, could promote TET adsorption. This finding was ascribed to the formed modified char-TET-Cu(II) or modified char-Cu(II)-TET complex. However, under the same initial TET concentration, the TET adsorption capacity decreased when the Cu(II) concentration increased to over 25 mg/L due to a greater competitive adsorption between them.

Temperature

Thermodynamic studies related to organics removal by chars are very important to have a better understanding of the involved reaction types. In this context, the effect of temperature on DO30 removal efficiencies by four chars generated from hornbeam sawdust was investigated at 20°C, 37°C, 47°C, 57°C, and 67°C (Ates & Tezcan Un, 2013). During these experiments, the initial DO30 concentration, adsorbent dosage, and aqueous pH were fixed at 50 mg/L, 2.4 g/L, and 2, respectively. They showed that for all tested chars, the dye removal efficiencies increase with increasing temperature. For instance, for the char produced at 800°C, this percentage increased from 37% to 59% when the temperature rose from 20°C to 67°C indicating that this process is endothermic. Moreover, Mahmoud et al. (2012) showed that MB adsorption efficiency increased when increasing the temperature from 30°C to 50°C, which confirms that the adsorption process is endothermic. Furthermore, the entropy and the Gibbs free energy values showed that the adsorption reaction is spontaneous and that an increasing randomness and disorder of the solid/solute interface at the char surface occurred during the adsorption process.

Concerning pesticides, Liu et al. (2015) studied the impact of temperature (10°C, 18°C, 27°C, and 40°C) on atrazine adsorption by six chars derived from the pyrolysis at 450°C of SBB, CSB, RSB, PMB, CMB, and PgMB. They showed higher removal capacities at higher temperatures, suggesting that the adsorption process is an endothermic process (Fig. 8). They suggested that the increase in the adsorbed atrazine amounts with increasing temperature could be explained by the increase in the availability of active sorption sites and the adsorbent becoming more desolvated, and also by the increase in frequency at which atrazine collided with the present functional groups on the char surfaces.

Globally, the same trend was observed by Yi et al. (2016) when studying the removal of levofloxacin (antibiotic) by adsorption on rice husk and woodchip chars. The impact of the main parameters cited in the sections above on the sorption of selected organic pollutants from aqueous solutions is summarized in Table 2.

7.3.2 Involved mechanisms

The involved mechanisms in the removal of organic compounds from aqueous solutions by chars depend on both pollutant properties and the chars' surface chemistry (Rosales et al., 2017). In general, high temperatures increase the aromaticity, hydrophobicity, and surface area of chars, which result in better removal efficiencies (Tan et al., 2015). The adsorption mechanisms include (Fig. 9): (1) pore filling, which is a function of micropore and mesopore volumes (Inyang & Dickenson, 2015; Tan et al., 2015), (2) diffusion, which occurs through diffusion into pores (e.g., during pore filling), or into the organic matter of the noncarbonized fractions, (3) partitioning, which concerns the amorphous carbon phase corresponding to low pyrolysis temperatures (Zhang, Sun, et al., 2013), (4) hydrophobic interaction, which occurs on char surfaces with low hydration energy, (5) aromatic-π and cation-π interactions depending on the π-electron systems present, (6) electrostatic interactions regarding the adsorption of ionic and ionizable organic compounds and where the aqueous pH and the pH_{ZPC} of the adsorbent play a key role, and (7) hydrogen bonding, which is related to polar organic compounds (Chen et al., 2015).

The adsorption of organic compounds occurs in the presence of different mechanisms (Inyang & Dickenson, 2015; Rosales et al., 2017). Zhang, Sun, Ren, Min, and Zhang (2018) studied the involved mechanisms in the adsorption of three pesticides by 24 chars derived from pyrolysis at various temperatures of maize straw and pig manure. They showed that for low pyrolysis temperatures (<500°C), besides hydrophobic partition, cation-π interactions and hydrogen bonding are the main involved mechanisms. However, for temperatures higher than 500°C, the involved mechanisms included pore filling and cation-π interactions.

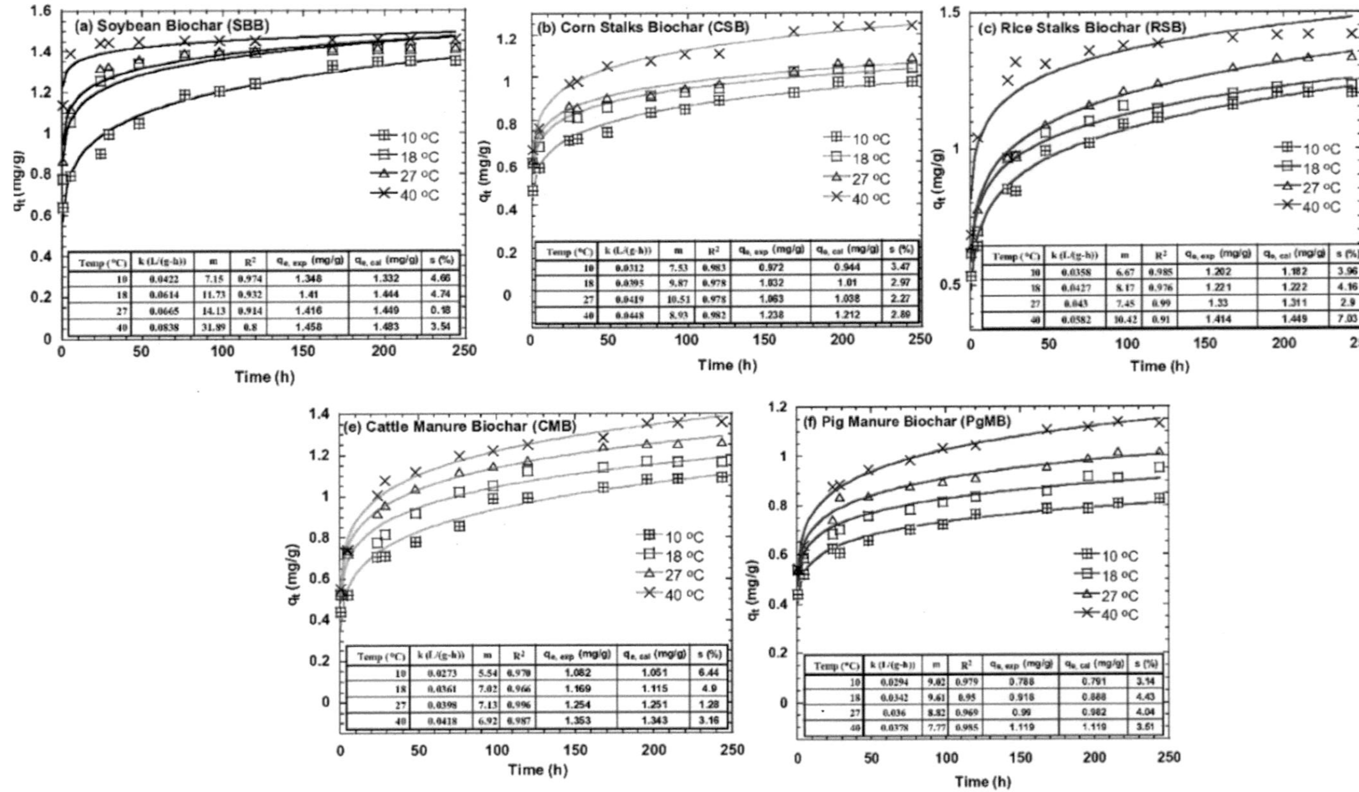

Fig. 8 Effect of temperature on adsorption capacity of six char types. Solid lines are best fit of the modified Freundlich model. (A) Soybean Biochar (SBB). (B) Corn Stalks Biochar (CSB). (C) Rice Stalks Biochar (RSB). (E) Cattle Manure Biochar (CMB). (F) Pig Manure Biochar (PgMB). Reprinted with permission from Liu, N., Charrua, A. B., Weng, C. H., Yuan, X., Ding, F. (2015). Characterization of biochars derived from agriculture wastes and their adsorptive removal of atrazine from aqueous solution: a comparative study. *Bioresource Technology, 198*, 55–62. Copyright 2018 Elsevier.

Table 2 Adsorption characteristics of various organics onto raw and modified chars

Feedstock	Pretreatment	Pyrolytic conditions	Posttreatment	SSA (m^2/g)	Pollutant	Adsorption batch conditions	Equilibrium time (h)	Kinetic model	Isotherm	QL (mg/g)[a] or K_f ($mg^{1-(1/n)}$ $L^{1/n}$ g^{-1})[b]	References
Dyes											
Korean cabbage waste	–	T = 500°C for 1 h Gradient = 10°C/min	–	11.4	Congo red Crystal violet	Dose = 1 g/L T = 30°C C_0 = 50–5000 mg/L pH = 7 150 rpm	120 180	PSO PSO	Langmuir Langmuir	95.81[a] 1304[a]	Sewu, Boakye, and Woo (2017)
Rice straw	–		–	34.7			90 288	PSO PSO	Freundlich Freundlich	190.8[a] 620[a]	
Woodchip	–		–	<0.1			120 240	PSO PSO	Langmuir Langmuir	110[a] 195.6[a]	
Municipal solid waste	–	T = 400°C for 30 min Gradient = 3°C/min	Activation with KOH	–	Methylene blue	Dose = 2 g/L T = 30°C C_0 = 50 mg/L pH = 6.5 150 rpm	200	PSO	–	24.75[a]	Sumalinog, Capareda, and de Luna (2018)
Safflower seed	Feedstock turned into a press cake material	500°C Gradient = 50°C/min	Activation: chemical with $ZnCl_2$	801.5	Methylene blue	Dose = 4 g/L T = 30°C C_0 = 50–500 mg/L pH = 6.5 120 rpm	–	–	Langmuir	128.21[a]	Angin et al. (2013)
Ordinary weed	–	500°C for 1 h Gradient = 5° C/min	Treatment with HCl	40.46	Methylene blue	Dose = 2 g/L T = 25°C C_0 = 50–400 mg/L pH = 7.4 120 rpm	450 min	PSO	Langmuir	161.29[a]	Güzel, Sayğılı, Akkaya Sayğılı, Koyuncu, and Yılmaz (2017)

Continued

Table 2 Continued

Corn straw	Preheated at 105°C	600°C for 120 min Gradient = 10°C	Steam activation with NH_3 at 800°C/1–3 h	418.7	Acid orange Methylene blue	Dose = 0.5 g/L T = 25°C C_0 = 75–300 mg/L pH = 11 150 rpm	25 h	PSO PSO	Freundlich Freundlich	203.1[a] 420.8[a]	Lian et al. (2016)
Sawdust and sugarcane straw	HCl treatment	800°C for 1 h Gradient = 5°C/min	–	590	Methylene blue Indigo carmine	Dose = 1 g/L T = 20°C C_0 = 0–1.1 mmol/L pH = 6 No rpm	30 min	PSO PSO	Langmuir Langmuir	13.11[a] 16.32[a]	de Mendonça et al. (2017)
Bamboo		T = 1173°C for 2 h No gradient	–	327	Acid blue 25 Acid yellow 117 Methylene blue	Dose = 1 g/L T = 30°C C_0 = – pH = 7 150 rpm	21 days	–	Redlich–Peterson	16.91[a] 22.23[a] 319.11[a]	Mui et al. (2010)
Pesticides											
Pig manure	–	T = 700°C for 2 h No gradient	Acid treatment (HCl and HF)	218.1	Carbaryl Atrazine	Dose = 1.25 g/L T = 20–24°C C_0, atrazyl = 1–40 mg/L C_0, carbanyl = 2–30 mg/L pH = 6.5 150 rpm	170 h	–	Freundlich Freundlich	2.85[b] 2.77[b]	Zhang, Sun, et al. (2013)
Corn straw	–	T = 600°C for 6 h No gradient	HCl demineralization	329.0	Simazine	No dose T = 25°C C_0 = 0.006–3.3 mg/L pH = 6.0 120 rpm	7 days	–	Dual model	3.4[a]	Zhang et al. (2011)

Tea waste	Washed and dried at 60°C for 48 h	700°C for 3 h Gradient = 7° C/min	–	342.22	Carbofuran	Dose = 1 g/L 50 mg/L pH = 5.0 100 rpm	700 min	PSO	–	10.74[a]	Vithanage et al. (2016)
Rice husk			–	377.00	Carbofuran		500 min	PSO	–	23.86[a]	
Rice straw	Dried at 60°C for 24 h	600°C for 1 h Gradient = 3° C/min	–	220.2	Atrazine	Dose = 1 g/L $T = 25°C$ $C_0 = 1–10$ mg/L pH = 5.0 100 rpm	24 h	–	Freundlich	1.68[b]	Mandal and Singh (2017)
Rice straw			–	192.3	Imidacloprid		24 h	–	Freundlich	3.41[b]	
Eucalyptus bark	Dried at 60°C for 24 h	600°C/1 h 3°C/min	–	188.2	Atrazine Imidacloprid	Dose = 1 g/L $T = 25°C$ $C_0 = 1–10$ mg/L pH = 5.0 100 rpm	14 h	Elovich	Freundlich	0.51[b] 0.39[b]	Mandal et al. (2017)
Corn cob			–	242.1			24 h	Elovich	Freundlich	0.53[b] 0.18[b]	
Bamboo chips			–	246.7			24 h	Elovich	Freundlich	0.32[b] 0.28[b]	
Rice husk			–	159.1			24 h	Elovich	Freundlich	0.44[b] 0.74[b]	
Rice straw			–	220.2			24 h	PSO	Freundlich	1.36[b] 1.70[b]	
Rice straw			Treatment with H_3PO_4	192.3			24 h	Elovich	Freundlich	2.71[b] 3.14[b]	
Soybean	–	450°C for 2 h No gradient	–	17.5	Atrazine	Dose = 10 g/L $T = 27°C$ $C_0 = 5–35$ mg/L pH = 6.5 115 rpm	10 days	–	Freundlich	1.37[b]	Liu et al. (2015)
Corn stalk	–		–	19.6			10 days	–	Freundlich	0.73[b]	
Rice stalk	–		–	25.8			10 days	–	Freundlich	1.11[b]	
Poultry manure	Dried for 14 days		–	15.4			10 days	–	Freundlich	0.93[b]	
Cattle manure	Dried for 14 days		–	13.5			10 days	–	Freundlich	0.89[b]	
Pig manure	Dried for 14 days		–	13.4			10 days	–	Freundlich	0.58[b]	

Continued

Table 2 Continued

Woodchips	–	600°C for 4h 5°C/min	Washed with distilled water and dried at 105°C	361	2,4-D	Dose = 1.25 g/L $T = 25°C$ $C_0 = 0–100 mg/L$ pH = 7.0 90–100 rpm	14 days	–	Freundlich	0.72[a]	Kearns et al. (2014)
	–	600°C for 4 days 5°C/min		516		Dose = 0.04 g/L $T = 25°C$ $C_0 = 0–100 mg/L$ pH = 7.0 90–100 rpm	14 days	–	Freundlich	21.3[a]	
Pharmaceuticals											
Eucalyptus wood	–	400°C for 2h Gradient = 11°C/min	H_3PO_4 treatment	–	Sulfamethazine Sulfamethoxazole Sulfathiazole Chloramphenicol	80 mg/L $T = 25°C$ $C_0 = 0.25–20 mg/L$ pH = 4.25 120 rpm	1800 min	PSO PSO PSO PFO	Langmuir Langmuir Langmuir Langmuir	20.71[a] 28.29[a] 45.19[a] 21.35[a]	Boshir et al. (2017)
Coconut, peanut, walnut shells		$T = 500°C$ for 1.5h Gradient = –	Ferric ion treatment	365	Carbamazepine Tetracycline	Dose = 0.2 mg/L $T = 25°C$ $C_0 = 5–60 mg/L$ pH = 6.0 170 rpm	20h	–	Langmuir	62.7[a]	Shan et al. (2016)
Granular activated char				486			Over 36h			135.1[a]	
Sawdust	Ferric/zinc ion treatment	$T = 600°C$ for 2h Gradient = 5°C/min	–	–	Tetracycline	Dose = 4 g/L $T = 25°C$ $C_0 = 40–150 mg/L$ pH = 6 No rpm	1500 min	Elovich and two compartment models	Langmuir	83[a]	Zhou, Liu, Xiang, et al. (2017)

Burcucumber-derived char	–	$T = 700°C$ for 2 h Gradient = 7° C/min	Steam activation	7.1	Sulfamethazine	Dose = 1 g/L $T = 25°C$ $C_0 = 2.5–50$ mg/L pH = 3 100 rpm	72 h	–	Temkin	37.73[a]	Vithanage et al. (2014)
Maize straw	–	$T = 600°C$ for 1.5 h No gradient	–	–	Oxytetracycline	Dose = 2 g/L $T = 25°C$ $C_0 = 0.02–0.4$ mmol/L pH = 5.5 15 rpm	–	–	Freundlich	26.12[a]	Jia et al. (2016)
Olive stones	Pretreatment with H_2SO_4	$T = 550°C$ for 1 h Gradient = –	–	83.72	Diclofenac	Dose = 5 g/L $T = 23°C$ $C_0 = 5–60$ mg/L pH = 4.2 500 rpm	20 min	PSO	BET	11.00[a]	Larous and Meniai (2016)

BET, Brunauer-Emmett-Teller; *PFO*, pseudo-first order model; *PSO*, pseudo-second-order model; *QL*, Langmuir's maximum adsorption capacity.
[a]Langmuir adsorption capacity.
[b]Freundlich coefficient.

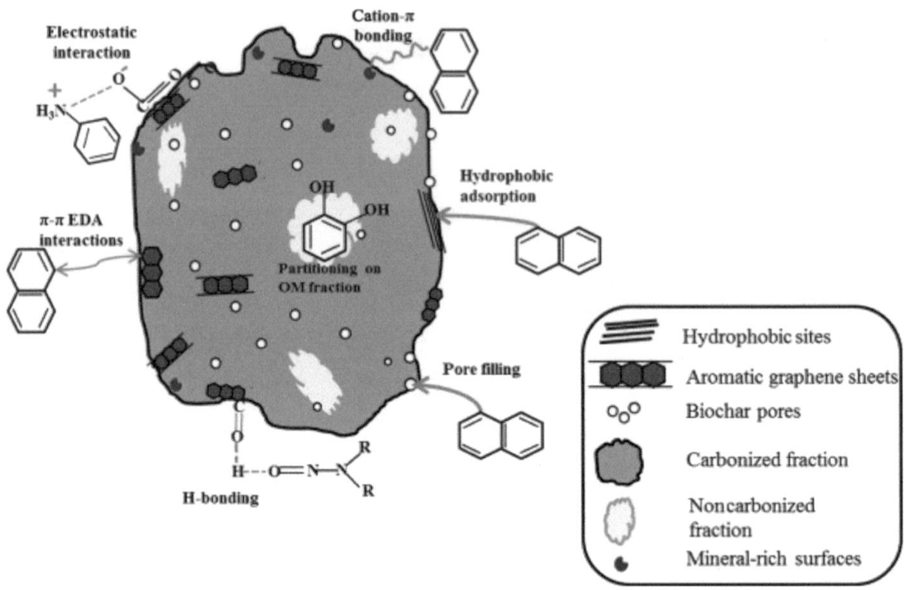

Fig. 9 Sorption mechanisms for the uptake of organic contaminants on char.
Reprinted with permission from Tan, X., Liu, Y., Zeng, G., Wang, X., Hu, X., Gu, Y., et al. (2015). Application of biochar for the removal of pollutants from aqueous solutions. *Chemosphere, 125*, 70–85. Copyright 2018 Elsevier.

7.4 Use of biomass-derived chars for gaseous effluents treatment

Biomass-derived chars are proposed as an alternative for activated carbons, for the adsorption of gaseous pollutants, due to their lower cost and less harmful effect on the environment. Experiments on gaseous pollutants adsorption on char could be carried out in dynamic or static conditions. Details of such experimental procedures are described in the review of Bamdad et al. (2018). Many parameters influence the adsorption affinity of chars toward gaseous pollutants.

7.4.1 Nature of biomass feedstock

The biomass type affects the adsorption properties of its derived char. Bamdad et al. (2018) found that the degree of carbonization is higher for SW compared to HW. Brunauer-Emmett-Teller (BET) surface areas depend also on the carbonized biomass type. Under the same pyrolysis temperature, 600°C, char derived from pine wood exhibits a surface area of $312\,m^2/g$, while the one derived from hickory is

256 m^2/g. These values are largely higher than those of herbaceous plant chars (0.2, 2.2, 15, 61, 168, and 182 m^2/g, respectively, for alfalfa, cotton, SG, corn, straw, rice husk, and wheat straw) (Zhang, Gao, Creamer, et al., 2017). Bamdad and Hawboldt (2016) performed characterization studies on three different woody biomasses, SW (shaving, bark [balsam fir]), an HW (ash wood), and a metal organic framework (MOF). They have evidenced that all chars were basic (pH 8.9–10.7), whereas MOFs are acidic. Biomass-derived chars possess higher carbon contents and more aromatic functional groups than MOF. These properties could confer on chars higher adsorption efficiencies toward acidic gases (CO_2, H_2S, etc.).

Lu and Lu (2010) reported higher adsorption efficiencies for SO_2 and NO on char prepared from wheat straws, pyrolyzed at 600°C, than on the ones prepared from rice straw and cotton straw. Zhang, Gao, Zheng, et al. (2017) found for a char derived from sugar beet tailings pyrolyzed at 600°C a lower surface area (2.6 m^2/g) than for the one derived from surgarcane bagasse treated in the same conditions (388.3 m^2/g). They stated that the wide variation of surface area between chars issued from different biomasses is correlated with the various contents of cellulose, hemicellulose, and lignin. In general, biomasses with higher lignin contents result in char with higher surface area and porosity. Furthermore, Zhang, Gao, Zheng, et al. (2017) obtained higher sorption capacities for acetone, cyclohexane, and toluene on the adsorbent with the higher surface area.

7.4.2 Physical and chemical treatment of char

The char characteristics and consequently its adsorption performances are influenced by the nature of the biomass and by the operating conditions during pyrolysis. It has been demonstrated (Chen, Zhou, & Zhu, 2008; Kim, Kim, Cho, & Choi, 2012; Zhang, Gao, Creamer, et al., 2017 and references therein; Zhang, Gao, Zheng, et al., 2017) that the pyrolysis temperature has more influence than the other factors. Indeed, increasing the pyrolysis temperature results in:

- higher surface area of char (Lu & Lu, 2010; Ouyang, Xu, Song, & Jiao, 2013; Zhang, Gao, Zheng, et al., 2017). During the pyrolysis process, large amounts of gases are released from the biomass resulting in high pore volume. However, some authors reported a decrease in the porosity with the increase in pyrolysis temperature (Zhang, Gao, Creamer, et al., 2017 and references therein; Chen, Zhang, Chen, Yang, & Chen, 2017);
- higher carbonization and increased ash contents in char;
- net decrease of oxygen and hydrogen content leading to the formation of hydrophobic char, efficient in the adsorption of nonpolar molecules such as CO_2 and hydrophobic VOCs. In other words, pyrolysis temperature influences the nature and relative amount of the chemical functional groups present on its surface (Chen et al., 2017; Iberahim, Sethupathi, & Bashir, 2017; Ouyang et al., 2013; Zhang, Gao, Zheng, et al., 2017).

Moreover, Zhang, Gao, Zheng, et al. (2017) stated that the decomposition of cellulose and hemicellulose into molecules with lower molecular weights results in the increase in the surface areas upon increasing the pyrolysis temperature at a range lower than

450°C. At higher temperatures, the transition from layered cellulose carbon to amorphous char, which takes place with the increase in pyrolysis temperature, is responsible for the development of higher surface areas.

Creamer, Gao, and Zhang (2014) noted that the surface area of a sugarcane bagasse increased from 5.20 to 388.30 m^2/g by increasing the pyrolysis temperature, respectively, from 300°C to 600°C, and that of hickory wood increased from 0.10 to 401.00 m^2/g for the same increase of pyrolysis temperature. In general, higher surface areas were associated with higher CO_2 adsorption capacities. For example, at 25°C, the adsorption capacity of sugarcane bagasse pyrolyzed at 300°C is about 35 mg/g, whereas the adsorption capacity of sugarcane bagasse pyrolyzed at 600°C is above 60 mg/g (Fig. 10) (Creamer et al., 2014).

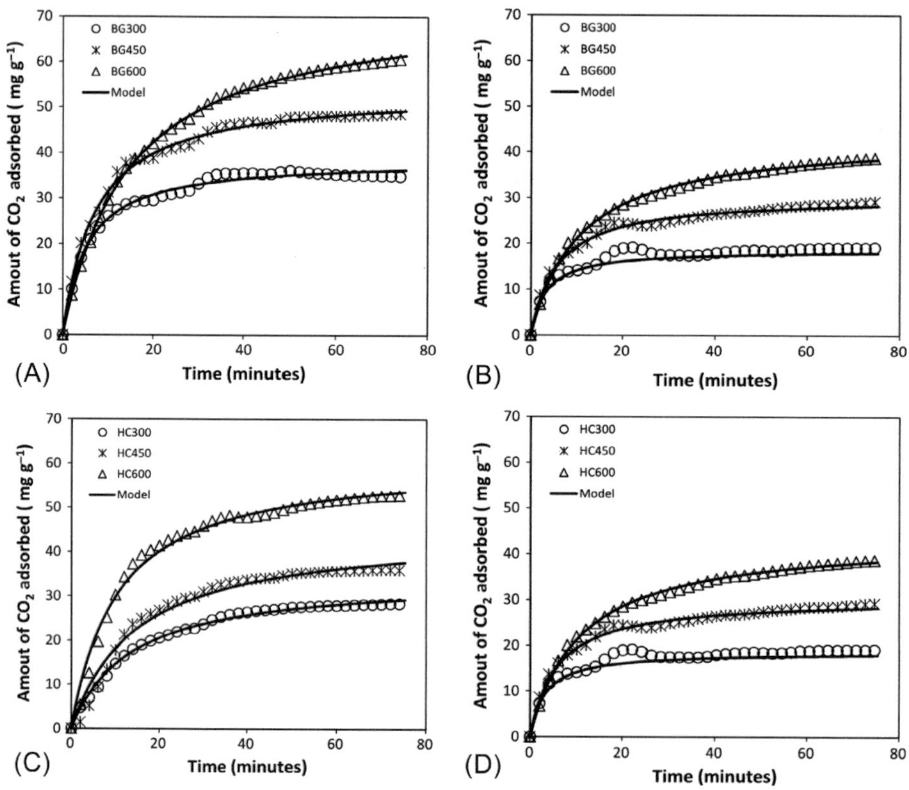

Fig. 10 Adsorption kinetics of CO_2 onto (A) sugarcane bagasse (BG) chars at 25°C, (B) sugarcane BG chars at 75°C, (C) hickory chip (HC) chars at 25°C, (D) HC chars at 75°C. BG and HC were pyrolyzed under nitrogen at 300°C, 450°C, and 600°C.

Zhang, Gao, Zheng, et al. (2017) compared the adsorption of three different VOCs, acetone, cyclohexane, and toluene, on different chars derived from bamboo, surgarcane bagasse, Brazilian pepper wood, sugar beet tailings, and hickory wood pyrolyzed at 300°C, 450°C, and 600°C. The higher sorption capacities of acetone were in line with its higher polarity compared to the two other studied VOCs. In addition, the sorption capacity of a given VOC depends on the surface area and on the content of the noncarbonized organic matter (NOM), both factors being controlled by the pyrolysis temperature. Increasing the pyrolysis temperature is expected to increase the surface area and to decrease the NOM. A compromise should be found. After five adsorption/desorption cycles, acetone sorption decreased due to the incomplete desorption that resulted from the formation of strong bonds between adsorbate and adsorbent and/or because of pore plugging.

In addition, Iberahim et al. (2017) investigated SO_2 adsorption on char issued from palm oil mill sludge pyrolyzed at different heating temperatures (300–800°C), heating rates (10–20°C/min), and holding times (60–120 min). The highest adsorption capacity, ~10 mg/g, was obtained for pyrolysis at 400°C with a heating rate of 15°C/min and a holding time of 90 min. The functional groups present on the surface of char determined mainly the ability to adsorb SO_2. The amount of these groups decreased with increasing temperature and hence lower pyrolysis temperatures resulted in higher SO_2 adsorption capacities. Lu and Lu (2010) evidenced that biomass fast pyrolysis (wheat straw, rice straw, and cotton straw) led to chars with higher adsorption capacities for SO_2 and NO than the chars prepared by slow pyrolysis (10°C/min).

In addition, Shen et al. (2017) studied the influence of ultrasound treatment on the properties of chars. In fact, chars prepared from tea waste pyrolyzed at 700°C and modified with NH_4Cl were subjected to magnetic stirring and then to ultrasound treatment (Shen et al., 2017). The latter was beneficial to mercury adsorption due to the increase in the surface area and the micropore volume by ultrasonication. Ultrasound treatment is thought to increase the dispersion, uniform distribution, and penetration of NH_4Cl species in the char. Ultrasound treatment could also remove the impurity present within the char. Furthermore, after pyrolysis at 400°C, Liu and Huang (2018) treated coffee waste by ammoxidation followed by activation with KOH. An important increase in the SSAs and pore volumes was observed due to activation by KOH. CO_2 adsorption capacity was enhanced due to this improvement in the textural properties and also due to the presence of N species and more pyrrolic nitrogen (active adsorption sites) on the char surface. The highest sorption capacity was noted for the sample treated with melamine. At 35°C, this sample exhibited a sorption capacity of 117.5 mg/g and this value was rather stable (110 mg/g), even after 10 cycles of adsorption/desorption (Fig. 11).

Moreover, Delgado et al. (2017) investigated char obtained via torrefaction of nonrecycled cardboard. Different physical treatment parameters (torrefaction temperature 250–300°C and residence time 60–120 min) and chemical treatment with KOH were surveyed. It was proven that a more stable char is obtained at higher torrefaction temperatures and residence times. CH_4 adsorption was studied at 30°C and atmospheric pressure with CH_4 partial pressures lower than 0.45 kPa. CH_4 adsorption capacity

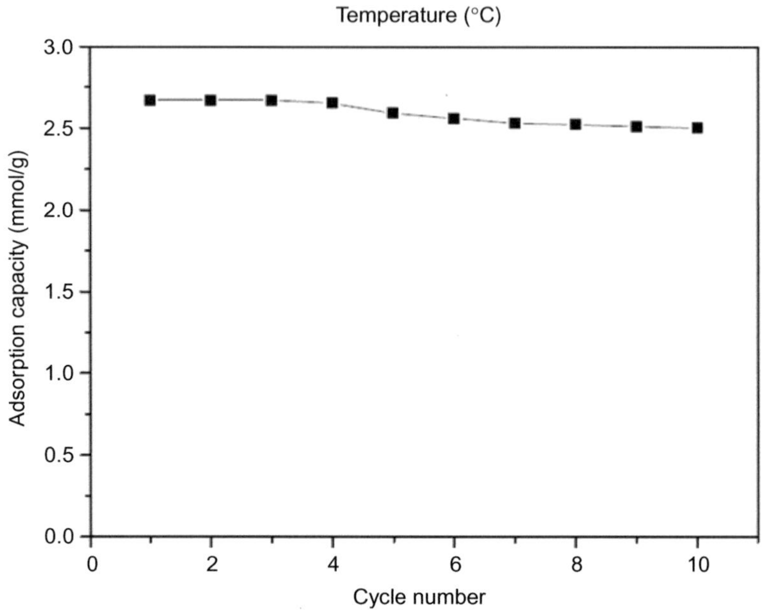

Fig. 11 Cyclic runs of a CO_2 adsorption test on coffee waste treated with melamine. Reprinted with permission from Liu, S. -H., Huang, Y. -Y. (2018). Valorization of coffee grounds to biochar-derived adsorbents for CO_2 adsorption. *Journal of Cleaner Production, 175*, 354–360. Copyright 2018 Elsevier.

increased with torrefaction time and temperature and decreased with chemical treatment. In general, low SSAs and pore volumes were obtained (respectively 2–6 m^2/g and 0.0008–0.023 cm^3/g). The highest adsorption capacity, 5.76 mg CH_4 per gram of sample, was for the char torrefied during 90 min at 300°C without treatment with KOH. Hence, chemical treatment of char with KOH was not beneficial for CH_4 adsorption. Indeed, CH_4 adsorption is influenced by the presence of functional groups on the surface of the adsorbent. CH_4 is a nonpolar molecule. Consequently, the presence of C—C bonds on the surface favors its adsorption, whereas the oxygen-based bonds, O—H, C—O, C=O, etc. confer on char a hydrophilic character resulting in lower affinity toward CH_4 and thus lower adsorption capacity.

In addition, sewage sludge pyrolyzed for 1 h at 700°C showed better formaldehyde adsorption after activation by both KOH and ammonia (Lee et al., 2011) than without activation. Surface basicity as well as oxygen and nitrogen functional groups generated by such activation are responsible for adsorption capacity improvement.

Furthermore, NO_2 adsorption in wet and dry conditions was investigated in coffee industry waste pyrolyzed at 500°C and 800°C (Nowicki, Skibiszewska, & Pietrzak, 2013). The chars pyrolyzed at 800°C showed adsorption capacities four times as high as those of their homologous pyrolyzed at 500°C. Activation by KOH led to an increase in the surface area from 9–12 to 2076 (for the one pyrolyzed at 500°C) and 1553 m^2/g (for the one pyrolyzed at 800°C). In dry conditions, the most effective adsorbent was the one pyrolyzed at 800°C and activated by KOH, with an NO_2 sorption capacity of about 44.5 mg/g. Similar effects of pyrolysis temperature and KOH activation were found on chars derived from waste tires and treated in the same way (Hofman & Pietrzak, 2011), but with lower surface areas and NO_2 adsorption capacities.

Another example is found in the work of Zhou, Liu, and Pan (2017) who investigated the effect of K_2FeO_4 loading (0.05, 0.1, and 0.3 wt%) on char derived from wheat straw pyrolyzed under nitrogen at 800°C. K_2FeO_4 presence increased the removal of mercury vapor (Fig. 12). Lower uptake was obtained for raw char (Fig. 12). The increase in BET surface area was mainly the reason behind this enhancement. However, an optimal content of K_2FeO_4 is necessary because the increase in the content from 0.1 to 0.3 wt% led to a decrease in the efficiency of the removal of Hg as a result of the decrease in the BET area, micropore volume, and total pore volume. Shen et al. (2017) studied mercury vapor adsorption on waste tea pyrolyzed under nitrogen at 60 mL/min at 500°C for 2 h, then activated by steam and impregnated with different contents of NH_4Cl to lead to 0, 1, 3, 5, 7, and 9 wt% of Cl and subjected to ultrasonic treatment. They found that the effect of NH_4Cl was beneficial to mercury removal, the optimum content being 5—7 wt% of Cl. Beyond 7 wt %, no further improvement was observed due to the pore blockage of the char.

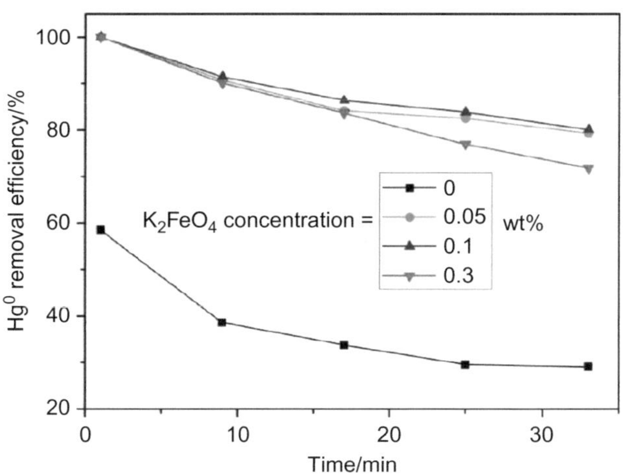

Fig. 12 Effect of K_2FeO_4 loading on Hg^0 removal efficiency. Experimental conditions—sorbent mass: 0.3 g, total gas flow: 1.2 L/min, humidity: 90%, SO_2 concentration: 1000 ppm, NO concentration: 400 ppm, O_2 content: 10%, reaction temperature: 50°C, Hg^0 inlet concentration: 40 μg/m^3.
Reprinted with permission from Zhou, J., Liu, Y., Pan, J. (2017). Removal of elemental mercury from flue gas using wheat straw chars modified by K_2FeO_4 reagent. *Environmental Technology, 38*, 3047–3054. Copyright 2018 Taylor & Francis Group.

7.4.3 Adsorption experimental conditions

7.4.3.1 Adsorption temperature and gas concentration

In general, increasing the adsorption temperature led to a decrease in the amount of the pollutant gas adsorbed. However, this observation depends on the range of temperature studied and on the experimental conditions. In some works, increasing the adsorption temperature led to an increase in the amount of adsorbed gas but a further increase in the temperature led to a decrease in the adsorption capacity.

In this context, Nguyen and Lee (2016) investigated a char produced by slow pyrolysis at 450°C of chicken manure, doped with nitrogen via an HNO_3 and NH_3 chemical treatment—to increase the number of amine functions on the surface—and converted into beads to make easier the collection and regeneration of the adsorbent. The beads exhibited SSAs of ~300–325 m^2/g with a CO_2 adsorption capacity of about 448.8 mg/g at 20°C, which is significantly higher than that of commercial activated carbon and other adsorbents such as MOFs. CO_2 adsorption capacity was improved by decreasing the adsorption temperature. Furthermore, after 10 adsorption/desorption cycles, bead recovery was about 85% of the initial adsorption capacity. In addition, increasing the adsorption temperature from 30°C to 70°C during CO_2 adsorption on sawdust pyrolyzed at different temperatures (450°C, 750°C, and 850°C), and modified or not with monoethanolamine, resulted also in a decrease in the adsorption capacities (Madzaki, Ghani, Rebitanim, & Alias, 2016). A similar trend was obtained in the work of Creamer et al. (2014) for the two CO_2 adsorption temperatures, 25°C and 75°C, on chars issued from surgarcane bagasse and hickory wood pyrolyzed at 300°C, 450°C, and 600°C (Fig. 10). Moreover, analogous tendencies were also noticed for other gases. For instance, increasing the adsorption temperature from 20°C to 40°C led to the decrease in the acetone sorption capacities by 73%–98% on different chars (bamboo, sugarcane bagasse, Brazilian pepper wood, sugar beet tailings, and hickory wood) pyrolyzed at 300°C, 450°C, and 600°C (Zhang, Gao, Zheng, et al., 2017). Lu and Lu (2010) also noted a decrease in SO_2 and NO adsorption capacities on chars prepared from cotton straw and pyrolyzed at 600°C, when the temperature of adsorption increased from 75°C to 120°C.

Shen et al. (2017) investigated three adsorption temperatures during the adsorption of Hg on char derived from tea waste, pyrolyzed at 500°C, and modified with different contents of NH_4Cl. The order of removal efficiency was 120°C > 180°C > 80°C. This reveals that the removal of Hg necessitates an optimal range of temperature. This temperature dependency suggests that Hg removal occurs according to physical and chemical adsorptions, the former being favored at low temperature and the latter at high temperature, the best efficiency being reached at 120°C by Shen et al. (2017). In addition, on char derived from wheat straw, pyrolyzed at 800°C, and activated by K_2FeO_4, an increase in the adsorption temperature from 30°C to 50°C increased the Hg^0 removal efficiency from 55.3% to 81.9% due to the enhancement of the rate of Hg^0 chemisorption (Fig. 13) (Zhou, Liu, & Pan, 2017). However, a further increase in the temperature from 50°C to 90°C resulted in a decrease in the removal efficiency to 39.1% (Fig. 13). Indeed, mercury removal takes place via physisorption followed

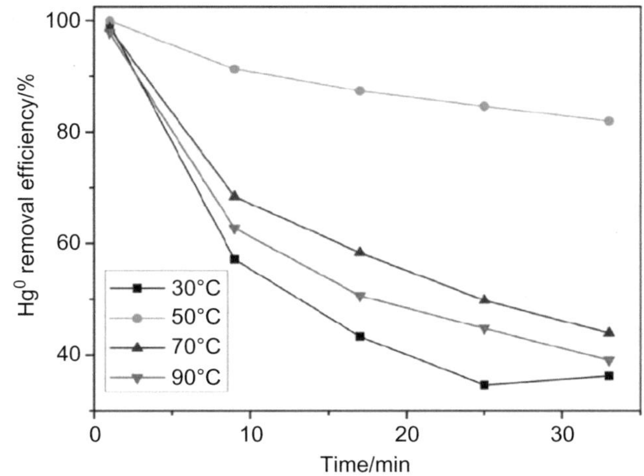

Fig. 13 Effect of the reaction temperature on Hg^0 removal efficiency. Experimental conditions—sorbent mass: 0.3 g, total gas flow: 1.2 L/min, humidity: 90%, SO_2 concentration: 1000 ppm, NO concentration: 400 ppm, O_2 content: 10%, Hg^0 inlet concentration: 40 μg/m^3. Reprinted with permission from Zhou, J., Liu, Y., Pan, J. (2017). Removal of elemental mercury from flue gas using wheat straw chars modified by K_2FeO_4 reagent. *Environmental Technology, 38*, 3047–3054. Copyright 2018 Taylor & Francis Group.

by chemisorption during which Hg^0 is oxidized into Hg^{2+}. Increasing the temperature decreases the rate of the physisorption process, followed by lowering of the removal efficiency (Zhou, Liu, & Pan, 2017).

On the other hand, gaseous pollutants removal efficiencies by raw or modified chars seem to be very dependent on their inlet concentrations. Indeed, Zhou, Liu, and Pan (2017) reported that when the inlet concentration of Hg^0 increased from 20 to 80 μg/m^3, Hg^0 uptake decreased from 99.7% to 71.8% (of the initial mercury concentration) (Fig. 14). It seems that the increase in the ratio of mol Hg^0/number active sites per time unit resulted in a decrease in the removal efficiencies due to the lower availability of the active sites whose number is constant and limited. Conversely, Nguyen and Lee (2016) found that CO_2 adsorption capacity was improved by increasing CO_2 concentration. In addition, the increase in CH_4 partial pressures resulted in an increase in its adsorption capacity on chars prepared by torrefaction of nonrecycled cardboard (Delgado et al., 2017).

7.4.3.2 Coexisting pollutants

The simultaneous presence of different pollutant gases affects the adsorption capacities of char toward each gas. The effect depends on the experimental conditions and on the type of interactions between adsorbate (gases) and adsorbent. Nguyen and Lee (2016) noted a high selectivity (∼80%) toward CO_2 adsorption under different

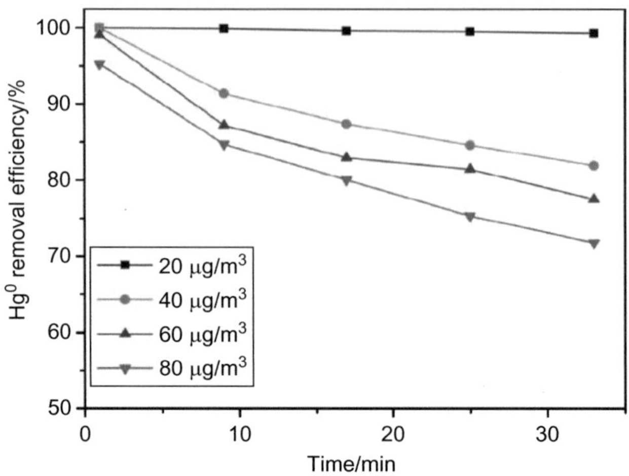

Fig. 14 Effect of the Hg^0 inlet concentration on Hg^0 removal efficiency. Experimental conditions—sorbent mass: 0.3 g, total gas flow: 1.2 L/min, humidity: 90%, SO_2 concentration: 1000 ppm, NO concentration: 400 ppm, O_2 content: 10%, reaction temperature: 50°C. Reprinted with permission from Zhou, J., Liu, Y., Pan, J. (2017). Removal of elemental mercury from flue gas using wheat straw chars modified by K_2FeO_4 reagent. *Environmental Technology, 38*, 3047–3054. Copyright 2018 Taylor & Francis Group.

N_2/CO_2 mixtures. A selectivity of 74% was obtained under the same mixture by Liu and Huang (2018). Furthermore, Sadasivam and Reddy (2015) investigated several chars derived from wood in CH_4 adsorption. They reported that the increase in temperature and the presence of water led to a decrease in the amount of CH_4 adsorbed (Fig. 15). The adsorption capacity of char was related to its textural (total porosity, surface area, effective particle size, etc.) and chemical (initial moisture, elemental, and fixed carbon contents) properties. The inhibiting effect of water was due to the similar size of water and methane molecules (0.306 nm diameter for water and 0.38 nm for CH_4), rendering the penetration of methane into the char pores more difficult. In addition, even if the CH_4 adsorption capacities were found to be higher for a granulated activated carbon than for the different chars studied, the inhibition effect of water on CH_4 adsorption was lower for chars (Fig. 15). In parallel, it was also noted that higher water adsorption capacity on activated carbon materials compared to nonactivated ones is ascribed to the chemical affinity between water and the functional groups existing on the surface of the activated matrixes (oxygenated sites). Conversely, in the case of nonactivated carbon, the molecular interactions between carbon and water are relatively weaker as compared to interactions between two water molecules.

Sethupathi et al. (2017) compared the capacities of four chars derived from perilla leaf, soybean stover, Korean oak (*Quercus dentata*), and Japanese oak (known as emperor oak) in the simultaneous adsorption of H_2S and CO_2 in the presence of CH_4, for their application in biogas purification. The first two chars were obtained by pyrolysis at 700°C (7°C/min), whereas the third char was obtained by pyrolysis at 400°C and the fourth one at 500°C. The chars were suitable for the biogas

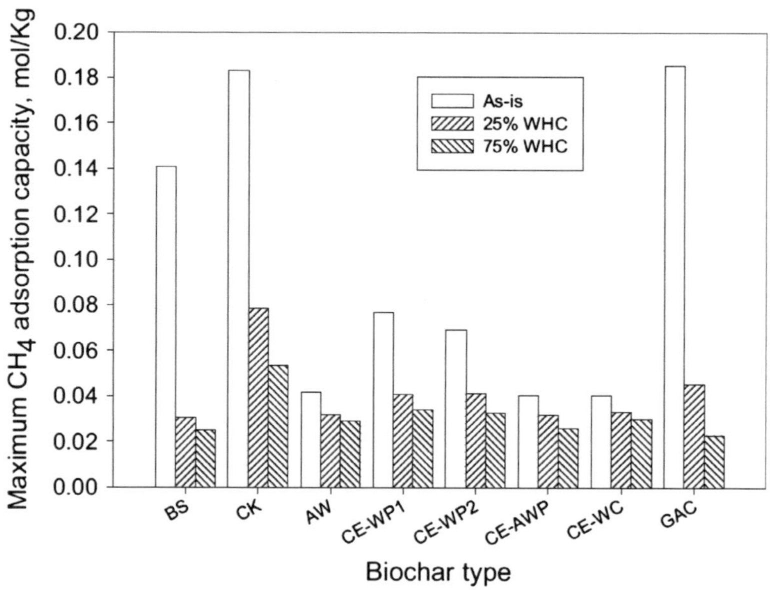

Fig. 15 Effect of moisture on maximum CH_4 adsorption capacity of different chars and granulated activated carbon (GAC).
Reprinted with permission from Sadasivam, B. Y., Reddy, K. R. (2015). Adsorption and transport of methane in biochars derived from waste wood. *Waste Management, 43*, 218–229. Copyright 2018 Elsevier.

purification industry. CH_4 was not adsorbed whereas H_2S and CO_2 were adsorbed. In addition, H_2S was more adsorbed than CO_2. The local pH within the pores was one of the factors, besides the surface area, responsible for H_2S adsorption. Indeed, the lowest pH observed for the Japanese oak could inhibit the dissociation of H_2S limiting the oxidation of sulfur leading thus to lower adsorption of H_2S on this material. Adsorption of CO_2 was correlated to the contents of N-groups on the surface. Adsorption of both CO_2 and H_2S is a physisorption. While physisorption of H_2S is promoted by a higher local pH, that of CO_2 is promoted by the interaction with the basic nitrogen chemical groups. Competition on the adsorption sites could also be responsible for the higher adsorption of H_2S compared to that of CO_2.

Also, H_2O adsorption on char prepared from olive stone was not significantly influenced by the presence of CO_2. However CO_2 adsorption capacities were decreased by H_2O adsorption and a further decrease was observed for higher H_2O contents (Plaza, Durán, Querejeta, Rubiera, & Pevida, 2016).

In addition, Nowicki et al. (2013) prepared char by pyrolyzing coffee waste at 500°C and 800°C, and then activation by KOH or CO_2. They investigated the adsorption of NO_2 in the presence of water. The highest adsorption capacity, 84.1 mg/g, was found for the char pyrolyzed at 800°C and not activated. The samples activated by CO_2 also showed high adsorption capacities, 65.8 (pyrolysis at 500°C) and 76.0 mg/g (pyrolysis at 800°C). It is worth noting that these capacities are higher than the highest capacity found on these samples in dry conditions, 44.5 mg/g, for the samples pyrolyzed at 800°

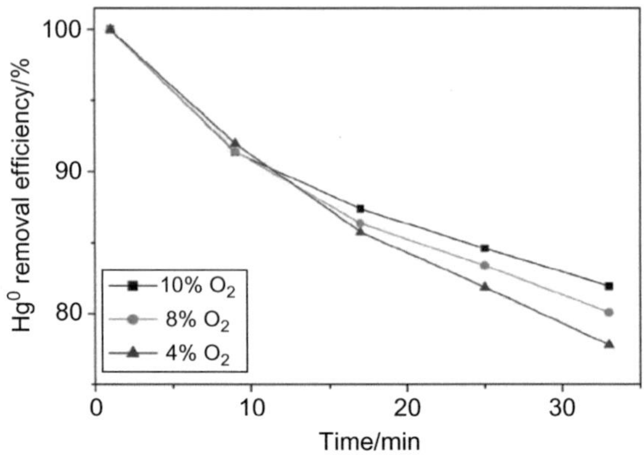

Fig. 16 Effect of O_2 content on Hg^0 removal efficiency. Experimental conditions—sorbent mass: 0.3 g, total gas flow: 1.2 L/min, humidity: 90%, SO_2 concentration: 1000 ppm, NO concentration: 400 ppm, reaction temperature: 50°C, Hg^0 inlet concentration: 40 μg/m^3. Reprinted with permission from Zhou, J., Liu, Y., Pan, J. (2017). Removal of elemental mercury from flue gas using wheat straw chars modified by K_2FeO_4 reagent. *Environmental Technology, 38*, 3047–3054. Copyright 2018 Taylor & Francis Group.

C and treated with KOH. The good sorption capacities in the presence of water, despite the low surface areas of the samples, are ascribed to the high content of basic oxygen functional groups and mineral substances that interact with NO_2; this interaction is enhanced in the presence of water. Lu and Lu (2010) studied the adsorption of NO and SO_2 on char prepared from wheat straw. NO adsorption was enhanced by the presence of SO_2, whereas SO_2 adsorption capacity was weakened by NO. SO_2 promoted the formation of chemicals between NO and chars. However, since some char sites are occupied by NO, SO_2 is less adsorbed.

Furthermore, Zhou, Liu, and Pan (2017) investigated the adsorption of mercury vapor Hg^0 in the presence of O_2, NO, and SO_2. The char used was derived from wheat straw modified by K_2FeO_4. Increasing the O_2 concentration from 4 to 10 vol% increases the uptake of mercury from 77.5% to 81.9% (Fig. 16) due to the synergistic effects of O_2 and Fe_2O_3. Moreover, the presence of O_2 may regenerate the lattice oxygen and replenish the oxygen functional groups on the adsorbent surface. An amount of 400 ppm NO greatly improved mercury removal (Fig. 17) due to the formation of $Hg(NO_3)_2$. However, the introduction of 1000 ppm SO_2 with the 400 ppm NO resulted in a decrease in the uptake of Hg^0 (Fig. 17) due to the competition between Hg and SO_2 on the adsorption sites. The effect of oxygen content was also investigated by Shen et al. (2017) during mercury adsorption on waste tea pyrolyzed and treated with NH_4Cl. These authors showed that an increase of oxygen content from 0 to 5 vol% resulted in an increase in the mercury removal efficiency, whereas a further increase in oxygen to 10 vol% decreased this efficiency. It seems that oxygen promotes the chemisorption of Hg but high concentrations of O_2 could lead to a competitive adsorption between O_2 and Hg. Shen et al. (2017) introduced 300 ppm NO and proved the beneficial effect of this gas due to the formation of $Hg(NO_3)_2$. A further increase in the concentration of NO was also beneficial. However, these scientists evidenced

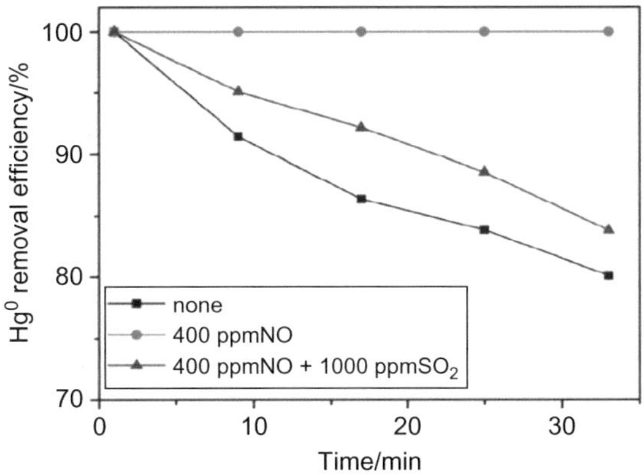

Fig. 17 Effect of gas mixture on Hg^0 removal efficiency. Experimental conditions—sorbent mass: 0.3 g, total gas flow: 1.2 L/min, humidity: 90%, O_2 content: 10%, reaction temperature: 50°C, Hg^0 inlet concentration: 40 μg/m^3.
Reprinted with permission from Zhou, J., Liu, Y., Pan, J. (2017). Removal of elemental mercury from flue gas using wheat straw chars modified by K_2FeO_4 reagent. *Environmental Technology, 38*, 3047–3054. Copyright 2018 Taylor & Francis Group.

that the introduction of 100 ppm of SO_2 was beneficial due to the formation of $HgSO_4$. It seems that, at lower contents, adsorption of SO_2 is not competitive with that of Hg. Shen et al. (2017) reported that the presence of 4 vol% H_2O slightly inhibited Hg removal probably because of the competition between H_2O and Hg.

7.4.3.3 Space velocity and particle size

The higher is the residence time (low space velocities), the higher are the removal efficiencies by chars. For instance, the decrease in space velocity has exhibited a positive effect on mercury removal (Shen et al., 2017). With a space velocity of 110,000 h^{-1}, the removal efficiency was 100% and it was maintained stable during 240 min. The efficiency decreased to values lower than 70% for a space velocity of 340,000 h^{-1}.

Concerning the particle size, Lu and Lu (2010) investigated SO_2 adsorption on char with different particle sizes. The char tested was prepared by fast pyrolysis at 600°C of wheat straw. Particle sizes higher than 0.425 mm gave lower adsorption capacities than particle sizes lower than 0.125 mm. Decreasing particle size increased the adsorption performance. Indeed, large particles present lower surface areas than small ones and thus will lead to lower adsorption efficiencies.

7.4.4 Involved mechanisms assessment and kinetics

Char surfaces are heterogeneous and vary in composition mainly with the raw feedstocks and treatment process. Thus, it is difficult to give one mechanism for the adsorption process on char. As concluded from the works cited above, in general, increasing the pyrolysis temperature results in an increase in the surface areas and porosities. Above a given pyrolysis temperature (higher than 700°C in most of the references), these textural properties will decrease. Activation and doping increase the porosity and the amount of functional groups present on the surface; this favors gas adsorption (Chen et al., 2017). It was reported by Nowicki et al. (2013) that the high pH values and the high number of oxygen functional groups are responsible for the high affinity to gases with acidic character. The basic chemical properties of a surface are suitable for the adsorption of acidic gases CO_2, H_2S, SO_2, NO_x, etc., whereas the acidic functions are convenient for the basic gases (Chen et al., 2017; Hervy et al., 2018; Sethupathi et al., 2017; Xu, Cao, Zhao, & Sun, 2014). Furthermore, the pore structure should match the particle size of the adsorbates (Chen et al., 2017; Hervy et al., 2018).

For example, adsorption of CO_2 is enhanced by physical properties such as high surface area and pore volume (Plaza, González, Pevida, & Rubiera, 2014; Plaza, González, Pis, Rubiera, & Pevida, 2014; Creamer et al., 2014; Li et al., 2016; Madzaki et al., 2016) and chemical properties such as N-containing groups (primary and secondary amines, azo N=N, aliphatic C—N) and also by the presence of aliphatic C—O and hydroxyl (Creamer et al., 2014; Nguyen & Lee, 2016; Sethupathi et al., 2017; Zhang et al., 2015). The presence of metal oxyhydroxide or hydroxide such as AlOOH or $Mg(OH)_2$ and of a metal oxide such as Fe_2O_3 was also found to be beneficial for CO_2 adsorption, the effect of AlOOH being better than the two others (Creamer, Gao, & Wang, 2016). In parallel, the surface basicity due to the large amounts of metal components and the generation of oxygen (due to KOH) and nitrogen (due to ammonia) functional groups resulting from the activation of char derived from sewage sludge by KOH and ammonia was responsible for the increase of formaldehyde adsorption (Lee et al., 2011). When the char was treated with steam, KOH, and ammonia, an enhancement of the uptake of formaldehyde was observed. This is ascribed to the amounts of oxygen (due to KOH) and nitrogen (due to ammonia) functional groups, as well as the SSA of the char that increased upon these activation processes (Lee et al., 2011).

Many studies (He, Xia, Wang, Pan, & Fang, 2011; Kanjanarong et al., 2016; Lehmann et al., 2011; Shang, Shen, Liu, Chen, & Xu, 2013; Xu et al., 2014) suggested, during H_2S adsorption on char, the oxidation of sulfur into elemental sulfur or into different sulfur oxide species (SO_2, SO_3, SO_4^{2-}, etc.) due to a reaction between adsorbed H_2S and oxygen of the functional groups existing on the char surface and/or water present. Indeed, moisture presence in char promoted H_2S removal (Xu et al., 2014). These authors reported that, on the char surface, complete oxidation of H_2S into SO_4^{2-} takes place, whereas inside the pores, only oxidation of H_2S into sulfur occurs. In an atmosphere poor in oxygen (for example, inside the pores of the adsorbent), sulfur of H_2S was oxidized into elemental sulfur. It was also noted that, upon H_2S adsorption on sewage sludge, SO_4^{2-} was mainly present as $CaSO_4$, whereas on pig manure, it was a soluble sulfate, such as K_2SO_4 and Na_2SO_4 (Xu et al., 2014). The increase in H_2S uptake could also be due to the chemical bonding between H_2S and COOH or OH functional groups that are present on the char surface (Shang et al., 2013). Some of the species formed, such as sulfuric acid, were reported to decrease the adsorption capacity of the adsorbent-activated carbon (Yan, Liang, Tsen, & Tay, 2002) or char (Hervy et al., 2018; Xu et al., 2014). Sethupathi et al. (2017) reported that lower pH is responsible for lower H_2S adsorption due to the limited possibility of H_2S to be dissociated and oxidized into more adsorbable sulfur species. In fact, acidification of the chars decreases their uptake for acidic gases such as CO_2 and H_2S. Strong alkalinities as well as the inorganic minerals originally present in the char play a crucial role in H_2S adsorption and oxidation (Hervy et al., 2018; Xu et al., 2014). These properties make natural char an adsorbent with interesting properties for H_2S removal, since it does not need further chemical activation by alkali or inorganic compounds (Xu et al., 2014). Hervy et al. (2018) investigated the performance of two chars, one issued from wood pellets and the other from a mixture of 50% food waste and 50% coagulation/flocculation sludge. These chars were pyrolyzed at 700°C and activated only by oxygen or steam. Mineral species (Ca, Fe) already present in char played an important role in the formation of metal sulfides and metal sulfates. Furthermore, activated chars were efficient in H_2S uptake due to their large SSA, alkaline pH, basic *O*-containing groups, and structural defects in graphene-like sheets (Hervy et al., 2018). It seems also that pores, with sizes lower than 70 Å, are important adsorption sites. The presence of CO_2 did not influence H_2S uptake, while the presence of O_2 decreased it due to the formation of sulfuric acid (Hervy et al., 2018). A char composed of 80% woodchips (spruce, pine, and fir) and of 20% residue of anaerobic digester, pyrolyzed at 600°C, was effective for the removal of H_2S (more than 98%) with various initial concentrations (105, 510, and 1020 ppmv). The high pH (7.98), the high moisture content of the char (80%–85%), and the presence of carboxylic and hydroxide radical groups were the main factors responsible for efficient H_2S removal (Kanjanarong et al., 2016).

Similarly to H_2S, mercury was also supposed to be oxidized during adsorption. Indeed, during the process of mercury removal by wheat straw char modified by K_2FeO_4 after pyrolysis at 800°C, it was assumed that Hg^0 was oxidized into Hg^{2+} by FeO_4^{2-} and Fe_2O_3. Hg^{2+} is more subjected to adsorption than Hg^0 (Zhou, Liu, & Pan, 2017). Shu, Lu, and He (2013) prepared char by the pyrolysis at 600°C of mulberry twig followed by its activation by steam, H_2O_2, $ZnCl_2$, or NaCl. The order of mercury Hg^0 adsorption capacity was: chloride > H_2O_2 > steam, with $ZnCl_2$ being better than NaCl. The highest adsorption capacity was 29.55 μg/g. Besides the enhancement of the physical properties, the improvement with H_2O_2 was related to its oxidizing properties and to the formation of surface oxygen functional groups. With chloride, it is thought that the formation of a C—Cl bond is followed by the formation of $[HgCl_4]^{2-}$. In both cases, mercury was oxidized, with the oxidation of mercury improving its adsorption (as stated above). Shen et al. (2017) investigated mercury adsorption on char derived from tea waste and impregnated with NH_4Cl. They evidenced that Hg^0 was oxidized upon chemisorption into Hg^{2+} and that the action of NH_4^+ and Cl^- with char will lead to the formation of —C—NH_2—Cl or —C—NH_2—Cl_2 bonds that break to form the radical Cl or the C—Cl bond. The further collision between radical Cl and Hg will oxidize Hg and reduce Cl leading to the formation of $HgCl_2$. Shen et al. (2017) also evidenced the conversion of C=O (with the action of H^+) to C—O—H and the further formation of the radical OH. The reaction of the latter with Hg results in the formation of $Hg(OH)_2$ and then of HgO upon the removal of H_2O.

Conversely, Seredych, Bashkova, Pietrzak, and Bandosz (2010) reported a reduction of NO_2 during its adsorption. They used char modified with silver nanoparticles and showed that reduction of NO_2 occurred during its adsorption due to the presence of metal and metal oxides sites on the surface. Metallic silver acts as an anchoring site for NO_2, forming chelates such as Ag_2—O—N—O or Ag—O_2—NO.

In addition, the molar ratios of O/C and H/C can be used to indicate, respectively, the polarity and aromaticity of char (Zhang, Gao, Creamer, et al., 2017 and references therein). These ratios have significant influence on the affinity of chars toward gaseous or liquid molecules and thus its adsorption capacity. For example, CH_4 adsorption is influenced by the presence of functional groups on the surface of the adsorbent. CH_4 is a nonpolar molecule. Consequently, the presence of C—C bonds on the surface favors its adsorption, whereas the oxygen-based bonds, O—H, C—O, C=O, etc., confer on char a hydrophilic character resulting in lower affinity toward CH_4 and thus lower adsorption capacity.

It was also stated that the adsorption of organic compounds on carbonaceous materials is governed by five types of interactions (Zhang, Gao, Creamer, et al., 2017 and references therein): hydrophobic ($\pi \rightarrow \pi$ bonds), hydrogen bond, van der Waals, covalent, and electrostatic interactions. For VOC adsorption, a compromise should be found between the increase in the surface area and the content of noncarbonized matter that does not vary in the same way with the increase in pyrolysis temperature (Zhang, Gao, Zheng, et al., 2017). For example, adsorption of trichloroethylene on pine needle chars produced at 700°C was more efficient than on those produced at 300°C (Ahmad et al., 2013), due to the increase in hydrophobicity and surface area, as well as the decrease of char polarity. Studies on the adsorption of nonpolar aromatics such as naphthalene and polar aromatics such as nitrobenzene and *m*-nitrobenzene on pine needle chars pyrolyzed at different temperatures, 100–700°C, evidenced also that higher pyrolysis temperatures result in higher adsorption capacities (Chen et al., 2008). In parallel, six different chars were prepared from *Miscanthus* straw pellets, oil seed rape straw pellets, rice husk, soft wood pellets, wheat straw pellets, and sewage sludge by pyrolysis under nitrogen at 550°C (Zhou, Richard, Ferronato, Chovelon, & Sleiman, 2018). These materials were tested in gaseous ozone removal. Even after 24 h exposure, ozone was completely adsorbed since no ozone was detected in the outlet of the reactor. Also, no VOCs were detected suggesting that no volatile by-products were formed that could be due to the oxidation of char by ozone. The tested chars were more efficient than a tested soil. It was inferred that ozone uptake is mainly related to the carbon content in the material. Surface area as well as pore size and volume also play a role.

On the other hand, adsorption of most of the pollutant gases on char was found to be a physisorption. This fact suggests an easy and low-cost regeneration of char adsorbents. For instance, adsorption of CO_2 was found to be mainly a physisorption (Chang, Chuang, Gray, & Soong, 2003; Plaza, González, Pevida, et al., 2014; Plaza, González, Pis, et al., 2014; Plaza et al., 2016; Creamer et al., 2016, 2014; Madzaki et al., 2016; Sethupathi et al., 2017), i.e., weak interactions (van der Waals type) exist between CO_2 and char. Furthermore, due to its physisorption, CO_2 is easily desorbed at a temperature as low as 120°C (Creamer et al., 2016, 2014). Adsorption of H_2S, like that of CO_2, is a physisorption (Kanjanarong et al., 2016). While physisorption of H_2S is promoted by a higher local pH, that of CO_2 is promoted by the interactions with the basic nitrogen chemical groups. In addition, Zhang, Gao, Creamer, et al. (2017) found that the adsorption of VOCs on different chars is rather physical and exothermic. It was also reported that NO and SO_2 are physisorbed on chars (Lu & Lu, 2010). Adsorption of CH_4 on char was also found to be a physical one (physisorption) (Delgado et al., 2017). This is advantageous for agricultural and environmental applications due to the easy desorption of CH_4, when needed, without the need of many regenerating treatments. Conversely, mercury adsorption takes place via a physisorption followed by a chemisorption during which Hg^0 is oxidized into Hg^{2+} as reported by Zhou, Liu, and Pan (2017) and Shen et al. (2017).

Due to the various types of chars, many isotherms were proposed to describe the adsorption process behavior. Four types of isotherms were mainly suggested for acidic gases adsorption on chars: Langmuir, Freundlich, Toth, and Sips (Bamdad et al., 2018). The Langmuir model assumes homogeneous sites on the surface and constant energy of adsorption. Freundlich assumes heterogeneous sites and energy of adsorption that varies with the surface coverage. The Toth isotherm is an empirical one developed to improve the Langmuir one to better describe heterogeneous systems. The Sips isotherm is a combination of the Langmuir and Freundlich ones. In most of the cases, Freundlich describes best the adsorption of acidic gases (CO_2, H_2S, etc.) (Bamdad et al., 2018) and of CH_4 (Delgado et al., 2017) on char since it fits best to heterogeneous surfaces such as char ones. However, Sadasivam and Reddy (2015) found that the Langmuir isotherm represented CH_4 adsorption on different chars. In parallel, the Toth model represented well the adsorption behavior of CO_2 on char issued from olive stone and almond shell (Plaza, González, Pevida, et al., 2014; Plaza, González, Pis, et al., 2014), whereas in a ternary mixture (H_2O, CO_2, N_2), CO_2 adsorption corresponded to the ideal adsorbed solution theory (Plaza et al., 2016). It is to be noted that H_2O adsorption on char derived from olive stone fits well with the extended cooperative multimolecular sorption model (Plaza et al., 2016).

Moreover, a pseudo-first-order kinetic model fits well for the adsorption of H_2S on char (Kanjanarong et al., 2016), whereas a pseudo-second-order model was found for CO_2 adsorption on char (Creamer et al., 2016, 2014). In addition, the kinetics of VOC adsorption on different chars is rather pseudo-second order, suggesting that the mechanism of adsorption is composed of more than one rate-determining step (Zhang, Gao, Creamer, et al., 2017). In parallel, Lagergren's second-order model represents the kinetic data of CH_4 adsorption on chars and activated carbon (Luo et al., 2011; Sadasivam & Reddy, 2015). The evaluation of the extent of molecular exchange at the interface between the adsorbate in the gas phase and the adsorbent was done by the parameters of the pseudo-second-order kinetic model.

It is also worth noting that molecular modeling simulation of adsorption on carbon-based adsorbents was performed. In the review of Bamdad et al. (2018), two potential models, fluid-fluid and solid-fluid interaction, as well as two simulation methods, the Vienna Ab-initio Simulation Package and Grand Canonical Monte Carlo, used to simulate adsorption of acidic gases on carbon adsorbents, were described. Indeed, modeling is helpful in reducing the number of experiments to be performed, and thus in reducing time and cost, as well as explaining some observations and elucidating some mechanisms.

Work done on adsorption of gaseous pollutants onto biomass-derived chars is summarized in Table 3.

Table 3 Adsorption characteristics of various gases onto biomass-derived chars

Feedstock	Conditions	Pollutant	Adsorption capacity (mg/g)	Remarks	References
Rice husk	Pyrolysis at 600°C (0.5 h), treatment with hydrofluoric acid and ammonia Adsorption temperature: 30°C	CO_2	3427.6	Beneficial effect of high surface area and high N contents on the surface	Zhang et al. (2015)
Raw cottonwood	Addition of salts of Mg, Fe, and Al and pyrolysis at 600°C under nitrogen for 3 h Mg content: 0–5.74 wt% Fe content: 0–3.7 wt% Al content: 0–0.8 wt% Adsorption at 25°C on 20 mg adsorbent under 50 mL/min of CO_2		27.79–71.05 The highest value is obtained for 0.8 wt% Al	Physisorption 90%–99% of adsorbed CO_2 was desorbed at 120°C, indicating the ease and low-cost regeneration of biochar Second order Beneficial role of surface area and of metal hydroxides, oxyhydroxides, and oxides ($Mg(OH)_2$, AlOOH, and Fe_2O_3), the Al-based species being the most beneficial	Creamer et al. (2016)
Mesquite	Pyrolysis at 450°C (4 h), one-step activation with KOH Adsorption at 25°C and 30 bar		1150	High surface areas and pore volumes Selectivity toward CO_2 in CO_2/CH_4 mixtures was studied	Li et al. (2016)
Sawdust	Pyrolysis at 450°C, 750°C, or 850°C Activation or not with monoethanolamine (5–7 wt%) Adsorption under 100 mL/min CO_2 at 30°C or 70°C on 10 mg adsorbent		Highest capacity: 47.5 for the sawdust pyrolyzed at 850°C with adsorption temperature of 30°C	Increasing the adsorption temperature decreases the adsorption capacity Physisorption Increasing pyrolysis temperature increased the surface area The effect of surface area was predominant over that of functional groups leading to higher capacities for the sawdust pyrolyzed at 850°C not treated with monoethanolamine, the functional groups blocking the pores	Madzaki et al. (2016)

Camphor, bamboo, rice hull	Pyrolysis at 400°C (10°C/min) 5 h, absence of oxygen Adsorption temperature 25°C, adsorption pressure 1 bar, 10–50 μL H_2S/L	H_2S	7.82	Rice hull biochar exhibited higher adsorption capacity than the others	Shang et al. (2013)
Sewage sludge and pig manure	Slow pyrolysis Adsorption temperature 25°C, adsorption pressure 1 bar Dynamic: 10 g of sewage sludge or 5.5 g of pig manure, 0.5 mL/min of 1% H_2S in air Static: 0.1 g biochar, 20 mL glass bottle, 1% H_2S Moisture content of biochar: 0 or 25, or 100 wt%		Pig manure: 59.6 (without moisture) to 65.4 (with 25 wt% moisture) Sewage sludge: 43.97 (without moisture) to 47.5 (with 25 wt% moisture)	Pig manure gave higher adsorption capacity than sewage sludge Presence of moisture in biochar improved H_2S adsorption capacities A mechanism was proposed for H_2S adsorption	Xu et al. (2014)
80% woodchips (spruce, pine, and fir) and 20% residue of anaerobic digester	Pyrolysis at 600°C Initial concentrations of H_2S: 105, 510, and 1020 ppmv		273.2 ± 1.9	Factors that improved H_2S uptake: high pH (7.98), high moisture content of biochar (80%–85%), presence of carboxylic and hydroxide radical groups	Kanjanarong et al. (2016)
Perilla leaf Soybean stover Korean oak (*Quercus dentata*) Japanese oak (known as emperor oak)	Pyrolysis at 700°C (7°C/min) for perilla leaf and soybean stover Pyrolysis at 400°C for Korean oak Pyrolysis at 500°C for Japanese oak Adsorption at 50°C under 300 mL/min of 0.3% H_2S, 40% CO_2, 59.7% CH_4, and 20% relative humidity. 0.5 g adsorbent with 0.1 g borosilicate glass fiber	CO_2 and H_2S in the presence of CH_4 (biogas purification)	Highest adsorptions obtained for perilla leaf CH_4: 1.584 H_2S: 18.258 CO_2: 101.728	No noticeable adsorption for CH_4 due to unsuitable size of carbon particles Higher adsorption for H_2S than for CO_2 Interaction between CO_2 and N-containing chemical groups Physisorption of H_2S and CO_2	Sethupathi et al. (2017)

Continued

Table 3 Continued

Olive stone	Activation at 400–500°C under air Adsorption temperature 23°C, 50°C, or 70°C, 1.4 bar, 140 mL/min of 2% water and 0 or 14% CO_2 (N_2 balance)	CO_2 and H_2O in the presence of N_2 (postcombustion)		H_2O adsorption was not significantly influenced by CO_2, whereas CO_2 adsorption capacities were decreased by H_2O adsorption, higher decrease was observed for higher H_2O contents Adsorption of water fitted with the extended cooperative multimolecular sorption model Adsorption of CO_2 in the ternary mixtures followed the ideal adsorbed solution theory	Plaza et al. (2016)
Cardboard	Torrefaction at 300°C during 60 and 90 min under atmospheric pressure, activation or not with KOH Adsorption of CH_4 at 30°C, 0.4 kPa, 150 mL/min	CH_4	3.53–5.70	Adsorption capacity increased with torrefaction time and decreased with chemical treatment by KOH Freundlich isotherm	Delgado et al. (2017)
Hardwood (seven types)	2, 5, 8, 10, 15, and 20% CH_4 in (CH_4/CO_2), without and with 25% and 75% moisture, at 25°C, 35°C, and 45°C and CH_4 gas pressure 0.15–1 kPa		0.8–13.92	CH_4 adsorption decreases with the increase in temperature and increases with the increase in CH_4 concentration Presence of moisture decreased the CH_4 adsorption capacity; this decrease was lower for biochar than for activated carbon The high microporosity and greater availability of surface sites in biochars could have potentially accommodated the adsorption of excessive amounts of methane within a short time period Lagergren's second-order model best represents the adsorption kinetic	Sadasivam and Reddy (2015)

Sewage sludge	Pyrolysis 1 h under 50 mL/min of nitrogen at 700°C, followed by an activation with steam KOH and/or ammonia 70 mg adsorbent, 20°C, 1 ppm formaldehyde (nitrogen balance)	Formaldehyde	–	Surface basicity and oxygen and nitrogen functional groups generated by activation by KOH and ammonia improved the adsorption capacity	Lee et al. (2011)
Bamboo, surgarcane bagasse, Brazilian pepper wood, sugar beet tailings, and hickory wood	Pyrolysis under nitrogen at 300°C, 450°C, and 600°C Adsorption: dynamic at 20°C	Acetone	7.06–91.16	Increasing the adsorption temperature to 40°C decreases the sorption capacity The sorption capacity depends on the surface area and the content of noncarbonized matter, which depends on pyrolysis temperature Pseudo-second-order Physical adsorption	Zhang, Gao, Zheng, et al. (2017)
		Toluene	12.63–62.91		
		Cyclohexane	5.58–69.33		
Palm oil mill sludge	Pyrolysis under N_2 at different heating temperatures (300–800°C), heating rates (10–20°C/min), and holding times (60–120 min) Adsorption at atmospheric pressure and room temperature, under 300 mL/min of 300 ppm SO_2 in N_2	SO_2	The highest: ~10 for pyrolysis at 400°C (15°C/min, 90 min)	Study of the influence of pyrolysis conditions on SO_2 adsorption Best conditions: 400°C, 15°C/min, 90 min The decrease in the amount of functional groups with increasing pyrolysis temperature led to lower SO_2 adsorption capacities	Iberahim et al. (2017)
Coffee waste	Pyrolysis at 500 and 800°C (10°C/min), under argon, activation by KOH or CO_2 Test under 0.450 L/min of 0.1% NO_2 with or without 70% humidity	NO_2	In the absence of water: 6.3–44.5 In the presence of water: 12.6–84.1	The NO_2 adsorption capacity was enhanced in the presence of humidity because of the improvement of the interaction between NO_2 and the surface functional group	Nowicki et al. (2013)

Continued

Table 3 Continued

Miscanthus straw pellets, oil seed rape straw pellets, rice husk, soft wood pellets, wheat straw pellets, sewage sludge	Pyrolysis 12–15 min at 550°C Adsorption: 0.1 g of biochar, 320 mL/min 55 ppbv ozone in air	O_3	No ozone was detected in the outlet of the reactor even after 24 h exposure	Ozone removal was mainly related to carbon content Surface area, pore volume, and pore size could also play a role No volatile by-product is formed	Zhou, Richard, et al. (2018)
Waste tea	Pyrolysis under nitrogen (60 mL/min) at 500°C for 2 h, then activation by steam and impregnation with different contents of NH_4Cl to lead to 0%, 1%, 3%, 5%, 7%, and 9% of Cl and ultrasonic treatment Total gas flow 850 mL/min of 0%, 5%, or 10% O_2, 60 μg/m^3 Hg, at different temperatures (80°C, 120°C, and 180°C) and different space velocities	Hg^0	The highest value noted was the sample with 5 wt% Cl and subjected to ultrasonication: 1.079	Beneficial effect of Cl, oxygen but with optimal contents Beneficial effect of ultrasound treatment due to the enhancement of the textural properties Beneficial effect of NO and SO_2 Negative effect of water presence Low space velocities are better than higher ones Optimal adsorption temperature: 120°C	Shen et al. (2017)

7.5 Conclusions

In this chapter, we focused on the recent developments regarding the environmental valorization of chars for the removal of pollutants from aqueous and gaseous phases. Based on an extended scientific literature review, it can be concluded that chars could be produced from a wide variety of feedstocks and their pollutants/adsorptive characteristics depend mainly on the used pyrolysis conditions. By controlling these thermal factors and adopting strategies for their functionalizing, chars with selective and also high pollutant adsorption capacities can be produced. To this end, physical and chemical activation of chars as well as the creation of char-based composites through their surface embedding with specific materials (iron, manganese, hematite, chitosan, etc.) have been performed to enhance their capacities to remove some inorganic and/or organic pollutants from aqueous and gaseous phases. Thanks to these modifications, the modified chars developed well-established microporous structures, high SSAs, and also more functionalized surfaces adapted to the pollutant to be removed. Regarding this modification process, important challenging aspects should be tackled such as the reduction of the consumed heating energy during the activation process as well as the minimization of the used amounts of chemical reagents.

The analysis of the mineral and organic pollutants' adsorption mechanisms shows that depending on the nature of the used char and also the selected pollutant, various kinds of interactions could jointly take place, including ion exchange, physical sorption, electrostatic attraction, precipitation, and hydrophobic interactions. The determination of the contribution of each mechanism in the overall adsorption process is a promising and innovative research topic.

The majority of the research work on the use of pristine and especially modified chars for pollutants removal from aqueous and gaseous effluents highlighted that chars can be considered as attractive, promising, and efficient adsorbents. This conclusion was justified by both their important adsorption capacities and also their environmental and economic benefits since they could be regarded also as environmentally sustainable substitutes for commercially activated carbons. However, modifying char leads generally to chemical and physical changes that promote pollutants adsorption but at the same time could increase toxicity. Special attention should be paid to the management of these pollutants' loaded materials to avoid any negative impacts on the environment in general and on aquatic organisms in particular.

Finally, beyond batch adsorption experiments, dynamic studies by using large columns or continuous stirring tank reactors during the investigation of the real application of pristine or modified chars in large demonstration-scale adsorption/filtration tests are crucial to close the current related knowledge gaps. Furthermore, the testing of chars in large-scale pilots in combination with other technologies such as advanced oxidation processes or disinfection methods could be an interesting future research option. Lastly, data on the adsorption capacities of chars in removing mineral and organic pollutants from real wastewaters and/or drinking water and/or real gas mixtures are recommended to gain a better understanding of the effect of water or air quality on the removal efficiency of pollutants by different chars.

References

Abdolali, A., Guo, W. S., Ngo, H. H., Chen, S. S., Nguyen, N. C., & Tung, K. L. (2014). Typical lignocellulosic wastes and by-products for biosorption process in water and wastewater treatment: a critical review. *Bioresource Technology*, *160*, 57–66. https://doi.org/10.1016/j.biortech.2013.12.037.

Agrafioti, E., Kalderis, D., & Diamadopoulos, E. (2014). Arsenic and chromium removal from water using biochars derived from rice husk, organic solid wastes and sewage sludge. *Journal of Environmental Management*, *133*, 309–314. https://doi.org/10.1016/j.jenvman.2013.12.007.

Ahmad, M., Lee, S. S., Rajapaksha, A. U., Vithanage, M., Zhang, M., Cho, J. S., et al. (2013). Trichloroethylene adsorption by pine needle biochars produced at various pyrolysis temperatures. *Bioresource Technology*, *13*, 615–622. https://doi.org/10.1016/j.biortech.2013.06.033.

Ahmad, M., Rajapaksha, A. U., Lim, J. E., Zhang, M., Bolan, N., Mohan, D., et al. (2014). Biochar as a sorbent for contaminant management in soil and water: a review. *Chemosphere*, *99*, 19–23. https://doi.org/10.1016/j.chemosphere.2013.10.071.

Angin, D., Altintig, E., & Köse, T. E. (2013). Influence of process parameters on the surface and chemical properties of activated carbon obtained from biochar by chemical activation. *Bioresource Technology*, *148*, 542–549. https://doi.org/10.1016/j.biortech.2013.08.164.

Ates, F., & Tezcan Un, U. (2013). Production of char from hornbeam sawdust and its performance evaluation in the dye removal. *Journal of Analytical and Applied Pyrolysis*, *103*, 159–166. https://doi.org/10.1016/j.jaap.2013.01.021.

Bamdad, H., & Hawboldt, K. (2016). Comparative study between physicochemical characterization of biochar and metal organic frameworks (MOFs) as gas adsorbents. *Canadian Journal of Chemical Engineering*, *94*, 2114–2120. https://doi.org/10.1002/cjce.22595.

Bamdad, H., Hawboldt, K., & MacQuarrie, S. (2018). A review on common adsorbents for acid gases removal: focus on biochar. *Renewable and Sustainable Energy Reviews*, *81*, 1705–1720. https://doi.org/10.1016/j.rser.2017.05.261.

Bogusz, A., Oleszczuk, P., & Dobrowolski, R. (2015). Application of laboratory prepared and commercially available biochars to adsorption of cadmium, copper and zinc ions from water. *Bioresource Technology*, *196*, 540–549. https://doi.org/10.1016/j.biortech.2015.08.006.

Boshir, M., Zhou, J. L., Hao, H., Guo, W., & Hasan, A. (2017). Competitive sorption affinity of sulfonamides and chloramphenicol antibiotics toward functionalized biochar for water and wastewater treatment. *Bioresource Technology*, *238*, 306–312. https://doi.org/10.1016/j.biortech.2017.04.042.

Cha, J. S., Park, S. H., Jung, S. C., Ryu, C., Jeon, J. K., Shin, M. C., et al. (2016). Production and utilization of biochar: a review. *Journal of Industrial and Engineering Chemistry*, *40*, 1–15. https://doi.org/10.1016/j.jiec.2016.06.002.

Chang, A. C. C., Chuang, S. S. C., Gray, M., & Soong, Y. (2003). In-situ infrared study of CO_2 adsorption on SBA-15 grafted with γ-(aminopropyl) triethoxysilane. *Energy and Fuels*, *17*, 468–473. https://doi.org/10.1021/ef020176h.

Chen, X., Chen, G., Chen, L., Chen, Y., Lehmann, J., McBride, M. B., et al. (2011). Adsorption of copper and zinc by biochars produced from pyrolysis of hardwood and corn straw in aqueous solution. *Bioresource Technology*, *102*, 8877–8884. https://doi.org/10.1016/j.biortech.2011.06.078.

Chen, Y., Zhang, X., Chen, W., Yang, H., & Chen, H. (2017). The structure evolution of biochar from biomass pyrolysis and its correlation with gas pollutant adsorption performance. *Bioresource Technology*, *246*, 101–109. https://doi.org/10.1016/j.biortech.2017.08.138.

Chen, C., Zhou, W., & Lin, D. (2015). Sorption characteristics of N-nitrosodimethylamine onto biochar from aqueous solution. *Bioresource Technology*, *179*, 359–366. https://doi.org/10.1016/j.biortech.2014.12.059.

Chen, B., Zhou, D., & Zhu, L. (2008). Transitional adsorption and parition of nonpolar and polar aromatic contaminants by biochars of pine needles with different pyrolytic temperatures. *Environmental Science & Technology*, *42*, 5137–5143. https://doi.org/10.1021/es8002684.

Choong, T. S. Y., Chuah, T. G., Robiah, Y., Gregory Koay, F. L., & Azni, I. (2007). Arsenic toxicity, health hazards and removal techniques from water: an overview. *Desalination*, *217*, 139–166. https://doi.org/10.1016/j.desal.2007.01.015.

Creamer, A. E., Gao, B., & Wang, S. (2016). Carbon dioxide capture using various metal oxyhydroxide-biochar composites. *Chemical Engineering Journal*, *283*, 826–832. https://doi.org/10.1016/j.cej.2015.08.037.

Creamer, A. E., Gao, B., & Zhang, M. (2014). Carbon dioxide capture using biochar produced from sugarcane bagasse and hickory wood. *Chemical Engineering Journal*, *249*, 174–179. https://doi.org/10.1016/j.cej.2014.03.105.

Cui, X., Fang, S., Yao, Y., Li, T., Ni, Q., Yang, X., et al. (2016). Potential mechanisms of cadmium removal from aqueous solution by Canna indica derived biochar. *Science of the Total Environment*, *562*, 517–525. https://doi.org/10.1016/j.scitotenv.2016.03.248.

de Mendonça, F. G., da Cunha, I. T., Soares, R. R., Tristão, J. C., & Lago, R. M. (2017). Tuning the surface properties of biochar by thermal treatment. *Bioresource Technology*, *246*, 28–33. https://doi.org/10.1016/j.biortech.2017.07.099.

Delgado, B., López González, D., Godbout, S., Lagacé, R., Giroir-Fendler, A., & Avalos Ramirez, A. (2017). A study of torrefied cardboard characterization and applications: composition, oxidation kinetics and methane adsorption. *Science of the Total Environment*, *593–594*, 406–417. https://doi.org/10.1016/j.scitotenv.2017.03.119.

Deng, J., Liu, Y., Liu, S., Zeng, G., Tan, X., Huang, B., et al. (2017). Competitive adsorption of Pb(II), Cd(II) and Cu(II) onto chitosan-pyromellitic dianhydride modified biochar. *Journal of Colloid and Interface Science*, *506*, 355–364. https://doi.org/10.1016/j.jcis.2017.07.069.

Essandoh, M., Kunwar, B., Pittman, C. U., Mohan, D., & Mlsna, T. (2015). Sorptive removal of salicylic acid and ibuprofen from aqueous solutions using pine wood fast pyrolysis biochar. *Chemical Engineering Journal*, *265*, 219–227. https://doi.org/10.1016/j.cej.2014.12.006.

García-jaramillo, M., Cox, L., Knicker, H. E., Cornejo, J., Spokas, K. A., & Hermosín, M. C. (2015). Characterization and selection of biochar for an efficient retention of tricyclazole in a flooded alluvial paddy soil. *Journal of Hazardous Materials*, *286*, 581–588. https://doi.org/10.1016/j.jhazmat.2014.10.052.

Goswami, R., Shim, J., Deka, S., Kumari, D., Kataki, R., & Kumar, M. (2016). Characterization of cadmium removal from aqueous solution by biochar produced from Ipomoea fistulosa at different pyrolytic temperatures. *Ecological Engineering*, *97*, 444–451. https://doi.org/10.1016/j.ecoleng.2016.10.007.

Güzel, F., Sayğılı, H., Akkaya Sayğılı, G., Koyuncu, F., & Yılmaz, C. (2017). Optimal oxidation with nitric acid of biochar derived from pyrolysis of weeds and its application in removal of hazardous dye methylene blue from aqueous solution. *Journal of Cleaner Production*, *144*, 260–265. https://doi.org/10.1016/j.jclepro.2017.01.029.

Hameed, B. H., & El-Khaiary, M. I. (2008). Equilibrium, kinetics and mechanism of malachite green adsorption on activated carbon prepared from bamboo by K_2CO_3 activation and subsequent gasification with CO_2. *Journal of Hazardous Materials*, *157*, 344–351. https://doi.org/10.1016/j.jhazmat.2007.12.105.

Han, Y., Boateng, A. A., Qi, P. X., Lima, I. M., & Chang, J. (2013). Heavy metal and phenol adsorptive properties of biochars from pyrolyzed switchgrass and woody biomass in correlation with surface properties. *Journal of Environmental Management*, *118*, 196–204. https://doi.org/10.1016/j.jenvman.2013.01.001.

Harikishore Kumar Reddy, D., & Lee, S.-M. (2014). Magnetic biochar composite: facile synthesis, characterization, and application for heavy metal removal. *Colloids and Surfaces A: Physicochemical and Engineering Aspects*, *454*, 96–103. https://doi.org/10.1016/j.colsurfa.2014.03.105.

Harvey, O. R., Herbert, B. E., Rhue, R. D., & Kuo, L. J. (2011). Metal interactions at the biochar-water interface: energetics and structure-sorption relationships elucidated by flow adsorption microcalorimetry. *Environmental Science & Technology*, *45*, 5550–5556. https://doi.org/10.1021/es104401h.

He, R., Xia, F. F., Wang, J., Pan, C. L., & Fang, C. R. (2011). Characterization of adsorption removal of hydrogen sulfide by waste biocover soil, an alternative landfill cover. *Journal of Hazardous Materials*, *186*, 773–778. https://doi.org/10.1016/j.jhazmat.2010.11.062.

Hervy, M., Minh, D. P., Gérente, C., Weiss-Hortala, E., Nzihou, A., Villot, A., et al. (2018). H_2S removal from syngas using wastes pyrolysis chars. *Chemical Engineering Journal*, *334*, 2179–2189. https://doi.org/10.1016/j.cej.2017.11.162.

Hofman, M., & Pietrzak, R. (2011). Adsorbents obtained from waste tires for NO_2 removal under dry conditions at room temperature. *Chemical Engineering Journal*, *170*, 202–208. https://doi.org/10.1016/j.cej.2011.03.054.

Holkar, C. R., Jadhav, A. J., Pinjari, D. V., Mahamuni, N. M., & Pandit, A. B. (2016). A critical review on textile wastewater treatments: possible approaches. *Journal of Environmental Management*, *182*, 351–366. https://doi.org/10.1016/j.jenvman.2016.07.090.

Iberahim, N., Sethupathi, S., & Bashir, M. J. K. (2017). Optimization of palm oil mill sludge biochar preparation for sulfur dioxide removal. *Environmental Science and Pollution Research*. https://doi.org/10.1007/s11356-017-9180-5. 13 pages.

Inyang, M., & Dickenson, E. (2015). The potential role of biochar in the removal of organic and microbial contaminants from potable and reuse water: a review. *Chemosphere*, *134*, 232–240. https://doi.org/10.1016/j.chemosphere.2015.03.072.

Inyang, M., Gao, B., Yao, Y., Xue, Y., Zimmerman, A. R., Pullammanappallil, P., et al. (2012). Removal of heavy metals from aqueous solution by biochars derived from anaerobically digested biomass. *Bioresource Technology*, *110*, 50–56. https://doi.org/10.1016/j.biortech.2012.01.072.

Jang, J., Mirana, W., Divine, S. D., Nawaz, M., Shahzad, A., Woo, S. H., et al. (2018). Rice straw-based biochar beads for the removal of radioactive strontium from aqueous solution. *Science of the Total Environment*, *615*, 698–707. https://doi.org/10.1016/j.scitotenv.2017.10.023.

Jellali, S., Diamantopoulos, E., Haddad, K., Anane, M., Durner, W., & Mlayah, A. (2016). Lead removal from aqueous solutions by raw sawdust and magnesium pretreated biochar: experimental investigations and numerical modelling. *Journal of Environmental Management*, *180*, 439–449. https://doi.org/10.1016/j.jenvman.2016.05.055.

Jia, M., Wang, F., Jin, X., Song, Y., Bian, Y., Boughner, L. A., et al. (2016). Metal ion—oxytetracycline interactions on maize straw biochar pyrolyzed at different temperatures. *Chemical Engineering Journal*, *304*, 934–940. https://doi.org/10.1016/j.cej.2016.05.064.

Kanjanarong, J., Giri, B. S., Jaisi, D. P., Oliveira, F. R., Boonsawang, P., Chaiprapat, S., et al. (2016). Removal of hydrogen sulfide generated during anaerobic treatment of sulfate-laden wastewater using biochar: evaluation of efficiency and mechanisms. *Bioresource Technology*, *234*, 115–121. https://doi.org/10.1016/j.biortech.2017.03.009.

Kearns, J. P., Wellborn, L. S., Summers, R. S., & Knappe, D. R. U. (2014). 2,4-D adsorption to biochars: effect of preparation conditions on equilibrium adsorption capacity and comparison with commercial activated carbon literature data. *Water Research*, *62*, 20–28. https://doi.org/10.1016/j.watres.2014.05.023.

Kılıc, M., Kırbıyık, I., Cepeliogullar, O., & Pütün, A. E. (2013). Adsorption of heavy metal ions from aqueous solutions by bio-char a by-product of pyrolysis. *Applied Surface Science*, *283*, 856–862. https://doi.org/10.1016/j.apsusc.2013.07.033.

Kim, K. H., Kim, J.-Y., Cho, T.-S., & Choi, J. W. (2012). Influence of pyrolysis temperature on physicochemical properties of biochar obtained from the fast pyrolysis of pitch pine (Pinus rigida). *Bioresource Technology*, *118*, 158–162. https://doi.org/10.1016/j.biortech.2012.04.094.

Kim, W. K., Shim, T., Kim, Y. S., Hyun, S., Ryu, C., Park, Y. K., et al. (2013). Characterization of cadmium removal from aqueous solution by biochar produced from a giant Miscanthus at different pyrolytic temperatures. *Bioresource Technology*, *138*, 266–270. https://doi.org/10.1016/j.biortech.2013.03.186.

Larous, S., & Meniai, A. (2016). Adsorption of Diclofenac from aqueous solution using activated carbon prepared from olive stones. *International Journal of Hydrogen Energy*, *41*, 10380–10390. https://doi.org/10.1016/j.ijhydene.2016.01.096.

Lee, J. Y., Park, S. H., Jeon, J.-K., Yoo, K.-S., Kim, S.-S., & Park, Y.-K. (2011). The removal of low concentration formaldehyde over sewage sludge char treated using various methods. *Korean Journal of Chemical Engineering*, *28*(7), 1556–1560. https://doi.org/10.1007/s11814-011-0007-7.

Lehmann, J., Rillig, M. C., Thies, J., Masiello, C. A., Hockaday, W. C., & Crowley, D. (2011). Biochar effects on soil biota—a review. *Soil Biology and Biochemistry*, *43*, 1812–1836. https://doi.org/10.1016/j.soilbio.2011.04.022.

Li, Y., Ruan, G., Jalilov, A. S., Tarkunde, Y. R., Fei, H., & Tour, J. M. (2016). Biochar as a renewable source for high-performance CO_2 sorbent. *Carbon*, *107*, 344–351. https://doi.org/10.1016/j.carbon.2016.06.010.

Lian, F., Cui, G., Liu, Z., Duo, L., Zhang, G., & Xing, B. (2016). One-step synthesis of a novel N-doped microporous biochar derived from crop straws with high dye adsorption capacity. *Journal of Environmental Management*, *176*, 61–68. https://doi.org/10.1016/j.jenvman.2016.03.043.

Lin, L., Jiang, W., & Xu, P. (2017). Comparative study on pharmaceuticals adsorption in reclaimed water desalination concentrate using biochar: impact of salts and organic matter. *Science of the Total Environment*, *601–602*, 857–864. https://doi.org/10.1016/j.scitotenv.2017.05.203.

Liu, N., Charrua, A. B., Weng, C. H., Yuan, X., & Ding, F. (2015). Characterization of biochars derived from agriculture wastes and their adsorptive removal of atrazine from aqueous solution: a comparative study. *Bioresource Technology*, *198*, 55–62. https://doi.org/10.1016/j.biortech.2015.08.129.

Liu, S.-H., & Huang, Y.-Y. (2018). Valorization of coffee grounds to biochar-derived adsorbents for CO_2 adsorption. *Journal of Cleaner Production*, *175*, 354–360. https://doi.org/10.1016/j.jclepro.2017.12.2076.

Lu, F., & Lu, P. (2010). Experiment study on adsorption characteristics of SO_2, NOx by biomass chars. In: *2010 International Conference on Digital Manufacturing & Automation 1* (pp. 682–685). https://doi.org/10.1109/ICDMA.2010.322.

Lu, K., Yang, X., Shen, J., Robinson, B., Huang, H., Liu, D., et al. (2014). Effect of bamboo and rice straw biochars on the bioavailability of Cd, Cu, Pb and Zn to Sedum plumbizincicola. *Agriculture, Ecosystems and Environment*, *191*, 124–132. https://doi.org/10.1016/j.agee.2014.04.010.

Lu, H., Zhang, W., Yang, Y., Huang, X., Wang, S., & Qiu, R. (2012). Relative distribution of Pb^{2+} sorption mechanisms by sludge-derived biochar. *Water Research*, *46*, 854–862. https://doi.org/10.1016/j.watres.2011.11.058.

Luo, J., Liu, Y., Jiang, C., Chu, W., Jie, W., & Xie, H. (2011). Experimental and modeling study of methane adsorption on activated carbon derived from anthracite. *Journal of Chemical & Engineering Data*, *56*, 4919–4926. https://doi.org/10.1016/j.biortech.2010.12.010.

Lyu, H., Gao, B., He, F., Zimmerman, A. R., Ding, C., Tang, J., et al. (2018). Experimental and modeling investigations of ball-milled biochar for the removal of aqueous methylene blue. *Chemical Engineering Journal*, *335*, 110–119. https://doi.org/10.1016/j.cej.2017.10.130.

Madzaki, H., Ghani, W. A. W. A. B. K., Rebitanim, N. Z., & Alias, A. B. (2016). Carbon dioxide adsorption on sawdust biochar. *Procedia Engineering*, *148*, 718–725. https://doi.org/10.1016/j.proeng.2016.06.591.

Mahmoud, D. K., Salleh, M. A. M., Karim, W. A. W. A., Idris, A., & Abidin, Z. Z. (2012). Batch adsorption of basic dye using acid treated kenaf fibre char: equilibrium, kinetic and thermodynamic studies. *Chemical Engineering Journal*, *181–182*, 449–457. https://doi.org/10.1016/j.cej.2011.11.116.

Mandal, A., & Singh, N. (2017). Optimization of atrazine and imidacloprid removal from water using biochars: designing single or multi-staged batch adsorption systems. *International Journal of Hygiene and Environmental Health*. https://doi.org/10.1016/j.ijheh.2017.02.010.

Mandal, A., Singh, N., & Purakayastha, T. J. (2017). Characterization of pesticide sorption behaviour of slow pyrolysis biochars as low cost adsorbent for atrazine and imidacloprid removal. *Science of the Total Environment*, *577*, 376–385. https://doi.org/10.1016/j.scitotenv.2016.10.204.

Mlayah, A., & Jellali, S. (2015). Study of continuous lead removal from aqueous solutions by marble wastes: efficiencies and mechanisms. *International journal of Environmental Science and Technology*, *12*, 2965–2978. https://doi.org/10.1007/s13762-014-0715-8.

Mondal, S., Aikat, K., Siddharth, K., & Sarkar, K. (2017). Optimizing ranitidine hydrochloride uptake of Parthenium hysterophorus derived N-biochar through response surface methodology. *Process Safety and Environmental Protection*, *107*, 388–401. https://doi.org/10.1016/j.psep.2017.03.011.

Mui, E. L. K., Cheung, W. H., Valix, M., & Mckay, G. (2010). Dye adsorption onto char from bamboo. *Journal of Hazardous Materials*, *177*, 1001–1005. https://doi.org/10.1016/j.jhazmat.2010.01.018.

Nguyen, M.-V., & Lee, B.-K. (2016). A novel removal of CO_2 using nitrogen doped biochar beads as a green adsorbent. *Process Safety and Environmental Protection*, *104*, 490–498. https://doi.org/10.1016/j.psep.2016.04.007.

Nowicki, P., Skibiszewska, P., & Pietrzak, R. (2013). NO_2 removal on adsorbents prepared from coffee industry waste materials. *Adsorption*, *19*, 521–528. https://doi.org/10.1007/s10450-013-9474-y.

Ouyang, S., Xu, S., Song, N., & Jiao, S. (2013). Coconut shell-based carbon adsorbents for ventilation air methane enrichment. *Fuel*, *113*, 420–425. https://doi.org/10.1016/j.fuel.2013.06.004.

Peng, H., Gao, P., Chu, G., Pan, B., Peng, J., & Xing, B. (2017). Enhanced adsorption of Cu(II) and Cd(II) by phosphoric acid-modified biochars. *Environmental Pollution*, *229*, 846–853. https://doi.org/10.1016/j.envpol.2017.07.004.

Plaza, M. G., Durán, I., Querejeta, N., Rubiera, F., & Pevida, C. (2016). Experimental and simulation study of adsorption in postcombustion conditions using a microporous biochar. 2. H_2O, CO_2, and N_2 adsorption. *Industrial and Engineering Chemistry Research*, *55*, 6854–6865. https://doi.org/10.1021/acs.iecr.6b01720.

Plaza, M. G., González, A. S., Pevida, C., & Rubiera, F. (2014). Influence of water vapor on CO_2 adsorption using a biomass-based carbon. *Industrial and Engineering Chemistry Research*, *53*, 15488–15499. https://doi.org/10.1021/ie500342q.

Plaza, M. G., González, A. S., Pis, J. J., Rubiera, F., & Pevida, C. (2014). Production of microporous biochars by single-step oxidation: effect of activation conditions on CO_2 capture. *Applied Energy*, *114*, 551–562. https://doi.org/10.1016/j.apenergy.2013.09.058.

Rafatullah, M., Sulaiman, O., Hashim, R., & Ahmad, A. (2010). Adsorption of methylene blue on low-cost adsorbents: a review. *Journal of Hazardous Materials*, *177*, 70–80. https://doi.org/10.1016/j.jhazmat.2009.12.047.

Rosales, E., Meijide, J., Pazos, M., & Sanromán, M. A. (2017). Challenges and recent advances in biochar as low-cost biosorbent: from batch assays to continuous-flow systems. *Bioresource Technology*, *246*, 176–192. https://doi.org/10.1016/j.biortech.2017.06.084.

Rufford, T. E., Smart, S., Watson, G. C. Y., Graham, B. F., Boxall, J., Diniz da Costa, J. C., et al. (2012). The removal of CO_2 and N_2 from natural gas: a review of conventional and emerging process technologies. *Journal of Petroleum Science and Engineering*, *94–95*, 123–154. https://doi.org/10.1016/j.petrol.2012.06.016.

Sadasivam, B. Y., & Reddy, K. R. (2015). Adsorption and transport of methane in biochars derived from waste wood. *Waste Management*, *43*, 218–229. https://doi.org/10.1016/j.wasman.2015.04.025.

Sekulić, M., Pap, S., Stojanović, Z., Bošković, N., Radonić, J., & Šolević Knudsen, T. (2018). Efficient removal of priority, hazardous priority and emerging pollutants with Prunus armeniaca functionalized biochar from aqueous wastes: experimental optimization and modeling. *Science of the Total Environment*, *613–614*, 736–750. https://doi.org/10.1016/j.scitotenv.2017.09.082.

Seredych, M., Bashkova, S., Pietrzak, R., & Bandosz, T. J. (2010). Interactions of NO_2 and NO with carbonaceous adsorbents containing silver nanoparticles. *Langmuir*, *26*, 9457–9464. https://doi.org/10.1021/la101175h.

Sethupathi, S., Zhang, M., Rajapaksha, A. U., Lee, S. R., Nor, N. M., Mohamed, A. R., et al. (2017). Biochars as Potential Adsorbers of CH_4, CO_2 and H_2S. *Sustainability*, *9*, 121–130. https://doi.org/10.3390/su9010121.

Sewu, D. D., Boakye, P., & Woo, S. H. (2017). Highly efficient adsorption of cationic dye by biochar produced with Korean cabbage waste. *Bioresource Technology*, *224*, 206–213. https://doi.org/10.1016/j.biortech.2016.11.009.

Shan, D., Deng, S., Zhao, T., Wang, B., Wang, Y., Huang, J., et al. (2016). Preparation of ultrafine magnetic biochar and activated carbon for pharmaceutical adsorption and subsequent degradation by ball milling. *Journal of Hazardous Materials*, *305*, 156–163. https://doi.org/10.1016/j.jhazmat.2015.11.047.

Shang, G., Shen, G., Liu, L., Chen, Q., & Xu, Z. (2013). Kinetics and mechanisms of hydrogen sulfide adsorption by biochars. *Bioresource Technology*, *133*, 495–499. https://doi.org/10.1016/j.biortech.2012.01.114.

Shannon, M. A., Bohn, P. W., Elimelech, M., Georgiadis, J. G., Mariñas, B. J., & Mayes, A. M. (2008). Science and technology for water purification in the coming decades. *Nature*, *452*, 301–310. https://doi.org/10.1038/nature06599.

Shen, B., Tian, L., Li, F., Zhang, X., Xu, H., & Singh, S. (2017). Elemental mercury removal by the modified bio-char from waste tea. *Fuel*, *187*, 189–196. https://doi.org/10.1016/j.fuel.2016.09.059.

Shu, T., Lu, P., & He, N. (2013). Mercury adsorption of modified mulberry twig chars in a simulated flue gas. *Bioresource Technology*, *136*, 182–187. https://doi.org/10.1016/j.biortech.2013.02.087.

Sizmur, T., Fresno, T., Akgül, G., Frost, H., & Moreno-Jiménez, E. (2017). Biochar modification to enhance sorption of inorganics from water. *Bioresource Technology*, *246*, 34–47. https://doi.org/10.1016/j.biortech.2017.07.082.

Son, E. B., Poo, K. M., Chang, J. S., & Chae, K. J. (2018). Heavy metal removal from aqueous solutions using engineered magnetic biochars derived from waste marine macro-algal biomass. *Science of the Total Environment*, *615*, 161–168. https://doi.org/10.1016/j.scitotenv.2017.09.171.

Sousa, J. C. G., Ribeiro, A. R., Barbosa, M. O., Pereira, M. F. R., & Silva, A. M. T. (2018). A review on environmental monitoring of water organic pollutants identified by EU guidelines. *Journal of Hazardous Materials*, *344*, 146–162. https://doi.org/10.1016/j.jhazmat.2017.09.058.

Sumalinog, D. A. G., Capareda, S. C., & de Luna, M. D. G. (2018). Evaluation of the effectiveness and mechanisms of acetaminophen and methylene blue dye adsorption on activated biochar derived from municipal solid wastes. *Journal of Environmental Management*, *210*, 255–262. https://doi.org/10.1016/j.jenvman.2018.01.010.

Tan, X. f., Liu, Y. g., Gu, Y. l., Xu, Y., Zeng, G. m., Hu, X. j., et al. (2016). Biochar-based nano-composites for the decontamination of wastewater: a review. *Bioresource Technology*, *212*, 318–333. https://doi.org/10.1016/j.biortech.2016.04.093.

Tan, X., Liu, Y., Zeng, G., Wang, X., Hu, X., Gu, Y., et al. (2015). Application of biochar for the removal of pollutants from aqueous solutions. *Chemosphere*, *125*, 70–85. https://doi.org/10.1016/j.chemosphere.2014.12.058.

Uchimiya, M., Lima, I. M., Klasson, K. T., & Wartelle, L. H. (2010). Contaminant immobilization and nutrient release by biochar soil amendment: roles of natural organic matter. *Chemosphere*, *80*, 935–940. https://doi.org/10.1016/j.chemosphere.2010.05.020.

Vithanage, M., Mayakaduwa, S. S., Herath, I., & Sik, Y. (2016). Kinetics, thermodynamics and mechanistic studies of carbofuran removal using biochars from tea waste and rice husks. *Chemosphere*, *150*, 781–789. https://doi.org/10.1016/j.chemosphere.2015.11.002.

Vithanage, M., Upamali, A., & Tang, X. (2014). Sorption and transport of sulfamethazine in agricultural soils amended with invasive-plant-derived biochar. *Journal of Environmental Management*, *141*, 95–103. https:/doi.org/10.1016/j.jenvman.2014.02.030.

Wan, S., Wu, J., Zhou, S., Wang, R., Gao, B., & He, F. (2018). Enhanced lead and cadmium removal using biochar-supported hydrated manganese oxide (HMO) nanoparticles: behavior and mechanism. *Science of the Total Environment*, *616–617*, 1298–1306. https://doi.org/10.1016/j.scitotenv.2017.10.188.

Wang, S., Gao, B., Li, Y., Mosa, A., Zimmerman, A. R., Ma, L. Q., et al. (2015). Manganese oxide-modified biochars: preparation, characterization, and sorption of arsenate and lead. *Bioresource Technology*, *181*, 13–17. https://doi.org/10.1016/j.biortech.2015.01.044.

Wang, H., Gao, B., Wang, S., Fang, J., Xue, Y., & Yang, K. (2015). Removal of Pb(II), Cu(II), and Cd(II) from aqueous solutions by biochar derived from $KMnO_4$ treated hickory wood. *Bioresource Technology*, *197*, 356–362. https://doi.org/10.1016/j.biortech.2015.08.132.

Wang, Y., & Liu, R. (2018). H_2O_2 treatment enhanced the heavy metals removal by manure biochar in aqueous solutions. *Science of the Total Environment*, 628–629. 1139–1148 https:/doi.org/10.1016/j.scitotenv.2018.02.137.

Xu, X., Cao, X., Zhao, L., & Sun, T. (2014). Comparison of sewage sludge- and pig manure-derived biochars for hydrogen sulfide removal. *Chemosphere*, *111*, 296–303. https://doi.org/10.1016/j.chemosphere.2014.04.014.

Xu, R. k., Xiao, S. c., Yuan, J. h., & Zhao, A. z. (2011). Adsorption of methyl violet from aqueous solutions by the biochars derived from crop residues. *Bioresource Technology*, *102*, 10293–10298. https://doi.org/10.1016/j.biortech.2011.08.089.

Xu, X., Zhao, Y., Sima, J., Zhao, L., Mašek, O., & Cao, X. (2017). Indispensable role of biochar-inherent mineral constituents in its environmental applications: a review. *Bioresource Technology*, *241*, 887–899. https://doi.org/10.1016/j.biortech.2017.06.023.

Yan, R., Liang, D. T., Tsen, L., & Tay, J. H. (2002). Kinetics and mechanisms of H_2S adsorption by alkaline activated carbon. *Environmental Science & Technology*, *36*, 4460–4466. https://doi.org/10.1021/es0205840.

Yang, Y., Lin, X., Wei, B., Zhao, Y., & Wang, J. (2014). Evaluation of adsorption potential of bamboo biochar for metal-complex dye: equilibrium, kinetics and artificial neural network modeling. *International journal of Environmental Science and Technology*, *11*, 1093–1100. https://doi.org/10.1007/s13762-013-0306-0.

Yi, S., Gao, B., Sun, Y., Wu, J., Shi, X., Wu, B., et al. (2016). Removal of levofloxacin from aqueous solution using rice-husk and wood-chip biochars. *Chemosphere*, *150*, 694–701. https://doi.org/10.1016/j.chemosphere.2015.12.112.

Yu, C.-H., Huang, C.-H., & Tan, C.-S. (2012). A review of CO_2 capture by absorption and adsorption. *Aerosol and Air Quality Research*, *12*, 745–769. https://doi.org/10.4209/aaqr.2012.05.0132.

Zhang, X., Gao, B., Creamer, A. E., Cao, C., & Li, Y. (2017). Adsorption of VOCs onto engineered carbon materials: a review. *Journal of Hazardous Materials*, *338*, 102–123. https://doi.org/10.1016/j.jhazmat.2017.05.013 0304-3894.

Zhang, X., Gao, B., Zheng, Y., Hu, X., Creamer, A. E., Annable, M. D., et al. (2017). Biochar for volatile organic compound (VOC) removal: sorption performance and governing mechanisms. *Bioresource Technology*, *245*, 606–614. https://doi.org/10.1016/j.biortech.2017.09.025.

Zhang, J., Liu, J., & Liu, R. (2014). Effects of pyrolysis temperature and heating time on biochar obtained from the pyrolysis of straw and lignosulfonate. *Bioresource Technology*, *176*, 288–291. https://doi.org/10.1016/j.biortech.2014.11.011.

Zhang, W., Mao, S., Chen, H., Huang, L., & Qiu, R. (2013). Pb(II) and Cr(VI) sorption by biochars pyrolyzed from the municipal wastewater sludge under different heating conditions. *Bioresource Technology*, *147*, 545–552. https://doi.org/10.1016/j.biortech.2013.08.082.

Zhang, C., Shan, B., Tang, W., & Zhu, Y. (2017). Comparison of cadmium and lead sorption by Phyllostachys pubescens biochar produced under a low-oxygen pyrolysis atmosphere. *Bioresource Technology*, *238*, 352–360. https://doi.org/10.1016/j.biortech.2017.04.051.

Zhang, P., Sun, H., Ren, C., Min, L., & Zhang, H. (2018). Sorption mechanisms of neonicotinoids on biochars and the impact of deashing treatments on biochar structure and neonicotinoids. *Environmental Pollution*, *234*, 812–820. https://doi.org/10.1016/j.envpol.2017.12.013.

Zhang, P., Sun, H., Yu, L., & Sun, T. (2013). Adsorption and catalytic hydrolysis of carbaryl and atrazine on pig manure-derived biochars: impact of structural properties of biochars. *Journal of Hazardous Materials*, *245*, 217–224. https://doi.org/10.1016/j.jhazmat.2012.11.046.

Zhang, G., Zhang, Q., Sun, K., Liu, X., Zheng, W., & Zhao, Y. (2011). Sorption of simazine to corn straw biochars prepared at different pyrolytic temperatures. *Environmental Pollution*, *159*, 2594–2601. https://doi.org/10.1016/j.envpol.2011.06.012.

Zhang, X., Zhang, S. H., Yang, H. P., Shao, J., Chen, Y. Q., Feng, Y., et al. (2015). Effects of hydrofluoric acid pre-deashing of rice husk on physicochemical properties and CO_2 adsorption performance of nitrogen-enriched biochar. *Energy*, *91*, 903–910. https://doi.org/10.1016/j.energy.2015.08.028.

Zhou, Q., Liao, B., Lin, L., Qiu, W., & Song, Z. (2018). Adsorption of Cu(II) and Cd(II) from aqueous solutions by ferromanganese binary oxide–biochar composites. *Science of the Total Environment*, *615*, 115–122. https://doi.org/10.1016/j.scitotenv.2017.09.220.

Zhou, J., Liu, Y., & Pan, J. (2017). Removal of elemental mercury from flue gas using wheat straw chars modified by K_2FeO_4 reagent. *Environmental Technology*, *38*, 3047–3054. https://doi.org/10.1080/09593330.2017.1287222.

Zhou, Y., Liu, X., Xiang, Y., Wang, P., Zhang, J., Zhang, F., et al. (2017). Modification of biochar derived from sawdust and its application in removal of tetracycline and copper from aqueous solution: adsorption mechanism and modelling. *Bioresource Technology*, *245*, 266–273. https://doi.org/10.1016/j.biortech.2017.08.178.

Zhou, L., Richard, C., Ferronato, C., Chovelon, J.-M., & Sleiman, M. (2018). Investigating the performance of biomass-derived biochars for the removal of gaseous ozone, adsorbed nitrate and aqueous bisphenol A. *Chemical Engineering Journal*, *334*, 2098–2104. https://doi.org/10.1016/j.cej.2017.11.145.

Applications in catalysis

8

Simona Bennici, Claire Courson*[†]
*University of Strasbourg, University of Upper Alsace, Institute of Materials Science of Mulhouse (IS2M—UMR CNRS 7361), Mulhouse, France, [†]Univ Strasbourg, CNRS, ICPEES UMR 7515, ECPM, Strasbourg, France

Chapter Outline

8.1 Introduction

During the last century, most chemical production processes had been characterized by an unconsidered exploitation of natural resources, very often ignoring the environmental consequences. Fortunately, nowadays there is a renewed consciousness for the necessity to reduce the human impact on the environment, push industries to develop sustainable production paths, and switch their practices toward cleaner processes that are less resource demanding. These advances represent the heart of so-called "sustainable development" that aims to reach a compromise between present-day production needs and the preservation of the ability of the next generations to meet their needs.

The development of new eco-friendly industrial processes and the optimization of the use of natural resources by minimizing wastes is a challenging objective. In particular, the need to upgrade the existing processes of the chemical and petrochemical industries toward a greener approach passes also through the development of new eco-compatible catalysts. The valorization of biosourced wastes such as chars, produced by thermal treatments, creates very promising materials that can be applied as catalysts or catalytic supports in various applications.

Char and Carbon Materials Derived from Biomass. https://doi.org/10.1016/B978-0-12-814893-8.00008-0

The design, preparation, and characterization of such materials represent very challenging fields due to the required multidisciplinary approach (physical chemistry, analytical chemistry, material science, process engineering, etc.) to answer the various fundamental and applicative issues that restrain their implementation in existing or new processes.

In this context, this chapter addresses the use of chars issued from biomass feedstocks as environmentally friendly catalysts or catalyst supports. Investigation of the surface, structural, morphological, and redox properties of the chars will be related to their catalytic behavior in various reactions. Moreover, the choice of biomass sources, chars, and catalyst preparation methods, as well as the chars' posttreatments, will be considered. The chapter aims to elucidate the dependence of the catalytic performance of the various reactions on the physicochemical properties of char-based catalysts and to identify the required characteristics necessary for the future of chars' development in the heterogeneous catalysis field.

8.2 Chars as catalyst supports

Among the various catalyst supports, biomass-derived chars are a versatile and environmentally friendly option. Some of chars' features such as large specific surface area, high porosity, high electron conductivity, and the possibility to relatively easily tailor their surface chemical properties make them particularly interesting as catalyst supports. Even if the main role of a catalytic support is to efficiently disperse the active phase (generally metal or metal oxides), thus improving the catalytic site's accessibility and stability, other contributions of the support have also to be considered. The catalytic support can present intrinsic catalytic activity and the capacity to interact with other chemical components of the catalysts, thus creating new active sites. Moreover, the interaction between the support and the active phase can produce synergetic effects that positively contribute to the overall catalytic activity. Different biomass sources can be used to produce chars by pyrolysis performed at different temperatures and with substoichiometric oxygen content. The cellulose, hemicellulose, and lignin contents differ for the various biomass feedstocks, thus producing chars with specific surface and morphological properties. Indeed, by modulating the thermal treatment temperature and increasing temperature rate (Lobos et al., 2016; Prati et al., 2018; Qian, Kumar, Zhang, Bellmer, & Huhnke, 2015), the three polysaccharides can be selectively decomposed giving rise to chars with different porous structures (Keiluweit, Nico, Johnson, & Kleber, 2010; Yang, Yan, Chen, Lee, & Zheng, 2007). Moreover, the char production yield also strongly depends on the feedstock composition (Aqsha, Tijani, Moghtaderi, & Mahinpey, 2017; Qian et al., 2013).

Certain specific properties (which will be detailed in the next sections) derive from the parent biomass used to produce the char, and might have a strong impact on the catalytic properties of the final material. As an example, Lobos et al. (2016) showed that chars derived from fast pyrolysis of cellulose at relatively low temperature (350°C) present high carbon content that is a positive characteristic when the char is used as a catalyst support for electro-oxidation reactions. Similar observations have been reported (Muley, Henkel, Abdollahi, Marculescu, & Boldor, 2016) on samples obtained by fast pyrolysis of lignin that displayed a higher amount of ash and fixed carbon when compared to those obtained by sawdust and cellulose. The char yield diminished for all biomasses when the temperature was increased from 500°C to 700°C. However, the cellulose char yields decrease much lower than those of the other two biomasses. This indicates that cellulose is more thoroughly broken down at lower reaction temperatures than lignin and sawdust. Moreover, the morphology of the obtained chars can be characterized by enhanced Brunauer-Emmett-Teller (BET) surface area and porosity, independently of the parent biomass source (Liu, Charrua, Weng, Yuan, & Ding, 2015). The decisive role of the carbon surface chemistry on catalytic behavior has been recognized for some time (Matos, Bernardo, & Fonseca, 2017; Rodríguez-Reinoso, 1998). More recently, biomass-derived chars have been identified as possible substitutes for conventional carbons (Xiong et al., 2017) due to the possibility of tuning their surface chemistry and porosity to match the required features for specific catalytic applications. The main properties characterizing chars as catalysts or catalyst supports are as follows.

8.2.1 Acid/base properties

Surface acid/base properties are the key factors that govern the adsorption ability of solid materials toward the different molecules that can control their performance in various applications, such as separation processes and catalysis.

The surface chemistry, and consequently the acid/base properties, of chars is related to the presence of heteroatoms such as oxygen, nitrogen, hydrogen, phosphor, and sulfur on the surface. These elements can be introduced from the biomass precursor during an activation process or by posttreatment on the already formed char. Interacting with the carbon surface, the heteroatoms can then form surface complexes (shown in Fig. 1 (Figueiredo, 2013)) that characterize the basic, acid, or neutral features of the solid material. Surface functionalities such as carboxylic acid, phenols, anhydrides, lactones, lactols, and phenols are acidic, while carbonyl and ether functionalities are neutral. On the other hand, chromene, quinone, pyrone, and nitrogen-containing groups are basic functionalities (Figueiredo, 2013; Rivera-Utrilla et al., 2011; Shafeeyan, Daud, Houshmand, & Shamiri, 2010).

The adsorption properties that are correlated to catalyst selectivity are strongly dependent on the surface acid/base properties of the char used as catalyst support.

The surface chemistry of chars can be modified by various treatments to create new basic/acid sites by modifying the nature and quantity of the surface functionalities. Often, the acidic character is enhanced by oxidative treatments of the chars that increase the amount of oxygen-containing surface groups. The treatments can be performed in the liquid (Belhachemi et al., 2009; Belhachemi, Jeguirim, Limousy, & Addoun, 2014; Figueiredo, Pereira, Freitas, & Orfao, 1999; Limousy, Ghouma, Ouederni, & Jeguirim, 2017; Moreno-Castilla et al., 1995; Rodríguez-Reinoso, Molina-Sabio, & González, 1997) or gas phase (Chiang, Chiang, & Huang, 2002; Figueiredo et al., 1999; Ghouma et al., 2015; Valdés, Sanchez-Polo, Rivera-Utrilla, & Zaror, 2002) with various oxidizing agents. The most common reactants used in the gas phase are oxygen, ozone, and nitrogen oxides, while in the liquid phase, nitric acid and hydrogen peroxide are those more widely applied. The choice of

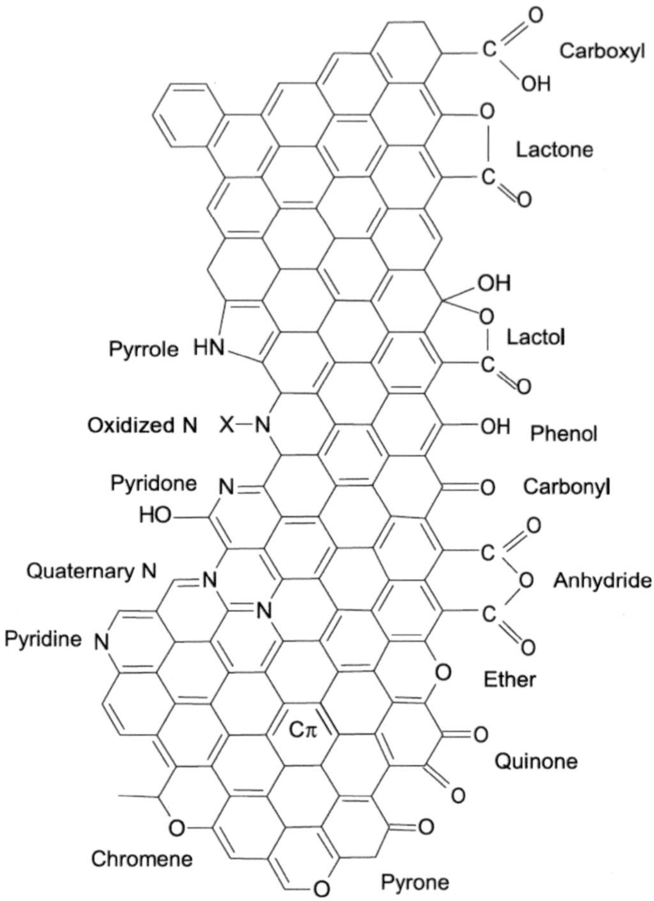

Fig. 1 Nitrogen and oxygen surface functional groups on activated carbon.
Reproduced with permission from Figueiredo, J. L., & Pereira, M. F. R. (2010). The role of surface chemistry in catalysis with carbons. *Catalysis Today*, *150*(1), 2–7. Copyright 2018, Elsevier.

performing a pretreatment in the gas or liquid phase is connected both to the desired type of oxygen-containing functionalities and to the extent of the modification of the textural properties of the char. Indeed, oxidation in the liquid phase is generally performed at more moderate reaction conditions (lower temperature and shorter reaction time) than those carried out in the gas phase. Liquid phase treatment creates surface oxygen functionalities like carboxylic groups, without having a strong impact on the textural properties (in particular char porosity (Moreno-Castilla et al., 1995)). Differently, during gas phase oxidation treatments (which create only low amounts of hydroxyl and carbonyl groups), the reaction time and temperature are more extreme, and can deeply modify the char porous structure by opening, extending, and creating pores (Samant, Gonçalves, Freitas, Pereiraand, & Figueiredo, 2004). As an example of liquid phase treatment, Moreno-Castilla et al. (1995) reported on the treatment of a series of activated carbons issued from almond shells by using three different chemicals (HNO_3, H_2O_2, and $(NH_4)_2S_2O_8$). They noticed that if carboxyl groups were created in each case, ether and ketone groups were magnified after treatment with hydrogen peroxide. Moreover, the $(NH_4)_2S_2O_8$ treatment surprisingly created stronger acidity than that observed after HNO_3 treatment.

Concerning the gas-phase treatments, and to improve the catalytic performance of woody chars, Ducousso, Weiss-Hortala, Nzihou, and Castaldi (2015) performed a gas-phase treatment with oxygen. Even if the process diminished the overall quantity of carboxylic groups (which are the strongest acid sites), oxygenation promoted the formation of weak acid sites, but did not have any impact on the amount of basic sites like pyrone and quinone functionalities (as shown in Fig. 2).

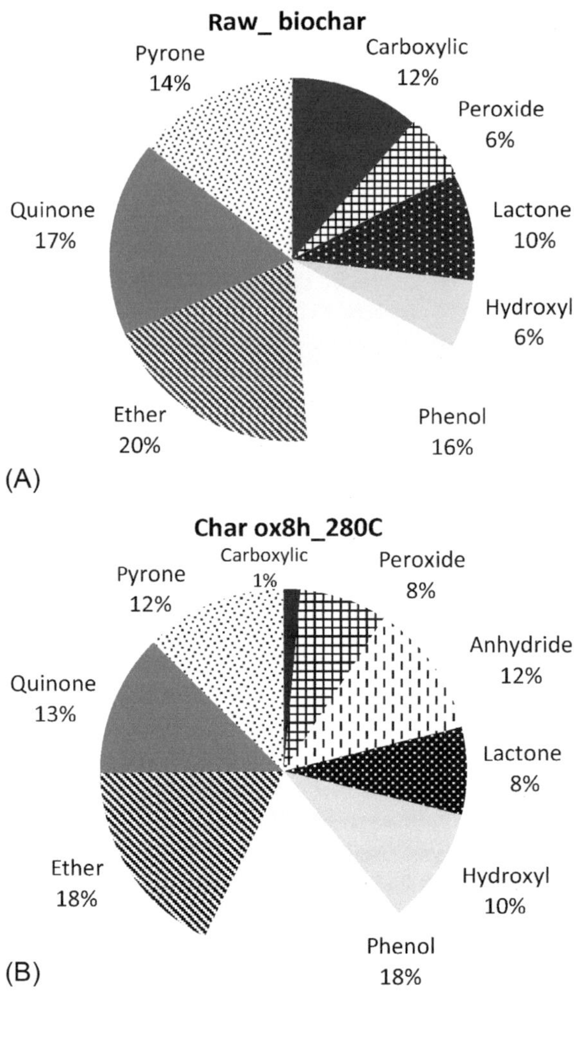

Fig. 2 Oxygenated group distribution in at.% of raw chars (A) and chars oxygenated for 8 h at 280°C (B).
Reproduced with permission from Ducousso, M., Weiss-Hortala, E., Nzihou, A., & Castaldi, M. J. (2015). Reactivity enhancement of gasification biochars for catalytic applications. *Fuel*, *159*, 491–499. Copyright 2018, Elsevier.

Among the others, Chiang et al. increased the phenolic and carboxylic groups on the surface of an activated carbon issued from coconut shells by treatment with ozone at room temperature (Chiang et al., 2002). Ozone treatment of chars was also used by Kastner et al. (2012) to introduce weak phenolic and lactonic acid sites. Besides the treatment with oxidizing agents, sulfonation is another effective method to create acid catalyst and catalyst supports. By treating the chars with H_2SO_4 and SO_3, strong surface sulfonic acid sites and weak carboxyl and phenolic hydroxyl acid groups were created, as reported by the same authors (Kastner et al., 2012). More recently, the acidity of chars, used as catalysts for the transformation of cellulose to 5-hydroxymethylfurfural (HMF), was enhanced by using ionic liquids. The chlorine-based ionic liquid had been grafted onto sulfonated chars to allow substitution of the chlorine by fluorine through ionic exchange with F-containing acids (Zhang et al., 2017).

Even if used less widely, a basic catalyst can be applied in specific reactions such as aldolization. Two types of basic sites can be identified on chars. The first kind of basic site is related to the presence of certain specific functionalities, such as chromene, ketone, pyrone, and amines, while the second is associated with the presence of oxygen-free Lewis basic sites on the graphene layers.

To improve the basic character of chars, both heat and chemical treatments can be applied. Heat treatment consists of processing the sample at high temperature under inert (i.e., helium, nitrogen) or reducing (i.e., hydrogen) gases; the acidic surface groups are selectively eliminated, thus enhancing the basicity ratio. Nitrogenation to form amine groups is the chemical treatment generally applied to increase the chars' surface basicity. Once exposed to high temperature for a long time, the formed amine is transformed into amide, then into imine, and then into imide. Finally, imide is transformed to nitrile and pyridine species (Fig. 1) (Kapteijn, Moulijn, Matzner, & Boehm, 1999; Mangun, Benak, Daley, Economy, & Foster, 2001; Pietrzak, 2009). As an example, activated carbons issued from olive stones and date pits were treated at high temperature under helium (Silvestre-Albero, Silvestre-Albero, Sepulveda-Escribano, & Rodriguez-Reinoso, 2009) and nitrogen (Belhachemi & Addoun, 2011) flows, respectively; in both cases the strongest carboxylic acidic groups were removed resulting in an improved basic character.

Modification of the surface acidity/basicity of biomass-derived chars by metals/metal oxides addition has not yet been widely investigated and put in practice as have other activation methods such as oxidation, sulfonation, etc. previously described. Indeed, to share between the carbon surface species' contribution and the metal/metal oxide sites' contribution to the surface acid/base character of a char-based composite is not an easy task. However, some authors identified the possibility of creating Lewis sites by metal impregnation. Such an objective was reached by Liu, Yang, Liu, and Li (2015) when preparing SnO_2-Co_3O_4/C catalysts used in corn cob hydrolysis into furfural. The authors showed that the acid properties of the char support can be modified when Sn^{4+} and Co^{2+} species are introduced in the material; as shown by pyridine adsorption infrared spectroscopy, the Brønsted acid sites were converted into Lewis acid sites (Fig. 3).

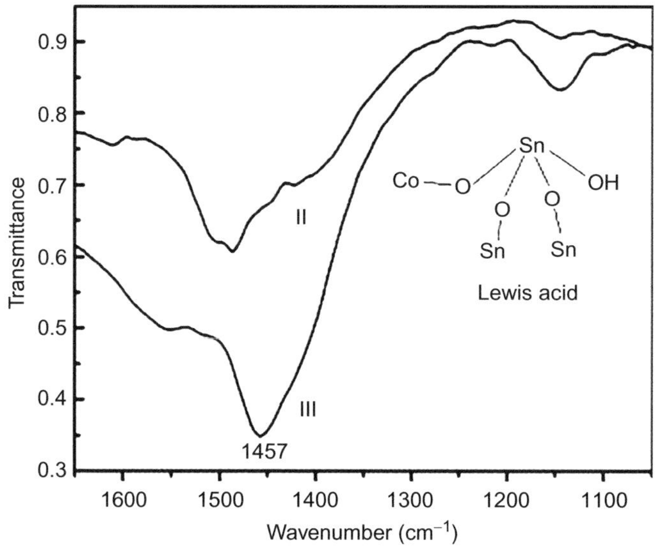

Fig. 3 Formation of Lewis acids: comparison of char (II) and SnO_2-Co_3O_4/C catalyst (III) through Fourier transform infrared spectra after pyridine adsorption.
Reproduced with permission from Liu, Q. Y., Yang, F., Liu, Z. H., & Li, G. (2015). Preparation of SnO2–Co3O4/C biochar catalyst as a Lewis acid for corncob hydrolysis into furfural in water medium. *Journal of Industrial and Engineering Chemistry*, *26*, 46–54. Copyright 2018, Elsevier.

8.2.2 Porosity/texture

Textural properties are very important for char reactivity. They depend on the char source and the preparation method and can be improved by chemical treatment, gas activation, and metal impregnation.

It is well known that the biomass precursor has a great influence on the texture and porosity of the elaborated char (Guerrero, Ruiz, Alzueta, Bilbao, & Millera, 2005; Kambo & Dutta, 2015; Lee, Kim, & Kwon, 2017; Xiong et al., 2017). In fact, a great variety of char sources can be considered, from agricultural residues to sewage sludge with various components (cellulose, hemicellulose, and lignin) distribution. However, because the thermal liability of these components is different (Yang et al., 2007), their pyrolysis creates several molecular arrangements (Abdullah et al., 2017), including cross-linking, depolymerization, and fragmentation (Cha et al., 2016), leading to various crystallinities (Keiluweit et al., 2010).

Depending on the nature of the biomass source, different pore sizes ranging from microporosity to meso- and macroporosity can be obtained (Dabrowski, Podkoscielny, Hubicki, & Barczak, 2005). Several authors have reported on the textural properties of chars (Ioannidou & Zabaniotou, 2007; Lee et al., 2017; Ronsse, Hecke, Dickinson, & Prins, 2013; Streat, Patrick, & Camporro Perez, 1995) derived from various biomasses, or after different treatments (chemical or physical). Table 1 summarizes the influence of biomass origin on these properties before any treatment.

Table 1 Specific surface area and porosity of chars derived from the pyrolysis of various biomass sources.

Raw material	Char preparation	BET surface area ($m^2\ g^{-1}$)[a]	Pore volume ($cm^3\ g^{-1}$)	Micropore volume ($cm^3\ g^{-1}$)[a]	Average pore diameter (nm)	References
Brown macroalga	1 h at 500°C, N_2 (50 mL min^{-1})	Under detection limit	–	–	–	Cho et al. (2013)
Woodchips	1 h at 400°C, N_2 (1–2 L min^{-1})	Under detection limit	–	–	–	Kastner et al. (2012)
Powdered peanuts hulls	1 h at 400°C, N_2 (1–2 L min^{-1})	1.1	0.001	–	1.2	Kastner et al. (2012)
Pelletized peanut hulls	1 h at 400°C, N_2 (1–2 L min^{-1})	1–4	–	–	–	Kastner et al. (2012)
Burcucumber plant	2 h at 700°C, air-limited supply	2.3	0.008	–	6.78	Rajapaksha et al. (2016)
Safflower seed	10 min at 600°C (100°C min^{-1}), N_2 (100 mL min^{-1})	4.9	0.91	–	–	Onay (2007)
Safflower seed	10 min at 600°C (800°C min^{-1}), N_2 (100 mL min^{-1})	6.7	1.73	–	–	Onay (2007)
Pelletized pine residues	1 h at 500°C, N_2 (1–2 L min^{-1})	10	0.005	–	0.96	Kastner et al. (2012)
Sewage sludge	1 h at 500°C, N_2 (50 mL min^{-1})	17.9	0.018	–	4.06	Cha et al. (2010)
Pinewood	2 h at 700°C (10°C min^{-1}), N_2 (10 mL min^{-1})	29	1.020	0.13	1.48	Liu, Zhang, and Wu (2010)
Sewage sludge	2 h at 900°C (5°C min^{-1}), N_2	44	–	–	–	Yuan, Yuan, Wang, Tang, and Zhou (2013)

Table 1 Continued

Almond shell	1 h at 600°C, N_2 (150 mL min^{-1})	42	0.094	0.023 (0.305)	–	Gonzalez, Roman, Encinar, and Martinez (2009)
Olive stone	1 h at 600°C, N_2 (150 mL min^{-1})	53	0.036	0.028 (0.399)	–	Gonzalez et al. (2009)
Radiata pine	5 min at 950°C (20°C s^{-1}), N_2	57 (290)	–	–	–	Cetin, Gupta, and Moghtaderi (2005)
Pectin	10 min at 450°C, N_2 (220 mL min^{-1})	70	–	–	–	Sharma, Wooten, Baliga, and Hajaligol (2001)
Corn cob	800°C (10°C min^{-1}), N_2 (200 mL min^{-1})	91	0.022	–	0.96	Chang, Chang, and Tsai (2000)
Rice straw	1 h at 500°C, N_2 (50 mL min^{-1})	139.5	0.092	–	2.64	Cha et al. (2010)
Giant *Miscanthus*	1 h at 500°C (10°C min^{-1}), N_2 (1.2 L min^{-1})	181	–	–	–	Shim, Yoo, Ryu, Park, and Jung (2015)
Microcrystalline cellulose from cotton	20 min at 500°C, N_2 (60 mL min^{-1})	199	0.026	–	3.0	Lobos et al. (2016)
Almond tree pruning	1 h at 600°C, N_2 (150 mL min^{-1})	204	0.118	0.097 (0.426)	–	Gonzalez et al. (2009)
Walnut shell	1 h at 600°C, N_2 (150 mL min^{-1})	280	0.160	0.138 (0.458)	–	Gonzalez et al. (2009)
Tea waste	2 h at 700°C, air-limited supply	342.2	0.022	–	1.76	Rajapaksha et al. (2016)
Corn cob	900°C (10°C min^{-1}), N_2 (200 mL min^{-1})	549	0.251	0.216	1.44	Chang et al. (2000)
Eucalyptus	1 h at 600°C (10°C min^{-1}), N_2 (1000 mL min^{-1})	(570)	–	(0.218)	–	Guerrero et al. (2005)

[a]Measured by N_2 adsorption and by CO_2 adsorption (between brackets when available).

The values reported in Table 1 are those measured by N_2 adsorption, and even if they do not to correctly evaluate microporosity they are still largely used. The values of BET surface area obtained from the CO_2 adsorption isotherms are indicated between brackets when available.

Currently, chars can be produced by several techniques. However, selection of the suitable preparation method depends on the feedstock nature (wet or dry) and the required properties of char for the specific application (Kambo & Dutta, 2015). The more common preparation methods are slow pyrolysis under limited oxygen or an oxygen-free environment (char/pyrochar) and hydrothermal carbonization (Table 2) under high pressure (hydrochar; Liu et al., 2010). The char preparation method has great influence on char porosity and morphology. In fact, conditions such as the reactive atmosphere, temperature, pressure, and treatment duration can vary and lead to a rough surface and porous structure through different reactions (Liu et al., 2010) and the formation of different intermediate compounds (Kambo & Dutta, 2015).

8.2.2.1 *Influence of char chemical treatment*

Chemical activation can be differentiated in prepyrolysis- and postpyrolysis-type treatments, i.e., a chemical treatment applied to the parent biomass or to the char, respectively. In the former, the raw material (generally wood or peat) is impregnated with a dehydrating reagent such as H_3PO_4, $ZnCl_2$, or KOH (or K_2CO_3), then the impregnated char is pyrolyzed and washed to remove the activating agent. In the latter method, chars or hydrochars are impregnated with (or mixed with) an acid (H_3PO_4, H_2SO_4, $ZnCl_2$) or, more frequently, a base (KOH, NaOH), then washed to remove the excess activating agent, and dried and/or pyrolyzed again.

Numerous studies have reported the interest of pretreatment procedures able to improve the textural properties of various biomass-derived chars by treatment with H_3PO_4 (Table 3), $ZnCl_2$ (Table 4), KOH (Table 5), and K_2CO_3 (Table 6).

Posttreatment (on chars or hydrochars instead of that performed on the biomass) is easier to put in place (smaller volumes and amount of minerals), therefore several researchers preferred this procedure to improve the textural properties of chars or hydrochars produced from various biomasses by acidic or basic treatments.

The acidic treatments performed after pyrolysis (i.e., on char or hydrochar) are generally used to modify char surface acidity and have rarely a positive effect on their textural properties (Table 7). The basic treatments performed after pyrolysis are used to improve the textural properties of chars and hydrochars and are associated with a second thermal treatment, generally performed under nitrogen between 600°C and 850°C (pyrolysis).

8.2.2.2 *Influence of char physical (gas) activation (during postsynthesis treatment)*

Pyro/hydrochars can be physically activated after their preparation. Various physical activation procedures exist, including steam and CO_2 activations as well as their mixture.

Table 2 Specific surface area and porosity of hydrochars derived from the hydrothermal carbonization of various biomass sources.

Raw material	Hydrochar preparation	BET surface area ($m^2\,g^{-1}$)	Pore volume ($cm^3\,g^{-1}$)	Micropore volume ($cm^3\,g^{-1}$)	Average pore diameter (nm)	References
Walnut shells	20 h at 220°C, $50\,g_{biomass}\,L_{water}^{-1}$	31	0.072	0.011	–	Román et al. (2013)
Cellulose	4 h at 250°C, $40\,g_{biomass}\,L_{water}^{-1}$	30	–	–	–	Sevilla and Fuertes (2009)
Cellulose	6 h at 230°C, N_2 ($50\,mL\,min^{-1}$)	28	0.369	–	–	Mumme et al. (2011)
Sunflower stem	20 h at 220°C, $50\,g_{biomass}\,L_{water}^{-1}$	27	0.051	0.016	–	Román et al. (2013)
Olive stone	20 h at 220°C, $50\,g_{biomass}\,L_{water}^{-1}$	22	0.025	0.013	–	Román et al. (2013)
Pinewood	20 min at 300°C ($10°C\,min^{-1}$), $40\,g_{biomass}\,L_{water}^{-1}$	21	0.140	0.090	0.86	Liu et al. (2010)
Digestate	6 h at 190°C, N_2 ($50\,mL\,min^{-1}$), $42.3\,g_C\,L^{-1}$	12	0.096	–	–	Mumme et al. (2011)
Solid waste of palm oil mill processing	20 min at 250°C	8	0.043	–	3.77	Parshetti, Kent Hoekman, and Balasubramanian (2013)

Table 3 H_3PO_4 pretreatment effect on textural properties measured by N_2 adsorption.

Raw material	Conditions	BET surface area ($m^2\,g^{-1}$)	Pore volume ($cm^3\,g^{-1}$)	Micropore volume ($cm^3\,g^{-1}$)	Average pore diameter (nm)	References
Effect of weight acid/biomass ratio						
Corn cob	50% solution of H_3PO_4 Pyrolysis at $T=500°C$	From 39 to 960	From 0.075 to 0.629	–	From 4.88 to 1.31	El-Hendawy, Samra, and Girgis (2001)
Cellulose	Weight acid/biomass ratio = 25% Pyrolysis at $T=500°C$	From 199 to 557	From 0.0264 to 0.2200	–	From 3.0 to 2.4	Lobos et al. (2016)
Lignin	Weight acid/lignin ratio ranging from 0.7 to 1.4	From 600 to 1100	From 0.3 to 0.5	From 0.25 to 0.40	–	Fierro, Torné-Fernandez, and Celzard (2006)
Effect of activation temperature						
Lignin	Activation temperature ranging from 400°C to 600°C	From 1000 to 1300	From 0.5 to 0.7	From 0.4 to 0.5	–	Fierro et al. (2006)
Effect of steam activation						
Cotton stalks	Weight acid/biomass ratio = 1.5/1 Pyrolysis at $T=600°C$ Steam activation at 800°C/1 h	From 516 to 1955	From 0.08 to 0.48	–	–	Shen, Chen, Yue, and Li (2015)
Effect of impregnation time						
Lignin	Impregnation time ranging from 1 to 48 h	From 1300 to 956	–	–	–	Fierro et al. (2006)

Table 4 $ZnCl_2$ pretreatment effect on textural properties measured by N_2 adsorption.

Raw material	Conditions	BET surface area ($m^2\ g^{-1}$)	Pore volume ($cm^3\ g^{-1}$)	Micropore volume ($cm^3\ g^{-1}$)	References
Effect of $ZnCl_2$/biomass weight ratio					
Cherry stones	Pyrolysis $T=500°C$	From 330 to 1566	–	–	Olivares-Marin, Fernandez-Gonzalez, Macias-Garcia, and Gomez-Serrano (2006)
Paulownia wood	Impregnated ratio = 4 Pyrolysis $T=400°C$	From 257 to 2620	–	–	Yorgun, Vural, and Demiral (2009)
Corn cob	From 0.75 to 1.75 Pyrolysis $T=500°C$ Impregnation time = 0.5 h	From 651 to 1410	From 0.33 to 0.70	–	Tsai and Lee (1997)
Cattle-manure compost	From 0.25 to 1.50 Pyrolysis $T=400°C$ Impregnation time = 0.5 h	From 600 to 2170	From 0.38 to 1.2	–	Qian, Machida, and Tatsumoto (2007)
Cattle-manure compost	From 1.5 to 2.5 Pyrolysis $T=400°C$ Impregnation time = 0.5 h	From 2170 to 2048	From 1.2 to 1.7	–	Qian et al. (2007)
Paulownia wood	From 1 to 4 Pyrolysis $T=400°C$ Impregnation time = 0.5 h	From 977 to 2736	From 0.516 to 1.387	From 0.27 to 0.69	Yorgun et al. (2009)
Effect of pyrolysis temperature ($ZnCl_2$/biomass weight ratio = 1, impregnation time = 0.5 h)					
Corn cob	Pyrolysis T from 400°C to 600°C	From 535 to 750	From 0.28 to 0.37	–	Tsai and Lee (1997)
Corn cob	Pyrolysis T from 600°C to 800°C	From 750 to 400	From 0.37 to 0.19	–	Tsai and Lee (1997)

Continued

Table 4 Continued

Effect of pyrolysis time ($ZnCl_2$/biomass weight ratio = 1)					
Corn cob	Impregnation time from 0.5 to 1 h Pyrolysis T = 400°C	From 535 to 613	From 0.28 to 0.31	–	Tsai and Lee (1997)
Corn cob	Impregnation time from 1 to 4 h Pyrolysis T = 600°C	From 820 to 744	From 0.40 to 0.38	–	Tsai and Lee (1997)

Table 5 KOH pretreatment effect on textural properties measured by N_2 adsorption.

Raw material	Conditions	BET surface area ($m^2\ g^{-1}$)	Pore volume ($cm^3\ g^{-1}$)	Micropore volume ($cm^3\ g^{-1}$)	References
Effect of KOH/precursor ratio					
Cassava peel	Impregnation weight ratio KOH/precursor from ½ to 5/2 Pyrolysis T = 750°C Pyrolysis time = 1 h	From 1027 to 1605	From 0.434 to 0.691	From 0.390 to 0.513	Sudaryanto, Hartono, Irawaty, Hindarso, and Ismadji (2006)
Effect of pyrolysis temperature					
Rice straw	Impregnation weight ratio KOH/precursor = 4/1 Pyrolysis T from 500°C to 800°C	From 370 to 900	From 0.25 to 0.46	From 0.12 to 0.30	Oh and Park (2002)
Cassava peel	Impregnation ration KOH/precursor = 1/1 Pyrolysis T from 450°C to 750°C Pyrolysis time = 1 h	From 972 to 1378	From 0.421 to 0.583	From 0.383 to 0.393	Sudaryanto et al. (2006)

Table 5 Continued

Effect of pyrolysis time					
Cassava peel	Impregnation weight ratio KOH/ precursor = 1/1 Pyrolysis T = 650°C Pyrolysis time from 1 to 3 h	From 1154 to 1183	From 0.519 to 0.520	–	Sudaryanto et al. (2006)

Table 6 K_2CO_3 pretreatment effect on textural properties measured by N_2 adsorption.

Raw material	Conditions	BET surface area ($m^2\ g^{-1}$)	Pore volume ($cm^3\ g^{-1}$)	Micropore volume ($cm^3\ g^{-1}$)	References
Effect of K_2CO_3/precursor weight ratio					
Almond shell	Impregnation K_2CO_3/precursor weight ratio from 0 to 2 Pyrolysis T = 600°C Pyrolysis time = 2 h	From 104 to 2111	–	–	Bevia, Rico, and Gomis (1984)
Black elutrilithe (coal waste)	Impregnation K_2CO_3/precursor weight ratio from 0.75 to 1.5 Pyrolysis T = 850°C Pyrolysis time = 3 h	From 776 to 856	From 0.469 to 0.523	From 0.244 to 0.234	Deng, Yue, and Gao (1997)
Effect of pyrolysis temperature					
Almond shell	Impregnation K_2CO_3/precursor weight ratio = 0.25 Pyrolysis T from 500°C to 800°C Pyrolysis time = 2 h	From 367 to 746	–	–	Bevia et al. (1984)

Continued

Table 6 Continued

Black elutrilithe (coal waste)	Impregnation K_2CO_3/precursor weight ratio = 1 Pyrolysis *T* from 500°C to 1000°C Pyrolysis time = 3 h	From 59 to 423	From 0.074 to 0.219	From 0 to 0.159	Deng et al. (1997)
Almond shell	Impregnation K_2CO_3/precursor weight ratio = 1 Pyrolysis *T* from 500°C to 800°C Pyrolysis time = 1 h	From 250 to 1200	–	–	Hayashi, Horikawa, Takeda, Muroyama, and Ani (2002)
Coconut shell	Impregnation K_2CO_3/precursor weight ratio = 1 Pyrolysis *T* from 500°C to 800°C Pyrolysis time = 1 h	From 700 to 1700	–	–	Hayashi et al. (2002)
Effect of pyrolysis time					
Almond shell	Impregnation K_2CO_3/precursor weight ratio of 0.25 Pyrolysis *T* = 600°C Pyrolysis time from 80 to 480 min	From 664 to 796	–	–	Bevia et al. (1984)
Black elutrilithe (coal waste)	Impregnation K_2CO_3/precursor weight ratio = 1 Pyrolysis *T* = 850°C Pyrolysis time from 1 to 3 h	From 726 to 858	From 0.433 to 0.512	From 0.224 to 0.258	Deng et al. (1997)

Table 7 Acidic chemical posttreatment effect on textural properties measured by N_2 adsorption.

Raw material	Conditions	BET surface area ($m^2 g^{-1}$)	Pore volume ($cm^3 g^{-1}$)	Micropore volume ($cm^3 g^{-1}$)	Average pore diameter (nm)	References
Chemical activation with H_3PO_4						
Sewage sludge char	H_3PO_4/char = 10/1 Following thermal treatment: $T = 450°C$, 4 h, $N_2 = 100\,mL\,min^{-1}$,	<5	–	–	–	Ros et al. (2006)
Beer waste HTC	H_3PO_4/HTC = unspecified Following thermal treatment: $T = 600°C$, 1 h, N_2	1073	0.978	0.101	–	Hao, Björkman, Lilliestråle, and Hedin (2013)
Chemical activation with H_2SO_4						
Hardwood char	H_2SO_4/char = 18.4 Following thermal treatment: $T = 70°C$, 1 h, air	5.8	–	–	–	Dehkhoda, West, and Ellis (2010)
Peanut hull char	H_2SO_4/char = 3 Following thermal treatment: $T = 110°C$, 12 h, air	242 (1–4 before activation)	–	–	–	Kastner et al. (2012)
Peanut hull char	H_2SO_4/char = 3 Following thermal treatment: $T = 110°C$, 12 h, air	338	0.18	–	1.1	Kastner et al. (2012)

Continued

Table 7 Continued

Rice husk char	H_2SO_4/char = 18 Following thermal treatment: $T = 80°C$, 24 h, air	4	–	–	7.7	Li, Zheng, Chen, and Zhu (2014)
Douglas fir woodchips char	H_2SO_4/char = 27.6 Following thermal treatment: $T = 105°C$, 4 h, air	3.5	–	–	–	Dong et al. (2015)
Chemical activation with $ZnCl_2$						
Groundnut shell char (pyrolysis $T = 400°C$)	$ZnCl_2$/char = 1.75 Following thermal treatment: $T = 650°C$, 15 min, N_2	1200	–	–	–	Malik, Ramteke, and Wate (2007)
Effect of weight acid/biomass ratio						
Safflower seed char	$ZnCl_2$/char from 1/1 to 4/1 Following thermal treatment: $T = 900°C$, N_2	From 620 to 802	From 0.330 to 0.393	From 0.223 to 0.284	From 2.13 to 1.96	Angin (2013)
Effect of activation temperature						
Safflower seed char	$ZnCl_2$/char = 4/1 Following thermal treatment: from 600°C to 900°C, N_2	From 249 to 802	From 0.151 to 0.393	From 0.077 to 0.284	From 2.42 to 1.96	Angin (2013)

HTC, hydrothermal carbonization.

Table 8 Textural characteristics of chars from various biomasses measured by N_2 or CO_2 (between brackets) adsorption.

Raw biomass	BET surface area ($m^2\ g^{-1}$)	Micropore volume ($cm^3\ g^{-1}$)	Mesopore volume ($cm^3\ g^{-1}$)
Walnut shell	280	0.138 (0.458)	0.022
Almond tree pruning	204	0.097 (0.426)	0.021
Almond shell	42	0.023 (0.305)	0.071
Olive stone	53	0.028 (0.399)	0.008

Steam activation was for the first time performed by using a stream of moisture-laden nitrogen (Streat et al., 1995) after the carbonization step. High BET surface areas were obtained (1000–1700 $m^2\ g^{-1}$ for coconut shell and wood char, respectively) after activation at high temperature (800–900°C). Improvement in micropore volume percentage was particularly highlighted (El-Hendawy et al., 2001). The positive effect of steam activation (under various steam contents) was largely confirmed on various chars, as sewage sludge and rice straw (Cha et al., 2010), broiler litter manure (Uchimiya, Lima, Klasson, & Wartelle, 2010), brown alga (Cho et al., 2013), mulberry twig (Shu, Lu, & He, 2013), *Miscanthus* (Shim et al., 2015), and burcucumber and tea waste (Rajapaksha et al., 2016).

The influence of biomass type was studied (Gonzalez et al., 2009) on walnut shell, almond tree pruning, almond shell, and olive stone (Table 8).

Steam activation usually increases the BET surface area, but leads to different micropore extents. In fact, the activated char from almond tree pruning presented broader pore size distribution (greater volumes of meso- and macropores) than the activated chars derived from other residues, leading to highly microporous carbons (Table 8). This different behavior seems to be linked to the higher lignocellulosic composition of almond tree pruning and to the higher hemicellulose fraction of almond shell. The thermal decomposition rate of the biomass, depending on the lignocellulosic fraction, plays a determinant role on the porosity of the activated carbon produced. The pyrolysis temperature appears to have no effect (Uchimiya et al., 2010) below 700°C, but is an important parameter (Iwazaki, Yang, Obinata, Sugimoto, & Takasu, 2010) at temperatures higher than 900°C, leading to a decrease in the BET surface area and micropore volume percentage.

The influence of the steam activation temperature was reported by different authors (Azargohar & Dalai, 2008; Chang et al., 2000; El-Hendawy et al., 2001) and identified an optimum temperature for obtaining the highest BET surface area and percentage of micropore volume. This parameter appears to be related to the mass ratio S/C and to the activation time (Azargohar & Dalai, 2008) (Fig. 4).

The effect of steam activation duration is difficult to understand. Indeed, high BET surface area and total pore volume depend on both temperature and duration time of the steam activation step, and to share between the two contributions is not an easy matter (Azargohar & Dalai, 2008; Buah & Kuma, 2012; Chang et al., 2000; Fan,

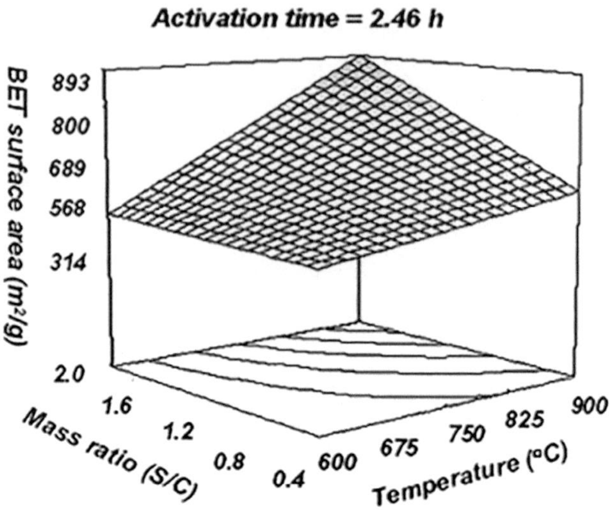

Fig. 4 Three-dimensional plot of BET surface area model of physically activated carbon. Reproduced with permission from Azargohar, R., & Dalai, A. K. (2008). Steam and KOH activation of biochar: experimental and modeling studies. *Microporous and Mesoporous Materials*, *110*, 413–421. Copyright 2018, Elsevier.

Marshall, Daugaard, & Brown, 2004; Ncibi et al., 2009; Youssef, Radwan, Abdel-Gawad, & Singer, 2005).

Few studies have examined the CO_2 activation of chars (Muradov, Fidalgo, Gujar, Garceau, & Ali, 2012) for catalysis application. An increase in activation temperature up to 900°C was reported (Chang et al., 2000; Zhang et al., 2004) to be positive on the BET surface area and total pore volume.

The effect of CO_2 activation duration was also studied (Chang et al., 2000; Zhang et al., 2004) and revealed a negative effect in the majority of cases (Zhang et al., 2004), but an optimum value of about 1 h was determined in other cases (Chang et al., 2000).

With respect to the micropore volume percentage, it appears that this fundamental property also depends on the activation temperature and time (Chang et al., 2000; Zhang et al., 2004) and on the type of biomass used (corn cobs (Chang et al., 2000), oak, corn hulls, and corn stover (Zhang et al., 2004)). The influence of CO_2 activation was also investigated (Hao et al., 2013) on various hydrochar types (horse manure, grass cuttings, beer waste, and biosludge). The authors reported different behaviors: grass cuttings led to a higher BET surface area and micropore volume, horse manure to the best total pore volume and lowest micropore volume percentage, and beer waste and biosludge showed intermediate behaviors.

A comparison of these two procedures (Chang et al., 2000; Rodríguez-Reinoso, Molina-Sabio, & González, 1995) indicates that the activation rate is higher for steam activation, whereas the specific surface area and pore size were higher for CO_2 activation. However, a difference was observed (Rodríguez-Reinoso et al., 1995) due to the formation of a lower micropore volume with steam than with CO_2 treatment.

8.2.3 Redox

Biomass-derived char displays higher redox properties than the parent biomass, and its redox properties may originate from both organic and inorganic constituents. Klüpfel, Keiluweit, Kleber, and Sander (2014) performed an in-depth study of the redox properties of various chars prepared from grass and woody biomasses, through the variation of pyrolysis thermosequence and temperature. The redox behavior of the obtained chars was then systematically investigated by means of mediated electrochemical analysis. The chars obtained at the highest pyrolysis temperature were those showing enhanced electron donor/acceptor properties. Phenolic groups were identified to be the main electron donating species, while quinones and polycondensed aromatic groups were identified more as oxidants (electron acceptor) (Klüpfel et al., 2014; Yuan et al., 2017).

Yuan et al. (2017), in a review article, summarized the different char synthesis methodologies and some specific technics for analyzing redox properties. The redox sites present on the char prepared at different pyrolysis temperatures are schematized in Fig. 5.

As previously described, the surface functionalities present on the chars' surface can affect the synthesis of supported catalysts. The group of Louloudi (Simaioforidou, Papastergiou, Margellou, Petrakis, & Louloudi, 2017) compared the immobilization of a nonheme Mn(II) complex onto commercially activated carbons (Mn^{II}-L_3imid@Cox) and pyrolytic carbon char (Mn^{II}-L_3imid@PCox). The catalyst prepared on the pyrolytic char presented equivalent performances (in terms of selectivity toward epoxides) than that prepared on commercially activated carbon in the alkene oxidation reaction by H_2O_2, in the presence of CH_3COONH_4, and even with higher kinetics (higher turnover frequency, Fig. 6), showing the promising application of chars as redox catalyst supports.

8.3 Chars as catalysts

The porous properties of catalysts are fundamental in catalysis, where the internal diffusion of the reactive molecules (favored by an enhanced porosity) and the amount of accessible active sites (on the catalyst surface) are crucial. Here, the ash content can also have a big impact on catalytic activity. In particular, certain accumulator plants (Lu et al., 2017), once pyrolyzed, give rise to chars that naturally contain metal oxides dispersed in the carbon structure and that can potentially act as catalysts in various reactions. More commonly, the percentage of inorganics such as sodium, magnesium, calcium, potassium, silicium, and phosphorus, concentrated in the chars after pyrolysis, strongly depends on the biomass source (Carpenter, Westover, Czernik, & Jablonski, 2014), and they can act as active sites in the field of biomass valorization for chemical production (Xiong et al., 2017), bio-oil upgrading (Li et al., 2014), tar reforming (Anis & Zainal, 2011; Bhandari, Kumar, Bellmer, & Huhnke, 2014; Klinghoffer, Castaldi, & Nzihou, 2015), hydrogen production (Yao et al., 2016), and also in electrochemical conversion and to a lesser extent in pollution control as in the selective catalytic reduction of nitrogen oxides (SCR-NO_x) (Wu, Nahil, Sun, et al., 2012).

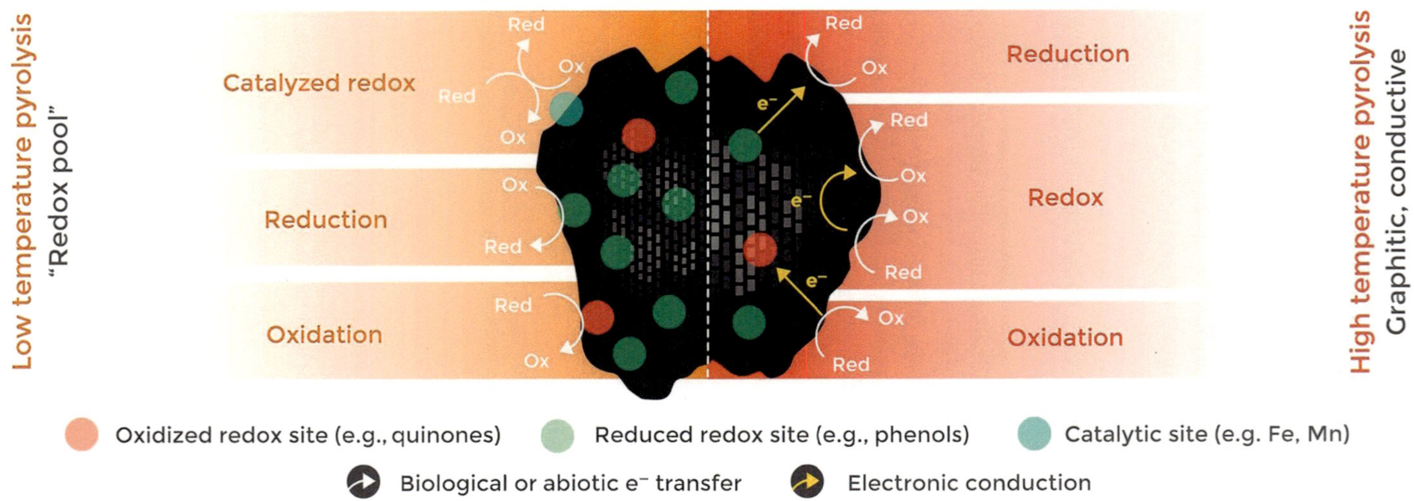

Fig. 5 Char redox behavior.
Reproduced with permission from Yuan, Y., Bolan, N., Prevoteau, A., Vithanage, M., Biswas, J. K., Ok, Y. S., et al. (2017). Applications of biochar in redox-mediated reactions. *Bioresource Technology*, *246*, 271–281. Copyright 2018, Elsevier.

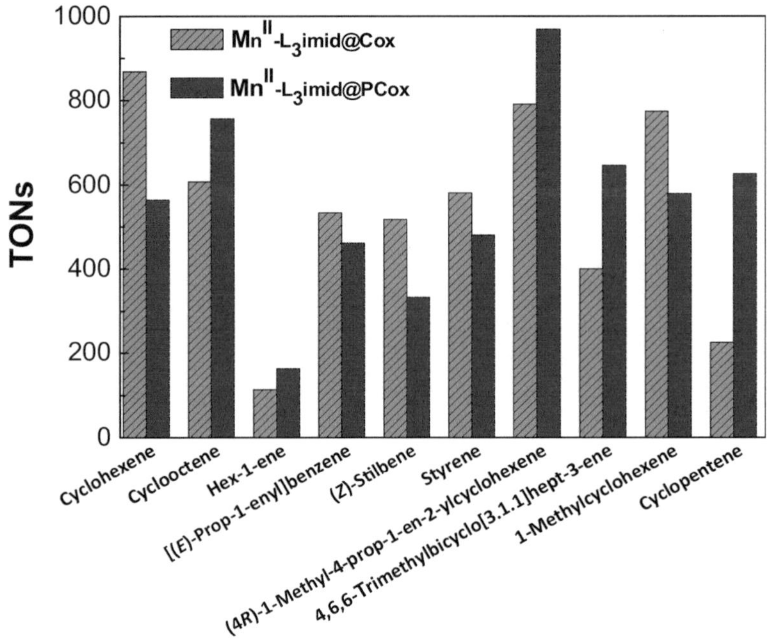

Fig. 6 Total turnover numbers (TONs) for alkene epoxidations catalyzed by Mn^{II}-L_3imid@Cox and Mn^{II}-L_3imid@PCox in the presence of H_2O_2.
Reproduced with permission from Simaioforidou, A., Papastergiou, M., Margellou, A., Petrakis, D., & Louloudi, M. (2017). Activated vs. pyrolytic carbon as support matrix for chemical functionalization: efficient heterogeneous non-heme Mn(II) catalysts for alkene oxidation with H2O2. *Journal of Molecular Catalysis A: Chemical*, *426*, 516–525. Copyright 2018, Elsevier.

8.3.1 Preparation methods

Metal addition can be also performed before or after char synthesis (presynthesis and postsynthesis methods, respectively) and can be achieved by impregnation, mixing, or sol-gel methods.

8.3.1.1 Presynthesis metal addition

The simplest and widely used method for biomass upgrading as a catalyst precursor is impregnation. Ni (from nitrate) is often used as the catalytic active phase, particularly for biomass pyrolysis and/or gasification (Richardson, Blin, Volle, Motuzas, & Julbe, 2010; Shen et al., 2014; Shen & Yoshikawa, 2014).

The pH of the solution used for impregnation is an important parameter to optimize both metal loading and dispersion on the biomass matrix (Richardson et al., 2010). When the pH is close to the pH of zero point of charge of char, nickel adsorption increases due to the negative charge density established on the biomass surface (woodchips) that favors the electrostatic attraction between the positive sorbate species and the adsorbent wood particles. Ni crystallite size depends on nickel loading, and was smaller than 4 nm for Ni content lower than 10 wt% (Fig. 7).

An impregnation method under reduced pressure (7 mbar) was also investigated (Richardson, Motuzas, Julbe, Volle, & Blin, 2013). This method permitted the

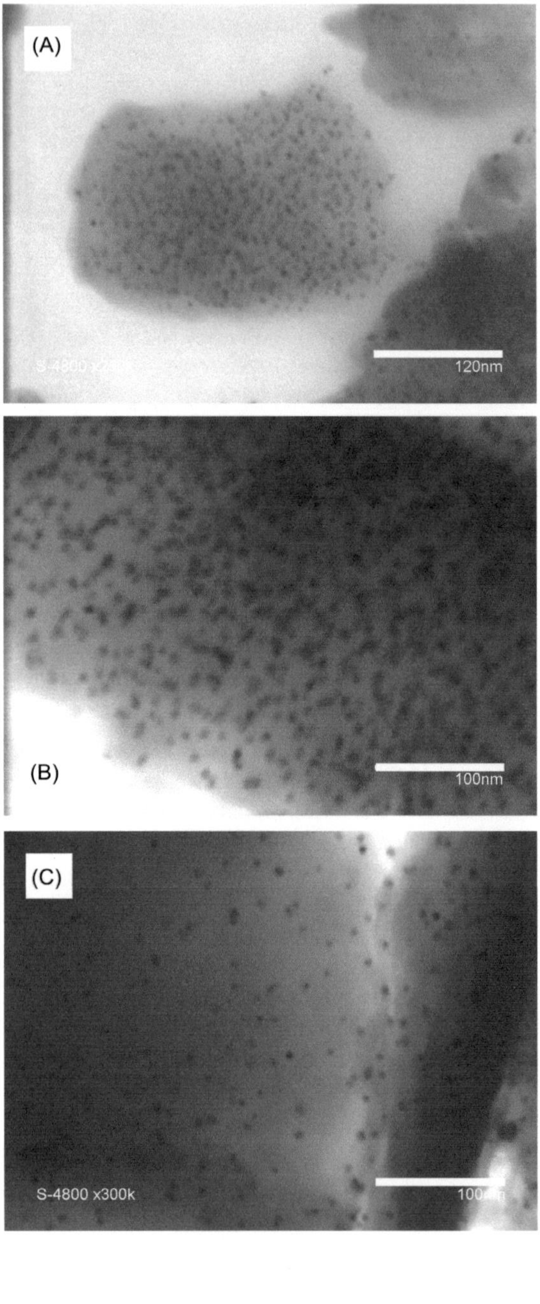

Fig. 7 STEM observations of (A) wood-Ni^{1} (9.9 wt% nickel content), (B) wood-$Ni^{0.5}$ (8.5 wt% nickel content), and (C) wood-$Ni^{0.1}$ (2.5 wt% nickel content), after pyrolysis at around 700°C. Reproduced with permission from Richardson, Y., Blin, J., Volle, G., Motuzas, J., & Julbe, A. (2010). In situ generation of Ni metal nanoparticles as catalyst for H2-rich syngas production from biomass gasification. *Applied Catalysis A: General*, *382*, 220–230. Copyright 2018, Elsevier.

generation of strong biomass-metal cation interaction, thanks to the formation of inner-sphere complexes with the oxygenated functions at the lignocellulosic surface. The same authors (Richardson et al., 2013) also investigated the use of a microwave-assisted polyol method to prepare Ni^0 char and to improve the catalytic properties. They succeeded in inserting Ni^0 into the biomass feedstock by ultrasound treatment and then depressurized it during biomass impregnation.

Iron is another transition metal often used in tar reforming and associated with char by the impregnation method (Zhang, Asadullah, Dong, Tay, & Li, 2013). The catalytic activity for tar reforming was identified to be linked to iron concentration but not directly proportional to the surface area and pore volume.

Tin can be coimpregnated with transition metals such as cobalt, copper, or iron starting from their chloride or hydroxide forms (Liu, Yang, et al., 2015) or mixed with them (Yu et al., 2016) to catalyze the hydrolysis of biomass.

8.3.1.2 Postsynthesis metal addition

To obtain char-based catalysts, impregnation of the metal active phase is the most used method. Applications of such catalytic materials are varied, such as the hydrodechlorination of chlorobenzene (Wu, Xu, & Ohnishi, 2005), the hydrogenation of benzene (Wojcieszak, Zielinski, Monteverdi, & Bettahar, 2006), and, more recently, the hydrodeoxygenation and hydrodenitrogenation of bio-oil on microalgae char (Nguyen, Pham, & Do, 2016) or biomass gasification for H_2 production (Yao et al., 2016). Ni nitrate and acetate are the most common precursors (Wojcieszak et al., 2006).

The effect of the support was studied by Wu et al. (2005) through the comparison of activated carbon derived from coconut shell (activated carbon) with classical catalyst supports such as γ-Al_2O_3 and SiO_2 in the chlorobenzene hydrodechlorination reaction. The high catalytic activity is due to the enhanced dispersion of nickel on the Ni/AC catalyst surface.

The effect of Ni salt (nitrate or acetate) was investigated in benzene hydrogenation by Wojcieszak et al. (2006). The authors reported that the reduction of nickel species occurs at lower temperature for the catalyst prepared with nitrate (450°C) compared to the one prepared with acetate (600°C), indicating a better Ni-support (activated carbon) interaction for the latter. On the contrary, the benzene conversion was largely higher for the catalyst prepared starting from the nitrate precursor (100% at 420°C) than for the catalyst prepared from acetate (20% at the same temperature).

The interest in metal addition and the influence of biomass feedstock (wheat straw, rice husk, cotton stalk) were validated (Yao et al., 2016) for the production of H_2 from biomass. The addition of Ni metal appeared to be more important than the morphology and structure of the catalyst for H_2 and gas production, and also the alkaline and alkaline earth metals content. To prepare a low-cost catalyst for tar reforming, Wang, Yuan, and Ji (2011) proposed a mechanical mixing of wood char and NiO. Tar removal in biomass gasification was proportional to the Ni loading (below 20 wt%) and led to 33 vol% of H_2 in the produced gas with 15 wt% of Ni compared to 28 vol% without Ni.

Iron was often reported to upgrade chars to catalysts for various applications such as dye removal (Yang, Zhu, Zhang, & Xu, 2008), phenol removal (Rey, Faraldos, Casas, Bahamonde, & Rodriguez, 2009; Zhang, Li, Li, & Hu, 2015), Fischer-Tropsch synthesis (Yan et al., 2013), or tar reforming (Kastner, Mani, & Juneja, 2015). The impregnation of iron nitrate remains the more common method, but iron pentacarbonyl (Rey et al., 2009) was also investigated as precursor. Iron pentacarbonyl obtained a homogeneous Fe distribution, but a catalyst with lower stability than that prepared with iron nitrate. However, rice husk pyrolytic char was impregnated (Zhang et al., 2015) with ferric nitrate solution, dried, and then activated under N_2 atmosphere at various temperatures between 400°C and 850°C to study the impact of the temperature. The optimum activation temperature was identified at 750°C leading to the highest BET surface area (480 $m^2 g^{-1}$), to an improved iron oxide particle (C-Fe-750-1.0) distribution in the porous structure, and enhanced dispersion on iron oxide on the surface of the activated carbon (smooth surface) during the impregnation process.

Yan et al. (2013) used a char from fast pyrolysis of pinewood as the support material to develop a carbon-encapsulated iron nanoparticle catalyst for the Fischer-Tropsch synthesis of biomass-derived syngas to liquid hydrocarbons. It presented very interesting results in terms of activity and stability due to the strong metal-carbon support interaction created during the thermal treatment at 1000°C, leading to the formation of very stable core-shell nanocomposites. The total conversion of toluene (studied as a tar model compound) was reported by Kastner et al. (2015) at 800°C on Fe/chars containing more than 10 wt% of Fe, while only 25% of toluene conversion was obtained on the bare char. At lower temperatures (600 and 700°C), toluene conversion was proportional to iron loading (up to 19 wt%). Iron and nickel were compared and their synergetic effect studied (Shen et al., 2014) in tar reforming. The highest performance (93% conversion) was obtained with Ni-Fe (equimolar)/char prepared by impregnation of Ni and Fe nitrates on rice husk-derived char.

Manganese was also largely used in low-temperature selective catalytic reduction (SCR) of NO. The effect of Mn impregnation was different for the various studied chars (rice straw and sludge) and the changes in specific surface area and pore volume were not dramatic (Cha et al., 2010). Cerium can be associated with Mn in char-based catalysts for low-temperature SCR of NO (Singh et al., 2013), by the coimpregnation method. Chemical activation by H_3PO_4 and physical activation by steam were also investigated to improve the char's catalytic properties (Shen et al., 2015; Singh et al., 2013).

Other metals such as Pt, Cu, and Ru were studied in applications such as the electrocatalytic oxidation of methanol and ethanol (Fan et al., 2015; Lobos et al., 2016). Pt chloride was impregnated on a hydrochar derived from pectin (Fan et al., 2015). Ru chloride and Cu sulfate were coimpregnated on chars derived from untreated and acid-treated cellulose (Lobos et al., 2016).

In the photocatalytic degradation field, the association of TiO_2 (from titanium isopropoxide (Kim & Kan, 2016) or tetrabutyl orthotitanate (Li, Zhang, Yu, & Yin, 2007)) and char was performed by the sol-gel method. Comparison with TiO_2 alone proved that the char/TiO_2 presented a much higher adsorption of sulfamethoxazole in water than the commercial TiO_2 powder, due to the hydrophobic interaction between the char and the sulfamethoxazole (Kim & Kan, 2016). The spherical-shaped TiO_2 particles, well dispersed in the char matrix, and the low crystallite size of TiO_2 (below 40 nm) created anticalcination effects (Li et al., 2007) (Fig. 8).

Shen and Yoshikawa (2014) proposed a comparison of pre- and postsynthesis addition of Ni (nitrate) by impregnation of rice husk and rice husk char. They concluded that the BET surface area of the catalyst prepared by postsynthesis was largely lower ($87\,m^2\,g^{-1}$) than that prepared by presynthesis ($135\,m^2\,g^{-1}$) for the same Ni content. Furthermore, the addition of $NaBH_4$ in the impregnation solution increases the BET surface area, improves the char porosity, and promotes the generation of metallic nickel (Ni^0) from the reduction of nickel oxides. Finally, the catalyst prepared by presynthesis in the presence of $NaBH_4$ led to the best textural properties and reactivity in tar reforming, particularly for the main organic compounds such as toluene, benzene, naphthalene, phenol, etc.

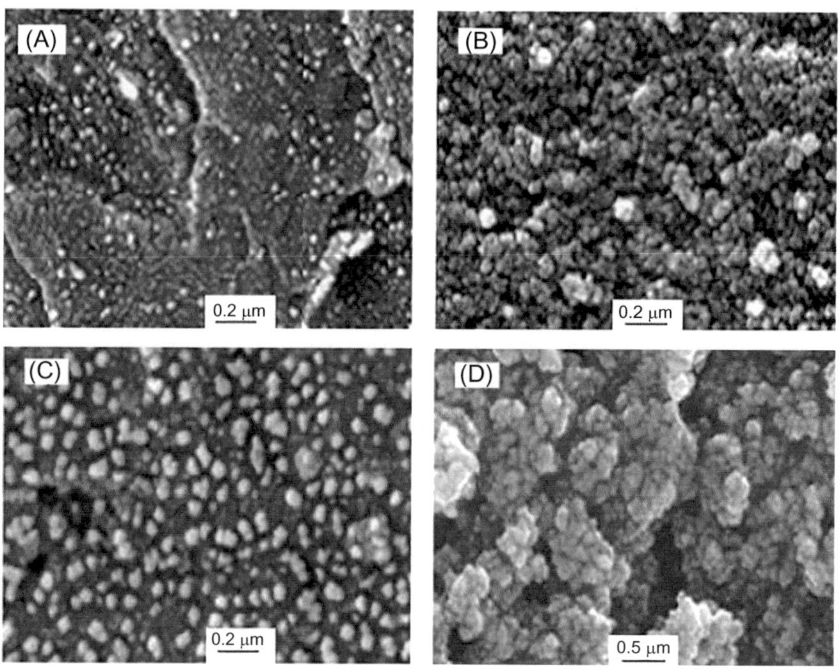

Fig. 8 SEM micrographs of samples prepared at different treatment temperatures: (A) TiO_2/activated carbon at 400°C, (B) TiO_2 at 400°C, (C) TiO_2/activated carbon at 500°C, and (D) TiO_2 at 500°C.
Reproduced with permission from Li, Y., Zhang, S., Yu, Q., & Yin, W. (2007). The effects of activated carbon supports on the structure and properties of TiO2 nanoparticles prepared by a sol–gel method. *Applied Surface Science*, *253*, 9254–9258. Copyright 2018, Elsevier.

8.3.2 Reactions

8.3.2.1 Biochemical conversion (valorization) of biomass

The synthesis of various platform chemicals and biofuels (Fig. 9) is largely studied from lignocellulosic biomass (instead of fossil resources) to meet the energy, food, and materials needs of a rising global population.

Valorization of biomass or its biopolymers (cellulose, hemicellulose, and lignin) needs several transformation reactions, including hydrolysis, isomerization, dehydration or rehydration, hydrogenation, and oxidation that can be performed with biomass-derived char-based catalysts.

Hydrolysis

Large efforts have been deployed to develop hydrolysis processes of glucose from cellulose, such as catalysis with mineral acids (sulfuric acid, hydrogen chloride, and phosphoric acid), enzymatic hydrolysis, hydrolysis in supercritical water, and solid catalysts. Hydrolysis with sulfuric acid is the only process developed in relatively large scale, due to the lower cost and higher reaction rate compared to enzymatic hydrolysis and other treatments. The replacement of strong acids by solid acidic catalysts (such as zeolites, solid superacids, sulfonated carbon nanostructures, sulfated zirconia) was largely studied due to many advantages such as the low corrosion power and reusability (Lazaridis et al., 2015).

The catalytic hydrolysis of cellulose with carbon solid acids prepared by sulfonation of partially carbonized crystalline cellulose is similar to saccharification using concentrated H_2SO_4 (Yamaguchi et al., 2009). The apparent activation energy for the hydrolysis of cellulose into glucose using the carbon-based catalyst is smaller than for sulfuric acid (Suganuma et al., 2008). The carbon catalyst can hydrolyze not only pure crystalline cellulose, but also natural lignocellulosic reactants, such as eucalyptus or straw tips, in the same way that mineral acid catalysts can (Yamaguchi et al., 2009). Strong acidic SO_3H groups bonded to the carbon material function as active sites capable of breaking the strong hydrogen bonds and hydrolyzing the β-1,4 glycosidic bonds of cellulose or water-soluble β-1,4 glucan; a good conversion of cellulose into glucose was obtained with these catalysts (Yamaguchi et al., 2009).

Biomass-derived char activated by chemical treatment (by H_2SO_4) showed enhanced catalytic performances compared to sulfuric acid (Li, Gu, Bjornson, & Muthukumarappan, 2013), with higher hydrolysis activity and xylose and glucose selectivity. The char solid acid also generated fewer soluble lignin-derived products than sulfuric acid. These catalytic performances are attributed to the ability of the material to adsorb β-1,4 glucan. Furthermore, the char solid acid catalyst can be readily separated from the saccharide solution after reaction and reused in the reaction without loss in activity (Suganuma et al., 2008). A char solid acid catalyst (peanut hulls or pinewood char chemically activated by H_2SO_4) was compared to a wood-based activated carbon (Ormsby, Kastner, & Miller, 2012). The char presented a specific surface area significantly lower than the activated carbon (365 vs. 1391 $m^2\,g^{-1}$, respectively), but a hydrolysis rate significantly higher (85% conversion of xylan for

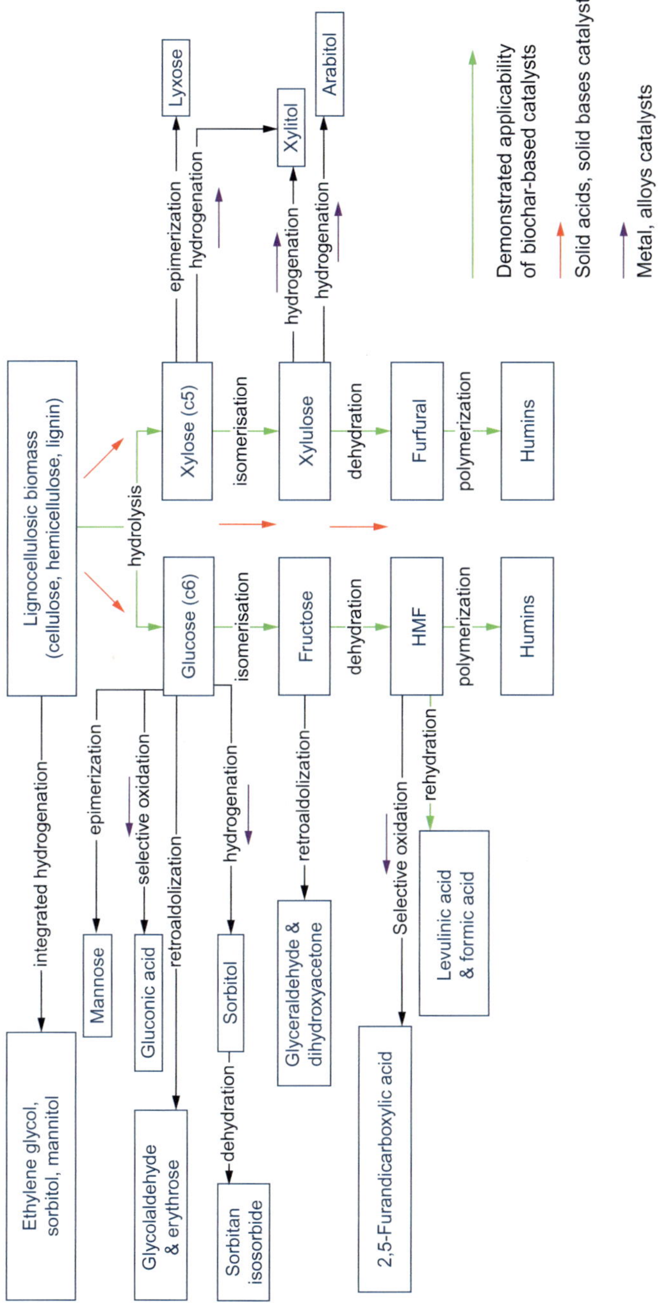

Fig. 9 Schematic presentation of biochemical processes in lignocellulosic biomass conversion. *HMF*, 5-hydroxymethylfurfural.
Based on Xiong, X., Yu, I. K. M., Cao, L., Tsang, D. C. W., Zhang, S., & Ok, Y. S. (2017). A review of biochar-based catalysts for chemical synthesis, biofuel production, and pollution control. *Bioresources Technology*, *246*, 254–270. Copyright 2018, Elsevier.

the char and 57% for the activated carbon). Compared to a commercially sulfonated styrene-divinylbenzene macroreticular resin (Amberlyst-15 resin), with relatively low surface area ($38\,m^2\,g^{-1}$) and limited pore structure, the char hydrolysis rate was significantly higher at low temperature (90–110°C), because of the lower adsorption capacity of Amberlyst-15 resin for xylan. Regarding the hydrolysis of cellobiose (Do Couto Fraga et al., 2016), it was reported that a relationship existed between the concentration of sulfonic acid groups and the higher activity of char chemically activated by H_2SO_4. The hydrolysis of holocellulose (cellulose and hemicelluloses) conversion into sugars was investigated (Wang, Jiang, Xu, Feng, & Wang, 2016), using sulfonated char catalysts derived from microcrystalline cellulose, hemicellulose, lignin, and lignin-rich residues. The sulfonated char catalysts, derived from lignin and residues, possess a graphene-like fused ring structure with —SO_3H, —COOH, and phenolic —OH groups. The synergetic combination of the fused ring structures and multifunctional groups in the sulfonated char catalysts is identified to be linked to the remarkable hydrolysis performance of holocellulose.

Isomerization

Xylose/xylulose and glucose/fructose are important chemical intermediates (Zhang, Wilson, & Lee, 2016). The *isomerization* of xylose or glucose was largely explored with solid Lewis acids, hydrotalcite solid bases, and metallosilicate solid bases. The use of char for this step is seldom reported, but it was demonstrated (Liu, Yang, et al., 2015) that in sulfonated char and mineral-impregnated (SnO_2-Co_3O_4) char catalyst, the multivalent metal (Lewis site) facilitates the isomerization of xylose, and the Brønsted acidity promotes the hydrolysis of the glycosidic bond.

Dehydration

The *dehydration* step (for the production of furfural or derivatives products) of C5-C6 sugars is conventionally catalyzed by liquid acids (such as H_2SO_4, HCl, H_3PO_4, oxalic, and maleic acid in aqueous solution), but they cause corrosion, toxicity, and difficulties for separation and product recovery (Zhang, Wilson, & Lee, 2016). Solid acid catalysts (such as zeolites, sulfonic acid derivative solid acids, sulfated metal oxides, metal phosphates, composite metal and nonmetal oxides, and heteropolyacids) are then largely studied as alternatives. The dehydration step was recently studied with sulfonated char for xylose dehydration to furfural (Sairanen, Vilonen, Karinen, & Lehtonen, 2013), or on a liquor of prehydrolyzed corn cob (Deng et al., 2016), and for the dehydration of fructose to HMF (Zhao et al., 2016). Furfural selectivity level was found to be higher over the functionalized carbon catalyst (chemically activated by H_2SO_4 and/or HNO_3) compared to other solid acid catalysts such as TiO_2 and H-mordenite zeolite (Sairanen et al., 2013). The combination of oxygen-containing and sulfur-containing surface sites on carbon seems to be the reason for the highest furfural selectivity. A sulfonated carbon-based solid acid catalyst bearing —SO_3H, —OH, and —COOH groups synthesized by a two-step hydrothermal method (Zhao et al., 2016) exhibited a high acid density that accelerated the reaction and improved the selectivity to HMF. An excessive amount of acid sites and high temperature resulted in the slight decrease in HMF selectivity due to the further condensation of HMF into by-products. The effectiveness of sulfonated char catalyst was demonstrated (Cao et al., 2018), directly correlated to the density of strong/weak Brønsted acidity (—SO_3H, —COOH, and —OH groups), and inversely correlated to humins content on the surface.

Rehydration

Rehydration is an important step leading to levulinic acid, a key platform chemical of polymers and fuels (Xiong et al., 2017). Even if char-based catalysts were not used for the production of levulinic acid, their modification to form acidic sites could be efficient for this application.

Hydrogenation

A metallic component (such as Raney Ni and Ru, Pt, Co, Cu, and Rh) can catalyze the *hydrogenation* of glucose, as well as the hydrogenolysis of glucose oligomers (Lazaridis et al., 2015; Zhang, Wilson, & Lee, 2016) to form sorbitol. Lazaridis et al. (2015) studied Pt- and Ru-based micro/mesoporous activated carbon (Norit SAE SUPER) with various contents of metal and different metal pretreatments (as a reduction by H_2 or $NaBH_4$). Although both Ru- and Pt-based catalysts were very active, the Pt-containing catalyst was also very selective toward sorbitol. Ru/ZSM-5 and Pt/activated carbon appear to be the best catalysts for glucose and xylose hydrogenation (Zhang et al., 2016). Pt was also associated with activated carbon cloths (Perrard et al., 2007) by metal cation exchange. It exhibited high activity in D-glucose hydrogenation and higher yields in D-sorbitol ($>$99.5%) than the Pt catalysts prepared on carbon powder or extrudates.

The one-pot/direct transformation of cellulose into sugar alcohols (such as sorbitol, glycerol, etc.) by combined *hydrolysis/hydrogenolysis* reactions is of great interest and needs bifunctional catalysts with acidic functionalities for cellulose hydrolysis, and metallic active sites for the hydrogenation of glucose and the hydrogenolysis of glucose oligomers (Lazaridis et al., 2015).

Oxidation

Selective *oxidation* of glucose was widely reported on metal catalysts (Zhang et al., 2013). To improve the performance of Pd and Au catalysts, new supports such as metal oxides, polymers, zeolites, carbon, resin, cellulose, etc., supported by bi- and trimetallic alloys (Zhang, Wilson, & Lee, 2016), have been investigated. Gold on activated carbon is recognized as one of the more active catalysts for this reaction, showing high productivity and easy recovery, compared to the biochemical route in the presence of enzymes (Comotti, Della Pina, Falletta, & Rossi, 2006).

Bio-oil upgrading (esterification and transesterification)

Transesterification concerns the reaction between a triglyceride and an alcohol in the presence of a catalyst, which produces a complex mixture of fatty acid methyl esters and glycerol as follows (Eq. 1).

$$
\begin{array}{lcccccc}
CH_2\text{–}OOC\text{–}R_1 & & & & R_1\text{–}COO\text{–}R' & & CH_2\text{–}OH \\
| & & & \text{Catalyst} & & & | \\
CH\text{–}OOC\text{–}R_2 & + & 3R'OH & \Leftrightarrow & R_2\text{–}COO\text{–}R' & + & CH\text{–}OH \\
| & & & & & & | \\
CH_2\text{–}OOC\text{–}R_3 & & & & R_3\text{–}COO\text{–}R' & & CH_2\text{–}OH \\
 & & & & & & \\
\text{Glyceride} & & \text{Alcohol} & & \text{Esters} & & \text{Glycerol}
\end{array}
\tag{1}
$$

In the last decade, the transesterification of various oils was studied to produce biodiesel, with a benign environmental footprint. Consequently, much research have been focused on solid base catalysts, such as calcium oxide, for transesterification of vegetable oils with methanol (Dong et al., 2015; Kouzu, Kasuno, Tajika, Yamanaka, & Hidaka, 2008). During the reaction, CaO is transformed into calcium diglyceroxide, due to the combination with glycerol (a by-product), without loss in activity and stability. Calcium ethoxide was also proposed (Liu, Plao, Wang, & Zhu, 2008) as catalyst for the transesterification of soybean oil to biodiesel with methanol and ethanol and was demonstrated to be better than CaO catalyst.

More recently, char with sulfonic acid groups was prepared as heterogeneous catalyst for transesterification of canola oil with methanol (Yu, Dehkhoda, & Ellis, 2011). The reaction yield was dependent on both catalyst surface area and total acid density. Calcium oxide-based catalyst, derived from palm kernel shell gasification residue (Bazargan, Kostic, Stomenkovic, Veljkovic, & McKay, 2015), valorized the most difficult fraction of the solid waste rich in calcium (50 wt% of Ca). The various biological sources for the preparation of CaO catalysts and their applications in biodiesel production were reviewed by Shan, Zhao, Lv, Yuan, and Yao (2016). A char physically activated by steam, then chemically activated by KOH, was also proposed (Dhawane, Kumar, & Halder, 2015) as catalyst for the transesterification of *Hevea brasiliensis* oil into biodiesel. The optimum conditions were identified as 1.5 h at 350°C for the physical activation of char, leading to the best surface and porosity properties (surface area of $812\,m^2\,g^{-1}$ and total pore volume of $0.325\,cm^3\,g^{-1}$). Then, the activated char was treated with various KOH/char ratios. The optimum catalyst loading (3.5 wt%) and methanol-to-oil ratio (1:15) were the most important parameters. A review of biomass-derived heterogeneous catalyst for sustainable biodiesel production was recently proposed (Abdullah et al., 2017). It appears that heterogeneous catalysts, capable of overcoming the problems connected with the use of homogeneous catalysts, were mostly derived from nonrenewable resources and presented high cost and low stability. Recently, heterogeneous catalysts, derived from biological wastes, and that offer several advantages (renewable resources, nontoxic, reusable, high catalytic activity, stability in both acidic and basic conditions, and high water tolerance properties), were proposed to provide a greener synthesis route for biodiesel production. Thus, a high biodiesel yield of 98.6% was obtained by using a peat char-supported catalyst in the transesterification of palm oil (Wang, Zhao, Shan, Wang, & Yuan, 2017). This good result was related to the very high total basicity of the catalyst. The transesterification of waste cooking oils was evaluated in a microwave reactor with a sulfonated char and the increase in $—SO_3H$ groups on the char surface was identified as being responsible for the higher fatty acid methyl ester

Table 9 Biodiesel production versus total acidity of different catalysts.

Catalyst	Total acid density (meq/g)	%FAME yield
Char derived from oat hull by slow pyrolysis at 600°C	not detected	0.5 ± 0.24
Char sulfonated with concentrated H_2SO_4 in a microwave reactor at 100°C for 30 min	3.74 ± 0.66	27.87 ± 0.10
Char sulfonated with concentrated H_2SO_4 in a microwave reactor at 140°C for 30 min	7.00 ± 1.18	72 ± 0.61

From Gonzalez, M. E., Cea, M., Romero-Hermoso, L., Hidalgo, P., Meier, S., Benito, N., et al. (2017). Functionalization of biochar derived from lignocellulosic biomass using microwave technology for catalytic application in biodiesel production. *Energy Conversion and Management*, *137*, 165–173.

(FAME) yield (Gonzalez et al., 2017) as reported in Table 9, where %FAME yield is the fatty acid methyl esters yield.

The esterification of fatty acids can also be catalyzed using solid acid catalysts as sulfonated chars or activated carbons (Kastner et al., 2012). This study demonstrated that, even if the sulfonation of the char by H_2SO_4 clearly increases the surface area and the porosity, the acid density is lower than that obtained by SO_3 sulfonation. Wood-based activated carbon sulfonated with SO_3 leads to the highest esterification activity and stability, due to improved particle strength, enhanced hydrophobicity, high surface area, and sulfonic acid group density. The activity loss detected for the sulfonated char was related to water adsorption, particle attrition, and strong acid site leaching. However, a sulfonated char was reported to be a stable catalyst for the preesterification of refined microalgae oil (Dong et al., 2015), exhibiting higher catalytic activity than Amberlyst-15. The activities of green sulfonated carbon-based catalyst from carbonized coffee residue were also higher than that of Amberlyst-15 (Ngaosuwan, Goodwin, & Prasertdham, 2016). A high catalytic activity and catalyst stability for esterification were achieved with this high surface area catalyst, which presented strong (—SO_3H) and weak (carboxylic and phenolic) acid site densities.

The transesterification of vegetable oils and the esterification of free fatty acids can be catalyzed on the same sulfonated catalyst, obtained with concentrated sulfuric or fuming sulfuric acid (Dehkhoda et al., 2010). The catalyst with the highest surface area and acid density (sulfonated for 15 h with fuming sulfuric acid) showed the highest catalytic activity in biodiesel production from canola oil and methanol. The catalyst surface area and porosity were optimized by chemical activation with KOH and sulfonation by fuming sulfuric acid (Dehkhoda & Ellis, 2013). A solid acid catalyst was prepared by sulfonating pyrolyzed rice husk with concentrated sulfuric acid. It was then compared to typical solid acid catalyst (Amberlyst-15) in the simultaneous esterification and transesterification of waste cooking oil to produce biodiesel (Li et al., 2014). The biomass-based catalyst led to higher free fatty acid conversion and FAME yield with an excellent stability compared to Amberlyst-15 (Fig. 10).

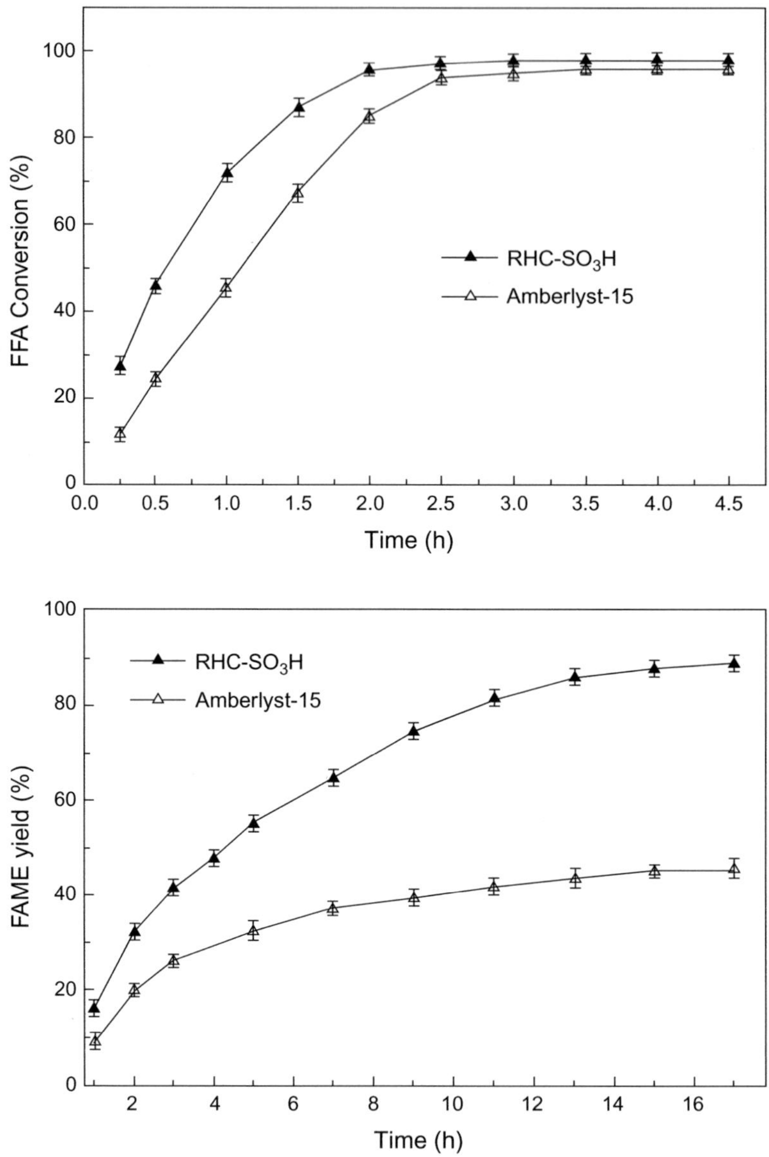

Fig. 10 Esterification activity of sulfonated pyrolyzed rice husk (RHC-SO_3H) and Amberlyst-15 (A) and FAME yield from esterification and transesterification catalyzed by RHC-SO_3H and Amberlyst-15 (B). *FFA*, free fatty acid.
Reproduced with permission from Li, M., Zheng, Y., Chen, Y., & Zhu, X. (2014). Biodiesel production from waste cooking oil using a heterogeneous catalyst from pyrolyzed rice husk. *Bioresource Technology*, *154*, 345–348. Copyright 2018, Elsevier.

8.3.2.2 Thermochemical conversion (valorization) of biomass

To convert biomass feedstocks into syngas, for the production of high added-value molecules (platform molecules, fine chemicals, biofuels, etc.), pyrolysis and gasification techniques have been developed during the last few decades (Ammendola, Cammisa, Chirone, Lisi, & Ruoppolo, 2012; Nino, Maria, Khaled, & Mohand, 2013). During these processes, tar is also produced as by-product, and catalytic methods have been developed to convert it into valuable products or additional syngas.

Tar reforming/cracking (syngas production)

Catalytic tar reforming is used to transform the heterogeneous hydrocarbon mixture deriving from gasification and pyrolysis into syngas. In addition to conventional catalysts (as transition-metal catalysts, alkali-metal catalysts, and zeolites (Anis & Zainal, 2011; Guan, Kaewpanha, Hao, & Abudula, 2016)), biomass-derived chars have also been used as catalysts or catalyst supports since it is possible to combine them in the same gasifier used for the reaction, thus considerably decreasing the costs (Abu, Bramer, & Brem, 2008; Li et al., 2016; Wang et al., 2011; Widayatno et al., 2014).

Many authors have shown the high potential of bare chars in tar reforming. The minerals content and their distribution in the char are important features, contributing to the catalytic activity, as shown by Klinghoffer et al. (2015). Zhang, Wilson, and Lee (2016) reported on the use of chars, produced by the pyrolysis of mallee wood, as catalyst in tar reforming during the gasification process). The performance of two kinds of catalysts, char and acid-washed char, was compared in a pilot plant. During this investigation, both samples showed good activity in tar reforming, even if the acid-treated catalyst was slightly less efficient than the raw char.

Carbon structure and morphology are also important characteristics for interpreting char reforming catalytic activity. Song et al. (2014) compared the catalytic behavior of various chars, and focused their attention on the positive effect of the presence of defects and unstable chemical bonds in the chars' structure that acted as active sites able to anchor and transform tar.

Chars have also been successfully used as reforming catalyst supports. Syed et al. (2016) prepared char-supported nickel and cobalt catalysts tested in the steam reforming of toluene, used as a tar model molecule. The char precursor was constituted of palm kernel shell and its stability was shown to be good up to 850°C, a temperature at which the agglomeration of the metal particles, the loss of alkali and alkaline earth metal, and the coke deposition take part. Shen and Yoshikawa (2014) performed in situ tar conversion with very high tar conversion (about 93%) during gasification by using a rice husk char as support for nickel and iron species.

Shen reported on tar removal by chars combining adsorption and catalytic cracking/reforming. A newly integrated concept, in which saturated chars, modified during adsorption of heavy metals or tars, were then used as char-supported metallic catalysts for tar conversion, was designed (Shen, 2015).

Although the use of chars as catalyst supports is promising, many issues connected to fast deactivation have to be urgently solved. In fact, during the reforming process the chars lose oxygen species due to ring condensation (Min et al., 2013), and the metal species are then relocated and agglomerated. Metal dispersion then decreases, negatively impacting the catalytic activity. The active sites are also deactivated by the deposition of coke on the catalyst surface that poisons the metal sites, blocks the char pores, and reduces the surface area (Bhandari et al., 2014; Shen & Yoshikawa, 2014; Syed-Hassan & Fuadi, 2016).

Hydrocarbon/oil reforming (hydrogen production)

A further step in biomass valorization concerns the production of hydrogen, an energy vector that represents a sustainable alternative to fossil fuels. Hydrogen can be obtained by further reforming of the hydrocarbons and bio-oils produced during pyrolysis, gasification, and tar reforming processes.

Ma, Xiao, and Zhang (2017) reached a high hydrogen concentration (up to 89%) by using an alkali- and alkaline metallic-rich char as catalyst for the steam reforming of bio-oils for the production of a hydrogen-rich gas. Another char type (a hydrochar obtained starting from a macroalgae, the *Cladophora glomerata*) was applied for the catalytic hydrothermal gasification of the same algae. The presence of the hydrochar promoted hydrogen production in the gas phase when compared to the process carried out in the absence of a catalyst (Safari, Norouzi, & Tavasoli, 2016). Yao et al. (2016) used different chars as catalyst and catalyst support (Ni-based char) for biomass gasification and noticed the influence of the char parent biomass (commercially activated carbon, wheat straw char, rise husk char, and cotton stalk char) on hydrogen production; the Ni-based catalyst, prepared on cotton-derived char, showed the highest hydrogen production (around 64% H_2 concentration in the outstreaming gas), while the rice-derived char was less active (Yao et al., 2016) (Fig. 11).

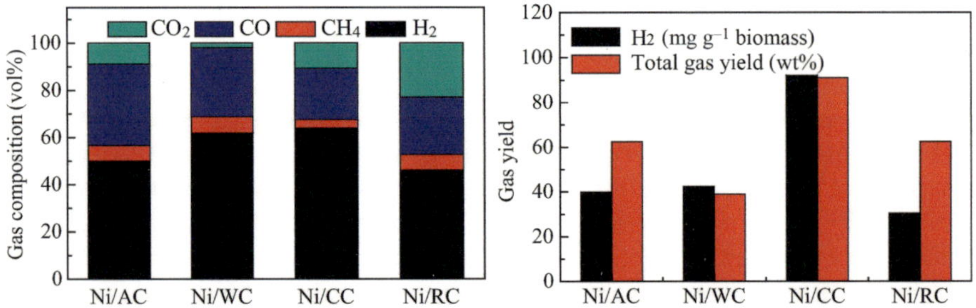

Fig. 11 Gas composition and gas yield of biomass with different Ni/char catalysts. *AC*, Commercially activated carbon; *CC*, cotton stalk char; *RC*, rise husk char; *WC*, wheat straw char.

Reproduced with permission from Yao, D., Hu, Q., Wang, D., Yang, H., Wu, C., Wang, X., et al. (2016). Hydrogen production from biomass gasification using biochar as a catalyst/support. Bioresource Technology, *216*, 159–164. Copyright 2018, Elsevier.

8.3.2.3 Electrochemical conversion

In the electrochemical conversion field, chars have also been studied as support to replace the high surface area carbon black originated from the fossil fuel industry, by a process with low efficiency (30% effective carbon conversion) and high energy consumption (Guilminot et al., 2008).

Char was used to support Pt or Fe species in proton-exchange membrane fuel cell electrocatalysts (Guilminot et al., 2008), in microbial fuel cells (MFCs) (Liu et al., 2015; Ma et al., 2014; Yuan et al., 2013) for the oxygen reduction reaction (ORR) (Wang et al., 2015), and in alcohol electro-oxidation (Fan et al., 2015; Lobos et al., 2016; Zhou, Wang, Ji, Feng, & Wang, 2014).

Oxygen reduction

The advantages of biomass-derived chars as metal support are the high surface area, enhanced microporosity, low cost (in particular when deriving from biomass wastes), and eventual high nitrogen content. In this context, an ultralight and highly porous pure cellulose material prepared from cellulose gels by drying in carbon dioxide in supercritical conditions (aerocellulose) was investigated (Guilminot et al., 2008) to support platinum. The metallic active phase was added following various procedures and compared to a commercial Pt/Vulcan XC72R catalyst in the ORR. Pt/aerocellulose materials exhibit ORR-specific activities comparable to those of the commercial catalysts. Chars are also used for the development of cathodes in MFCs and are associated with Fe_3O_4 (Ma et al., 2014). Ma et al. (2014) reported that waste cornstalks and pomelo skins can be used as carbon sources for the development of iron oxide nanocomposites/partly graphitized carbon (Fe_3O_4/PGC-CS and Fe_3O_4/PGC-PS). The improvement of the BET surface area, the electrical conductivity, and the catalyst's density were obtained by optimizing the pyrolysis temperature and by introducing Fe species, respectively. The formulation of the MFC led to a mixed composite (Fe_3O_4/PGC-CS:Fe_3O_4/PGC-PS = 1:1) cathode, generating the highest power density (also higher than that of a Pt/C cathode) and was stable during 18 cycles (Ma et al., 2014).

The low cost of this type of support also motivates investigations into the development of cathode catalysts of polymer electrolyte fuel cells for the ORR (Iwazaki, Obinata, Sugimoto, & Takasu, 2009). The catalytic activity of carbonized silk fibroin, which is free of metallic elements, for the ORR was greatly enhanced by steam activation (Iwazaki et al., 2010). It was demonstrated that the introduction of nitrogen (by a nitrogen-containing precursor such as NH_3) into carbon materials enhances the catalytic activity in the ORR. Sewage sludge chars were used for their abundant microporosity and their high nitrogen and iron contents (Yuan et al., 2013). In an MFC, the maximum power density was obtained from the cathode prepared with a char pyrolyzed at 900°C. The obtained power density was comparable to that of a conventional Pt cathode. Other research has proposed to develop nitrogen- and phosphorus-doped carbons by pyrolysis of cellulose, using ammonium phosphate as doping source, and to employ them as ORR catalysts in MFC cathodes (Liu, Zhou, et al., 2015). Wang et al. (2015) prepared N-carbon catalysts from okara (a cheap, nitrogen-rich, biomass precursor) using a simple synthesis method where the addition of $FeCl_3$ was performed at different steps. They established a correlation between N-species (pyridinic-N and graphitic-N) and ORR activity. The mesoporous N-carbon catalyst showed an ORR activity comparable to that of a commercial Pt/C catalyst.

Methanol and ethanol oxidation

Recent research has reported on carbon-derived materials, prepared by carbonization of Okara-Derived Carbon (ODC), and used as metal support for Pt and applied in alcohol electro-oxidation, due to their mesoporous structure and large specific area (Zhou et al., 2014). Higher electrocatalytic activity and resistance to poisoning of ODC than Pt/carbon catalyst were reported in methanol electro-oxidation (Zhou et al., 2014). Carbonaceous materials prepared by fast pyrolysis of cellulose (untreated and acid treated) were also studied as catalyst support for producing Cu-Ru@Pt core-shell nanoparticles applied in the electro-oxidation of methanol (Lobos et al., 2016) in acid media. The electrocatalytic performances observed for trimetallic catalysts supported on acid-treated cellulose were shown to be promising when compared to other electrocatalysts, with similar or higher Pt loading, on conventional carbon supports. Porous carbon spheres (PCSs), prepared via the hydrothermal carbonization and chemical activation route of pectin, were reported (Fan et al., 2015) to present a large specific surface area, a well-balanced micro/mesoporosity with narrow pore size distribution, and a high content of oxygen-containing functional groups. These properties are indeed very important for Pt electrocatalyst supports. A three-dimensional hybrid catalyst material exhibited high catalytic activity and stability toward both methanol and ethanol electro-oxidation in acidic media. Such behavior was attributed to the Pt nanowires (PtNWs) network growing on PCSs. Furthermore, CO-poisoning tolerance was promoted by PCS surface oxygen groups.

8.3.2.4 Pollution control (NO_x, ammonia, organics)

Little research has been published on the possible applications of char as catalyst or catalyst support for the reduction, oxidation, and removal of gaseous pollutants.

Biomass-derived char can be applied in the SCR-NO_x as reported by Wu et al. (2012), which used the residue of the pyrolysis of cotton stalk waste, once steam activated, as support for Ce and Mn. Various catalysts with different porous distributions were obtained. The authors observed that the extent of mesoporosity influenced the temperature at which the catalyst was active. The catalyst with the lower amount of mesopores showed the highest NO conversion at low temperature (180–240°C), while a bigger amount of mesopores corresponded to enhanced activity (55% NO conversion) at temperatures higher than 260°C. Cotton stalk-derived chars were also used by Singh et al. (2013) as support for Ce and Mn. Low catalytic activity was observed. It was attributed to both the low metal loading and the presence of inorganics (Na, K, Ca, and Mg) that neutralize the Brønsted acid active sites of the catalyst involved in the activation (by protonation) of the reducing agent (NH_3). Biomass-derived char seems to be more adapted than commercially activated carbons for application at low temperature; the higher presence of oxygen functional groups in Mn/Ce deposited on cotton char was shown to have a key role in SCR of NO at low temperature (Cha et al., 2010; Shen et al., 2015). The influence of the specific surface area, the amount of oxygen functional groups, the porosity, and the NO and NH_3 adsorption capacities have been taken into account in the SCR reaction performed on Mn-based catalysts supported on rice straw char (Cha et al., 2010).

Attempts to use chars as catalyst in the catalytic ozonation of ammonia are rarely reported (Kastner, Miller, Kolar, & Das, 2009). The use of chars in this process is potentially very profitable due to the low cost of biomass-derived catalysts, and the possibility of simultaneously reusing agriculture waste materials in agricultural operations that involve ammonia release.

8.4 Conclusions and perspectives

In the last few years, char has been identified as an environmentally friendly and low-cost catalyst and catalyst support in various reactions for syngas and bio-oil production, chemical synthesis, and pollutant treatments.

Much research has reported on the performance of such relatively new catalytic materials and compared them to more conventional catalysts. Unfortunately, often the selection of a certain char for a target reaction is not based on objective correlations between char properties and reactivity, but just on the accessibility of a certain char, due to the availability of the parent biomass. As a consequence, it is difficult to identify the properties directing a specific reaction and to produce more active char-based catalysts by applying a bottom-up approach. Amelioration of char catalytic properties by designing and producing appropriate chars (in terms of surface functionalities, hydrophobic/hydrophilic features, and pores distribution) will facilitate to apply chars to a larger extent and in a larger variety of reactions.

References

Abdullah, S. H. Y. S., Hanapia, N. H. M., Azida, A., Umara, R., Juahira, H., Khatoonc, H., et al. (2017). A review of biomass-derived heterogeneous catalyst for a sustainable biodiesel production. *Renewable and Sustainable Energy Reviews*, *70*, 1040–1051.

Abu, E. R. Z., Bramer, E. A., & Brem, G. (2008). Experimental comparison of biomass chars with other catalysts for tar reduction. *Fuel*, *87*, 2243–2252.

Ammendola, P., Cammisa, E., Chirone, R., Lisi, L., & Ruoppolo. (2012). Effect of sulphur on the performance of Rh–LaCoO3 based catalyst for tar conversion to syngas. *Applied Catalysis B: Environmental*, *11-18*, 113–114.

Angin, D. (2013). Effect of pyrolysis temperature and heating rate on biochar obtained from pyrolysis of safflower seed press cake. *Bioresource Technology*, *128*, 593–597.

Anis, S., & Zainal, Z. A. (2011). Tar reduction in biomass producer gas via mechanical, catalytic and thermal methods: a review. *Renewable and Sustainable Energy Reviews*, *15*, 2355–2377.

Aqsha, A., Tijani, M. M., Moghtaderi, B., & Mahinpey, N. (2017). Catalytic pyrolysis of straw biomasses (wheat, flax, oat and barley) and the comparison of their product yields. *Journal of Analytical and Applied Pyrolysis*, *125*, 201–208.

Azargohar, R., & Dalai, A. K. (2008). Steam and KOH activation of biochar: experimental and modeling studies. *Microporous and Mesoporous Materials*, *110*, 413–421.

Bazargan, A., Kostic, O. S., Stomenkovic, O. S., Veljkovic, V. B., & McKay, G. (2015). calcium oxide-based catalyst derived from palm kernel shell gasification residues for biodiesel production. *Fuel*, *150*, 519–525.

Belhachemi, M., & Addoun, F. (2011). Effect of heat treatment on the surface properties of activated carbons. *E-Journal of Chemistry*, *8*, 992–999.

Belhachemi, M., Rios, R. V. R. A., Addoun, F., Silvestre-Albero, J., Sepulveda-Escribano, A., & Rodriguez-Reinoso, F. (2009). Preparation of activated carbon from date pits: effect of the activation agent and liquid phase oxidation. *Journal of Analytical and Applied Pyrolysis*, *86*, 168–172.

Belhachemi, M., Jeguirim, M., Limousy, L., & Addoun, F. (2014). Comparison of NO2 removal using date pits activated carbon and modified commercialized activated carbon via different preparation methods: effect of porosity and surface chemistry. *Chemical Engineering Journal*, *253*, 121–129.

Bevia, F. R., Rico, D. P., & Gomis, A. F. M. (1984). Activated carbon from almond shells. Chemical activation. 1. Activating reagent selection and variables influence. *Industrial and Engineering Chemistry Product Research and Development*, *23*, 266–269.

Bhandari, P. N., Kumar, A., Bellmer, D. D., & Huhnke, R. L. (2014). Synthesis and evaluation of biochar-derived catalysts for removal of toluene (model tar) from biomass-generated producer gas. *Renewable Energy*, *66*, 346–353.

Buah, W. K., & Kuma, J. S. Y. (2012). Properties of activated carbon prepared from coconut shells in Ghana, Ghana. *Mining Journal*, *13*, 51–55.

Cao, L., Yu, I. K. M., Chen, S. S., Tsang, D. C. W., Wang, L., Xiong, X., et al. (2018). Production of 5-hydroxymethylfurfural from starch-rich food waste catalyzed by sulfonated biochar. *Bioresource Technology*, *252*, 76–82.

Carpenter, D., Westover, T. L., Czernik, S., & Jablonski, W. (2014). Biomass feedstock for renewable fuel production a review of the impacts of feedstock and pretreatment on the yield and product distribution of fast pyrolysis bio-oils vapors. *Green Chemistry*, *16*, 384–406.

Cetin, E., Gupta, R., & Moghtaderi, B. (2005). Effect of pyrolysis pressure and heating rate on radiata pine char structure and apparent gasification reactivity. *Fuel*, *84*, 1328–1334.

Cha, J. S., Choi, J. C., Ko, J. H., Park, Y. K., Park, S. H., Jeong, K. E., et al. (2010). The low temperature SCR of NO over rice straw and sewage sludge derived char. *Chemical Engineering Journal*, *156*, 321–327.

Cha, J. S., Park, S. H., Jung, S.-C., Ryu, C., Jeon, J.-K., Shin, M.-C., et al. (2016). Production and utilization of biochar: a review. *Journal of Industrial and Engineering Chemistry*, *40*, 1–15.

Chang, C.-F., Chang, C.-Y., & Tsai, W.-T. (2000). Effects of burn-off and activation temperature on preparation of activated carbon from corn Cob agrowaste by CO_2 and steam. *Journal of Colloid and Interface Science*, *232*, 45–49.

Chiang, H. L., Chiang, P. C., & Huang, C. P. (2002). Ozonation of AC and its effects on the adsorption of VOCs exemplified by methylethylketone and benzene. *Chemosphere*, *47*, 267–275.

Cho, H. J., Baek, K., Jeon, J.-K., Park, S. H., Suh, D. J., & Park, Y. K. (2013). Removal characteristics of copper by marine macro-algae-derived chars. *Chemical Engineering Journal*, *217*, 205–211.

Comotti, M., Della Pina, C., Falletta, E., & Rossi, M. (2006). Is the biochemical route always advantageous? The case of glucose oxidation. *Journal of Catalysis*, *244*, 122–125.

Dabrowski, A., Podkoscielny, P., Hubicki, Z., & Barczak, M. (2005). Adsorption of phenolic compounds by activated carbon—a critical review. *Chemosphere*, *58*, 1049–1070.

Dehkhoda, A. M., & Ellis, N. (2013). Biochar-based catalyst for simultaneous reactions of esterification and transesterification. *Catalysis Today*, *207*, 86–92.

Dehkhoda, A. M., West, A. H., & Ellis, N. (2010). Biochar based solid acid catalyst for biodiesel production. *Applied Catalysis A: General*, *382*, 197–204.

Deng, A., Lin, Q., Yan, Y., Li, H., Ren, J., Liu, C., et al. (2016). A feasible process for furfural production from the pre-hydrolysis liquor of corncob via biochar catalysts in a new biphasic system. *Bioresource Technology*, *216*, 754–760.

Deng, X., Yue, Y., & Gao, Z. (1997). Preparation and characterization of active carbon adsorbents for wastewater treatment from elutrilithe. *Journal of Colloid and Interface Science*, *192*, 475–480.

Dhawane, S. H., Kumar, T., & Halder, G. (2015). Central composite design approach towards optimization of flamboyant pods derived steam activated carbon for its use as heterogeneous catalyst in transesterification of *Hevea brasiliensis* oil. *Energy Conversion and Management*, *100*, 277–287.

Do Couto Fraga, A., Pontes Bittencourt Quitete, C., Loureiro Ximenes, V., Falabella Sousa-Aguiar, E., Fonseca, I. M., & Botelho Rego, A. M. (2016). Biomass derived solid acids as effective hydrolysis catalysts. *Journal of Molecular Catalysis A: Chemical*, *422*, 248–257.

Dong, T., Gao, D., Miao, C., Yu, X., Degan, C., Garcia-Pérez, M., et al. (2015). Two step microalgal biodiesel production using acidic catalyst generated from pyrolysis-derived bio-char. *Energy Conversion and Management*, *105*, 1389–1396.

Ducousso, M., Weiss-Hortala, E., Nzihou, A., & Castaldi, M. J. (2015). Reactivity enhancement of gasification biochars for catalytic applications. *Fuel*, *159*, 491–499.

El-Hendawy, A.-N. A., Samra, S. E., & Girgis, B. S. (2001). Adsorption characteristics of activated carbons obtained from corncobs. *Colloids and Surfaces A: Physicochemical and Engineering Aspects*, *180*, 209–221.

Fan, Y., Liu, P. F., Yang, Z. J., Jiang, T. W., Yao, K. L., Han, R., et al. (2015). Bi-functional porous carbon spheres derived from pectin as electrode material for supercapacitors and support material for Pt nanowires towards electrocatalytic methanol and ethanol oxidation. *Electrochimica Acta*, *163*, 140–148.

Fan, M., Marshall, W., Daugaard, D., & Brown, R. C. (2004). Steam activation of chars produced from oat hulls and corn stover. *Bioresource Technology*, *93*, 103–107.

Fierro, V., Torné-Fernandez, V., & Celzard, A. (2006). Kraft lignin as a precursor for microporous activated carbons prepared by impregnation with ortho-phosphoric acid: synthesis and textural characterization. *Microporous and Mesoporous Materials*, *92*, 243–250.

Figueiredo, J. L. (2013). Functionalization of porous carbons for catalytic applications. *Journal of Materials Chemistry A*, *1*, 9351–9364.

Figueiredo, J. L., Pereira, M. F. R., Freitas, M. M. A., & Orfao, J. J. M. (1999). Modification of the surface chemistry of activated carbons. *Carbon*, *37*, 1379–1389.

Ghouma, M., Jeguirim, S., Dorge, L., Limousy, C., Matei Ghimbeu, & Ouederni, A. (2015). Activated carbon prepared by physical activation of olive stones for the removal of NO2 at ambient temperature. *Comptes Rendus Chimie*, *18*, 63–74.

Gonzalez, M. E., Cea, M., Romero-Hermoso, L., Hidalgo, P., Meier, S., Benito, N., et al. (2017). Functionalization of biochar derived from lignocellulosic biomass using microwave technology for catalytic application in biodiesel production. *Energy Conversion and Management*, *137*, 165–173.

Gonzalez, J. F., Roman, S., Encinar, J. M., & Martinez, G. (2009). Pyrolysis of various biomass residues and char utilization for the production of activated carbons. *Journal of Analytical and Applied Pyrolysis*, *85*, 134–141.

Guan, G., Kaewpanha, M., Hao, X., & Abudula, A. (2016). Catalytic steam reforming of biomass tar: prospects and challenges. *Renewable and Sustainable Energy Reviews*, *58*, 450–461.

Guerrero, M., Ruiz, M. P., Alzueta, M. U., Bilbao, R., & Millera, A. (2005). Pyrolysis of eucalyptus at different heating rates: studies of char characterization and oxidative reactivity. *Journal of Analytical and Applied Pyrolysis*, *74*, 307–314.

Guilminot, E., Gavillon, R., Chatenet, M., Berthon-Fabry, S., Rigacci, A., & Budtova, T. (2008). New nanostructured carbons based on porous cellulose: elaboration, pyrolysis and use as platinum nanoparticles substrate for oxygen reduction electrocatalysis. *Journal of Power Sources*, *185*, 717–726.

Hao, W., Björkman, E., Lilliestråle, M., & Hedin, N. (2013). Activated carbons prepared from hydrothermally carbonized waste biomass used as adsorbents for CO2. *Applied Energy*, *112*, 526–532.

Hayashi, J., Horikawa, T., Takeda, I., Muroyama, K., & Ani, F. N. (2002). Preparing activated carbon from various nutshells by chemical activation with K2CO3. *Carbon*, *40*, 2381–2386.

Ioannidou, O., & Zabaniotou, A. (2007). Agricultural residues as precursors for activated carbon production—a review. *Renewable and Sustainable Energy Reviews*, *11*, 1966–2005.

Iwazaki, T., Obinata, R., Sugimoto, W., & Takasu, Y. (2009). High oxygen-reduction activity of silk-derived activated carbon. *Electrochemistry Communications*, *11*, 376–378.

Iwazaki, T., Yang, H., Obinata, R., Sugimoto, W., & Takasu, Y. (2010). Oxygen-reduction activity of silk-derived carbons. *Journal of Power Sources*, *195*, 5840–5847.

Kambo, H. S., & Dutta, A. (2015). A comparative review of biochar and hydrochar in terms of production, physico-chemical properties and applications. *Renewable and Sustainable Energy Reviews*, *45*, 359–378.

Kapteijn, F., Moulijn, J. A., Matzner, S., & Boehm, H. P. (1999). The development of nitrogen functionality in model chars during gasification in CO2 and O2. *Carbon*, *37*, 1143–1150.

Kastner, J. R., Mani, S., & Juneja, A. (2015). Catalytic decomposition of tar using iron supported biochar. *Fuel Processing Technology*, *130*, 31–37.

Kastner, J. R., Miller, J., Geller, D. P., Locklin, J., Keith, L. H., & Johnson, T. (2012). Catalytic esterification of fatty acids using solid acid catalysts generated from biochar and activated carbon. *Catalysis Today*, *190*, 122–132.

Kastner, J. R., Miller, J., Kolar, P., & Das, K. C. (2009). Catalytic ozonation of ammonia using biomass char and wood fly ash. *Chemosphere*, *75*, 739–744.

Keiluweit, M., Nico, P. S., Johnson, M. G., & Kleber, M. (2010). Dynamic molecular structure of plant biomass-derived black carbon (biochar). *Environmental Science & Technology*, *44*, 1247–1253.

Kim, J. R., & Kan, E. (2016). Heterogeneous photocatalytic degradation of sulfamethoxazole in water using a biochar-supported TiO2 photocatalyst. *Journal of Environmental Management*, *180*, 94–101.

Klinghoffer, N. B., Castaldi, M. J., & Nzihou, A. (2015). Influence of char composition and inorganics on catalytic activity of char from biomass gasification. *Fuel*, *157*, 37–47.

Klüpfel, L., Keiluweit, M., Kleber, M., & Sander, M. (2014). Redox properties of plant biomass-derived black carbon (biochar). *Environmental Science & Technology*, *48*, 5601–5611.

Kouzu, M., Kasuno, T., Tajika, M., Yamanaka, S., & Hidaka, J. (2008). Active phase of calcium oxide used as solid base catalyst for transesterification of soybean oil with refluxing methanol. *Applied Catalysis A: General*, *334*, 357–365.

Lazaridis, P. A., Karakouli, S., Delimitis, A., Coman, S. M., Parvulescu, V. I., & Triantafyllidis, K. S. (2015). D-Glucose hydrogenation/hydrogenolysis reactions on noble metal(Ru, Pt)/activated carbon supported catalysts. *Catalysis Today*, *257*, 281–290.

Lee, J., Kim, K. H., & Kwon, E. E. (2017). Biochar as a catalyst. *Renewable and Sustainable Energy Reviews*, *77*, 70–79.

Li, S., Gu, Z., Bjornson, B. E., & Muthukumarappan, A. (2013). Biochar based solid acid catalyst hydrolyze biomass. *Journal of Environmental Chemical Engineering*, *1*, 1174–1181.

Li, L., Song, Z., Zhao, X., Ma, C., Kong, X., & Wang, F. (2016). Microwave-induced cracking and CO2 reforming of toluene on biomass derived char. *Chemical Engineering Journal*, *284*, 1308–1316.

Li, Y., Zhang, S., Yu, Q., & Yin, W. (2007). The effects of activated carbon supports on the structure and properties of TiO2 nanoparticles prepared by a sol–gel method. *Applied Surface Science*, *253*, 9254–9258.

Li, M., Zheng, Y., Chen, Y., & Zhu, X. (2014). Biodiesel production from waste cooking oil using a heterogeneous catalyst from pyrolyzed rice husk. *Bioresource Technology*, *154*, 345–348.

Limousy, L., Ghouma, I., Ouederni, A., & Jeguirim, M. (2017). Amoxicillin removal from aqueous solution using activated carbon prepared by chemical activation of olive stone. *Environmental Science and Pollution Research*, *24*, 9993–10004.

Liu, N., Charrua, A. B., Weng, C.-H., Yuan, X., & Ding, F. (2015). Characterization of biochars derived from agriculture wastes and their adsorptive removal of atrazine from aqueous solution: a comparative study. *Bioresource Technology*, *198*, 55–62.

Liu, X., Plao, X., Wang, Y., & Zhu, S. (2008). Calcium ethoxide as a solid base catalyst for the transesterification of soybean oil to biodiesel. *Energy & Fuels*, *22*, 1313–1317.

Liu, Q. Y., Yang, F., Liu, Z. H., & Li, G. (2015). Preparation of SnO2–Co3O4/C biochar catalyst as a Lewis acid for corncob hydrolysis into furfural in water medium. *Journal of Industrial and Engineering Chemistry*, *26*, 46–54.

Liu, Z., Zhang, F.-S., & Wu, J. (2010). Characterization and application of chars produced from pinewood pyrolysis and hydrothermal treatment. *Fuel*, *89*, 510–514.

Liu, Q., Zhou, Y., Chen, S., Wang, Z., Hou, H., & Zhao, F. (2015). Cellulose-derived nitrogen and phosphorus dual-doped carbon as high performance oxygen reduction catalyst in microbial fuel cell. *Journal of Power Sources*, *273*, 1189–1193.

Lobos, M. L. N., Sieben, J. M., Comignani, V., Duarte, M., Volpe, M. A., & Moyano, E. L. (2016). Biochar from pyrolysis of cellulose: an alternative catalyst support for the electro-oxidation of methanol. *International Journal of Hydrogen Energy*, *41*, 10695–10706.

Lu, Q., Li, J., Chen, F., Liao, M., Lin, L., Tang, Y., et al. (2017). Effects of mutual intercropping on the cadmium accumulation in accumulator plants *Stellaria media*, *Malachium aquaticum*, and *Galium aparine*. *Environmental Monitoring and Assessment*, *189*, 622–632.

Ma, M., Dai, Y., Zou, J. L., Wang, L., Pan, K., & Fu, H. G. (2014). Synthesis of iron oxide/partly graphitized carbon composites as a high-efficiency and low-cost cathode catalyst for microbial fuel cells. *ACS Applied Materials & Interfaces*, *6*, 13438–13447.

Ma, Z., Xiao, R., & Zhang, H. (2017). Catalytic steam reforming of bio-oil model compounds for hydrogen-rich gas production using bio-char as catalyst. *International Journal of Hydrogen Energy*, *42*, 3579–3585.

Malik, R., Ramteke, D. S., & Wate, S. R. (2007). Adsorption of malachite green on groundnut shell waste based powdered activated carbon. *Waste Management*, *27*, 1129–1138.

Mangun, C. L., Benak, K. R., Daley, M. A., Economy, J., & Foster, K. L. (2001). Surface chemistry, pore sizes and adsorption properties of activated carbon fibers and precursors treated with ammonia. *Carbon*, *39*, 1809–1820.

Matos, I., Bernardo, M., & Fonseca, I. (2017). Porous carbon: a versatile material for catalysis. *Catalysis Today*, *285*, 194–203.

Min, Z., Zhang, S., Yimsiri, P., Wang, Y., Asadullah, M., & Li, C. Z. (2013). Catalytic reforming of tar during gasification. Part IV. Changes in the structure of char in the char-supported iron catalyst during reforming. *Fuel*, *106*, 858–863.

Moreno-Castilla, C., Ferro-Garcia, M. A., Joly, J. P., Bautista-Toledo, I., Carrasco-Marín, F., & Rivera-Utrilla, J. (1995). Activated carbon surface modifications by nitric acid, hydrogen peroxide, and ammonium peroxydisulfate treatments. *Langmuir*, *11*, 4386–4392.

Muley, P. D., Henkel, C., Abdollahi, K. K., Marculescu, C., & Boldor, D. (2016). A critical comparison of pyrolysis of cellulose, lignin, and pine sawdust using an induction heating reactor. *Energy Conversion and Management*, *117*, 273–280.

Mumme, J., Eckervogt, L., Pielert, J., Diakité, M., Rupp, F., & Kern, J. (2011). Hydrothermal carbonization of anaerobically digested maize silage. *Bioresource Technology*, *102*, 9255–9260.

Muradov, N., Fidalgo, B., Gujar, A. C., Garceau, N., & Ali, T. (2012). Production and characterization of *Lemna minor* bio-char and its catalytic application for biogas reforming. *Biomass and Bioenergy*, *42*, 123–131.

Ncibi, M. C., Jeanne-Rose, V., Mahjouba, B., Jean-Marius, C., Lambert, J., Ehrhardt, J. J., et al. (2009). Preparation and characterisation of raw chars and physically activated carbons derived from marine *Posidonia oceanica* (L.) fibres. *Journal of Hazardous Materials*, *165*, 240–249.

Ngaosuwan, K., Goodwin, J. G., Jr., & Prasertdham, P. (2016). A green sulfonated carbon-based catalyst derived from coffee residue for esterification. *Renewable Energy*, *86*, 262–269.

Nguyen, H. K. D., Pham, V. V., & Do, H. T. (2016). Preparation of Ni/biochar catalyst for hydrotreating of bio-oil from microalgae biomass. *Catalysis Letters*, *146*, 2381–2391.

Nino, L., Maria, P., Khaled, L., & Mohand, T. (2013). Biodiesel production from biomass gasification tar via thermal/catalytic cracking. *Fuel Processing Technology*, *106*, 776–783.

Oh, G. H., & Park, C. R. (2002). Preparation and characteristics of rice-straw-based porous carbons with high adsorption capacity. *Fuel*, *81*, 327–336.

Olivares-Marin, M., Fernandez-Gonzalez, C., Macias-Garcia, A., & Gomez-Serrano, V. (2006). Preparation of activated carbon from cherry stones by chemical activation with ZnCl2. *Applied Surface Science*, *252*, 5967–5971.

Onay, O. (2007). Influence of pyrolysis temperature and heating rate on the production of bio-oil and char from safflower seed by pyrolysis, using a well-swept fixed-bed reactor. *Fuel Processing Technology*, *88*, 523–531.

Ormsby, R., Kastner, J. R., & Miller, J. (2012). Hemicellulose hydrolysis using solid acid catalysts generated from biochar. *Catalysis Today*, *190*, 89–97.

Parshetti, G. K., Kent Hoekman, S., & Balasubramanian, R. (2013). Chemical, structural and combustion characteristics of carbonaceous products obtained by hydrothermal carbonization of palm empty fruit bunches. *Bioresource Technology*, *135*, 683–689.

Perrard, A., Gallezot, P., Joly, J.-P., Durand, R., Baljou, C., Coq, B., et al. (2007). Highly efficient metal catalysts supported on activated carbon cloths: a catalytic application for the hydrogenation of D-glucose to D-sorbitol. *Applied Catalysis A: General*, *331*, 100–104.

Pietrzak, R. (2009). XPS study and physico-chemical properties of nitrogen-enriched microporous activated carbon from high volatile bituminous coal. *Fuel*, *88*, 1871–1877.

Prati, L., Bergna, D., Villa, A., Spontoni, P., Bianchi, C. L., Hu, T., et al. (2018). Carbons from second generation biomass as sustainable supports for catalytic systems. *Catalysis Today*, *301*, 239–243.

Qian, K., Kumar, A., Patil, K., Bellmer, D., Wang, D., Yuan, W., et al. (2013). Effects of biomass feedstocks and gasification conditions on the physiochemical properties of char. *Energies*, *6*, 3972–3986.

Qian, K., Kumar, A., Zhang, H., Bellmer, D., & Huhnke, R. (2015). Recent advances in utilization of biochar. *Renewable and Sustainable Energy Reviews*, *42*, 1055–1064.

Qian, Q., Machida, M., & Tatsumoto, H. (2007). Preparation of activated carbons from cattle-manure compost by zinc chloride activation. *Bioresource Technology*, *98*, 353–360.

Rajapaksha, A. U., Vithanage, M., Lee, S. S., Seo, D. C., Tsang, D. C. W., & Ok, Y. S. (2016). Steam activation of biochars facilitates kinetics and pH-resilience of sulfamethazine sorption. *Journal of Soils and Sediments*, *16*, 889–895.

Rey, A., Faraldos, M., Casas, J. A., Bahamonde, A., & Rodriguez, J. J. (2009). Catalytic wet peroxide oxidation of phenol over Fe/AC catalysts: influence of iron precursor and activated carbon surface. *Applied Catalysis B: Environmental*, *86*, 69–77.

Richardson, Y., Blin, J., Volle, G., Motuzas, J., & Julbe, A. (2010). In situ generation of Ni metal nanoparticles as catalyst for H2-rich syngas production from biomass gasification. *Applied Catalysis A: General*, *382*, 220–230.

Richardson, Y., Motuzas, J., Julbe, A., Volle, G., & Blin, J. (2013). Catalytic investigation of in situ generated Ni metal nanoparticles for tar conversion during biomass pyrolysis. *Journal of Physical Chemistry C*, *117*, 23812–23831.

Rivera-Utrilla, J., Sanchez-Polo, M., Gomez-Serrano, V., Alvarez, P. M., Alvim-Ferraz, M. C. M., & Dias, J. M. (2011). Activated carbon modifications to enhance its water treatment applications. An overview. *Journal of Hazardous Materials*, *187*, 1–23.

Rodríguez-Reinoso, F. (1998). The role of carbon materials in heterogeneous catalysis. *Carbon*, *36*, 159–175.

Rodríguez-Reinoso, Molina-Sabio, M., & González, M. T. (1995). The use of steam and CO2 as activating agents in the preparation of activated carbons. *Carbon*, *33*, 15–23.

Rodríguez-Reinoso, Molina-Sabio, M., & González, M. T. (1997). Effect of oxygen surface groups on the immersion enthalpy of activated carbons in liquids of different polarity. *Langmuir*, *13*, 2354–2358.

Román, S., Valente Nabais, J. M., Ledesma, B., Gonzalez, J. F., Langinhas, C., & Titirici, M. M. (2013). Production of low-cost adsorbents with tunable surface chemistry by conjunction of hydrothermal carbonization and activation processes. *Microporous and Mesoporous Materials*, *165*, 127–133.

Ronsse, F., Hecke, S., Dickinson, D., & Prins, W. (2013). Production and characterization of slow pyrolysis biochar: influence of feedstock type and pyrolysis conditions. *GCB Bioenergy*, *5*, 104–111.

Ros, A., Lillo-Rodenas, M. A., Fuente, E., Montes-Moran, M. A., Martin, M. J., & Linares-Solano, A. (2006). High surface area materials prepared from sewage sludge-based precursors. *Chemosphere*, *65*, 132–140.

Safari, F., Norouzi, O., & Tavasoli, A. (2016). Hydrothermal gasification of *Cladophora glomerata* macroalgae over its hydrochar as a catalyst for hydrogen-rich gas production. *Bioresource Technology*, *222*, 232–241.

Sairanen, E., Vilonen, K., Karinen, R., & Lehtonen, J. (2013). Functionalized activated carbon catalysts in xylose dehydration. *Topics in Catalysis*, *56*, 512–521.

Samant, P. V., Gonçalves, F., Freitas, M. M. A., Pereiraand, M. F. R., & Figueiredo, J. L. (2004). Surface activation of a polymer based carbon. *Carbon*, *42*, 1321–1325.

Sevilla, M., & Fuertes, A. B. (2009). The production of carbon materials by hydrothermal carbonization of cellulose. *Carbon*, *47*, 2281–2289.

Shafeeyan, M. S., Daud, W. M. A. W., Houshmand, A., & Shamiri, A. (2010). A review on surface modification of activated carbon for carbon dioxide adsorption. *Journal of Analytical and Applied Pyrolysis*, *89*, 143–151.

Shan, R., Zhao, C., Lv, P., Yuan, H., & Yao, J. (2016). Catalytic applications of calcium rich waste materials for biodiesel: current state and perspectives. *Energy Conversion and Management*, *127*, 273–283.

Sharma, R. K., Wooten, J. B., Baliga, V. L., & Hajaligol, M. R. (2001). Characterization of chars from biomass-derived materials: pectin chars. *Fuel*, *80*, 1825–1836.

Shen, Y. (2015). Chars as carbonaceous adsorbents/catalysts for tar elimination during biomass pyrolysis or gasification. *Renewable and Sustainable Energy Reviews*, *43*, 281–295.

Shen, B., Chen, J., Yue, S., & Li, G. (2015). A comparative study of modified cotton biochar and activated carbon based catalysts in low temperature SCR. *Fuel*, *156*, 47–53.

Shen, Y., & Yoshikawa, K. (2014). Tar conversion and vapor upgrading via in situ catalysis using silica-based nickel nanoparticles embedded in rice husk char for biomass pyrolysis/gasification. *Industrial and Engineering Chemistry Research*, *53*, 10929–10942.

Shen, Y., Zhao, P., Shao, Q., Ma, D., Takahashi, F., & Yoshikawa, K. (2014). In-situ catalytic conversion of tar using rice husk char-supported nickel-iron catalysts for biomass pyrolysis/gasification. *Applied Catalysis B: Environmental*, *152–153*, 140–151.

Shim, T., Yoo, J., Ryu, C., Park, Y.-K., & Jung, J. (2015). Effect of steam activation of biochar produced from a giant Miscanthus on copper sorption and toxicity. *Bioresource Technology*, *197*, 85–89.

Shu, T., Lu, P., & He, N. (2013). Mercury adsorption of modified mulberry twig chars in a simulated flue gas. *Bioresource Technology*, *136*, 182–187.

Silvestre-Albero, A., Silvestre-Albero, J., Sepulveda-Escribano, A., & Rodriguez-Reinoso, F. (2009). Ethanol removal using activated carbon: effect of porous structure and surface chemistry. *Microporous and Mesoporous Materials*, *120*, 62–68.

Simaioforidou, A., Papastergiou, M., Margellou, A., Petrakis, D., & Louloudi, M. (2017). Activated vs. pyrolytic carbon as support matrix for chemical functionalization: efficient heterogeneous non-heme Mn(II) catalysts for alkene oxidation with H2O2. *Journal of Molecular Catalysis A: Chemical*, *426*, 516–525.

Singh, S., Nahil, M. A., Sun, X., Wu, C., Chen, J., Shen, B., et al. (2013). Novel application of cotton stalk as a waste derived catalyst in the low temperature SCR-deNOx process. *Fuel*, *105*, 585–594.

Song, Y., Wang, Y., Hu, X., Hu, S., Xiang, J., Zhang, L., et al. (2014). Effects of volatile–char interactions on in-situ destruction of nascent tar during the pyrolysis and gasification of biomass. Part I. Roles of nascent char. *Fuel*, *122*, 60–66.

Streat, M., Patrick, J. W., & Camporro Perez, M. J. (1995). Sorption of phenol and para-chlorophenol from water using conventional and novel activated carbons. *Water Research*, *29*, 467–472.

Sudaryanto, Y., Hartono, S. B., Irawaty, W., Hindarso, H., & Ismadji, S. (2006). High surface area activated carbon prepared from cassava peel by chemical activation. *Bioresource Technology*, *97*, 734–739.

Suganuma, S., Nakajima, K., Kitano, M., Yamagushi, D., Kato, H., Hayashi, S., et al. (2008). Hydrolysis of cellulose by amorphous carbon bearing SO3H, COOH, and OH groups. *Journal of the American Chemical Society*, *130*, 12787–12793.

Syed-Hassan, S. S. A., & Fuadi, F. A. (2016). Catalytic steam reforming of biomass tar model compound using nickel and cobalt catalysts supported on palm kernel shell char. *Journal of Chemical Engineering of Japan*, *49*, 29–34.

Tsai, W. T., & Lee, S. L. (1997). Preparation and characterization of activated carbons from corn cob. *Carbon*, *35*, 1198–1200.

Uchimiya, M., Lima, I. M., Klasson, K. T., & Wartelle, L. H. (2010). Contaminant immobilization and nutrient release by biochar soil amendment: roles of natural organic matter. *Chemosphere*, *80*, 935–940.

Valdés, H., Sanchez-Polo, M., Rivera-Utrilla, J., & Zaror, C. A. (2002). Effect of ozone treatment on surface properties activated carbon. *Langmuir*, *18*, 2111–2116.

Wang, K., Jiang, J., Xu, J., Feng, J., & Wang, J. (2016). Effective saccharification of holocellulose over multifunctional sulfonated char with fused ring structures under microwave irradiation. *RSC Advances*, *6*, 14164–14170.

Wang, R., Wang, H., Zhou, T., Key, J., Ma, Y., Zhang, Z., et al. (2015). The enhanced electrocatalytic activity of okara-derived N-doped mesoporous carbon for oxygen reduction reaction. *Journal of Power Sources*, *274*, 741–747.

Wang, D., Yuan, W., & Ji, W. (2011). Char and char-supported nickel catalysts for secondary syngas cleanup and conditioning. *Applied Energy*, *88*, 1656–1663.

Wang, S., Zhao, C., Shan, R., Wang, Y., & Yuan, H. (2017). A novel peat biochar supported catalyst for the transesterification reaction. *Energy Conversion and Management*, *139*, 89–96.

Widayatno, W. B., Guan, G. Q., Rizkiana, J., Hao, X. G., Wang, Z. D., Samart, C., et al. (2014). Steam reforming of tar derived from *Fallopia japonica* stem over its own chars prepared at different conditions. *Fuel*, *132*, 204–210.

Wojcieszak, R., Zielinski, M., Monteverdi, S., & Bettahar, M. M. (2006). Study of nickel nanoparticles supported on activated carbon prepared by aqueous hydrazine reduction. *Journal of Colloid and Interface Science*, *299*, 238–248.

Wu, C. F., Nahil, M. A., Sun, X., et al. (2012). Novel application of biochar from biomass pyrolysis for low temperature selective catalytic reduction. *Journal of the Energy Institute*, *85*, 236–239.

Wu, W., Xu, J., & Ohnishi, R. (2005). Complete hydrodechlorination of chlorobenzene and its derivatives over supported nickel catalysts under liquid phase conditions. *Applied Catalysis B: Environmental*, *60*, 129–137.

Xiong, X., Yu, I. K. M., Cao, L., Tsang, D. C. W., Zhang, S., & Ok, Y. S. (2017). A review of biochar-based catalysts for chemical synthesis, biofuel production, and pollution control. *Bioresource Technology*, *246*, 254–270.

Yamaguchi, D., Kitano, M., Suganuma, S., Nakajima, K., Kato, H., & Hara, M. (2009). Hydrolysis of cellulose by a solid acid catalyst under optimal reaction conditions. *Journal of Physical Chemistry C*, *113*, 3181–3188.

Yan, Q., Wan, C., Liu, J., Gao, J., Yu, F., Zhang, J., et al. (2013). Iron nanoparticles in situ encapsulated in biochar-based carbon as an effective catalyst for the conversion of biomass-derived syngas to liquid hydrocarbons. *Green Chemistry*, *15*, 1631–1640.

Yang, H., Yan, R., Chen, H., Lee, D. H., & Zheng, C. (2007). Characteristics of hemicellulose, cellulose and lignin pyrolysis. *Fuel*, *86*, 1781–1788.

Yang, N., Zhu, S., Zhang, D., & Xu, S. (2008). Synthesis and properties of magnetic Fe3O4-activated carbon nanocomposite particles for dye removal. *Materials Letters*, *62*, 645–647.

Yao, D., Hu, Q., Wang, D., Yang, H., Wu, C., Wang, X., et al. (2016). Hydrogen production from biomass gasification using biochar as a catalyst/support. *Bioresource Technology*, *216*, 159–164.

Yorgun, S., Vural, N., & Demiral, H. (2009). Preparation of high-surface area activated carbons from Paulownia wood by ZnCl2 activation. *Microporous and Mesoporous Materials*, *122*, 189–194.

Youssef, A. M., Radwan, N. R. E., Abdel-Gawad, I., & Singer, G. A. A. (2005). Textural properties of activated carbons from apricot stones. *Colloids and Surfaces A: Physicochemical and Engineering Aspects*, *252*, 143–151.

Yu, J. T., Dehkhoda, A. M., & Ellis, N. (2011). Development of biochar-based catalyst for transesterification of canola oil. *Energy and Fuels*, *25*, 337–344.

Yu, I. K. M., Tsang, D. C. W., Yip, A. C. K., Chen, S. S., Ok, Y. S., & Poon, C. S. (2016). Valorization of food waste into hydroxymethylfurfural: dual role of metal ions in successive conversion steps. *Bioresource Technology*, *219*, 338–347.

Yuan, Y., Bolan, N., Prevoteau, A., Vithanage, M., Biswas, J. K., Ok, Y. S., et al. (2017). Applications of biochar in redox-mediated reactions. *Bioresource Technology*, *246*, 271–281.

Yuan, Y., Yuan, T., Wang, D., Tang, J., & Zhou, S. (2013). Sewage sludge biochar as an efficient catalyst for oxygen reduction reaction in a microbial fuel cell. *Bioresource Technology*, *144*, 115–120.

Zhang, S., Asadullah, M., Dong, L., Tay, H.-L., & Li, C.-Z. (2013). An advanced biomass gasification technology with integrated catalytic hot gas cleaning. Part II: tar reforming using char as a catalyst or as a catalyst support. *Fuel*, *112*, 646–653.

Zhang, C., Cheng, Z., Fu, Z., Liu, Y., Yi, X., Zheng, A., et al. (2017). Effective transformation of cellulose to 5-hydroxymethylfurfural catalyzed by fluorine anion-containing ionic liquid modified biochar sulfonic acids in water. *Cellulose*, *24*, 95–106.

Zhang, H., Haba, M., Okumura, M., Akita, T., Hashimoto, S., & Toshima, N. (2013). Novel formation of Ag/Au bimetallic nanoparticles by physical mixture of monometallic nanoparticles in dispersions and their application to catalysts for aerobic glucose oxidation. *Langmuir*, *29*, 10330–10339.

Zhang, X., Li, Y., Li, G., & Hu, C. (2015). Preparation of Fe/activated carbon directly from rice husk pyrolytic carbon and its application in catalytic hydroxylation of phenol. *RSC Advances*, (7), 4984–4992.

Zhang, S., Song, Y., Song, Y. C., Yi, Q., Dong, L., Li, T. T., et al. (2016). An advanced biomass gasification technology with integrated catalytic hot gas cleaning. Part III: effects of inorganic species in char on the reforming of tars from wood and agricultural wastes. *Fuel*, *183*, 177–184.

Zhang, T., Walawender, W. P., Fan, L. T., Fan, M., Daugaard, D., & Brown, R. C. (2004). Preparation of activated carbon from forest and agricultural residues through CO2 activation. *Chemical Engineering Journal*, *105*, 53–59.

Zhang, X., Wilson, K., & Lee, A. F. (2016). Heterogeneously catalyzed hydrothermal processing of C5–C6 sugars. *Chemical Reviews*, *116*, 12328–12368.

Zhao, J., Zhou, C., He, C., Dai, Y., Jia, X., & Yang, Y. (2016). Efficient dehydration of fructose to 5-hydroxymethylfurfural oversulfonated carbon sphere solid acid catalysts. *Catalysis Today*, *264*, 123–130.

Zhou, T., Wang, H., Ji, S., Feng, H., & Wang, R. (2014). Synthesis of mesoporous carbon from Okara and application as electrocatalyst support. *Fuel Cells*, *14*, 296–302.

Further reading

Corcho-Corral, B., Olivares-Marın, M., Fernandez-Gonzalez, C., Gomez-Serrano, V., & Macias-Garcia, A. (2006). Preparation and textural characterisation of activated carbon from vine shoots (*Vitis vinifera*) by H3PO4—chemical activation. *Applied Surface Science*, *252*, 5961–5966.

Gao, Y., Wang, X., Wang, J., Li, X., Cheng, J., Yang, H., et al. (2013). Effect of residence time on chemical and structural properties of hydrochar obtained by hydrothermal carbonization of water hyacinth. *Energy*, *58*, 376–383.

Zhang, C., Geng, Z., Cai, M., Zhang, J., Liu, X., Xin, H., et al. (2013). Microstructure regulation of super activated carbon from biomass source corncob with enhanced hydrogen uptake. *IJHE*, *38*, 9243–9250.

Gas storage

9

Deneb Peredo-Mancilla[*,‡], *Imen Ghouma*[†], *Cecile Hort*[‡], *David Bessieres*[*], *Camélia Matei Ghimbeu*[†]
[*]CNRS/Total/University of Pau and Pays de l'Adour/E2S UPPA, Laboratory of Complex Fluids and Their Reservoirs-IPRA, Pau, France, [†]University of Strasbourg, University of Upper Alsace, Institute of Materials Science of Mulhouse (IS2M—UMR CNRS 7361), Mulhouse, France, [‡]University of Pau and Pays de l'Adour/E2S UPPA, Laboratory of Thermal Engineering, Energy and Processes-IPRA, Pau, France

Chapter Outline

9.1 Introduction

Gases play an important role in our daily lives. They are usually found in the form of gas mixtures with the most common being atmospheric air, which is made up from approximate volumes of 78.08% nitrogen (N_2), 20.95% oxygen (O_2), 0.93% argon (Ar), 0.03% carbon dioxide (CO_2), and traces of methane CH_4, helium (He), neon (Ne), krypton (Kr), xenon (Xe), and ozone (O_3) (Encyclopedia Britannica, 2018). Among them, CO_2, CH_4, and nitrogen oxides (NO_x) are a product of incomplete burning of fossil fuels (coal, oil, diesel, natural gas) and are the major contributors to global warming because they are emitted in higher amounts than other GHGs such as hydrofluorocarbons, perfluorocarbons, and hexafluorocarbon. As can be seen in Fig. 1, the atmospheric concentrations of these three gases have dramatically increased since preindustrial years ($\approx$1850), thus the reduction in their emissions has become the main target and has created the need for new technologies for the selective adsorption and storage of these pollutants (Core Writing Team et al., 2014).

Carbon dioxide is a major compound in the gaseous effluent of many industrial processes, with fossil fuel power plants being the most important source (40% of total CO_2). To reduce its emission, the main issues include increasing the power plants'

Char and Carbon Materials Derived from Biomass. https://doi.org/10.1016/B978-0-12-814893-8.00009-2

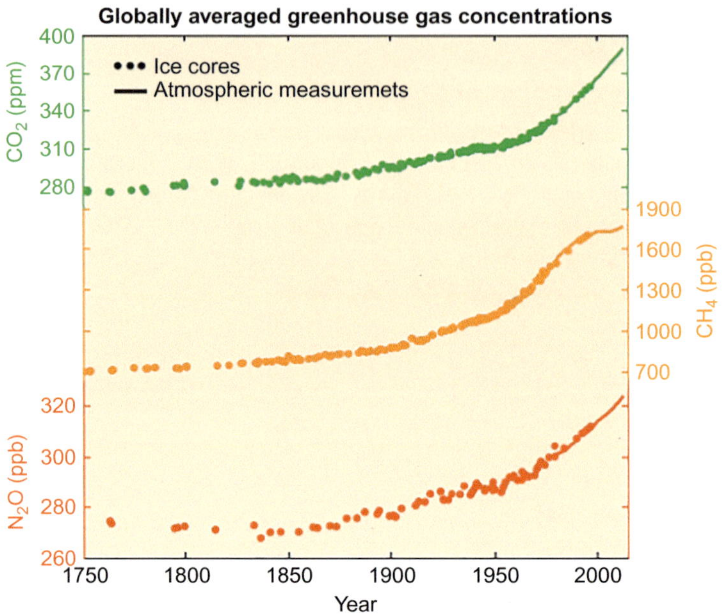

Fig. 1 Observed changes in atmospheric greenhouse gas concentrations between the years 1750 to 2011. Atmospheric concentrations of carbon dioxide (CO_2, *green*), methane (CH_4, *orange*), and nitrous oxide (N_2O, *red*). Data from ice cores (*symbols*) and direct atmospheric measurements (*lines*) are overlaid.
Reprinted from Core Writing Team, Pachauri, R. K., Meyer, L. A. (2014). *Climate Change 2014 Synthesis Report*. Contribution of Working Groups I, II and III to the Fifth Assessment Report of the Intergovernmental Panel on Climate Change.

efficiency, changing to nuclear and renewable energies, and the implementation of carbon capture and storage (CCS) systems (Alonso et al., 2017). CCS can be used as a promising strategy to reduce not only the power plants' emissions but it can also be employed in many industrial sectors such as the petrochemical industry, iron and steel refineries, and cement production.

There are many separation technologies available for the separation and storage of CO_2: amine scrubbing, water scrubbing, membrane separation, cryogenic separation, and biological and physical adsorption. Most of the commercial CCS systems use amine or water scrubbing but they have important drawbacks such as production of undesirable by-products, high energy regeneration, high corrosion, and low-cost efficiency. In this context, physical adsorption has the advantages of having low energy requirements, cost advantages, no use of water or additional chemicals, and it is easy to scale to a wide range of temperatures and pressures (Ben-Mansour et al., 2016).

For the case of methane, the principal anthropogenic sources are thought to be the production of livestock, natural gas production and distribution, emissions from landfills, and coal mining (Alonso et al., 2017). Due to methane's low volumetric energy density, its storage for use as a clean energy source involves the use of high pressures to obtain compressed natural gas (CNG), and low temperatures, down to 120 K, for liquefied natural gas (LNG). These methods have a very high energy output as well as various security issues that limit their use. The storage of methane by adsorption in a solid material, known as adsorbed natural gas (ANG), presents advantages over LNG and CNG in terms of energy efficiency and safety. An ANG storage system would have the ability to store great quantities of methane in small volumes under atmospheric conditions. To compete with other fuels (and other methane storage forms), the US Department of Energy (DOE) has established a CH_4 storage capacity of 180 v/v (volume of gas adsorbed at standard temperature and pressure: 298 K, 0.1 MPa per volume of the storage vessel) (Choi et al., 2016; Ma et al., 2008).

To mitigate climate change a number of strategies are being implemented in addition to CO_2 emission reduction and the use of methane as a clean energy source. Hydrogen (H_2) has a heating value three times higher than oil and its use as a form of energy is completely free of pollution making it one of the most promising alternatives to the burning of fossil fuels. Unfortunately, similar to methane, the transportation and storage of hydrogen have proven to be very challenging tasks. To be liquefied, it needs to be compressed below 33.2 K (H_2 critical temperature) in a high-energy demand process, and even when the compression process can be carried out the storage capacity of hydrogen is as low as 2 wt% even at high pressures (20 MPa). An H_2 storage target for use on vehicles of 5.5 wt%/40 g L^{-1} by 2017 and 7.5 wt%/70 g L^{-1} ultimately was established by the DOE (Office of Energy Efficiency & Renewable Energy, n.d.). To this end the storage of hydrogen using porous materials is one of the most promising techniques (Liu, Zou, & Zhao, 2016).

The development of porous adsorbent materials compatible with the adsorption of gases such as CH_4, CO_2, and H_2 has become a target issue for researchers, with the goals of finding cheap and *eco-friendly* precursor materials and implementing the synthesis procedures that will produce better yielding and the required properties of the material. There are a variety of porous materials used for the adsorption of gases such as zeolites (Cavenati, Grande, & Rodrigues, 2004; Chen & Ahn, 2014; Kacem, Pellerano, & Delebarre, 2015), carbon nanotubes (Gui, Yap, Chai, & Mohamed, 2013; Lee, Kang, Shim, Kim, & Moon, 2006), metal organic frameworks (MOFs) (Casco et al., 2015; Llewellyn et al., 2008; Zhou, Wu, Hartman, & Yildirim, 2007), and activated carbons (Buss, 1995; Byamba-Ochir, Shim, Balathanigaimani, & Moon, 2017; Munusamy, Somani, & Bajaj, 2015). Among these materials, activated carbons are the most widely used in the industry, due to their ease of synthesis as well as the possibility of their pore tailoring and chemical functionalization. In addition, activated carbons can be produced from biomass as the carbon source with the advantages of low cost, high availability, valorization of agricultural wastes, and development of high specific surface area adsorbency (Feroldi, Neves, Borba, & Alves, 2018).

In this chapter, the use of activated carbons for the adsorption of key gases with a focus on biomass-derived activated carbons is discussed. Particular attention is paid to the effect of textural (such as surface area and pore volume) and chemical (surface groups) properties upon the activated carbons' adsorption behavior.

9.2 Methane storage

Methane accounts for about 10% of total GHG emissions. Over 60% of total CH_4 is generated by human activities such as energy production, organic waste, and the raising of livestock, while the other 40% is emitted by natural sources like natural wetlands (EEA, 2014; EPA, 2016). As part of the global effort to mitigate its environmental impact, methane is considered as a clean energy in the forms of natural gas and biomethane. The global increase in the main GHGs from 1750 to 2011 is shown in Fig. 1.

Natural gas is considered an alternative fuel due to its environmental benefits; its combustion is relatively cleaner than that of other fossil fuels. Natural gas use results in the production and emission of fewer pollutants: 29% less carbon dioxide per joule delivered than oil and 44% less than coal (Natural Gas.org, n.d.). As a vehicle fuel, methane emits 15%–20% fewer GHGs than gasoline. The main component of natural gas is methane (approximately 90% and higher in most cases); it also contains a small portion of heavier hydrocarbons such as ethane (C_2H_6), propane (C_3H_8), and butane (C_4H_{10}). On the other hand, biogas is the product of the anaerobic fermentation of organic matter whose main constituents are CH_4 (50%–75%) and CO_2 (25%–45%) (Table 1). Biogas can be directly burned to generate electricity in a combined heat and power plant or alternatively it can be purified to produce biomethane (96% CH_4) (Feroldi et al., 2018).

The utilization of biomethane as an alternative energy source also presents a variety of environmental benefits, including decrease in GHG emissions, cheap organic waste recycling, and the formation of nitrogen-rich products that can be used as fertilizers. In addition, the similar properties of biomethane and methane make biomethane an alternative to replace or to be added to natural gas for vehicular engines. However, despite the benefits of using methane-rich fuels, their use is limited due to transportation and storage constraints. At normal temperature and pressure, methane is in the gaseous state, which results in a volumetric energy density of only 0.12% of that of gasoline. Table 2 shows the physical properties of methane.

Currently, CH_4 storage methods are performed by means of CNG or LNG. These methods involve high cost processes (cryogenics and compression) and special storage tanks. In addition, these methods present risks to safety due to the use of high

Table 1 Composition of natural gas and biogas in volume percentage (Feroldi et al., 2018)

Compound	Natural gas (%)	Biogas (%)
Methane (CH_4)	88.1	50–75
Carbon dioxide (CO_2)	0.8	25–45
Ethane (C_2H_6)	4.2	[a]
Nitrogen (N_2)	5.2	0–3
Hydrogen (H_2)	[a]	0–2
Hydrogen sulfide (H_2S)	[a]	**0–1**

[a]None detected.

Table 2 Physical properties of methane (National Center for Biotechnology Information, n.d.)

Molar mass ($g\,mol^{-1}$)	Boiling point (K)	Critical temperature (K)	Critical pressure (MPa)	Critical density ($g\,cm^{-3}$)
16.043	111.65	190.90	4.64	0.16

pressure and/or the refueling of cryogenic liquids that complicate their use in the vehicle sector. A promising alternative that will reduce the costs and safety risks consists of the storage of natural gas (or biomethane) at low pressure and normal temperature in the adsorbed form (ANG).

According to the DOE, to commercialize ANG technology a target of CH_4 storage capacity of 180 v/v at 35 MPa (volume of gas at standard pressure and temperature conditions per volume of the storage vessel) was established (Burchell & Rogers, 2000). In this context, the search for efficient adsorbent materials that can meet the DOE standards has become a major challenge.

The established value of 180 v/v corresponds to the energy density of CNG at a pressure of 16.3 MPa (Table 3). This value is considered as a reference target when searching for new adsorbents. However, the DOE under the program "Methane Opportunities for Vehicular Energy (MOVE)" demands an energy density for the ANG systems equivalent to CNG under 25 MPa (266 v/v) (Energy, U.S. Department of Energy: Washington, 2012).

It is of general consensus that an efficient adsorbent for the storage of methane must have a high specific surface area (from 1000 up to 3000 m^2/g), with a pore size distribution that ranges around 1.0–2.0 nm with a micropore volume of at least 85% of total pore volume (Biloe, Goetz, & Mauran, 2001). A different class of porous materials has been studied aiming to reach the DOE target, including activated carbons, ordered carbonaceous materials, zeolites, and MOFs.

Table 3 Energy densities of methane and conventional fluids (Kumar, Preuss, Titirici, & Rodríguez-Reinoso, 2017)

Pressure (MPa)	Volumetric storage capacity (v/v) at STP	Energy density (MJ/L)
CNG (15 MPa)	168	5.80
CNG (20 MPa)	222	7.68
CNG (25 MPa)	266	9.2
LNG (110 K and 0.1 MPa)	600	22.2
Gasoline	[a]	34.2
Diesel	[b]	**37.3**

1 L of CH_4 = 0.0345 MJ.
[a]Required density of methane in a CNG tank to make it equivalent to gasoline.
[b]Required density of methane in a CNG tank to make it equivalent to diesel: 1.0785 g/cm^3.

Among the available adsorbent materials, activated carbons present advantages that make them suitable materials for gas storage such as high surface area, large pore volume, and light weight. In particular, for the methane storage process they have the benefits of not being sensitive to humid conditions; they tend to be reasonably priced and have good adsorption properties at atmospheric pressure.

The adsorption capacity of activated carbons is highly related to the textural properties. They present a wide range of pore sizes that goes from the micropore to the macropore region and a surface area between 400 and 3000 $m^2 g^{-1}$. Furthermore, their textural properties can be controlled by various activation factors.

The CH_4 molecule is symmetrical and therefore it does not present a dipole or quadripole. It is difficult then to increase the storage capacity of the adsorbent by surface modification. However, the gas storage capacity of activated carbons can be enhanced by controlling the pore structures. For methane adsorption, it has been found that the presence of high microporosity with no macropore volume is optimal since the size of the methane molecules is equivalent to the size of the micropore. An average pore volume of at least 0.8 nm (diameter greater than two molecules of methane) is desired (Antoniou et al., 2014; Cracknell, Gordon, & Gubbins, 1993). In fact, since the interaction of the methane molecules with activated carbons is governed by van der Waals interactions, the attractive force of the pore walls of the adsorbent is a function of the distance z between the center of the methane molecule and the center of a carbon atom on the pore wall; this interaction can be described by the Lennar-Jones potential (Fig. 2).

The attraction force becomes stronger when z is equivalent to one methane molecular diameter (0.36 nm). At longer distances the attraction energy becomes too week and thus the formation of a second adsorbed layer is impeded; the rest of the methane molecules will fill the empty pore space with a low gas-like density.

A linear relationship between the micropore volume (up to 1.6 $cm^3 g^{-1}$) and the methane adsorption capacity of different activated carbons at 35 bar and 298 K (Fig. 3) could be established. However, a large portion of pores that have an optimal size of around 0.8 nm plays a major role in the adsorption of methane (Lozano-Castelló, Cazorla-Amorós, Linares-Solano, & Quinn, 2002).

Furthermore, it has been reported that the presence of pores larger than 0.8 nm is needed to facilitate the access and adsorption of the gas molecules. For example, García-Blanco et al. showed that the Maxorb activated carbon, with a surface area of 3090 $m^2 g^{-1}$ and a pore size distribution from pores between 0.5 and 3 nm, showed a micropore filling of only 59% when adsorbing methane at 298 K and 3500 kPa. However, even if the micropore filling was low, the narrow micropore (<0.8 nm) filling is total, indicating that methane is not only adsorbed in the narrow micropores but also in pores of a larger size (García Blanco, Vallone, Korili, Gil, & Sapag, 2016).

This need for larger pores for methane adsorption was also evidenced by Casco et al. when comparing a set of activated carbons with different pore volumes (Fig. 4) (Casco et al., 2015). Samples with a pure microporous network (F400 and Maxsorb) showed a maximum adsorption capacity at moderate working pressures (6–8 MPa). On the contrary, with samples that combine micropores with mesopores (pore size <2 nm) such as the activated carbons RGC30, LMA738, LMA405, and

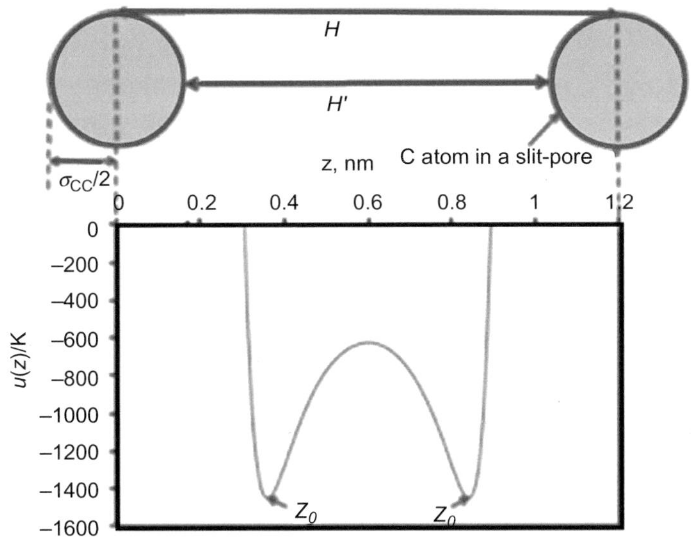

Fig. 2 Potential energy, $u(z)$, for a spherical methane Lennard-Jones site interacting with walls of a slit-pore of width $H = 1.2$ nm in terms of the distance z (distance between the center of the methane molecule and the center of a wall carbon atom).
Reprinted with permission from Kumar, K. V., Preuss, K., Titirici, M. M., Rodríguez-Reinoso, F. (2017). Nanoporous materials for the onboard storage of natural gas. *Chemical Reviews, 117*, 1796–1825. doi: 10.1021/acs.chemrev.6b00505. Copyright 2017 American Chemical Society.

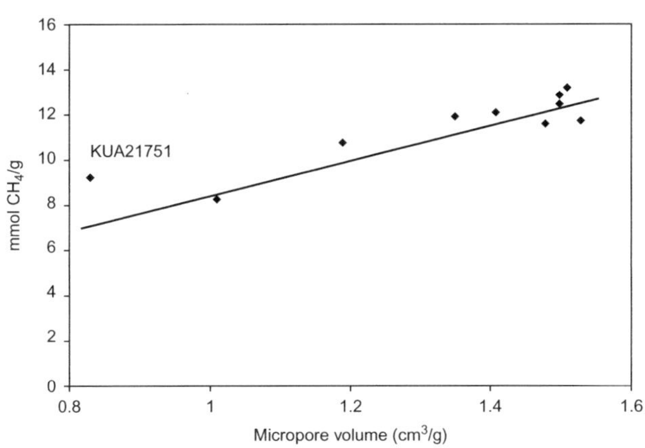

Fig. 3 Relationship of the methane adsorption capacity at 35 bar and 298 K with the micropore volume for a series of activated carbons calculated by N_2 adsorption isotherms.
Reprinted from Lozano-Castelló, D., Cazorla-Amorós, D., Linares-Solano, A., Quinn, D. F. (2002). Influence of pore size distribution on methane storage at relatively low pressure: preparation of activated carbon with optimum pore size. *Carbon, 40*, 989–1002. doi:10.1016/S0008-6223(01)00235-4 with permission from Elsevier.

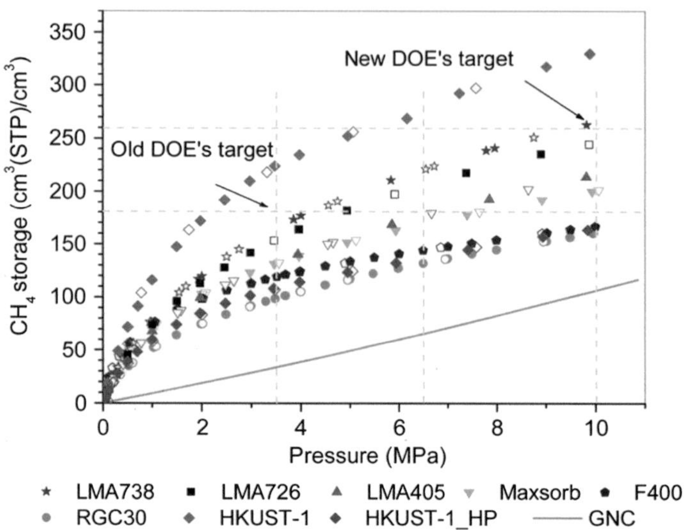

Fig. 4 Methane storage capacity for different activated carbons at 298 K and up to 10 MPa. Metal organic framework sample HKUST-1 has been included for the sake of comparison (*closed symbols* represent adsorption data, *open symbols* represent desorption data). Reprinted with permission from Casco, M. E., Martínez-Escandell, M., Gadea-Ramos, E., Kaneko, K., Silvestre-Albero, J., Rodríguez-Reinoso, F. (2015). High-pressure methane storage in porous materials: are carbon materials in the pole position? *Chemistry of Materials, 27*, 959–964. doi: 10.1021/cm5042524. Copyright 2015 American Chemical Society.

LMA726 there was a continuous increase in the adsorbed quantity. Furthermore, they found that the correct design of activated carbons can result in materials that meet both the DOE and DOE MOVE requirements set by the DOE.

As was previously mentioned, the feasibility of using ANG instead of CNG and LNG relies on the ability of the adsorbent to store a high amount of gas inside the fuel tank, which means that the porous adsorbent must present a high packing density ensuring that the storage capacity and the energy density on a volumetric basis is high. Since the microporosity of activated carbons is created by removal of carbon atoms during the activation process, after a certain microporosity volume further activation is followed by the creation of macropores translating into a useless volume for gas storage and thus a decrease in the packing density. Hence, carbons with high CH_4 uptake might result in a lower volumetric storage capacity of methane than some with lower surface area. In this sense, it is useful to express the adsorption capacities in terms of volumetric quantities instead of gravimetric basis (Fig. 5) (Menon & Komarneni, 1998).

Along with a high micropore volume, the activated carbon is required to have a high packing density, thus presenting a high energy density on a volumetric basis. Optimization of the packing density can be done by means of improving the bulk density, monolith preparation, and particle size design. It is strongly related to the

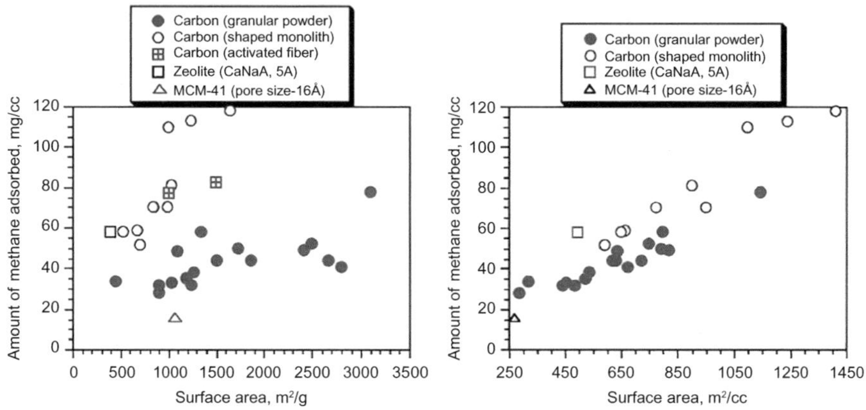

Fig. 5 Effect of expressing the methane adsorption capacity in terms of gravimetric surface ($m^2\ g^{-1}$) area versus volumetric surface area ($m^2\ cm^{-3}$) for different porous materials at 3.447 MPa and 298 K. A suitable adsorbent for the adsorbed natural gas storage presents a maximum volumetric surface area.
Reprinted by permission from Springer Menon, V. C., Komarneni, S. (1998). Porous adsorbents for vehicular natural gas storage: a review. *Journal of Porous Materials, 5*, 43–58. doi:10.1023/A:1009673830619. Copyright 1988.

activation degree of the adsorbent; the higher the surface area and pore volume are, the lower the packing density is. Byamba-Ochir et al. showed the importance of the activation conditions for the preparation of activated carbons for methane storage. The authors found that the BET surface area increased with an increase in the ratio of activating agent and the Mongolian raw anthracite; however, the packing density gradually decreases and so does the volumetric capacity (Table 4) (Byamba-Ochir et al., 2017).

To further investigate the effect of the adsorbent design parameters and the volumetric storage capacity, Kumar et al. plotted the volumetric adsorption capacity versus the product of the packing density and the specific surface area of the adsorbents (Fig. 6) finding a logarithmic relationship. They determined that to achieve the volumetric methane adsorption capacity of 180 v/v, the adsorbents must have a surface

Table 4 Effect of activating agent/Mongolian raw anthracite ratio in volumetric methane adsorption capacity (Byamba-Ochir et al., 2017)

Carbon monoliths	S_{BET} (m^2/g)	v/v ($cm^3_{CH4}/cm^3_{activated\ carbon}$)
PMAC2/1-C3-65	622	43.6
PMAC1/1-C3-65	845	60.1
PMAC1/2-C3-65	1460	162.2
PMAC1/3-C3-65	1757	147.8

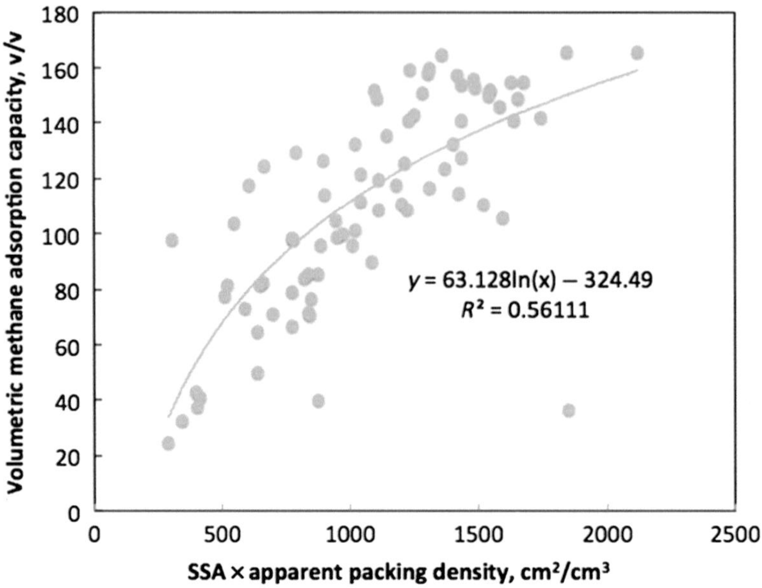

Fig. 6 Volumetric adsorption capacity (v/v) of carbon structures versus the product of the BET specific surface area (SSA) and packing density.
Reprinted with permission from Kumar, K. V., Preuss, K., Titirici, M. M., Rodríguez-Reinoso, F. (2017). Nanoporous materials for the onboard storage of natural gas. *Chemical Reviews, 117*, 1796–1825. doi: 10.1021/acs.chemrev.6b00505. Copyright 2017 American Chemical Society.

area of 2463 m^2 cm^{-3} and a packing density of about 1.2 g cm^{-3}. These are just guidelines since consideration of pore volume and distribution must be made (Kumar et al., 2017).

One important advantage of using activated carbons as adsorbent materials for gas separation and storage is the possibility of using different organic materials rich in carbon contents as precursors with coal and wood being the most employed ones. However, in recent years there has been an increasing interest in the production of activated carbon from agricultural wastes. The use of these types of material as precursors can reduce the pressure on mines and forests, while valorizing products that otherwise result in waste production with economic and environmental impacts (Mozammel, Masahiro, & Bhattacharya, 2002).

Examples of such precursor materials include rubber seed coat, olive seed, coconut shell, oil palm fiber, and almond shell (Arami-Niya, Wan Daud, Mjalli, & Abnisa, 2012). Some of the results obtained using low-cost lignocellulosic biomass activated carbons for methane adsorption and storage are described in Table 5.

Activated carbons can be optimized for methane storage by testing several activation conditions, such as activation method, activation temperature, activation agent, and activation agent-to-char ratio. Bagher and Abedi studied the effect of activation conditions on a corn cob-based activated carbon, showing that the methane adsorption capacity can be increased from a low 25 v/v to up to 120 v/v at 298 K and 3.44 MPa

Table 5 Biowaste precursors used to prepare activated carbons along with their specific surface area and CH_4 storage (González-García, 2018)

Precursors	Activating agent	Surface area (m^2 /g)	Storage capacity
Sugarcane molasses	KOH	2202	197.23 mg g^{-1}
Oil palm shell	$ZnCl_2$	870	13 cm^3 g^{-1}
Corn cob	KOH	3227	273.4 cm^3 g^{-1}
Corn cobs	KOH	1320	120 v/v
Palm shell	H_3PO_4		25.7 cm^3 g^{-1}
Olive stones	H_3PO_4	1014	4.69 mmol g^{-1}

when using optimal activation conditions:activation temperature of 823 K for 60 min and impregnation with KOH/char on a 1:1 ratio (Bagheri & Abedi, 2011). Using a similar optimization process, Policicchio et al. were able to obtain a CH_4 adsorption capacity of 150 v/v at 298 K and 3.5 MPa for a cellulose-based activated carbon (Policicchio et al., 2013). This optimization relies on the change of the textural parameters of the adsorbent (such as BET surface area and pore size distribution) upon different activation conditions. Ruiz et al. tested the effect of different activation temperatures, activation agent/precursor weight, and inner flow gas on the pore size distribution (Fig. 7) (Ruiz, Ferrera-Lorenzo, & Fuente, 2017). Overall the best methane storage capacities were found when the activation method produced samples with high BET surface area. In the case of physical activation methods, a higher activation degree resulted in the development of porosity, thus resulting in a higher surface area. In the case of chemical activation, a low activation temperature (between 700°C and 800°C) and a higher activation ratio/precursor resulted in a higher narrow micropore (<0.8 nm) volume.

All the hitherto present studies show that the new DOE MOVE requirements are difficult to achieve by activated carbons and their feasibility is questionable. Furthermore, using Monte Carlo simulations a maximum methane storage capacity of activated carbons has been predicted to be 209 v/v at 3.4 MPa (Matranga, Myers, & Glandt, 1992). Thus, this result implies that researchers working on the design of new materials for methane storage should reconsider this target and/or changing the operating conditions (i.e., using higher pressures).

9.3 Carbon dioxide storage

Reducing carbon dioxide emissions is a key factor to combat global warming since among the greenhouse gases, CO_2 is the one emitted in higher volume. It is estimated that in 2006, CO_2 alone accounted for 81.6% of total GHG emissions in the United States (United States Environmental Protection Agency (EPA), n.d.; Yamasaki, 2003). CO_2 atmospheric concentration has increased from a preindustrial value of 280 to 408 ppm in 2018 resulting in an increase of 1°C of the global temperature (Intergovernmental Panel on Climate Change (IPCC), 2007; NASA, 2018).

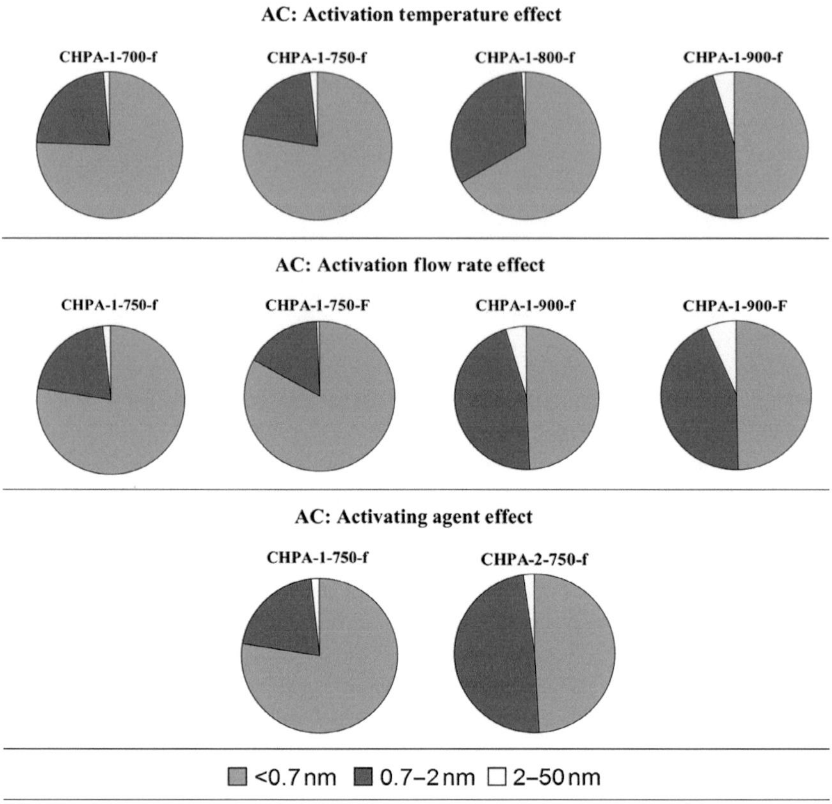

Fig. 7 Micro/mesoporosity volume distributions (in percentages) within chestnut shell activated carbons.
Reprinted from Ruiz, B., Ferrera-Lorenzo, N., Fuente, E. (2017). Valorisation of lignocellulosic wastes from the candied chestnut industry. Sustainable activated carbons for environmental applications. *Journal of Environmental Chemical Engineering, 5*, 1504–1515. doi:10.1016/j.jece.2017.02.028. Copyright (2017), with permission from Elsevier.

The main source of carbon dioxide is fossil fuel combustion followed by land-use change. Annual fossil fuel emissions increased from an average 6.4 GtC in the 1990s to 7.2 GtC per year in 2000–05 (1 GtC of carbon = 1 billion tonnes of carbon) (Plaza et al., 2008). Fossil fuel combustion is the principal source of anthropogenic carbon dioxide emissions (58% of global GHG emissions) (International Energy Agency, 2017).

CCS has been proposed as one of the most promising technologies to reduce carbon dioxide emissions in the short term, while carbon-free processes are implemented. It is believed that by the year 2050, CCS can reduce carbon dioxide emissions by 20% (Global CCS Institute, 2011). CSS is a three step process:

1. Separation of CO_2 from other emissions
2. CO_2 transportation to the storage site
3. Permanent storage

The transportation and storage steps are already developed technologies. It is the first stage of the CCS that has made its implementation difficult. The separation of the CO_2 accounts for approximately two-thirds of the total cost of a CCS (Intergovernmental Panel on Climate Change (IPCC), 2005; Li et al., 2011). In this sense, between the proposed carbon dioxide capture and separation options, postcombustion capture has been appointed as the most feasible to implement in power plants because it can be retrofitted to existing plants without modification of the already existing systems (Lee, Lee, Bhatia, & Mohamed, 2012).

The process of postcombustion can be done by many separation technologies such as physical adsorption, chemical adsorption, cryogenic separation, and membrane separation. Most of the carbon dioxide capture systems commercially available employ chemical absorption with alkaline amines. Despite method availability, remarkable disadvantages such as high energy requirements, oxidative degradation of the adsorbents, high corrosion, and low-cost efficiency are noted. Other possible technologies are cryogenic separation and membrane separation. However, cryogenic separation has the disadvantage of using a great quantity of energy that makes it too expensive, and the use of membranes is still at a very early development phase (Oh, 2010).

Adsorption technology involves two steps: the first step consists of carbon dioxide adsorption onto a porous adsorbent by the formation of chemical bonds (chemisorption) or by physical interactions such as van der Waals forces (physisorption). The second step is the regeneration of the adsorption column after system equilibration. Regeneration can be done by reducing the pressure (pressure swing adsorption) or by increasing the temperature (temperature swing adsorption).

The adsorption process using physisorbent materials is considered as the most cost-effective option for carbon dioxide separation due to its low energy requirements. In addition, it is a very noble technology that can be easily scaled to different plant sizes and used on a wide range of temperatures and pressures. The success of an adsorption separation technology relies on the quality of the employed adsorbent. It has been described that a successful adsorbent for carbon dioxide capture and storage should satisfy five main requirements (Sevilla, Falco, Titirici, & Fuertes, 2012):

1. Large CO_2 uptake
2. High sorption rate
3. Good selectivity between CO_2 and other competing gases
4. Easy regeneration
5. Low cost and high availability

The search for new adsorbent materials that better fulfill these requirements has become an active field of research. Adsorbents that have been studied for this purpose include activated carbons, zeolites, MOFs, and polymers.

It is important to mention here that development of better adsorbents for carbon dioxide separation would be beneficial not only to the CCS systems but also for the upgrading of biogas and natural gas. In this context, activated carbons are considered as a superior material for CO_2 capture and separation due to their hydrophobic character, lower cost, high thermal stability, and low regeneration energy requirements (Plaza, García, Rubiera, Pis, & Pevida, 2010).

Similar to the adsorption process of methane, the textural properties of activated carbons play a major role in adsorption behavior. CO_2 adsorption is enhanced by the presence of micropores of up to 1 nm. By comparing a number of porous carbons, Presser et al. showed that the volume of pores larger than 1 nm is not a major contributor to CO_2 adsorption. In fact, it is the volume of ultramicropores (size smaller than 0.8 nm) that has the higher influence upon CO_2 adsorption as proven by a higher correlation coefficient (R^2 of 0.9958) (Fig. 8B). Presser et al. reported that upon very low CO_2 partial pressures such as 0.01 MPa it is the pores with a size smaller than 0.5 nm (Fig. 8A) that are the main contributors (Presser, McDonough, Yeon, & Gogotsi, 2011). The dependency of the adsorption of carbon dioxide on ultramicropore volume was also observed by Yin et al. at 273, 283, and 293 K and pressures between 0.01 and 0.10 MPa (Yin, Liu, Liu, & Wu, 2013).

If ultramicropore volume has proven to influence the adsorption capacity of the activated carbons, no clear relationship between the BET surface area and the adsorption capacity has been reported for pressures up to 0.1 MPa (Presser et al., 2011;

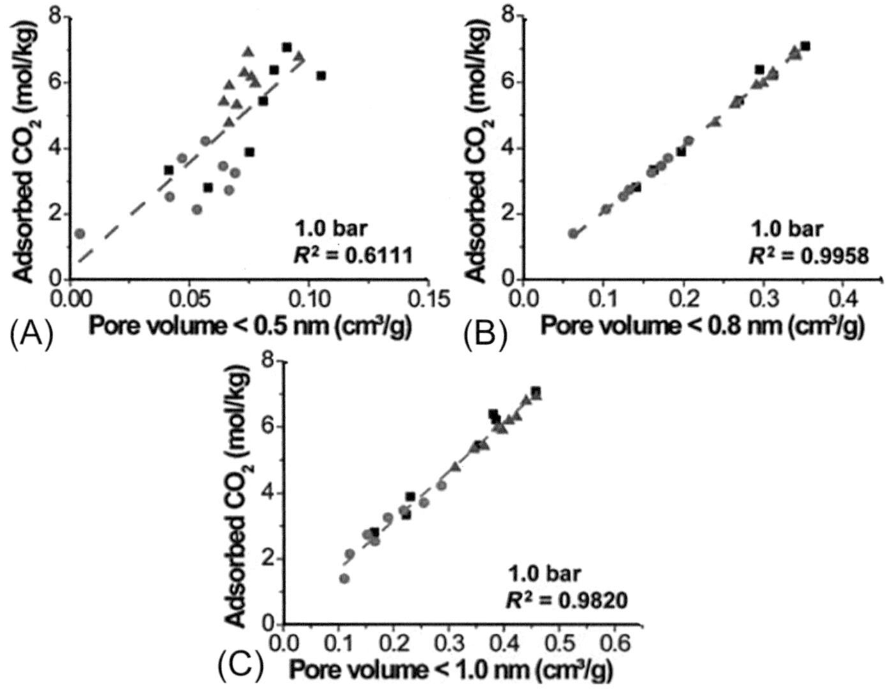

Fig. 8 CO_2 uptake at 273 K and 0.1 MPa for the volume of pores (A) pore volume < 0.5 nm, (B) pore volume < 0.8 nm, and (C) pore volume < 1.0 nm.
Reproduced from Presser, V., McDonough, J., Yeon, S. -H., Gogotsi, Y. (2011). Effect of pore size on carbon dioxide sorption by carbide derived carbon. *Energy & Environmental Science, 4*, 3059. doi:10.1039/c1ee01176f with permission of The Royal Society of Chemistry.

Table 6 Biowaste precursors used to prepare activated carbons along with their activation method, specific surface area, and CO_2 storage capacity (González-García, 2018)

Precursors	Activating agent	Surface area (m^2 g)	Storage capacity
Oil cake/walnut	CO_2	1207	34.72 mL g^{-1}
Coconut shell	CO_2	371	1.8 mmol g^{-1}
Almond shell	CO_2	862	2.7 mmol g^{-1}
Olive stone	CO_2	1215	3.1 mmol g^{-1}
Coffee residue	CO_2	593	2.4 mmol g^{-1}
Rice husk	$ZnCl_2$	927	1.3 mmol g^{-1}
Peanut shell	KOH	956	4.0 mmol g^{-1}
Sunflower seed	KOH	1790	4.6 mmol g^{-1}
Bamboo	KOH	1846	4.5 mmol g^{-1}
Palm stone	H_3PO_4	924	2.7 mmol g^{-1}

Serafin, Narkiewicz, Morawski, Wróbel, & Michalkiewicz, 2017; Yin et al., 2013). The potential theory indicates that CO_2 adsorption happens with micropore filling followed by capillary condensation and liquefaction with increasing pressure; it is expected then that at high pressure the BET surface area and total pore volume gain influence in CO_2 adsorption. Drage et al. reported that at a pressure of 4.1 MPa a linear relationship between the CO_2 adsorption capacity of activated carbons and the BET surface area can be found (Drage, Blackman, Pevida, & Snape, 2009).

Another important factor to be considered when designing a carbonaceous porous material for carbon dioxide adsorption is the source of the raw material from which it will be produced. One of the main advantages of activated carbons is their versatility. Any carbonaceous material can be converted into activated carbon with the main requirements being a high carbon content and low ash content; this raises the possibility of producing environmentally friendly activated carbons based on lignocellulosic biomass. In this context, several efforts have been made to produce activated carbons from lignocellulosic resources with the aim of CO_2 adsorption (Table 6).

The selection of raw materials prior to activated carbon production is a major concern for the gas storage suitability of the adsorbent. Olivares-Marín et al. showed the importance of the careful selection of the precursor. The authors compared the CO_2 adsorption of one sample of preconsumer carpet and two samples of post-consumer carpet under the same activation conditions finding that the presence of inorganic binders in the precursor can result in a lower micropore volume thus lowering the CO_2 adsorption capacity between the samples from 13.8 to 3.7 wt%. Hao et al. obtained similar findings when comparing activated carbons obtained from grass cuttings, house manure, organic waste from beer production, and biosludge from a wastewater treatment plant. The activated carbons produced from grass cuttings and horse manure showed higher CO_2 adsorption capacities due to a larger volume of ultramicropores (Hao, Björkman, Lilliestråle, & Hedin, 2013). Munusamy et al. showed that by using carboxyl methyl cellulose sodium salt and soluble starch as binders to convert a powdered mango seed shell-derived activated carbon into an extruded form they could enhance the maximum CO_2 equilibrium adsorption capacity from 14.3 to 21.3 wt% at 0.1 MPa and 273 K, which was explained by the formation of surface functionality by the binders on the extruded activated carbon (Munusamy et al., 2015).

In addition to the precursor nature, the optimal operating conditions of the activation process need to be determined since they can significantly affect the adsorption capacity of the activated carbon. There are two possible activation paths for producing activated carbons: chemical activation and physical activation. Chemical activation using KOH, H_3PO_4, or $ZnCl_2$ is usually preferred since it results in hierarchical porous structures with high surface areas; it also consumes less energy because the chemical activation temperature is between 600°C and 800°C versus 800°C and 1000°C for physical activation. Chemical activation has the additional advantage that it can be done in one stage where carbonization and impregnation of the precursor are done simultaneously, while physical activation involves a first step of carbonization and a second step where the material is activated in the presence of oxidants such as air, CO_2, and steam (Singh et al., 2017; Wang & Kaskel, 2012).

Song et al. compared different activation methods and conditions for corn stalk-based activated carbons. The activated carbons produced by physical activation at different activation concentrations or activation times presented surface areas lower than $500\,m^2\,g^{-1}$, while chemical activation resulted in slightly higher surface areas (up to $639.8\,m^2\,g^{-1}$); additionally, the chemical activated carbons also showed higher micropore volumes. These two factors resulted in higher CO_2 uptakes of chemical activated carbons than physical activated ones (Fig. 9). Between all the samples a maximum carbon dioxide adsorption capacity of 7.33 wt% was obtained (Song, Liao, Xiao, & Shen, 2015). Alabadi, Razzaque, Yang, Chen, and Tan (2015) were able to obtain highly porous activated carbons by chemical activation with KOH of a starch and gelatin 1:1 mix; a CO_2 capture capacity of $3.8\,mmol\,g^{-1}$ at 298 K and 0.1 MPa was obtained (Alabadi et al., 2015). Using also KOH activation to produce a pine nut shell-derived activated carbon, a high CO_2 uptake of $5.0\,mmol\,g^{-1}$ at 273 K and 0.1 MPa was determined by Deng et al. (2014). A CO_2 adsorption of $5.22\,mmol\,g^{-1}$ at 293 K and 0.1 bar, one of the highest reported CO_2 adsorption uptakes for carbonaceous materials, was obtained by KOH chemical activation of Coca-Cola as a source of biomass by Boyjoo et al. (2017).

Typical activated carbons show weak affinity for carbon dioxide having an average adsorption capacity of 5 wt% at low pressure (0.01 MPa) and 298 K (Lee & Park, 2015). This value, however, can be improved by chemical modification of the activated carbon's surface. One possibility is to introduce Lewis bases onto the activated carbon surface, which favors CO_2 adsorption due to its acidic properties. Nitrogen enrichment has been reported to be an effective way for introducing basic groups on the activated carbon. Meng et al. reported a CO_2 adsorption capacity of up to 17.7 wt% at 298 K and 0.01 MPa when treating a polypyrrole activated carbon with NaOH; they reported that the adsorption capacity was a function of both the nitrogen content and the micropore volume (Meng & Park, 2014). Other works reported the same dependence between carbon dioxide capacity and both textural properties and N content (Tiwari, Bhunia, & Bajpai, 2017; Yaumi, Bakar, & Hameed, 2018).

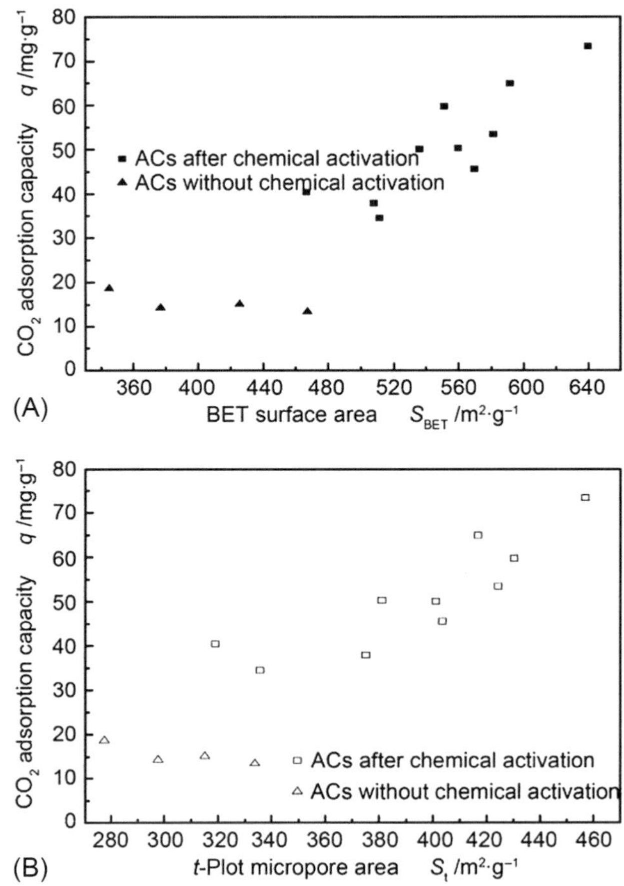

Fig. 9 Effect of (A) Brunauer-Emmett-Teller surface area and (B) *t*-plot micropore area on CO_2 adsorption capacity at 298 K for chemical activated (*squares*) and physical activated (*triangles*) carbons from corn stalk.
Reprinted from Song, T., Liao, J. M., Xiao, J., Shen, L. H. (2015). Effect of micropore and mesopore structure on CO2 adsorption by activated carbons from biomass. *Xinxing Tan Cailiao/New Carbon Materials, 30*, 156–166. doi:10.1016/S1872-5805(15)60181-0, Copyright (2015) with permission from Elsevier.

When introducing nitrogen functionalities the presence of oxygen surface groups is desirable, since they will act as anchoring sites for the nitrogen groups. Caglayan and Aksoylu studied the effect of different oxidation and heat treatments on the formation of N surface groups. They found that the CO_2 adsorption capacity was enhanced by Na_2CO_3 impregnation of the air-oxidized and nitric acid-oxidized activated carbon. They reported that sodium atoms (Na) have CO_2 sorption ability and that oxidation with HNO_3 provided anchoring sites for Na atoms (Caglayan & Aksoylu, 2013). Shafeeyan et al. compared the CO_2 adsorption capacity of 2 activated carbons (pristine and pre-oxidised) thermally treated with ammonia (Shafeeyan, Daud, Houshmand, &

Arami-Niya, 2011). The samples that were not preoxidized presented a higher BET surface area, pore volume, and micropore volume. This behavior is due to the thermal decomposition of functional groups inside the pores. Furthermore, when the sample is previously oxidized there is a blockage of the micropore's entrances by oxygen and also a collapse of some pore walls leading to a decrease in the BET surface area and the micropore volume. However, if the oxidation step is done at high temperature (800°C) the oxygen surface groups are decomposed and the textural properties are recovered (Fig. 10, sample OXA-800). They found a clear relationship between both micropore volumes and nitrogen content with the CO_2 adsorption capacity.

Seema et al. showed that upon sulfur (S) doping of microporous carbonaceous materials the CO_2 adsorption capacity can be increased. They found a correlation between the BET surface area, micropore volume, and oxidized sulfur surface groups with the carbon dioxide adsorption capacity. A maximum adsorption capacity of $4.5\,mmol\,g^{-1}$ was reached (Seema et al., 2014). Xiao et al. (2014) also reported an increase in the CO_2 adsorption capacity after S-doping of microporous carbons; they obtained a maximum adsorption capacity of $2.46\,mmol\,g^{-1}$ when an S content of 6.56% was present and of $1.69\,mmol\,g^{-1}$ when there were no sulfur atoms on the carbonaceous material.

It is, then, that even if the CO_2 adsorption capacity of activated carbons is greatly influenced by the BET surface area and the volume of narrow micropores (<0.8 nm), the surface chemistry can also play an important role in the search for better materials for CO_2 separation and storage. Different functionalities can be added to the surface of the adsorbent that will participate in the adsorption process through acid/base interactions and polarizability of the carbon dioxide molecules.

In addition to nitrogen and sulfur doping, activated carbons can be treated with metal oxides resulting in chemical reactions between the CO_2 molecules and the metal oxides at elevated temperatures. Published works include impregnation of activated carbons with alkaline metals such as Mg and Ca, and transition metals such as Cu, Co, Ni, Fe, Cr, and Zn (Somy, Mehrnia, Amrei, Ghanizadeh, & Safari, 2009; Son et al., 2005; Yong & Mata, 2001). Son et al. determined the CO_2 adsorption capacity of metal-impregnated activated carbons to decrease in the order of $Mg > Ca > Co = Cu > Ni$ (impregnated metal) (Son et al., 2005). Using Zn^{2+} as a promotor (Fig. 11), Somy et al. tested the adsorption behavior of activated carbons impregnated by Fe_2O_3 and Cr_2O, finding a 20% increase in the CO_2 adsorption capacity upon Cr_2O impregnation (compared with raw AC); meanwhile Fe_2O_3 impregnation was not effective (Somy et al., 2009). It has been reported that an acidic treatment prior to metal impregnation can further increase carbon dioxide adsorption; the acid groups will oxidize the samples and provide oxygen surface groups that can form metal complexes with the metal ions (Hosseini et al., 2015).

9.4 Hydrogen storage

One way to reduce GHG emissions is to design suitable adsorbents to efficiently remove gases like CO_2 and CH_4 from the atmosphere. In addition, renewable clean energy sources (water, wind, solar) are currently being developed to replace fossil

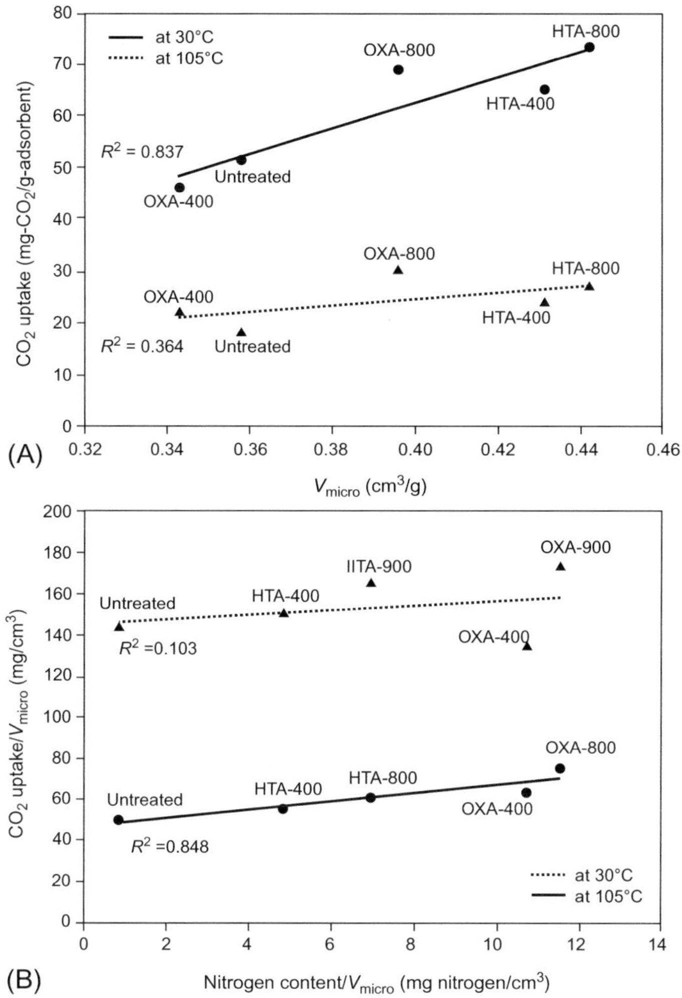

Fig. 10 Relationship between the CO_2 capture capacity and (A) micropore volume and (B) nitrogen content (N%). OXA-400 and OXA-800: preoxidized samples aminated at 400°C and 800°C, HTA-400 and HTA-800: heat-treated samples aminated at 400°C and 800°C without preoxidation.
Reprinted from Publication Shafeeyan, M. S., Daud, W. M. A. W., Houshmand, A., Arami-Niya, A. (2011). Ammonia modification of activated carbon to enhance carbon dioxide adsorption: effect of pre-oxidation. *Applied Surface Science, 257*, 3936–3942. doi:10.1016/j.apsusc.2010.11.127, Copyright (2011), with permission from Elsevier.

fuels in the long term. However, an important disadvantage of renewable energy sources is their intermittence, which requires efficient energy storage systems to store and release energy when needed. Hydrogen is one of the renewable resources that can help in addressing the growth of energy demand without contributing to global climate

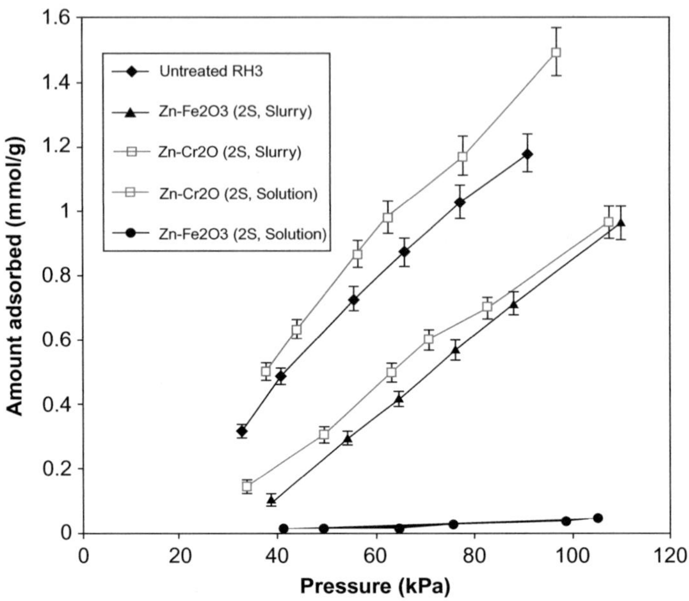

Fig. 11 CO_2 adsorption isotherms of Zn^{+2}–metal oxide samples prepared by solution and slurry impregnation methods.
Reprinted from Somy, A., Mehrnia, M. R., Amrei, H. D., Ghanizadeh, A., Safari, M. (2009). Adsorption of carbon dioxide using impregnated activated carbon promoted by Zinc. *International Journal of Greenhouse Gas Control, 3*, 249–254. doi:10.1016/j.ijggc.2008.10.003, Copyright (2009), with permission from Elsevier.

change. Hydrogen is an abundant element in fossil hydrocarbons and all organic matter. In its pure molecular form, H_2, hydrogen is a colorless, odorless, nontoxic gas that produces only water vapor when used in a fuel cell, making it a clean fuel. H_2 has the advantage of possessing the highest energy density among common fuels by weight. Unfortunately, gaseous H_2 has the lowest energy density by volume and presents a risk of explosion in contact with air. Therefore, developing safe, efficient, and inexpensive hydrogen storage systems with acceptable gravimetric and volumetric energy densities is still an objective for researchers to achieve (Andrews & Shabani, 2014).

At present, two methods are mainly used for the storage of large quantities of hydrogen. Storage in the gaseous state is done at a high pressure (700 bar) in bottles, which is rather heavy and bulky. Storage in the liquid state requires a temperature of 20 K and therefore expensive cooling devices along with significant losses by evaporation (boil-off phenomenon). Both methods present significant problems related to storage under pressure and the high explosive potential of hydrogen in air (above 4% by volume). The third solution is solid hydrogen storage, with significant safety and volume capacity advantages.

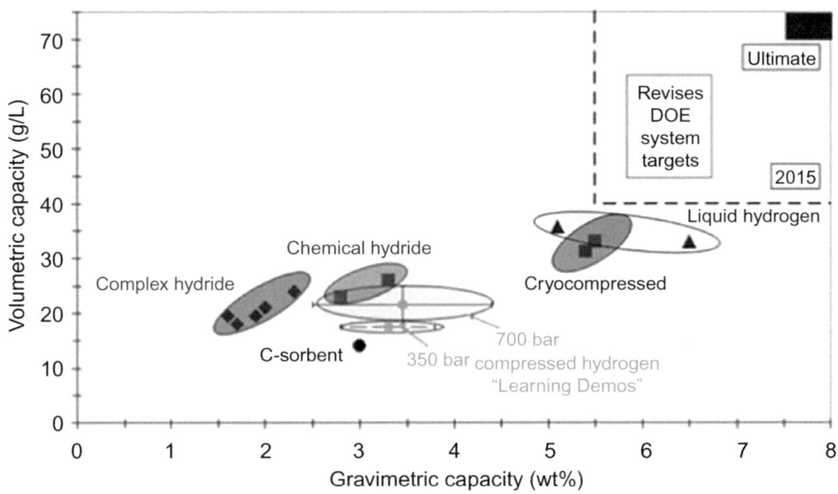

Fig. 12 Ranking of current hydrogen storage technologies by volumetric and gravimetric capacities showing the Department of Energy (DOE) 2015 and ultimate targets.
Reprinted from Durbin, D. J., Malardier-Jugroot, C. (2013). Review of hydrogen storage techniques for on board vehicle applications. *International Journal of Hydrogen Energy, 38*, 14595–14617. doi:10.1016/j.ijhydene.2013.07.058, Copyright (2013), with permission from Elsevier.

The DOE targeted a minimum gravimetric and volumetric requirement for onboard hydrogen storage systems. However, the current market and technology scenario has led to some revisions and changes in targets. In 2010, the targeted storage system had a gravimetric density of 4.5 wt% (1.5 kWh/kg) and volumetric density of 28 g/L (0.9 kWh/L), while for 2015 the targeted gravimetric density was changed to 5.5 wt% (1.8 kWh/kg), and the volumetric density was revised to 40 g H_2/L (1.3 kWh/L) (Md Arshad et al., 2016).

As highlighted in Fig. 12 (Durbin & Malardier-Jugroot, 2013), the current hydrogen storage technology does not meet the DOE standards. This may be related to fundamental limitations of each particular storage method: (1) gaseous H_2 requires a large volume to be stored, (2) liquid hydrogen requires low temperatures and evaporates easily, (3) metal hydrides are heavy and present thermodynamic hydrogen release limitation, and (4) adsorbent materials present low adsorption capacities at room temperature and pressure. Therefore, significant efforts are made to improve these storage techniques and to develop new storage methods.

Great interest has been focused on hydrogen physisorption on different adsorbents, including zeolites, MOFs, and carbon-based materials. Particularly, carbon adsorbents exhibit several advantages compared to other adsorbents, i.e., low cost, low densities, fast kinetics, and complete hydrogen reversibility (Fierro, Zhao, Izquierdo, Aylon, & Celzarda, 2010). Moreover, hydrogen sorption capacity can be comparable or even higher than MOFs and zeolites (Builes et al., 2011). Among the carbonaceous materials, activated carbons have been widely investigated as H_2 storage materials due to their easy preparation process combined with more developed porous structures when compared to nanostructured carbons (carbon nanotubes, carbon nanofibers, template carbons).

Biomass is widely employed as a precursor to obtain carbon mainly because of its eco-friendly nature, widespread availability, and high adsorption/desorption rates (Agarwal, Noh, Schwarz, & Davini, 1987; Masika & Mokaya, 2013). Several biomass precursors have been explored, i.e., cellulose, corncob, rice husk, olive bagasse, starch, etc. (Table 7); however, simple pyrolysis of biomass results in low surface area carbon materials and in most cases an activation step is usually used to develop the porosity. Generally, two types of activation procedure based on physical (CO_2 or water steam) and chemical (KOH, H_3PO_4, etc.) activation are used. In particular, as revealed by Tables 7–9 gathering literature data, chemical activation with KOH is the most used since it leads to highly microporous activated carbons with well-developed surface areas (up to $3700\,m^2\,g^{-1}$). In addition, chemical activation requires lower activation temperatures and shorter reaction times than physical activation; however, it requires a supplementary washing step to remove the impurities created during the activation process.

The development of the porosity is a consequence of carbon oxidation (etching) due to the redox reaction between potassium hydroxide and carbon with the formation of carbonates, release of volatiles (H_2, CO_x), and intercalation of K, according to the following equations (Choi & Park, 2015; Wang et al., 2009):

$$6KOH + 2C => 2K + 3H_2 + 2K_2CO_3 \tag{1}$$

$$K_2CO_3 + C => K_2O + 2CO \tag{2}$$

$$K_2O + C => 2K + CO \tag{3}$$

Other parameters such as carbon precursor type, activation agent type/amount, and activation temperature/time used during carbon preparation are the main factors determining porosity development and surface chemistry of the materials. Both parameters are of prime importance when designing efficient adsorbents for hydrogen storage applications and will be discussed further in more detail.

9.4.1 Influence of adsorption conditions (T, P)

Hydrogen storage studies are mainly focused at 77 K, due to the weak interaction between molecular hydrogen and solids. The amount of hydrogen stored at room temperature is very low, while it can consistently increase at low temperature, typically 77 K, that is, the temperature of liquid nitrogen. At elevated temperatures such as room temperature, the storage capacity was found to vary linearly with pressure, following Henry's law, while at 77 K the adsorption isotherm of all samples could be satisfactorily explained using the Langmuir model (Fig. 13).

Therefore, at room temperature the hydrogen adsorption capacity is very low. This trend can also be seen in Table 7, where several hydrogen adsorption capacities of

Table 7 Activated carbon derived from different biomasses along with their textural properties and hydrogen adsorption capacity at room temperature (398–303 K) and different pressures

Precursor	Activating agent	BET SSA (m^2/g)	$V\mu$ (cm^3/g)	H_2 adsorption capacity (wt%)	T (K)	P (bar)	References
Cellulose	KOH	3771	–	0.5	298	20	Blankenship Ii, Balahmar, and Mokaya (2017)
Olive pomace	KOH	1269	0.48	1.22	298	204	Heo and Park (2015)
Jute fibers	KOH	1224	0.43	1.2	303	40	Ramesh, Rajalakshmi, and Dhathathreyan (2017)
Tamarind seeds	KOH	1784	0.64	1.36	303	60	Ramesh, Rajalakshmi, and Dhathathreyan (2015)
Coffee shell	KOH	3149	–	0.91	298	140	Li, Li, Tan, et al. (2016)
Olive bagasse	CO_2	1185	0.45	0.63	298	200	Bader and Abdelmottaleb (2017)

Table 8 Activated carbon derived from different biomasses along with their textural properties and hydrogen adsorption capacity at 77 K and 1 bar

Precursor	Activating agent	BET SSA ($m^2 g^{-1}$)	$V\mu$ ($cm^3 g^{-1}$)	H_2 adsorption capacity (wt%)	*T* (K)	*P* (bar)	References
Fruit bunch	CO_2	687	0.297	1.97	77	1	Md Arshad et al. (2016)
Beer lees	KOH	1927	0.70	2.92	77	1	Balathanigaimani et al. (2018)
Rice husks	KOH	2682	0.792	2.85	77	1	Heo and Park (2015)
Cellulose	KOH	3771	–	3.9	77	1	Andrews and Shabani (2014)
Starch	CO_2	3281	1.1	2.3	77	1	Li, Li, Rao, Zhao, and Wu (2016)
Fungi	KOH	2137	0.87	2.4	77	1	Wang, Senkovska, Kaskel, and Liu (2014)
Corncob	KOH	3708	1.14	3.2	77	1	Liu, Zhang, Geng, and Cai (2014)
Chitosan	KOH	2840	1.06	2.88	77	1	Wróbel-Iwaniec, Díez, and Gryglewicz (2015)
Chitosan	KOH	2919	1.19	2.71	77	1	Wang et al. (2016)
Sucrose	KOH	1552	0.72	2.5	77	1	Choi and Park (2015)
Hemp stem	KOH	3078	1.13	3.18	77	1	Zhang, Gao, Chen, Hao, and Jin (2017)

Table 9 Activated carbon derived from different biomasses along with their textural properties and hydrogen adsorption capacity at low temperature, 77 K, and different high pressures

Precursor	Activating agent	BET SSA (m^2/g)	$V\mu$ (cm^3/g)	H_2 adsorption capacity (wt%)	*T* (K)	*P* (bar)	References
Corncob	KOH	3708	0.59	5.80	77	40	Wang, Gao, and Hu (2009)
Cellulose	KOH	3771	–	8.1	77	20	Blankenship Ii et al. (2017)
Olive pomace	KOH	1269	0.48	3	77	25	Bader and Ouederni (2017)
				6.11	77	209	
Activated carbon	KOH	3190	1.09	7.08	77	20	Wang et al. (2009)
Olive bagasse	CO_2	1185	0.45	2.59	77	25	Bader and Abdelmottaleb (2017)
				3.34	77	200	
Starch	CO_2	3281	1.1	6.9	77	20	Li, Li, Rao, et al. (2016)
Fungi	KOH	2137	0.87	5.0	77	20	Wang et al. (2014)
Corncob	KOH	3708	1.14	5.5	77	20	Liu et al. (2014)
Chitosan	KOH	2840	1.06	5.01	77	40	Wróbel-Iwaniec et al. (2015)
		2919	1.19	6.77	77	20	Wang et al. (2016)
Melaleuca bark	KOH	3170	1.07	3.18	77	10	Xiao et al. (2014)

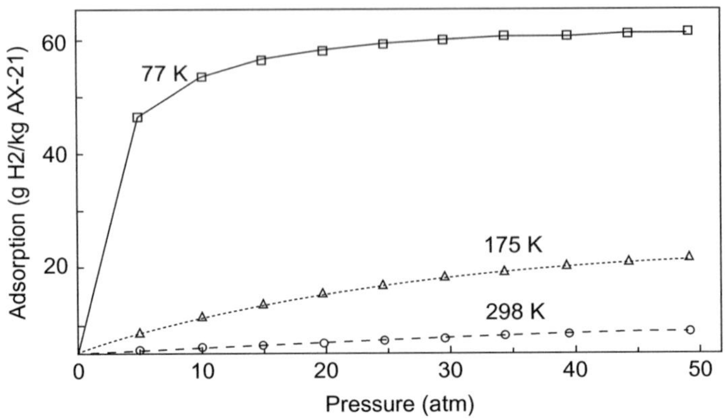

Fig. 13 Adsorption isotherms of H_2 on the activated carbons AX-21 at different temperatures. Reprinted from Chahine, R., Bose, T. K. (1994). Low-pressure adsorption storage of hydrogen. *International Journal of Hydrogen Energy, 19*, 161–164. doi:10.1016/0360-3199(94)90121-X, Copyright (1994), with permission from Elsevier.

several biomass-derived activated carbons are shown. It can be observed that the maximum achieved capacity at room temperature and high pressure is only slightly superior to 1 wt% even if the surface area of the carbon is very high (~3000 $m^2\ g^{-1}$).

Therefore, further studies are made in cryogenic conditions, i.e., 77 K and different pressures. It is therefore worth discussing the influence of porosity on adsorption capacity in such conditions.

9.4.2 Influence of porosity

Table 8 presents the hydrogen adsorption capacities at 77 K and 1 bar for activated carbons issued from several biomasses (beer lees, rice husks, starch hemp stem, etc.) activated with CO_2 or KOH. At first glance, it can be noticed that the hydrogen capacities are comprised between 1.97 and 3.2 wt% for activated carbon materials possessing specific BET surface areas ranging between 687 and 3771 $m^2\ g^{-1}$ and micropore volumes between 0.3 and 1.19 $cm^3\ g^{-1}$. To find evidence of any correlation between hydrogen adsorption capacity, specific surface area, and micropore volume, several data were collected from the references listed in Table 8 and plotted in Fig. 14.

As can be noticed, the capacity increases rather linearly from 1.5 to 3.2 wt% when increasing the surface area from 468 to 3078 $m^2\ g^{-1}$. A similar trend of capacity improvement is observed with the increase in microporous volume; however, in this case the data points are more scattered. This behavior may be due to the different pore size or surface chemistries of these materials, which also induce an effect on the adsorption capacity as will be presented later in the chapter.

Pressure has an important influence on hydrogen adsorption capacity. In Table 3, several literature data are collected at 77 K and high pressures (20–209 bar) for several activated carbons. The adsorption capacities are much higher than at atmospheric

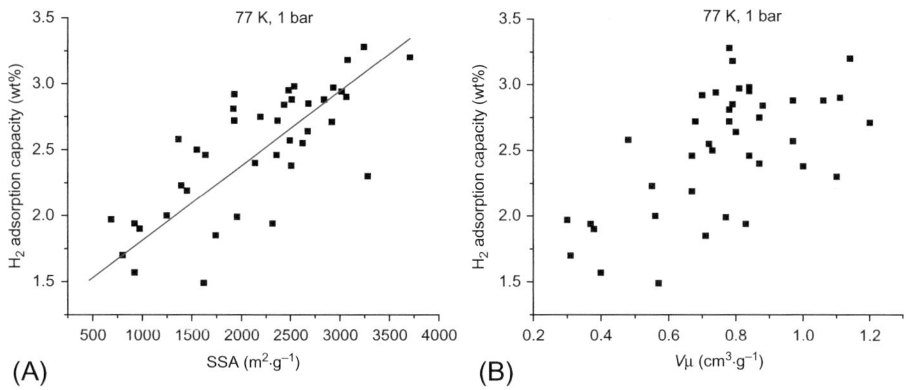

Fig. 14 Correlation between the hydrogen adsorption capacity at 77 K and 1 bar, and the specific surface area (SSA) (A) and microporous volume (B) (literature data are extracted from references listed in Table 8).

pressure and may reach 8.1 wt% by combining a high surface area as well. Nevertheless, if one plots the values from Table 3 references (Fig. 4) a linear relationship between the adsorption capacity, the specific surface area, and the microporous volume can be clearly seen. The capacity increases from 4.2 to 8.1 by increasing the BET surface from 1452 to 3771 $m^2 g^{-1}$.

Indeed, the experimental conditions (T, P) and the textural properties of carbon materials are the important parameters influencing hydrogen adsorption capacity. Both the specific surface area and the microporous volume affect the adsorption capacity, and generally an increase in the capacity with the increase in the textural values is observed but a linear correlation is difficult to obtain when gathering many different carbon materials. This may be mainly because the carbon materials exhibit different pore size and surface chemistries. If we consider a series of activated carbons derived from the same precursor, for instance chitosan biopolymer, Wróbel-Iwaniec et al. (2015) correlate the H_2 adsorption capacity at 77 K and 1 bar with BET surface area (S_{BET}), total pore volume (V_T), micropore volume (V_{DRN2}), and linear relationship was observed with a correlation coefficient (R^2) being the highest for the microporous volume (0.916). The same trend was observed at high pressure (40 bar) and in this case the correlation coefficient (R^2) is even higher than at low pressure (1 bar), i.e., 0.961 for microporous volume. This is in line with the observations made in Figs. 14 and 15 where several activated carbon hydrogen sorption capacities were presented. Similar observations showing the importance of micropores (pores smaller than 2 nm) on hydrogen adsorption have been shown by other authors as well (Schaefer, Fierro, Izquierdo, & Celzard, 2016; Wang et al., 2009; Yang, Liu, Li, Zhang, & Hao, 2012) (Fig. 16).

Further investigation led to the conclusion that H_2 storage capacity correlates better with the ultramicropore volume (volume of pore having size $<$ 0.7 nm) than the surface area of carbon materials. For example, Texier-Mandoki et al. (2004) showed that

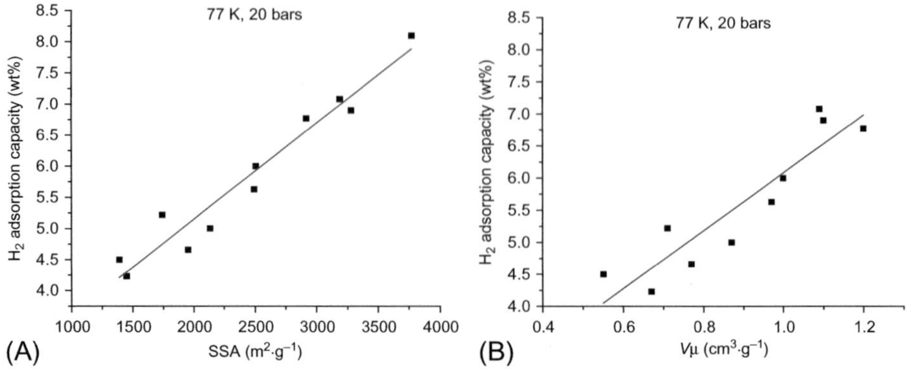

Fig. 15 Correlation between the hydrogen adsorption capacity at 77 K and 20 bar, and the specific surface area (SSA) (A) and microporous volume (B) (literature data are extracted from references listed in Table 9).

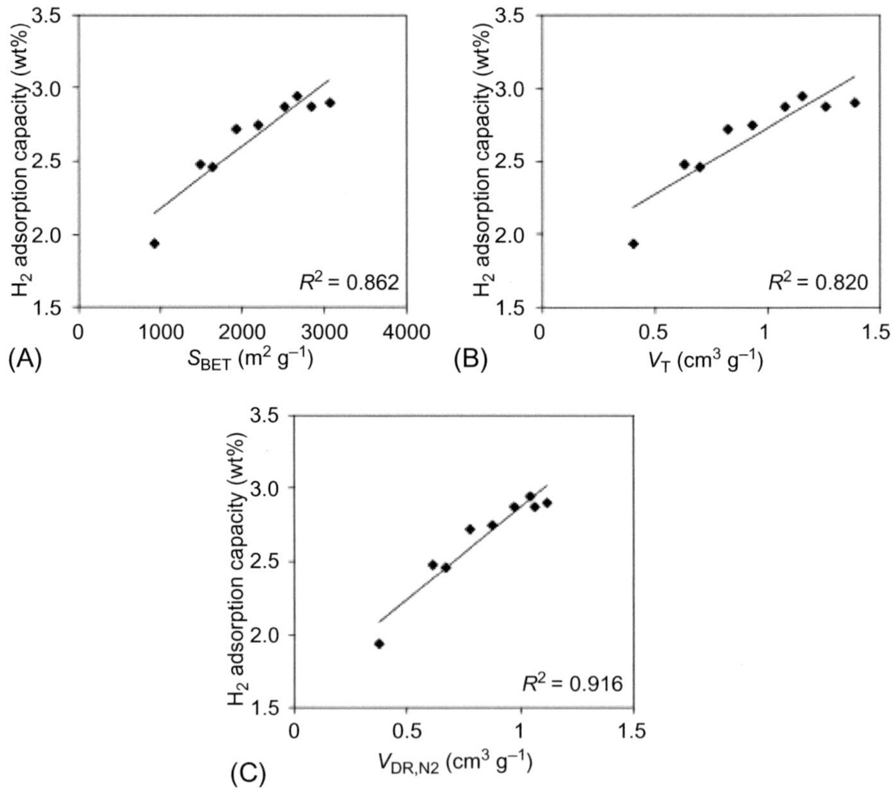

Fig. 16 The variation in hydrogen adsorbed at 77 K and 1 bar pressure with (A) BET surface area, (B) total pore volume, and (C) total micropore volume for activated carbons derived from chitosan.

Reprinted from Wróbel-Iwaniec, I., Díez, N., Gryglewicz, G. (2015). Chitosan-based highly activated carbons for hydrogen storage. *International Journal of Hydrogen Energy, 40*, 5788–5796. doi:10.1016/j.ijhydene.2015.03.034, Copyright (2015), with permission from Elsevier.

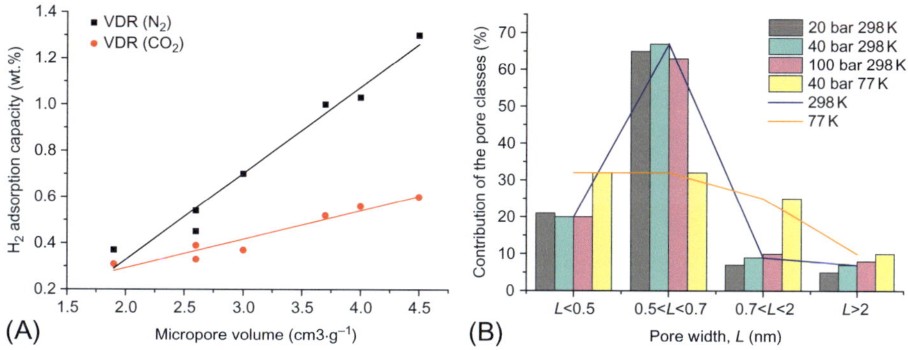

Fig. 17 H_2 adsorption capacity at 77 K and 10 bar versus N_2 and CO_2 micropore volume (A). Contribution of each pore diameter to the hydrogen up-take at different temperatures and pressures (B).
(A) Adapted from Texier-Mandoki, N., Dentzer, J., Piquero, T., Saadallah, S., David, P., Vix-Guterl, C. (2004). Hydrogen storage in activated carbon materials: role of the nanoporous texture. *Carbon, 42*, 2744–2747. doi:10.1016/j.carbon.2004.05.018, Copyright (2004), with permission from Elsevier. (B) Adapted from Schaefer, S., Fierro, V., Izquierdo, M. T., Celzard, A. (2016). Assessment of hydrogen storage in activated carbons produced from hydrothermally treated organic materials. *International Journal of Hydrogen Energy, 41*, 12146–12156. doi:10.1016/j.ijhydene.2016.05.086, Copyright (2016), with permission from Elsevier.

the hydrogen adsorption capacity of activated carbons is directly related to its CO_2 micropore volume (Fig. 17A). This work emphasizes the importance of the presence of a homogeneous narrow microporosity ($d_{pore} \leq 0.7$ nm) for hydrogen adsorption and the necessity to complete the N_2 adsorption characterizations of carbon materials by the CO_2 adsorption measurements at 273 K, to obtain better knowledge of the whole microporosity range of activated carbons. However, in most of the available literature data, N_2 adsorption is systematically used, while CO_2 adsorption is only marginally employed. To get a deeper understanding of the contribution of porosity to hydrogen capacity, Schaefer et al. (2016) gathered the adsorption capacity of different activated carbons at room temperature, 77 K, and different pressures. As seen in Fig. 17B, the main contribution to H_2 uptake is due to the narrowest micropores (0.5–0.7 nm). When increasing the pore width the hydrogen uptake gradually decreases. The authors also pointed out that the pore diameter through the overlap of the adsorption potentials is significantly less important at 77 K than at 298 K, a temperature at which the adsorption potential needs to be very high to allow hydrogen adsorption at the carbon surface.

The main conclusions drawn from these studies is that smaller pores ~0.7 nm are more convenient for hydrogen storage. Therefore, several studies have focused on the effect of pore size on hydrogen uptake in high surface area porous carbons and the optimum pore diameters for H_2 storage. However, Sethia and Sayari (2016) and Masika and Mokaya (2013) pointed out that in many studies the pore size varies in conjunction with pore volume or surface area, therefore, ambiguities may exist when studying the effect of pore size for such materials. In this sense, Sethia and Sayari

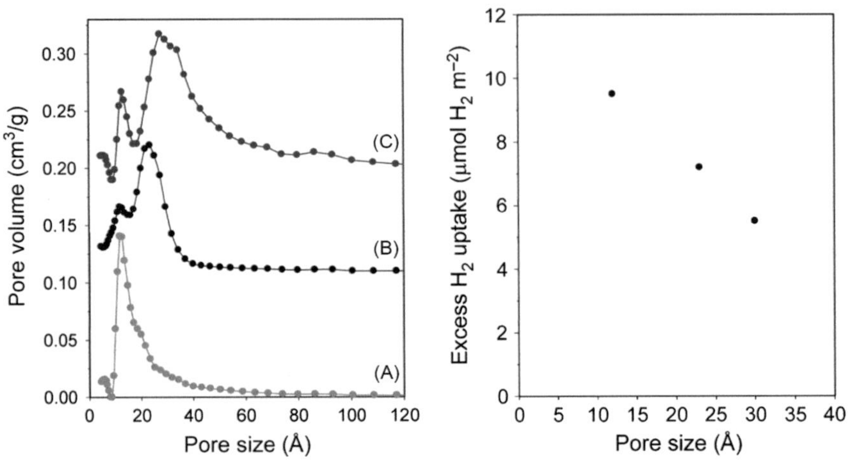

Fig. 18 Pore size distribution curves (*left*) of porous carbon materials along with their hydrogen adsorption uptake versus pore size (*right*).
Reprinted with permission from Masika, E., Mokaya, R. (2012). Hydrogen storage in high surface area carbons with identical surface areas but different pore sizes: direct demonstration of the effects of pore size. *Journal of Physical Chemistry C, 116*, 25734–25740. doi:10.1021/jp3100365, Copyright (2012) American Chemical Society.

(2016) concluded that the high hydrogen uptake is associated with the activated carbon having the most important cumulative volume of ultramicropores (<0.7). Two materials exhibiting identical total pore volume (0.64 vs 0.65 cm^3/g) and BET surface area (1312 vs 1342 m^2/g) were prepared and it was determined that they exhibited a 21.8% difference in hydrogen uptake. The authors attributed this significant decrease in hydrogen uptake to the occurrence of a broader pore size distribution with larger micropores (0.63 vs 0.59 nm) together with a 28.6% decrease in ultramicropore volume (0.21 vs 0.27 cm^3/g). Masika and Mokaya (2013) showed that the excess and total hydrogen uptake (at 77 K and 20 bar) of activated carbons with identical surface areas of 3340 m^2/g (Fig. 18 left) increased from 3.7 and 5.4 wt% (31 Å pore size sample) to 4.8 and 6.3 wt% (23 Å pore size sample), and to 6.3 and 7.3 wt% for a 12 Å pore size sample. The authors concluded that the excess hydrogen storage density ($\mu mol\, H_2\, m^{-2}$) of the carbons decreases linearly with pore size (Fig. 18 right).

9.4.3 *Influence of surface chemistry*

The interaction between adsorbed hydrogen and carbons is very weak and can be improved via functionalization of the surface with heteroatoms. Research has demonstrated that the surface chemical properties of activated carbons have an important role in their sorption capacity. The amount of hydrogen adsorbed on activated carbons depends on the surface groups present on activated carbons; however, the presented literature results are somehow contradictory. Agarwal et al. (1987) prepared activated

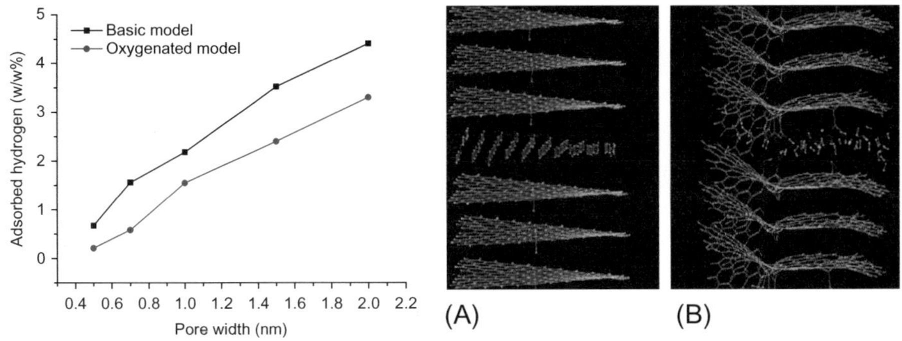

Fig. 19 Correlation between the adsorbed hydrogen amount for basic carbon (no oxygen functional groups) and oxygenated model (contains six types of oxygen groups) vs the pore size obtained by molecular dynamics simulation (*left*); hydrogen adsorption snapshots for the pore size of 0.7 nm (A) purely carbonaceous material and (B) oxygenated materials when the adsorption simulation is complete in the basic model (*right*).
Readapted (*left*) and reprinted (*right*) from Georgakis, M., Stavropoulos, G., Sakellaropoulos, G. P. (2007). Molecular dynamics study of hydrogen adsorption in carbonaceous microporous materials and the effect of oxygen functional groups. *International Journal of Hydrogen Energy*, *32*, 1999–2004. doi:10.1016/j.ijhydene.2006.08.040, Copyright (2007), with permission from Elsevier.

carbons with modified amounts of oxygen surface groups by thermal oxidation at different temperatures and for different time periods. The nature and amount of formed oxygen complexes were determined by Boehm titration, and it was found that the surface oxygen complexes affect the amount of hydrogen adsorbed on the surface, i.e., the amount of hydrogen adsorbed increases with the amount of surface acidic complexes. Concerning the influence of basic complex groups the authors did not find any trend with the amount of hydrogen adsorbed. Georgakis, Stavropoulos, and Sakellaropoulos (2007) used a theoretical approach via molecular dynamic simulation to understand the impact of oxygen functional groups on the hydrogen adsorption capacity. Six types of common oxygen groups found on porous carbon have been used (carboxyl, carbonyl, lactones, anhydrides, phenols, and quinones). The authors came to the conclusion that the oxygen functional groups induce a decrease in hydrogen adsorption capacity and this effect is more pronounced with the narrow pores (Fig. 19A). They accounted this behavior mainly to three factors, i.e., increase in weight due to oxygen, steric hindrance effects, and weaker oxygen-hydrogen interactions compared to carbon-hydrogen (Fig. 19B).

From an experimental point of view, Huang, Chen, Chen, and Huang (2010) found the same trend, i.e., hydrogen uptake decreases when the carbons present oxygen functional groups. The authors used activated carbon derived from litchi wood, which was postmodified by nitric acid or hydrogen peroxide at various temperatures to enable the formation of oxygen functional groups. In some cases, however, the drastic oxidation conditions lead not only to the introduction of functional groups but also to the

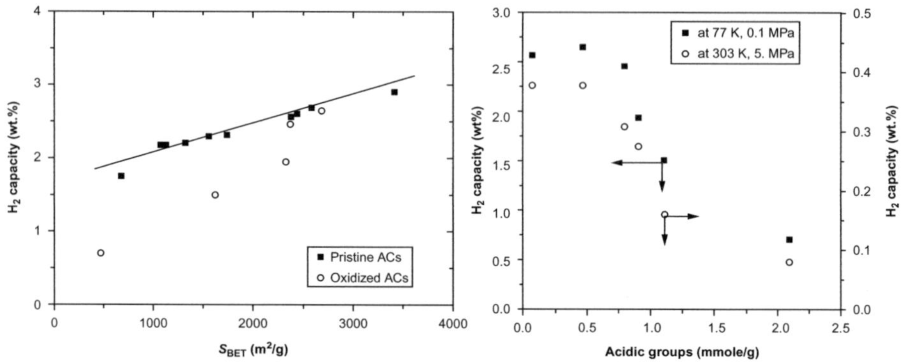

Fig. 20 Comparison of hydrogen capacity of the pristine activated carbon and oxidized activated carbon (*left*) and effect of surface acidic groups on hydrogen capacity measured at different conditions (*right*).
Reprinted from Huang, C. -C., Chen, H. -M., Chen, C. -H., Huang, J. -C. (2010). Effect of surface oxides on hydrogen storage of activated carbon. *Separation and Purification Technology, 3*, 291–295. doi:10.1016/j.seppur.2009.10.009, Copyright (2010), with permission from Elsevier.

porosity collapsing and therefore to the decrease in hydrogen uptake (Fig. 20A). The authors found that hydrogen capacity does not depend on the amount of basic groups, but rather on the amount of acidic groups. It has been observed as well that hydrogen capacity is significantly suppressed when increasing the acidic group amounts by more than 0.8 mmol/g (Fig. 20B); this fact is also related to the simultaneous decrease in the specific surface area (Fig. 20A).

Many other authors found that the amount of hydrogen adsorbed remarkably decreased by oxidation of the carbons (Bleda-Martínez, Pérez, Linares-Solano, Morallón, & Cazorla-Amorós, 2008; Takagi, Hatori, Yamada, Matsuo, & Shiraishi, 2004; Zhao, Xiao, Fletcher, & Thomas, 2005) and they accounted this to the decreased accessibility to aromatic C—C bonds due to pore blockage with the functional groups (Schimmel et al., 2003; Takagi et al., 2004). Some authors proved that the oxygen groups do not affect hydrogen uptake since no significant differences were detected by comparing their capacity with that of pristine carbons (Llorens & Pera-Titus, 2009; Tellez-Juarez et al., 2014). However, recent work performed by Blankenship Ii et al. (2017) claims an opposite behavior. By using hydrothermally treated cellulose further activated with KOH at different temperatures (600–800°C), highly porous carbon (2000–3700 $m^2\ g^{-1}$) containing high amounts of oxygen (17.9–20.6 wt%) were prepared. This combination of high surface area and high amount of functional oxygen groups was reached at 700°C of activation (CA-4700) and improved hydrogen uptake (Fig. 21A) significantly compared to the reference activated carbon AX21. To explain this behavior, the isosteric heat of hydrogen adsorption (Q_{st}) was measured (Fig. 21B) allowing an estimation of the strength of interaction between the hydrogen and carbon surface. As can be noticed, the isosteric heat of hydrogen adsorption is significantly

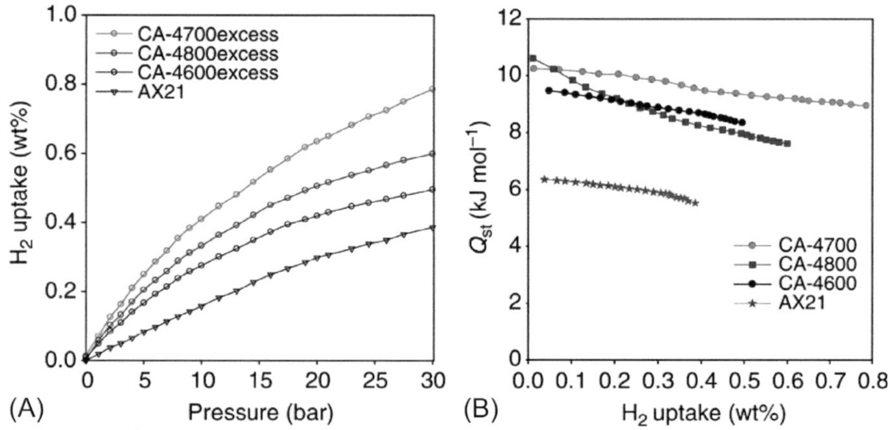

Fig. 21 Hydrogen storage capacity at room temperature (A) and isosteric heat of hydrogen adsorption (B) of activated carbons derived from cellulose acetate and a commercially reference activated carbon AX21.
Reprinted from Blankenship Ii, T. S., Balahmar, N., Mokaya, R. (2017). Oxygen-rich microporous carbons with exceptional hydrogen storage capacity. *Nature Communications, 8*, 1545. doi: 10.1038/s41467-017-01633-x.

much higher for the cellulose-derived activated carbon, CA (8–10 KJ/mol) compared to that of reference activated carbon, AX21 (5–6 KJ/mol). The authors accounted this difference not only to porosity but also to the presence of oxygen functional groups.

Taking into consideration all these works showing contradictory results, one can state that the effect of oxygen groups on hydrogen adsorption still needs to be investigated to better understand it. Moreover, when other functional groups are present on the carbon surface, such as nitrogen-based groups, the effect of these groups on hydrogen adsorption is still difficult to assess. However, there are only a few studies dedicated to the effect of nitrogen atoms on hydrogen storage performance. Wang et al. (2016) used chitosan precursor, containing nitrogen in its skeleton, to obtain N-doped porous carbon through an activation process. The amount of nitrogen could be varied from 6.5 to 0.53 wt% by increasing the activation temperature from 600°C to 800°C. The authors found that the high N-doping is beneficial for hydrogen adsorption at low pressure but it is detrimental at high pressure. It is worth mentioning that high nitrogen contents are always accompanied by significant oxygen contribution. However, the authors did not take into consideration this aspect, which may contribute as well to the observed hydrogen storage behavior.

Functionalization of carbon with metal-based nanoparticles such as Pd, Pt, Ni, etc. or with metal hydrides MgH_2, $LiBH_4$, $LiAlH_4$ may further improve the hydrogen uptake by chemisorptions and/or spillover effects (Fig. 22); however, this is a large topic that will not be developed here but can be overviewed in a recent review (Ren, Musyoka, Langmi, Mathe, & Liao, 2017).

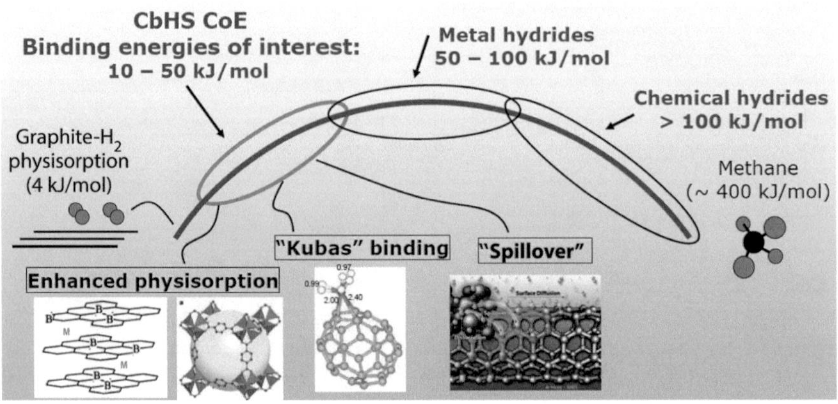

Fig. 22 Binding energies and the associated sorption mechanisms in different materials. Reprinted from Dillon, A. C., Parilla, P. A., Zhao, Y., Kim, Y. H., Gennett, T., Curtis, C., Heben, M. J., Blackburn, J. L., Gilbert, K. E. H., Alleman, J. L., Jones, K. M., Zhang, S. B. (n.d.) NREL activities in DOE Carbon-based Materials Center of Excellence.

9.5 Conclusions

From an environmental and economic point of view the use of biomass to prepare carbon materials with tunable features is demonstrated in many works. Many sources of biomass or biomass waste were used, and optimized activation conditions with physical or chemical agents to obtain materials with high specific surface area, high porous volume, tunable pore size characteristics, and various surface chemistries are highly desired for gas storage applications. For the three gases evaluated herein (CH_4, CO_2, and H_2), the adsorption capacities on carbon adsorbents show many similar trends. As demonstrated, adsorption capacity depends first on the available porosity, namely the specific surface area and in particular microporous volume, with the pore size with narrow distributions (0.5–0.8 nm) being likely more favorable for achieving higher gas adsorption uptakes. In addition to porosity, carbon functional groups also influence gas adsorption capacity. Oxygen, nitrogen, and sulfur were widely studied and their influence on gas uptake is not the same for all the gases. For CO_2 adsorption, oxygen seems beneficial to increase the gas uptake, while for hydrogen adsorption, oxygen proved to be detrimental. The impact of other heteroatoms such as nitrogen and sulfur was found more difficult to establish due to the complexity to design materials with similar texture and oxygen contents. The impact of morphology on gas adsorption is rarely addressed and deserves further investigations to elucidate its impacts on the materials' density, gas diffusion, and volumetric gas capacity. Besides the carbon characteristics that strongly influence gas adsorption capacity, the analysis condition such as temperature and pressure has significant importance. Due to the weak interactions of the gases with the carbon structure, the gas adsorption capacities at room temperature and atmospheric pressure or even high pressure remain very small to meet the requirements of DOE departments. They can, however, be strongly improved at low temperature (77 K) and high pressure, achieving in some cases the capacities required by the DOE. Nevertheless, such a solution involving low temperature/high pressures remains expensive and is not feasible for real-life application.

References

Agarwal, R. K., Noh, J. S., Schwarz, J. A., & Davini, P. (1987). Effect of surface acidity of activated carbon on hydrogen storage. *Carbon*, *25*, 219–226. https://doi.org/10.1016/0008-6223(87)90119-9.

Alabadi, A., Razzaque, S., Yang, Y., Chen, S., & Tan, B. (2015). Highly porous activated carbon materials from carbonized biomass with high CO_2 capturing capacity. *Chemical Engineering Journal*, *281*, 606–612. https://doi.org/10.1016/j.cej.2015.06.032.

Alonso, A., Moral-Vico, J., Abo Markeb, A., Busquets-Fité, M., Komilis, D., Puntes, V., et al. (2017). Critical review of existing nanomaterial adsorbents to capture carbon dioxide and methane. *Science of the Total Environment*, *595*, 51–62. https://doi.org/10.1016/j.scitotenv.2017.03.229.

Andrews, J., & Shabani, B. (2014). The role of hydrogen in a global sustainable energy strategy. *Wiley Interdisciplinary Reviews: Energy and Environment*, *3*, 474–489. https://doi.org/10.1002/wene.103.

Antoniou, M. K., Diamanti, E. K., Enotiadis, A., Policicchio, A., Dimos, K., Ciuchi, F., et al. (2014). Microporous and mesoporous materials methane storage in zeolite-like carbon materials. *Microporous and Mesoporous Materials*, *188*, 16–22. https://doi.org/10.1016/j.micromeso.2013.12.030.

Arami-Niya, A., Wan Daud, W. M. A., Mjalli, F. S., Abnisa, F., & Shafeeyan, M. S. (2012). Production of microporous palm shell based activated carbon for methane adsorption: modeling and optimization using response surface methodology. *Chemical Engineering Research and Design*, *90*, 776–784. https://doi.org/10.1016/j.cherd.2011.10.001.

Bader, N. Abdelmottaleb, O. CO_2 activation of olive bagasse for hydrogen storage. *Environmental Progress & Sustainable Energy 36*, 2017, 315–324. https://doi.org/10.1002/ep.12514.

Bader, N., & Ouederni, A. (2017). Functionalized and metal-doped biomass-derived activated carbons for energy storage application. *Journal of Energy Storage*, *13*, 268–276. https://doi.org/10.1016/j.est.2017.07.013.

Bagheri, N., & Abedi, J. (2011). Adsorption of methane on corn cobs based activated carbon. *Chemical Engineering Research and Design*, *89*, 2038–2043. https://doi.org/10.1016/j.cherd.2011.02.002.

Balathanigaimani, M. S., Haider, M. B., Jha, D., Kumar, R., Lee, S. J., Shim, W. G., et al. (2018). *Nanostructured Biomass Based Carbon Materials from Beer Lees for Hydrogen Storage [WWW Document]*. https://www.ingentaconnect.com/contentone/asp/jnn/2018/00000018/00000003/art00121 (Accessed 21 June 2018).

Ben-Mansour, R., Habib, M. A., Bamidele, O. E., Basha, M., Qasem, N. A. A., Peedikakkal, A., et al. (2016). Carbon capture by physical adsorption: materials, experimental investigations and numerical modeling and simulations—a review. *Applied Energy*, *161*, 225–255. https://doi.org/10.1016/j.apenergy.2015.10.011.

Biloe, S., Goetz, V., & Mauran, S. (2001). Characterization of adsorbent composite blocks for methane storage. *Carbon*, *39*, 1653–1662. https://doi.org/10.1016/S0008-6223(00)00288-8.

Blankenship Ii, T. S., Balahmar, N., & Mokaya, R. (2017). Oxygen-rich microporous carbons with exceptional hydrogen storage capacity. *Nature Communications*, *8*, 1545. https://doi.org/10.1038/s41467-017-01633-x.

Bleda-Martínez, M. J., Pérez, J. M., Linares-Solano, A., Morallón, E., & Cazorla-Amorós, D. (2008). Effect of surface chemistry on electrochemical storage of hydrogen in porous carbon materials. *Carbon*, *46*, 1053–1059. https://doi.org/10.1016/j.carbon.2008.03.016.

Boyjoo, Y., Cheng, Y., Zhong, H., Tian, H., Pan, J., Pareek, V. K., et al. (2017). From waste Coca Cola® to activated carbons with impressive capabilities for CO_2 adsorption and supercapacitors. *Carbon*, *116*, 490–499. https://doi.org/10.1016/j.carbon.2017.02.030.

Builes, S., Roussel, T., Ghimbeu, C. M., Parmentier, J., Gadiou, R., Vix-Guterl, C., et al. (2011). Microporous carbon adsorbents with high CO_2 capacities for industrial applications. *Physical Chemistry Chemical Physics*, *13*, 16063–16070. https://doi.org/10.1039/C1CP21673B.

Burchell, T., & Rogers, M. (2000). Low Pressure Storage of Natural Gas for Vehicular Applications. *SAE Tech. Pap.* doi: 10.4271/2000.

Buss, E. (1995). Gravimetric measurement of binary gas adsorption equilibria of methane-carbon dioxide mixtures on activated carbon. *Gas Separation and Purification*, *9*, 189–197. https://doi.org/10.1016/0950-4214(95)98226-B.

Byamba-Ochir, N., Shim, W. G., Balathanigaimani, M. S., & Moon, H. (2017). High density Mongolian anthracite based porous carbon monoliths for methane storage by adsorption. *Applied Energy*, *190*, 257–265. https://doi.org/10.1016/j.apenergy.2016.12.124.

Caglayan, B. S., & Aksoylu, A. E. (2013). CO_2 adsorption on chemically modified activated carbon. *Journal of Hazardous Materials*, *252–253*, 19–28. https://doi.org/10.1016/j.jhazmat.2013.02.028.

Casco, M. E., Martínez-Escandell, M., Gadea-Ramos, E., Kaneko, K., Silvestre-Albero, J., & Rodríguez-Reinoso, F. (2015). High-pressure methane storage in porous materials: are carbon materials in the pole position? *Chemistry of Materials*, *27*, 959–964. https://doi.org/10.1021/cm5042524.

Cavenati, S., Grande, C. A., & Rodrigues, A. E. (2004). Adsorption equilibrium of methane, carbon dioxide, and nitrogen on zeolite 13X at high pressures. *Journal of Chemical & Engineering Data*, *49*, 1095–1101. https://doi.org/10.1021/je0498917.

Chen, C., & Ahn, W. S. (2014). CO_2 adsorption on LTA zeolites: effect of mesoporosity. *Applied Surface Science*, *311*, 107–109. https://doi.org/10.1016/j.apsusc.2014.04.218.

Choi, P.-S., Jeong, J.-M., Choi, Y.-K., Kim, M.-S., Shin, G.-J., & Park, S.-J. (2016). A review: methane capture by nanoporous carbon materials for automobiles. *Carbon Letters*, *17*, 18–28. https://doi.org/10.5714/CL.2016.17.1.018.

Choi, Y.-K., & Park, S.-J. (2015). Preparation and characterization of sucrose-based microporous carbons for increasing hydrogen storage. *Journal of Industrial and Engineering Chemistry*, *28*, 32–36.

Core Writing Team, Pachauri, R. K., & Meyer, L. A. (2014). Climate Change 2014 Synthesis Report. In *Contribution of Working Groups I, II and III to the Fifth Assessment Report of the Intergovernmental Panel on Climate Change*. .

Cracknell, R. F., Gordon, P., & Gubbins, K. E. (1993). Influence of pore geometry on the design of microporous materials for methane storage. *The Journal of Physical Chemistry*, *97*, 494–499. https://doi.org/10.1021/j100104a036.

Deng, S., Wei, H., Chen, T., Wang, B., Huang, J., & Yu, G. (2014). Superior CO_2 adsorption on pine nut shell-derived activated carbons and the effective micropores at different temperatures. *Chemical Engineering Journal*, *253*, 46–54. https://doi.org/10.1016/j.cej.2014.04.115.

Drage, T. C., Blackman, J. M., Pevida, C., & Snape, C. E. (2009). Evaluation of activated carbon adsorbents for CO_2 capture in gasification. *Energy & Fuels*, *23*, 2790–2796. https://doi.org/10.1021/ef8010614.

Durbin, D. J., & Malardier-Jugroot, C. (2013). Review of hydrogen storage techniques for on board vehicle applications. *International Journal of Hydrogen Energy*, *38*, 14595–14617. https://doi.org/10.1016/j.ijhydene.2013.07.058.

EEA. (2014). *Trends and projections in Europe 2014, EEA Report 6/2014*. doi: 10.2800/93693.

Encyclopedia Britannica. (2018). *Air, Atmospheric Gas [WWW Document]. URL* https://www.britannica.com/science/air.

Energy, U.S. Department of Energy: Washington, D. (2012). *MOVE: Methane Oppurtuities for Vehicular Energy*. Advanced Research Projects Agency. http://arpa-e.energy.gov/?q=arpa-e-programs/move (Accessed 5 May 2018).

EPA (2016). *Inventory of U.S. Greenhouse Emissions and Sinks*.

Feroldi, M., Neves, A. C., Borba, C. E., & Alves, H. J. (2018). Methane storage in activated carbon at low pressure under different temperatures and flow rates of charge. *Journal of Cleaner Production*, *172*, 921–926. https://doi.org/10.1016/j.jclepro.2017.10.247.

García Blanco, A. A., Vallone, A. F., Korili, S. A., Gil, A., & Sapag, K. (2016). A comparative study of several microporous materials to store methane by adsorption. *Microporous and Mesoporous Materials*, *224*, 323–331. https://doi.org/10.1016/j.micromeso.2016.01.002.

Georgakis, M., Stavropoulos, G., & Sakellaropoulos, G. P. (2007). Molecular dynamics study of hydrogen adsorption in carbonaceous microporous materials and the effect of oxygen functional groups. *International Journal of Hydrogen Energy*, *32*, 1999–2004. https://doi.org/10.1016/j.ijhydene.2006.08.040.

Global CCS Institute. (2011). *Accelerating the Uptake of CCS: Industrial Use of Captured Carbon Dioxide.* .

González-García, P. (2018). Activated carbon from lignocellulosics precursors: a review of the synthesis methods, characterization techniques and applications. *Renewable and Sustainable Energy Reviews*, *82*, 1393–1414. https://doi.org/10.1016/j.rser.2017.04.117.

Gui, M. M., Yap, Y. X., Chai, S. P., & Mohamed, A. R. (2013). Multi-walled carbon nanotubes modified with (3-aminopropyl)triethoxysilane for effective carbon dioxide adsorption. *International Journal of Greenhouse Gas Control*, *14*, 65–73. https://doi.org/10.1016/j.ijggc.2013.01.004.

Hao, W., Björkman, E., Lilliestråle, M., & Hedin, N. (2013). Activated carbons prepared from hydrothermally carbonized waste biomass used as adsorbents for CO_2. *Applied Energy*, *112*, 526–532. https://doi.org/10.1016/j.apenergy.2013.02.028.

Heo, Y.-J., & Park, S.-J. (2015). Synthesis of activated carbon derived from rice husks for improving hydrogen storage capacity. *Journal of Industrial and Engineering Chemistry*, *31*, 330–334. https://doi.org/10.1016/j.jiec.2015.07.006.

Hosseini, S., Bayesti, I., Marahel, E., Eghbali Babadi, F., Chuah Abdullah, L., & Choong, T. S. Y. (2015). Adsorption of carbon dioxide using activated carbon impregnated with Cu promoted by zinc. *Journal of the Taiwan Institute of Chemical Engineers*, *52*, 109–117. https://doi.org/10.1016/j.jtice.2015.02.015.

Huang, C.-C., Chen, H.-M., Chen, C.-H., & Huang, J.-C. (2010). Effect of surface oxides on hydrogen storage of activated carbon. *Separation and Purification Technology*, *3*, 291–295. https://doi.org/10.1016/j.seppur.2009.10.009.

Intergovernmental Panel on Climate Change (IPCC). (2007). *Human and Natural Drivers of Climate Change*. https://www.ipcc.ch/publications_and_data/ar4/wg1/en/spmsspm-human-and.html (Accessed 25 March 2018).

Intergovernmental Panel on Climate Change (IPCC) (2005). *Carbon Dioxide Capture and Storage*. Cambridge University Press.

International Energy Agency (2017). *CO_2 emissions from fuel combustion*. Highlights.

Kacem, M., Pellerano, M., & Delebarre, A. (2015). Pressure swing adsorption for CO_2/N_2 and CO_2/CH_4 separation: Comparison between activated carbons and zeolites performances. *Fuel Processing Technology*, *138*, 271–283. https://doi.org/10.1016/j.fuproc.2015.04.032.

Kumar, K. V., Preuss, K., Titirici, M. M., & Rodríguez-Reinoso, F. (2017). Nanoporous materials for the onboard storage of natural gas. *Chemical Reviews*, *117*, 1796–1825. https://doi.org/10.1021/acs.chemrev.6b00505.

Lee, J. W., Kang, H. C., Shim, W. G., Kim, C., & Moon, H. (2006). Methane adsorption on multi-walled carbon nanotube at (303.15, 313.15, and 323.15) K. *Journal of Chemical & Engineering Data*, *51*, 963–967. https://doi.org/10.1021/je050467v.

Lee, S. Y., & Park, S. J. (2015). A review on solid adsorbents for carbon dioxide capture. *Journal of Industrial and Engineering Chemistry*, *23*, 1–11. https://doi.org/10.1016/j.jiec.2014.09.001.

Lee, Z. H., Lee, K. T., Bhatia, S., & Mohamed, A. R. (2012). Post-combustion carbon dioxide capture: evolution towards utilization of nanomaterials. *Renewable and Sustainable Energy Reviews*, *16*, 2599–2609. https://doi.org/10.1016/j.rser.2012.01.077.

Li, G., Li, J., Tan, W., Jin, H., Yang, H., Peng, J., et al. (2016). Preparation and characterization of the hydrogen storage activated carbon from coffee shell by microwave irradiation and KOH activation. *International Biodeterioration & Biodegradation*, *113*, 386–390. https://doi.org/10.1016/j.ibiod.2016.05.003.

Li, J. R., Ma, Y., McCarthy, M. C., Sculley, J., Yu, J., Jeong, H. K., et al. (2011). Carbon dioxide capture-related gas adsorption and separation in metal-organic frameworks. *Coordination Chemistry Reviews*, *255*, 1791–1823. https://doi.org/10.1016/j.ccr.2011.02.012.

Liu, J., Zou, R., & Zhao, Y. (2016). Recent developments in porous materials for H_2 and CH_4 storage. *Tetrahedron Letters*, *57*, 4873–4881. https://doi.org/10.1016/j.tetlet.2016.09.085.

Liu, X., Zhang, C., Geng, Z., & Cai, M. (2014). High-pressure hydrogen storage and optimizing fabrication of corncob-derived activated carbon. *Microporous and Mesoporous Materials*, *194*, 60–65. https://doi.org/10.1016/j.micromeso.2014.04.005.

Li, Y., Li, D., Rao, Y., Zhao, X., & Wu, M. (2016). Superior CO_2, CH_4, and H_2 uptakes over ultrahigh-surface-area carbon spheres prepared from sustainable biomass-derived char by CO_2 activation. *Carbon*, *105*, 454–462. https://doi.org/10.1016/j.carbon.2016.04.036.

Llewellyn, P. L., Bourrelly, S., Serre, C., Vimont, A., Daturi, M., Hamon, L., et al. (2008). High uptakes of CO_2 and CH_4 in mesoporous metals organic frameworks MIL-100 and MIL-101. *Langmuir*, 7245–7250. https://doi.org/10.1021/la800227x.

Llorens, J., & Pera-Titus, M. (2009). Influence of surface heterogeneity on hydrogen adsorption on activated carbons. *Colloids and Surfaces A: Physicochemical and Engineering Aspects*, *350*, 63–72. https://doi.org/10.1016/j.colsurfa.2009.08.035.

Lozano-Castelló, D., Cazorla-Amorós, D., Linares-Solano, A., & Quinn, D. F. (2002). Influence of pore size distribution on methane storage at relatively low pressure: preparation of activated carbon with optimum pore size. *Carbon*, *40*, 989–1002. https://doi.org/10.1016/S0008-6223(01)00235-4.

Masika, E., & Mokaya, R. (2013). Preparation of ultrahigh surface area porous carbons templated using zeolite 13X for enhanced hydrogen storage. *Progress in Natural Science: Materials International*, *23*, 308–316. https://doi.org/10.1016/j.pnsc.2013.04.007.

Ma, S., Sun, D., Simmons, J. M., Collier, C. D., Yuan, D., & Zhou, H. C. (2008). Metal-organic framework from an anthracene derivative containing nanoscopic cages exhibiting high methane uptake. *Journal of the American Chemical Society*, *130*, 1012–1016. https://doi.org/10.1021/ja0771639.

Matranga, K. R., Myers, A. L., & Glandt, E. D. (1992). Storage of natural-gas by adsorption on activated carbon. *Chemical Engineering Science*, *47*, 1569–1579. https://doi.org/10.1016/0009-2509(92)85005-v.

Md Arshad, S. H., Ngadi, N., Aziz, A. A., Amin, N. S., Jusoh, M., & Wong, S. (2016). Preparation of activated carbon from empty fruit bunch for hydrogen storage. *Journal of Energy Storage*, *8*, 257–261. https://doi.org/10.1016/j.est.2016.10.001.

Meng, L. Y., & Park, S. J. (2014). One-pot synthetic method to prepare highly N-doped nanoporous carbons for CO2adsorption. *Materials Chemistry and Physics*, *143*, 1158–1163. https://doi.org/10.1016/j.matchemphys.2013.11.016.

Menon, V. C., & Komarneni, S. (1998). Porous adsorbents for vehicular natural gas storage: a review. *Journal of Porous Materials*, *5*, 43–58. https://doi.org/10.1023/A:1009673830619.

Mozammel, H. M., Masahiro, O., & Bhattacharya, S. C. (2002). Activated charcoal from coconut shell using $ZnCl_2$ activation. *Biomass and Bioenergy*, *22*, 397–400. https://doi.org/10.1016/S0961-9534(02)00015-6.

Munusamy, K., Somani, R. S., & Bajaj, H. C. (2015). Breakthrough adsorption studies of mixed gases on mango (Mangifera indica L.) seed shell derived activated carbon extrudes. *Journal of Environmental Chemical Engineering*, *3*, 2750–2759. https://doi.org/10.1016/j.jece.2015.05.010.

NASA. (2018). *Carbon Dioxide 2018.* https://climate.nasa.gov/vital-signs/carbon-dioxide/ *(Accessed 25 March 2018).*

National Center for Biotechnology Information n.d. PubChem Compound Database; CID = 297 https://pubchem.ncbi.nlm.nih.gov/compound/297 (Accessed 19 April 2018).

Natural Gas.org n.d. Natural Gas and the Environment http://naturalgas.org/environment/naturalgas/ (Accessed 5 April 2018).

Office of Energy Efficiency & Renewable Energy n.d. DOE Technical Targets for Onboard Hydrogen Storage for Light-Duty Vehicles. https://www.energy.gov/eere/fuelcells/doe-technical-targets-onboard-hydrogen-storage-light-duty-vehicles

Oh, T. H. (2010). Carbon capture and storage potential in coal-fired plant in Malaysia—a review. *Renewable and Sustainable Energy Reviews*, *14*, 2697–2709. https://doi.org/10.1016/j.rser.2010.06.003.

Plaza, M. G., García, S., Rubiera, F., Pis, J. J., & Pevida, C. (2010). Post-combustion CO_2 capture with a commercial activated carbon: Comparison of different regeneration strategies. *Chemical Engineering Journal*, *163*, 41–47. https://doi.org/10.1016/j.cej.2010.07.030.

Plaza, M. G., Pevida, C., Arias, B., Fermoso, J., Arenillas, A., Rubiera, F., et al. (2008). Application of thermogravimetric analysis to the evaluation of aminated solid sorbents for CO_2 capture. *Journal of Thermal Analysis and Calorimetry*, *92*, 601–606. https://doi.org/10.1007/s10973-007-8493-x.

Policicchio, A., MacCallini, E., Agostino, R. G., Ciuchi, F., Aloise, A., & Giordano, G. (2013). Higher methane storage at low pressure and room temperature in new easily scalable large-scale production activated carbon for static and vehicular applications. *Fuel*, *104*, 813–821. https://doi.org/10.1016/j.fuel.2012.07.035.

Presser, V., McDonough, J., Yeon, S.-H., & Gogotsi, Y. (2011). Effect of pore size on carbon dioxide sorption by carbide derived carbon. *Energy & Environmental Science*, *4*, 3059. https://doi.org/10.1039/c1ee01176f.

Ramesh, T., Rajalakshmi, N., & Dhathathreyan, K. S. (2017). Synthesis and characterization of activated carbon from jute fibers for hydrogen storage. *Renewable Energy and Environmental Sustainability*, *2*, 4. https://doi.org/10.1051/rees/2017001.

Ramesh, T., Rajalakshmi, N., & Dhathathreyan, K. S. (2015). Activated carbons derived from tamarind seeds for hydrogen storage. *Journal of Energy Storage*, *4*, 89–95. https://doi.org/10.1016/j.est.2015.09.005.

Ren, J., Musyoka, N. M., Langmi, H. W., Mathe, M., & Liao, S. (2017). Current research trends and perspectives on materials-based hydrogen storage solutions: a critical review. *International Journal of Hydrogen Energy*, *42*, 289–311. https://doi.org/10.1016/j.ijhydene.2016.11.195.

Ruiz, B., Ferrera-Lorenzo, N., & Fuente, E. (2017). Valorisation of lignocellulosic wastes from the candied chestnut industry. Sustainable activated carbons for environmental applications. *Journal of Environmental Chemical Engineering*, *5*, 1504–1515. https://doi.org/10.1016/j.jece.2017.02.028.

Schaefer, S., Fierro, V., Izquierdo, M. T., & Celzard, A. (2016). Assessment of hydrogen storage in activated carbons produced from hydrothermally treated organic materials. *International Journal of Hydrogen Energy*, *41*, 12146–12156. https://doi.org/10.1016/j.ijhydene.2016.05.086.

Schimmel, H. G., Kearley, G. J., Nijkamp, M. G., Visser, C. T., de Jong, K. P., & Mulder, F. M. (2003). Hydrogen adsorption in carbon nanostructures: comparison of nanotubes, fibers, and coals. *Chemistry A European Journal*, *9*, 4764–4770. https://doi.org/10.1002/chem.200304845.

Seema, H., Kemp, K. C., Le, N. H., Park, S. W., Chandra, V., Lee, J. W., et al. (2014). Highly selective CO_2 capture by S-doped microporous carbon materials. *Carbon*, *66*, 320–326. https://doi.org/10.1016/j.carbon.2013.09.006.

Serafin, J., Narkiewicz, U., Morawski, A. W., Wróbel, R. J., & Michalkiewicz, B. (2017). Highly microporous activated carbons from biomass for CO_2 capture and effective micropores at different conditions. *Journal of CO_2 Utilization*, *18*, 73–79. https://doi.org/10.1016/j.jcou.2017.01.006.

Sethia, G., & Sayari, A. (2016). Activated carbon with optimum pore size distribution for hydrogen storage. *Carbon*, *99*, 289–294. https://doi.org/10.1016/j.carbon.2015.12.032.

Sevilla, M., Falco, C., Titirici, M.-M., & Fuertes, A. B. (2012). High-performance CO_2 sorbents from algae. *RSC Advances*, *2*, 12792–12797. https://doi.org/10.1039/c2ra22552b.

Shafeeyan, M. S., Daud, W. M. A. W., Houshmand, A., & Arami-Niya, A. (2011). Ammonia modification of activated carbon to enhance carbon dioxide adsorption: effect of preoxidation. *Applied Surface Science*, *257*, 3936–3942. https://doi.org/10.1016/j.apsusc.2010.11.127.

Singh, G., Lakhi, K. S., Kim, I. Y., Kim, S., Srivastava, P., Naidu, R., et al. (2017). Highly efficient method for the synthesis of activated mesoporous biocarbons with extremely high surface area for high-pressure CO_2 adsorption. *ACS Applied Materials & Interfaces*, *9*, 29782–29793. https://doi.org/10.1021/acsami.7b08797.

Somy, A., Mehrnia, M. R., Amrei, H. D., Ghanizadeh, A., & Safari, M. (2009). Adsorption of carbon dioxide using impregnated activated carbon promoted by Zinc. *International Journal of Greenhouse Gas Control*, *3*, 249–254. https://doi.org/10.1016/j.ijggc.2008.10.003.

Song, T., Liao, J. M., Xiao, J., & Shen, L. H. (2015). Effect of micropore and mesopore structure on CO_2 adsorption by activated carbons from biomass. *Xinxing Tan Cailiao/New Carbon Materials*, *30*, 156–166. https://doi.org/10.1016/S1872-5805(15)60181-0.

Son, S.-J., Choi, J.-S., Choo, K.-Y., Song, S.-D., Vijayalakshmi, S., & Kim, T.-H. (2005). Development of carbon dioxide adsorbents using carbon materials prepared from coconut shell. *Korean Journal of Chemical Engineering*, *22*, 291–297. https://doi.org/10.1007/BF02701500.

Takagi, H., Hatori, H., Yamada, Y., Matsuo, S., & Shiraishi, M. (2004). Hydrogen adsorption properties of activated carbons with modified surfaces. *Journal of Alloys and Compounds*, *385*, 257–263. https://doi.org/10.1016/j.jallcom.2004.03.139.

Tellez-Juarez, M. C., Fierro, V., Zhao, W., Fernandez-Huerta, N., Izquierdo, M. T., Reguera, E., et al. (2014). Hydrogen storage in activated carbons produced from coals of different ranks: effect of oxygen content. *International Journal of Hydrogen Energy*, *39*, 4996–5002. https://doi.org/10.1016/j.ijhydene.2014.01.071.

Texier-Mandoki, N., Dentzer, J., Piquero, T., Saadallah, S., David, P., & Vix-Guterl, C. (2004). Hydrogen storage in activated carbon materials: role of the nanoporous texture. *Carbon*, *42*, 2744–2747. https://doi.org/10.1016/j.carbon.2004.05.018.

Tiwari, D., Bhunia, H., & Bajpai, P. K. (2017). Synthesis of nitrogen enriched porous carbons from urea formaldehyde resin and their carbon dioxide adsorption capacity. *Journal of CO_2 Utilization*, *21*, 302–313. https://doi.org/10.1016/j.jcou.2017.08.002.

United States Environmental Protection Agency (EPA) n.d. Overview of Greenhouse Gases 2016.

Fierro, V., Zhao, W., Izquierdo, M. T., Aylon, E., & Celzarda, A. (2010). Adsorption and compression contributions to hydrogen storage in activated anthracites. *International Journal of Hydrogen Energy*, *35*, 9038–9045.

Wang, H., Gao, Q., & Hu, J. (2009). High hydrogen storage capacity of porous carbons prepared by using activated carbon. *Journal of the American Chemical Society*, *131*, 7016–7022. https://doi.org/10.1021/ja8083225.

Wang, J., & Kaskel, S. (2012). KOH activation of carbon-based materials for energy storage. *Journal of Materials Chemistry*, *22*, 23710. https://doi.org/10.1039/c2jm34066f.

Wang, J., Senkovska, I., Kaskel, S., & Liu, Q. (2014). Chemically activated fungi-based porous carbons for hydrogen storage. *Carbon*, *75*, 372–380. https://doi.org/10.1016/j.carbon.2014.04.016.

Wang, Z., Sun, L., Xu, F., Zhou, H., Peng, X., Sun, D., et al. (2016). Nitrogen-doped porous carbons with high performance for hydrogen storage. *International Journal of Hydrogen Energy*, *41*, 8489–8497. https://doi.org/10.1016/j.ijhydene.2016.03.023.

Wróbel-Iwaniec, I., Díez, N., & Gryglewicz, G. (2015). Chitosan-based highly activated carbons for hydrogen storage. *International Journal of Hydrogen Energy*, *40*, 5788–5796. https://doi.org/10.1016/j.ijhydene.2015.03.034.

Xiao, Y., Dong, H., Long, C., Zheng, M., Lei, B., Zhang, H., et al. (2014). Melaleuca bark based porous carbons for hydrogen storage. *International Journal of Hydrogen Energy*, *39*, 11661–11667. https://doi.org/10.1016/j.ijhydene.2014.05.134.

Yamasaki, A. (2003). An overview of CO_2 mitigation options for global warming—emphasizing CO_2 sequestration options. *Journal of Chemical Engineering of Japan*, *36*, 361–375.

Yang, R., Liu, G., Li, M., Zhang, J., & Hao, X. (2012). Preparation and N_2, CO_2 and H_2 adsorption of super activated carbon derived from biomass source hemp (Cannabis sativa L.) stem. *Microporous and Mesoporous Materials*, *158*, 108–116. https://doi.org/10.1016/j.micromeso.2012.03.004.

Yaumi, A. L., Bakar, M. Z. A., & Hameed, B. H. (2018). Melamine-nitrogenated mesoporous activated carbon derived from rice husk for carbon dioxide adsorption in fixed-bed. *Energy*, *155*, 46–55. https://doi.org/10.1016/j.energy.2018.04.183.

Yin, G., Liu, Z., Liu, Q., & Wu, W. (2013). The role of different properties of activated carbon in CO_2 adsorption. *Chemical Engineering Journal*, *230*, 133–140. https://doi.org/10.1016/j.cej.2013.06.085.

Yong, Z. O. U., & Mata, V. G. (2001). Adsorption of Carbon Dioxide on Chemically Modified High Surface Area Carbon-Based Adsorbents at High Temperature. *Adsorption*, 41–50. https://doi.org/10.1023/A:1011220900415.

Zhang, J., Gao, J., Chen, Y., Hao, X., & Jin, X. (2017). Characterization, preparation, and reaction mechanism of hemp stem based activated carbon. *Results in Physics*, *7*, 1628–1633. https://doi.org/10.1016/j.rinp.2017.04.028.

Zhao, X. B., Xiao, B., Fletcher, A. J., & Thomas, K. M. (2005). Hydrogen adsorption on functionalized nanoporous activated carbons. *The Journal of Physical Chemistry B*, *109*, 8880–8888. https://doi.org/10.1021/jp050080z.

Zhou, W., Wu, H., Hartman, M. R., & Yildirim, T. (2007). Hydrogen and methane adsorption in metal-organic frameworks: A high-pressure volumetric study. *Journal of Physical Chemistry C*, *111*, 16131–16137. https://doi.org/10.1021/jp074889i.

Further reading

Chahine, R., & Bose, T. K. (1994). Low-pressure adsorption storage of hydrogen. *International Journal of Hydrogen Energy*, *19*, 161–164. https://doi.org/10.1016/0360-3199(94)90121-X.

Dillon AC, Parilla PA, Zhao Y, Kim YH, Gennett T, Curtis C, Heben MJ, Blackburn JL, Gilbert KEH, Alleman JL, Jones KM,, Zhang SB, n.d. NREL activities in DOE Carbon-based Materials Center of Excellence.

Masika, E., & Mokaya, R. (2012). Hydrogen storage in high surface area carbons with identical surface areas but different pore sizes: direct demonstration of the effects of pore size. *Journal of Physical Chemistry C*, *116*, 25734–25740. https://doi.org/10.1021/jp3100365.

Supercapacitors (electrochemical capacitors)

10

Joanna Conder, Krzysztof Fic†, Camélia Matei Ghimbeu*,‡*
*University of Strasbourg, University of Upper Alsace, Institute of Materials Science of Mulhouse (IS2M—UMR CNRS 7361), Mulhouse, France, †Institute of Chemistry and Technical Electrochemistry, Poznan University of Technology, Poznan, Poland, ‡French Research Network on Electrochemical Energy Storage (RS2E), FR CNRS 3459, Amiens, France

Chapter Outline

10.1 Introduction

Energy storage is continuously attracting and maintaining scientists' attention because the immediate exploitation of the converted, harvested, or produced energy (from, e.g., coal, geothermal sources, sun, wind, or from other sources) can be problematic. Therefore, systems designed for energy accumulation and supply "on demand" are being sought (Miller, 2010). Rapid development of portable devices in the last few decades has stimulated intensive research into efficient power sources, mainly batteries and electrochemical capacitors (ECs) (Burke, 2007).

ECs, often called electric double-layer capacitors (EDLCs) or supercapacitors, are energy storage devices exploiting charge accumulation in the electric double-layer (EDL). This phenomenon is based on the weak electrostatic interactions of ions from the bulk of the electrolyte with the electrode surface (Simon, Gogotsi, & Dunn, 2014).

Char and Carbon Materials Derived from Biomass. https://doi.org/10.1016/B978-0-12-814893-8.00010-9

Unlike batteries, EDLCs store the charge in a physical manner, thus their energy density is moderate. At the same time, the lack of electrochemical reactions ensures very high power (response time up to 1 s) and very long cyclability (1,000,000 cycles).

These characteristics provide ECs with a link between conventional dielectric capacitors (high power) and batteries (high energy), as can be seen in the Ragone plot in Fig. 1.

ECs can be classified into EDLC with a purely capacitive charge storage mechanism and hybrid capacitors (HECs), merging capacitive and faradaic charge storage. The latter includes asymmetric capacitors (ASCs) and Li-ion capacitors. Nevertheless, this classification is not exhaustive (Sharma & Bhatti, 2010).

10.2 Basic principles of ECs

A charge storage mechanism in ECs is attributed to charging of the EDL, formed at the electrode/electrolyte interface (Béguin, Presser, Balducci, & Frackowiak, 2014). The first EDL model, proposed by Helmholtz, assumed a simple organization of ions at this charged surface followed by a linear potential drop with the distance from the electrode (Fig. 2). Later, Gouy-Chapman proposed a concept including the so-called diffuse layer with nonlinear potential decrease and solvent present at the proximity of the electrode. Finally, Stern, and later Graham, derived the model widely accepted today, which merges both previous concepts and reflects most likely the situation at the electrode/electrolyte interface.

The capacitance, C, (F) of this interface is directly proportional to the electrochemically accessible surface area A (m^2), electrical permittivity of free space (vacuum) ε_0, and dielectric constant ε_r, and inversely proportional to the thickness of the double layer d (m) (Kötz & Carlen, 2000; Lu, Béguin, & Frackowiak, 2013):

$$C = \frac{\varepsilon_0 \varepsilon_r A}{d} \tag{1}$$

For this reason, the increase in C is often a result of the development of the electrochemically accessible surface area of the electrode material or the ion size adjustment, both of which influence the thickness (d) of EDL. The electrochemically accessible surface area of the electrode should not be, however, confused with the specific surface area (SSA) determined from gas adsorption (N_2, CO_2, Ar) (Centeno, Sereda, & Stoeckli, 2011; Jagiello & Olivier, 2013; Stoeckli & Centeno, 2012). It should be considered as the surface in contact with the electrolyte, which is accessible for ions and depends, among other things, on pore size/diameter (Centeno et al., 2011; Chmiola, Yushin, Dash, & Gogotsi, 2006; Garcia-Gomez, Moreno-Fernandez, Lobato, & Centeno, 2015; Lobato, Suárez, Guardia, & Centeno, 2017; Vix-Guterl et al., 2005), surface functionalities, type of electrolyte used, and finally the size of the electrolyte ions.

The capacitance of the electrode, that is, the charge Q accumulated within given potential range U:

$$C = \frac{Q}{U} \tag{2}$$

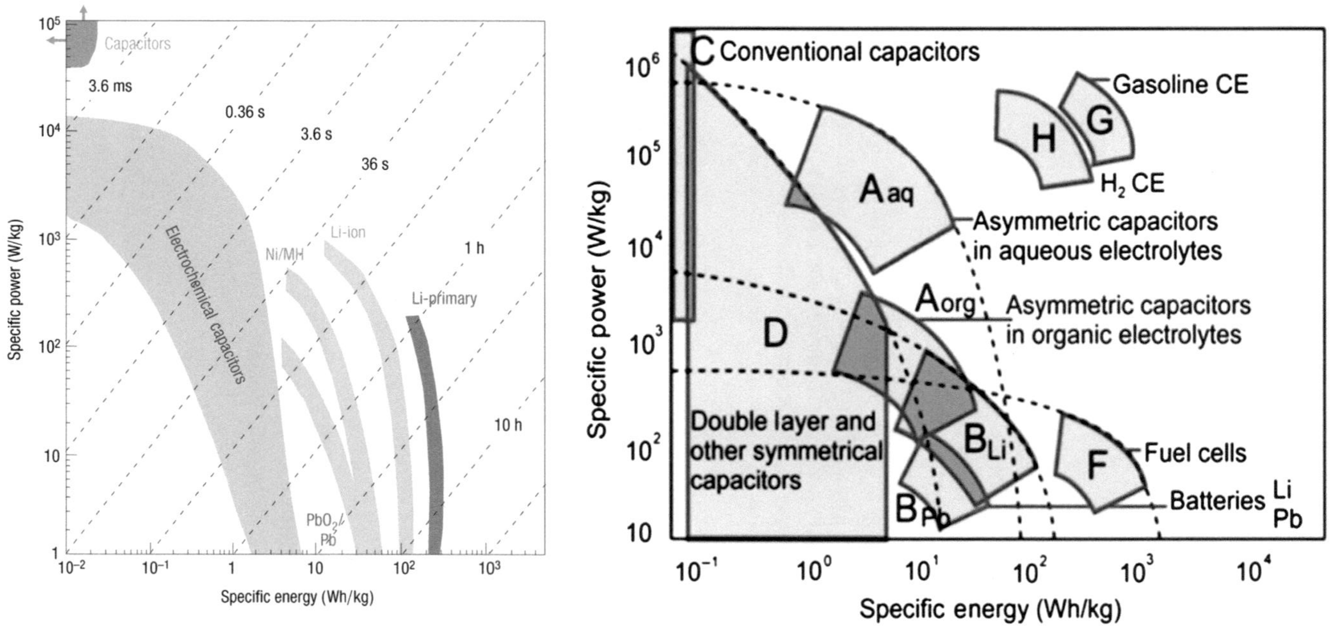

Fig. 1 (*Left*) Ragone plot for various energy storage systems and (*right*) a detailed comparison of various electrochemical capacitors (Dubal, Ayyad, Ruiz, & Gomez-Romero, 2015).
Reproduced with permission from Nature and The Royal Society of Chemistry.

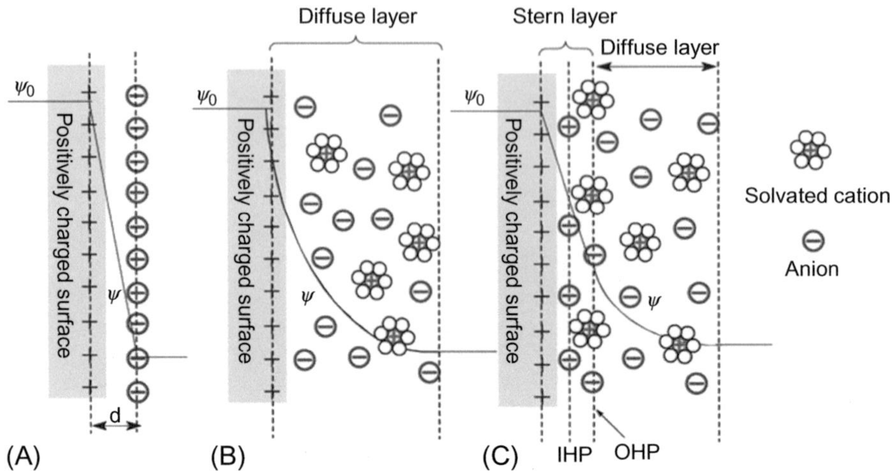

Fig. 2 Schematic representation of the electric double-layer at a positively charged electrode with various models: (A) the Helmholtz, (B) Gouy-Chapman, and (C) Stern models, showing the inner Helmholtz plane (IHP) and outer Helmholtz plane (OHP).
Reproduced with permission from The Royal Society of Chemistry, Zhang, L. L., & Zhao, X. S. (2009). Carbon-based materials as supercapacitor electrodes. *Chemical Society Reviews*, *38*(9), 2520–2531.

can be determined using different electrochemical techniques assessing either the kinetics (cycling voltammetry, electrochemical impedance spectroscopy) or the thermodynamics (galvanostatic and/or potentiostatic cycling). Other parameters such as self-discharge, leakage currents, and cyclability are also often determined (Andreas, 2015; Chen, Bai, Huang, & Li, 2014; García-Cruz, Ratajczak, Iniesta, Montiel, & Béguin, 2016; Kowal et al., 2011). It is widely accepted that the capacitor reaches its end of life either when the capacitance drops to 20% of the initial value or the internal resistance increases by 100% (Azaïs et al., 2007; He, Fic, Frckowiak, Novak, & Berg, 2016; Ratajczak, Jurewicz, & Béguin, 2014).

Although the capacitance is one of the most important parameters for capacitor evaluation, the energy density (Wh/L) or specific energy (Wh/kg) is a more useful parameter for final applications (Abbas, Babuchowska, Frąckowiak, & Béguin, 2016; Chae & Chen, 2012; François, Volker, Andrea, & Elzbieta, 2014; Gogotsi & Simon, 2011; Naoi, Nagano, & Naoi, 2013). For ECs, the amount of stored energy is directly proportional to the capacitance (C) and the voltage (U) squared:

$$E = \frac{1}{2} C_{\text{device}} U^2 \tag{3}$$

This equation should be used for the capacitance and voltage determination in a real, two-electrode cell configuration. It should not, however, be applied for calculations based on the three-electrode cell tests because these are far away from practical applications (excess of the active material and the electrolyte). In addition, because the two-electrode cell is composed of two capacitors connected in series, the final capacitance is the sum of the two individual electrode capacitances:

$$\frac{1}{C} = \frac{1}{C_+} + \frac{1}{C_-} \tag{4}$$

and thus is two times lower than that of a single electrode. Consequently, the final gravimetric capacitance (F/g) is four times lower than that of individual electrodes. Therefore, for calculating the energy for a symmetric capacitor one should use the following equation (derived from Eq. 3):

$$E = \frac{1}{8} C_{\text{electrode}} U^2 \tag{5}$$

However, it is strongly advised to calculate the energy density from the capacitance of the full device.

The maximum voltage of the capacitor is dictated by the type of the electrolyte used. Initially, aqueous (protic) and organic (aprotic) electrolytes were distinguished (Simon & Gogotsi, 2008). The recent tremendous development of ionic liquids, however, has led to considering them as a third class of electrolytes for ECs (Balducci, 2016; François et al., 2014). Solid-state electrolytes, (hydro)gels, and polyelectrolytes (Le Bideau, Viau, & Vioux, 2011) are also being increasingly employed.

The key characteristic of the electrolyte for ECs is the electrochemical stability window. For organic electrolytes, acetonitrile (ACN), or propylene carbonate (PC)-based formulations, the maximum working voltage reaches 2.7–2.8 V (Morita, Kaigaishi, Yoshimoto, Egashira, & Aida, 2006). The use of ionic liquids (Kurig et al., 2016; Thomberg et al., 2015), e.g., imidazolium, pyrrolidinium, and ammonium-based cations paired with BF_4^-, PF_6^-, and bis(trifluoromethanesulfonyl)imide ($TFSI^-$) anions (Balducci, 2016; François et al., 2014), instead of conventional electrolytes can reach extremely wide stability windows (up to 6 V). These results obtained in three-electrode cells with nonporous electrodes have, however, yet to be confirmed in two-electrode systems assembled with porous carbon electrodes. For the time being, it seems that 3 V is the practical limit. Aqueous electrolytes, the eco-friendly alternative to organic solvents and ionic liquids (Fic, Meller, Menzel, & Frackowiak, 2016; Johns, Roberts, Wakizaka, Sanders, & Owen, 2009; Long et al., 2011), are significantly more conductive than their nonaqueous counterparts (usually >80 mS/cm, sometimes even up to 600 mS/cm) and thus can enhance the power of the capacitor. In addition, assembling the capacitors with water-based electrolytes does not require thorough and expensive drying of the electrode materials and an atmosphere free from oxygen and moisture. Acidic, alkaline, and neutral electrolytes can all be used in combination with a stainless steel current collector, desired because of its low price. The electrochemical stability of these electrolytes is, however, limited by the decomposition of water at 1.23 V and thus is the major obstacle in the way to wide commercialization of these systems. Nonetheless, for ECs operating with neutral electrolytes (pH around 7), such as 0.5–2 M Li^-, Na^-, K_2SO_4 or 1–5 M Li^-,

Na^-, KNO_3, effective operating voltages up to 1.9 V have been reported (Azaïs et al., 2007; Bichat, Raymundo-Piñero, & Béguin, 2010; Chae & Chen, 2012; Chien, Wu, Rajkumar, & Hu, 2016; Fic, Lota, Meller, & Frackowiak, 2012; Fic, Meller, & Frackowiak, 2015; Gao, Demarconnay, Raymundo-Pinero, & Béguin, 2012).

By definition, the capacitance is independent of the potential/voltage range. This is, however, not true for porous electrodes, in which faradaic contributions to charge accumulation influence the final capacitance. Furthermore, the amount of stored charge depends also on the current load, because the porosity of the electrode impacts the transport of ions from the electrolyte bulk to the electrode surface. Incorporation of redox species brings pseudocapacitance into the system and boosts the capacitance (Brousse, Bélanger, & Long, 2015; Conway, Birss, & Wojtowicz, 1997; Guan, Yu, & Chen, 2016; Lu et al., 2013). This phenomenon has been primary reported for RuO_2 (Conway et al., 1997; Conway & Pell, 2003) and then studied for various transition metal oxides (TMOs). Recently, however, besides RuO_2, only MnO_2 can be accepted as a typical pseudocapacitive material. Apart from TMOs, iodide/iodine and bromide/bromine redox couples present in the electrolyte were also initially considered as a source of enormous capacitance (Lota, Fic, & Frackowiak, 2011; Lota & Frackowiak, 2009; Menzel, Fic, Meller, & Frackowiak, 2014; Yamazaki, Ito, Yamagata, & Ishikawa, 2012). Hydroquinone and its isomers (Pognon, Brousse, & Bélanger, 2011; Roldán et al., 2011; Shul & Belanger, 2016), methylene blue (Roldán, Granda, Menéndez, Santamaría, & Blanco, 2012), indigo carmine (Roldán et al., 2011), humic acids (Wasiński, Walkowiak, & Lota, 2014), and other species, were tested as well. However, because the performance of the capacitors with redox-active electrolytes is closer to high-power batteries one should take a special care in reporting the correct metrics, especially for capacitance estimations (Laheäär, Przygocki, Abbas, & Béguin, 2015). As for the electrode materials, activated carbons (ACs) with their versatile properties have been and still are the most popular materials for applications in ECs.

10.3 From biomass to capacitor electrode material

Ever since the late 1980s, when Nippon Electric Corporation launched the first EDLC on the market, ACs became an inherent part of this energy storage system, playing almost the first fiddle. Although for a long time these materials were considered as the "golden goose," recently they are more and more often seen as weak links limiting further progress of high-performance capacitors. This is mainly because of the expensive and often hazardous synthetic routes of ACs as well as the scarcely tunable pore structure dominated by micropores restricting the diffusion of the electrolyte ions (Deng, Li, & Wang, 2016; Wang, Li, Liu, Lu, & Cheng, 2008). Optimizing the pore structure of ACs is never an easy task and requires well-thought-out choice of both carbon precursor and its processing conditions. Besides the optimal pore size and geometry, the ideal candidate for the EC electrode material should have a relatively high SSA, electronic and ionic conductivities, and sufficient mechanical and chemical stability (Yu, Tetard, Zhai, & Thomas, 2015) (Scheme 1), the interplay of which is the focus of this chapter.

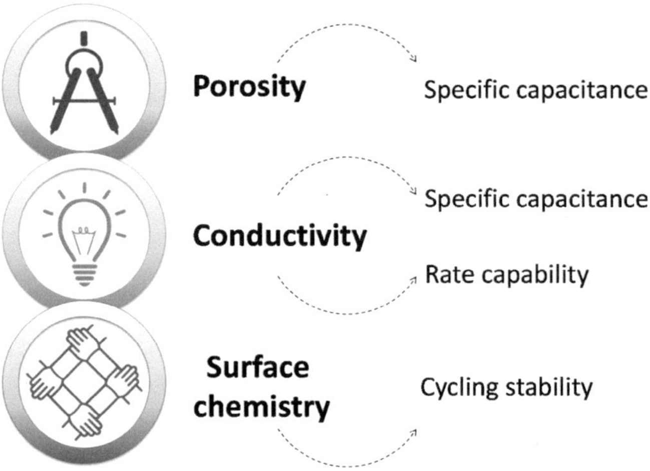

Scheme 1 Illustration summarizing the concept of the chapter, namely the interplay between the properties of the electrode material and those related to the performance of the capacitor.

In addition, it is arranged such that the subsequent parts constitute a guideline for the preparation of the biomass-derived carbon materials for different types of ECs, from classical EDLCs to their asymmetric and hybrid counterparts. This chapter not only leads through the process of preparation and optimization of the biomass-derived carbons for the specific application but also highlights the key parameters used to quantify their promise as cost-effective high-performance robust electroactive materials for capacitors.

10.3.1 Step 1: Pretreatment of the biomass crude

Various biomasses could be converted into carbonaceous materials and used for applications in ECs. On the one hand, coconut shells (Jain et al., 2013), olive pits (Redondo, Carretero-González, Goikolea, Ségalini, & Mysyk, 2015), pine cones (Karthikeyan et al., 2014), and other plant-based carbon-rich resources mainly composed of cellulose, lignin, and hemicellulose are a good starting point for electrode materials. On the other hand, bovine, porcine, and other animal bones built of apatite (Huang, Zhang, Huang, Wang, & Wei, 2011) could be used to prepare macro/mesoporous carbons. While the source of the biomass predestines the morphology (shape, particle size, and porosity) of the end-annealing products (Fig. 3), the primary chemical composition of the biomass, in particular the presence of impurities, has a big share in defining other physicochemical properties. Inorganic elements, e.g., calcium (Jenkins, Bakker, & Wei, 1996), potassium (Jenkins et al., 1996), silica (Ahiduzzaman & Sadrul Islam, 2016), and sodium (Jenkins et al., 1996), are considered undesired because their presence in to-be-annealed biomass often (1) significantly reduces the overall char yield (an exception can be found in Haddad et al., 2017) and promotes secondary reactions (Banks, Nowakowski, & Bridgwater, 2014), (2) impedes the formation of

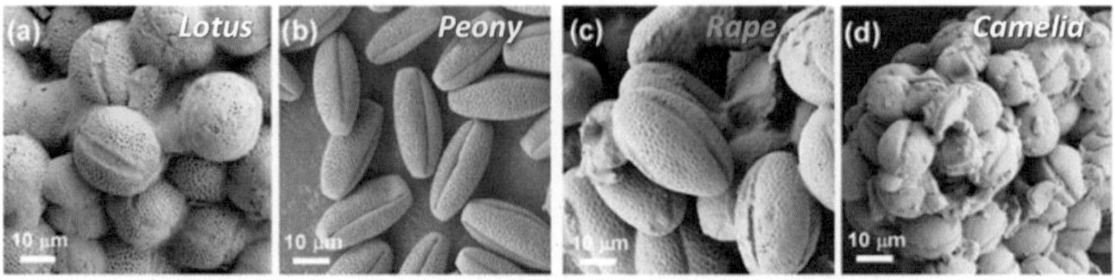

Fig. 3 SEM images of hydrochars derived from different pollens. (A) Lotus; (B) peony; (C) rape; and (D) camellia.
Reproduced with permission from John Wiley and Sons, Zhang, L., Zhang, F., Yang, X., Leng, K., Huang, Y., & Chen, Y. (2013). High-performance supercapacitor electrode materials prepared from various pollens. *Small*, *9*(8), 1342–1347.

pores (Ahiduzzaman & Sadrul Islam, 2016) and alters the surface chemistry of the resulting char (formation of glassy melt phases, fusions, and inclusions) (Jenkins et al., 1996), and (3) leads to a partial blocking of the micropores and thus restricted diffusion of the electrolyte ions therein (Wang et al., 2017).

Usually, washing the crude biomass only with water is sufficient to remove 90% of alkali metals unless they are present in high concentrations and bound to the organic structure (Banks et al., 2014). In this case, the use of strong acids, e.g., HCl (Tan & Wang, 2009) and HNO_3, instead of water is recommended because they hydrolyze cellulose and hemicellulose, therefore increasing the content of lignin (resistant to acid hydrolysis (Mayer, Apfelbacher, & Hornung, 2012)) and thus the overall char yield. While it removes metal ions from the biomass, strong acid washing alters the native polymer structure of the crude (promotes the development of the pores) and causes nonnegligible loss of mass (Park, Yun, & Park, 2005). In addition, the acid-washed biomass has to be thoroughly rinsed with water and subsequently dried to remove the remaining chlorine and other acid ions, and to reduce the moisture content before further treatment (Banks et al., 2014).

High moisture content may significantly decrease the efficiency of the "waste to wealth" conversion process because in some cases, mainly during the pyrolysis, it promotes the formation of liquid products (bio-oil) instead of char (Sarkar & Praveen, 2017). It also causes inhomogeneous heat transfer during charring, thereby adversely affecting the properties of the solid end-products. Therefore, besides drying in the classical conditions, elevated temperature and reduced pressure, other possibilities are currently being considered. These include mechanical dewatering of the biomass and adding a surfactant to the washing solution (Banks et al., 2014). In the latter case, the surface-active "additive" not only reduces the amount of moisture kept by the biomass but also removes most of the inorganic impurities, thereby trimming the operational cost (Banks et al., 2014).

10.3.2 Step 2: Thermal treatment

The choice of the biomass, and its extraction and/or pretreatment, commences the chains of consequences for the carbonaceous material. In addition, although there is no thermal treatment-memory effect, the properties of the char are closely intertwined with the annealing and processing conditions. Most common biomass carbonization methods include hydrothermal carbonization (HTC) and slow pyrolysis. HTC, often referred to as precarbonization or wet torrefaction, is increasingly being seen as a "waste to wealth" conversion route for biomass (Gallifuoco, Taglieri, Scimia, Papa, & Di Giacomo, 2017). Carried out in the presence of hot compressed water between 120°C and 250°C (Deng, Li, & Wang, 2016), HTC mimics natural coal formation but in a shorter time (BGU, 2017). HTC relies on complex chemical reactions of the biomass basic building blocks: cellulose, hemicellulose, and lignin, taking place at the interface of water and given biopolymer, and in its bulk (Deng, Li, & Wang, 2016). These reactions include, among others, aromatization, dehydration, hydrolysis, and polymerization (Funke & Ziegler, 2010). Overall, the materials ("hydrochars") produced through HTC are rich in oxygenated and other heteroatomic functional groups, and have a low level of aromaticity, rendering them good precursors for ACs(Funke & Ziegler, 2010). The combination of the starting material (also referred to as parent material) and other parameters, namely: (1) annealing temperature, (2) residence time (also referred to as dwell time), (3) concentration of the reactants, and (4) the presence/absence of the catalyst, plays a substantial role in producing HTC-derived AC precursors with desired chemical and physical characteristics (Funke & Ziegler, 2010). In particular, the choice of the annealing temperature comes to the fore when aiming at high-surface-area ACs with well-developed pore structure. As shown by Jain, Balasubramanian, and Srinivasan (2015) in the example of coconut shell-derived hydrochars, there is a threshold value, after which further increase in the HTC temperature does not result in higher yield of the oxygenated functional groups but quite the opposite—it leads to their decomposition to carbon di- and monoxide, methane, hydrogen, and other gaseous products (Williams & Onwudili, 2005).

Among the advantages of HTC are relatively low environmental impact and the carbon yield reaching the highest value of all biomass-based technologies, 95% (R3Water, n.d.). Besides that, unlike other methods, HTC does not require predrying of biomass, which inevitably increases the cost of the entire conversion process. It falls short, however, in adding value to the crude rich in lignin because the presence of this nonpolysaccharide during thermal treatment causes a so-called "shadowing effect," lowering the carbon content and the coalification degree of the hydrochar (Liu & Guo, 2015). In addition, high energy and water consumption always accompany this carbonization process, taking place in saturated steam and at high pressure (Singh & Kalia, 2017) thereby postponing its industrial implementation.

The high energy consumption and the "shadowing effect" are not a concern for pyrolysis that starts to mature as a commercial technology (Bauer, 2017). In combination with physical and/or chemical activation, pyrolysis constitutes the classical synthesis route for converting the majority of biowastes into value-added carbons for capacitors (Deng, Li, & Wang, 2016). Given the operating temperature range (500 and 1000°C) (Deng, Li, & Wang, 2016), this thermochemical conversion process is in particular suited for woody and herbaceous plants (Chen, 2014) and other lignin-rich biowastes (thermal decomposition of lignin occurs between 100°C and 900°C (Yang, Yan, Chen, Lee, & Zheng, 2007)) (Fig. 4). Pyrolysis of lignocellulosic materials yields three different products, namely biochar, bio-oil, and syn-gas, the relative

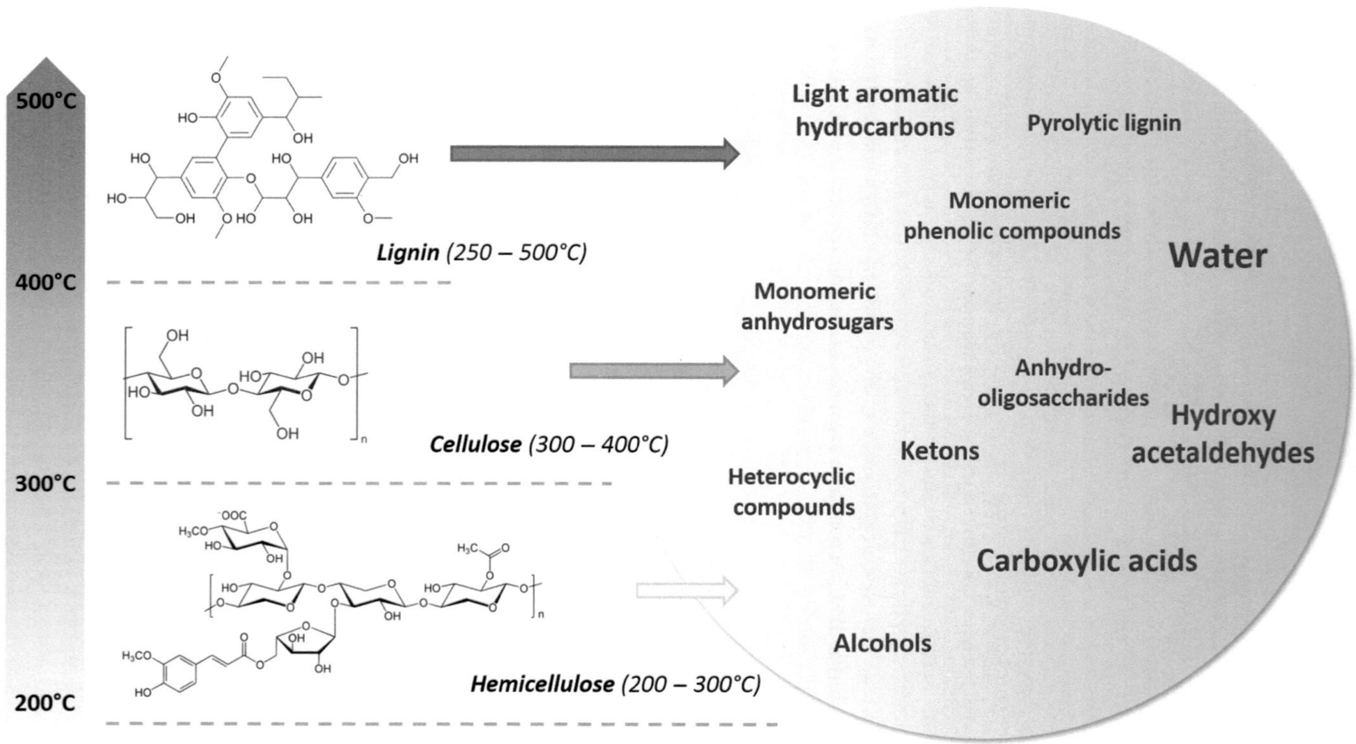

Fig. 4 Schematics illustrating decomposition of the biomass basic building blocks as a function of temperature.
Based on Arnold, S., Moss, K., Henkel, M., & Hausmann, R. (2017). Biotechnological perspectives of pyrolysis oil for a bio-based economy. *Trends in Biotechnology*, *35*(10), 925–936.

ratio of which depends, among other parameters, on the particle size of the raw feedstock, processing temperature, heating rate, and residence time. As shown by Demirbas (2004) in the example of olive husk, corncob, and tea waste, large particle size in combination with low pyrolysis temperature and slow heating rate promotes formation of carbon (Hernández-Montoya, García-Servin, & Bueno-López, 2012), significantly increasing fuel-to-feed efficiency. The latter is also related to the chemical composition of the raw feedstock. The higher the lignin content in the starting material, the higher the char yield after the pyrolysis, which in the given example leads to the following order: olive husk > tea waste > corncob.

The effect of pyrolysis temperature, residence time, and heating rate on the biomass-derived carbon was also studied by Lua, Lau, and Guo (2006) for oil palm shells with a lignin content similar to that of olive husk (around 54 wt%) (Yuliansyah & Hirajima, 2012). They confirmed that the three main textural properties of ACs (activated by CO_2), that is, the SSA, the micropore area, and volume, are already decided during thermal treatment. Moreover, materials with the most favorable characteristics were also obtained at lower temperatures, between 400°C and 600°C. Once 600°C was exceeded, further increase in temperature had an adverse effect on the forming char, causing softening and sintering of the high-molecular-weight volatiles. These in turn triggered the depolymerization of the melt and eventually led to an increase in char density. Consequently, ACs produced from these compact chars could not boast a well-developed micropore network and high SSA (Fig. 5, left).

Besides pyrolysis temperature, residence time is also an important parameter that can have its large share in defining the structure of the pyrolysis end-products (Fig. 5, right). The longer the residence time, the higher the release of volatile components, and thus the larger the micropore volume, unless the heat treatment at 600°C lasts longer than 2 h. Once this threshold is exceeded, the micropore area and micropore volume drop steeply, from almost 460 to 340 m^2/g and from 0.22 to $<0.16\,cm^3/g$, respectively, and the SSA decreases correspondingly.

In the pursuit of energy saving and cost reduction, other carbonization methods are currently being investigated. Among them, molten-salt carbonization, which can be considered as a derivative of HTC, continues to gather attention. In this method, thermochemical decomposition of the biomass takes place in a bath of molten salts or their eutectics (Nygård & Olsen, 2012), through which inert gas is passed at a temperature higher than 400°C. Alkali metal carbonates, halides, and nitrates are the essential ingredients of this bath. Not only do they act as catalysts accelerating gasification of the carbon with CO_2 and steam by cracking the links between cellulose, hemicellulose, and lignin (mainly glycosidic bonds) but they also play the role of heat carriers and suppliers, as well as reaction media. Molten salts are often employed for upgrading the biomass rich in cellulose. Lu, Hu, Yin, Xiao, and Wang (2016) reported a study of firewood-derived carbon prepared with and without the eutectic carbonate molten salts, Na_2CO_3 and K_2CO_3, in to-be-pyrolyzed feedstock. Although the yield of the carbon was significantly lower when molten salt was used for biomass conversion (11.6% vs 35.9%), the obtained char had an SSA and pore volume twice as big as the SSA and pore volume of its reference pyrolyzed only under argon (818 vs 463 m^2/g and 0.44 vs 0.22 cm^3/g).

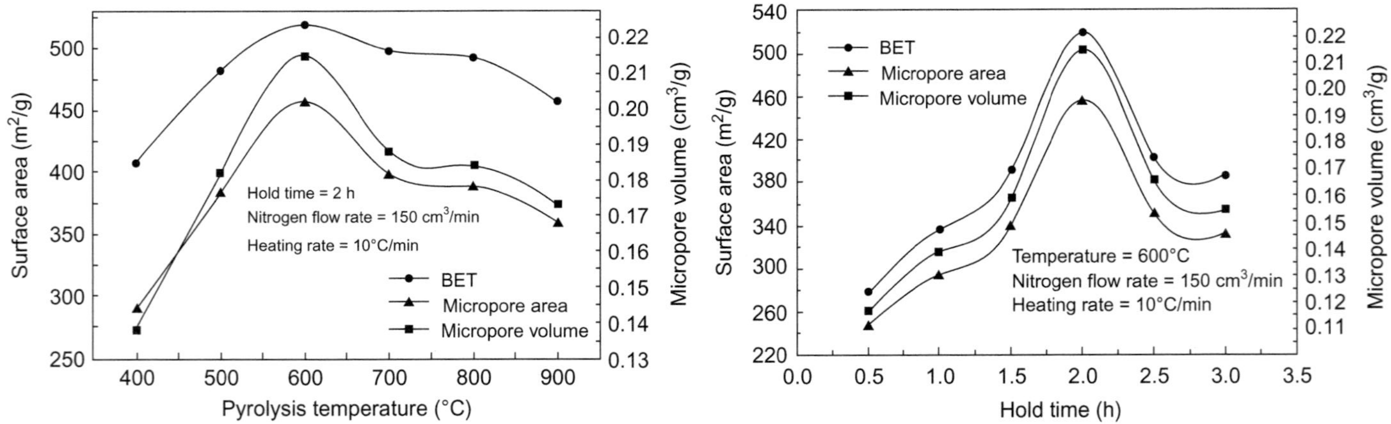

Fig. 5 (*Left*) Effect of pyrolysis temperature on the BET surface area, micropore area, and micropore volume of the activated carbons. (*Right*) Effect of pyrolysis hold time on the BET surface area, micropore area, and micropore volume of the activated carbons.
Reproduced with permission from Elsevier, Lua, A. C., Lau, F. Y., & Guo, J. (2006). Influence of pyrolysis conditions on pore development of oil-palm-shell activated carbons. *Journal of Analytical and Applied Pyrolysis*, *76*(1), 96–102.

10.3.3 Step 3: Activation

Usually after thermal treatment the SSA of the (hydro/pyro)chars does not exceed a few hundreds of m^2/g (Gallifuoco et al., 2017) and their pore structure is rather poorly developed (Lu & Zhao, 2017), therefore precluding these materials from attaining specific capacitances comparable to those of commercial capacitors. Fortunately, both the SSA and the pore structure can be enhanced through the activation of chars.

Activation can be carried out chemically, physically, or physiochemically. The first of the three, chemical activation, which is usually but not necessarily paired to carbonization (Fig. 6), is chosen for this purpose twice as often as physical activation because of: (1) lower temperature and shorter time required to activate the (hydro/pyro)char, (2) higher SSA of the end product (Fig. 7 vs Fig. 8), (3) better control of the textural properties, in particular the micropore size distribution, and (4) overall higher carbon yield. However, it always carries a risk of introducing inorganic impurities to the AC and requires thorough postwashing (Hernández-Montoya et al., 2012).

In a typical procedure, a biomass carbon precursor or its precarbonized counterpart is impregnated with chemical agent and subsequently thermally treated under an inert atmosphere (Hernández-Montoya et al., 2012). H_3PO_4 and $ZnCl_2$ are generally recommended for lignocellulosic materials, whereas K_2CO_3 and KOH, and other alkaline carbonates and hydroxides, are considered as more universal activating agents for

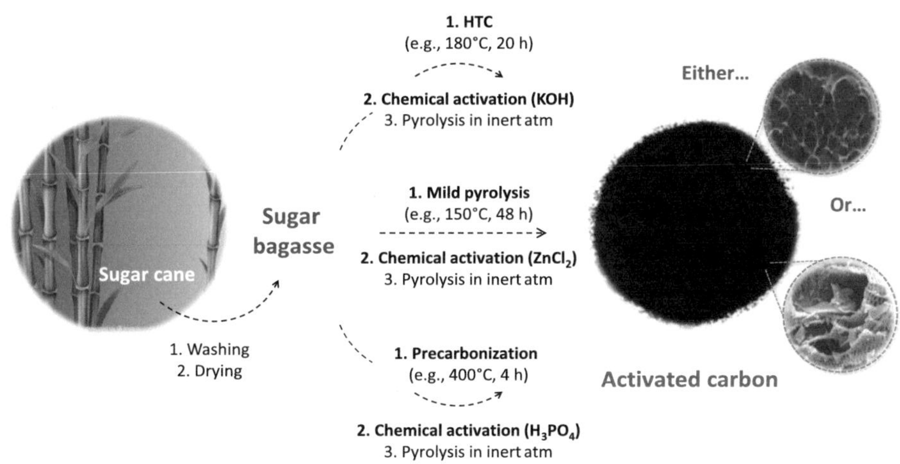

Fig. 6 Summary of different "waste to wealth" pathways for converting sugar cane into activated carbon. The sugar cane graphic was designed by Freepik (2018a). (*Top right*) SEM picture is reprinted in part with permission from Wahid, M., Puthusseri, D., Phase, D., & Ogale, S. (2014). Enhanced capacitance retention in a supercapacitor made of carbon from sugarcane bagasse by hydrothermal pretreatment. *Energy & Fuels*, *28*(6), 4233–4240. Copyright 2018 American Chemical Society. (*Bottom right*) SEM picture is reprinted in part from Adinaveen, T., Kennedy, L. J., Vijaya, J. J., & Sekaran, G. (2013). Studies on structural, morphological, electrical and electrochemical properties of activated carbon prepared from sugarcane bagasse. *Journal of Industrial and Engineering Chemistry*, *19*(5), 1470–1476, with the permission of Elsevier.

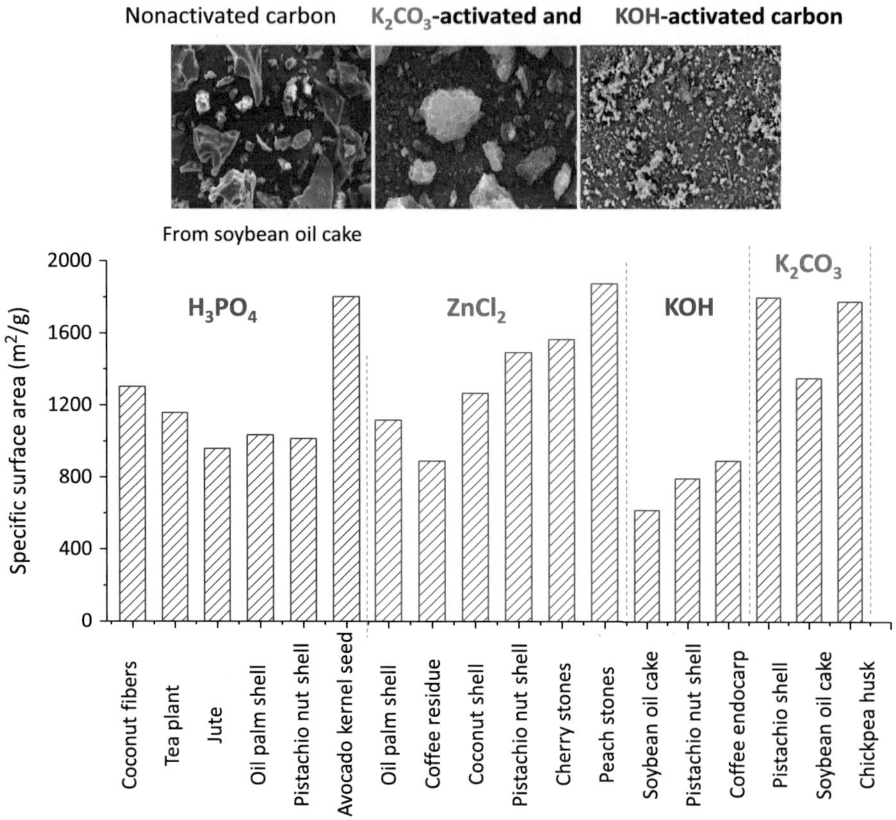

Fig. 7 (*Top*) SEM images of biochar and activated carbons. (*Bottom*) Comparison of the specific surface area of the carbons derived from different bioprecursors, activated chemically either with H_3PO_4 (Arami-Niya, Daud, & Mjalli, 2011; Elizalde-González, Mattusch, Peláez-Cid, & Wennrich, 2007; Phan et al., 2006; Yagmur, Ozmak, & Aktas, 2008), or $ZnCl_2$ (Arami-Niya et al., 2011; Azevedo et al., 2007; Boudrahem, Aissani-Benissad, & Aït-Amar, 2009; Caturla, Molina-Sabio, & Rodríguez-Reinoso, 1991; Kaghazchi, Kolur, & Soleimani, 2010; Olivares-Marín, Fernández-González, Macías-García, & Gómez-Serrano, 2006), or KOH (Lua & Yang, 2004; Nabais, Carrott, Ribeiro Carrott, Luz, & Ortiz, 2008; Tay, Ucar, & Karagöz, 2009), or K_2CO_3 (Hayashi, Horikawa, Muroyama, & Gomes, 2002; Hayashi, Horikawa, Takeda, Muroyama, & Nasir Ani, 2002; Tay et al., 2009).
(*Top*) Modified from Tay, T., Ucar, S., & Karagöz, S. (2009). Preparation and characterization of activated carbon from waste biomass. *Journal of Hazardous Materials*, *165*(1), 481–485; (*Bottom*) Based on Hernández-Montoya, V., García-Servin, J., & Bueno-López, J. I. (2012). Thermal treatments and activation procedures used in the preparation of activated carbons. In V. H. Montoya (Ed.), *Lignocellulosic precursors used in the synthesis of activated carbon—Characterization techniques and applications in the wastewater treatment*. InTech. (Chapter 2).

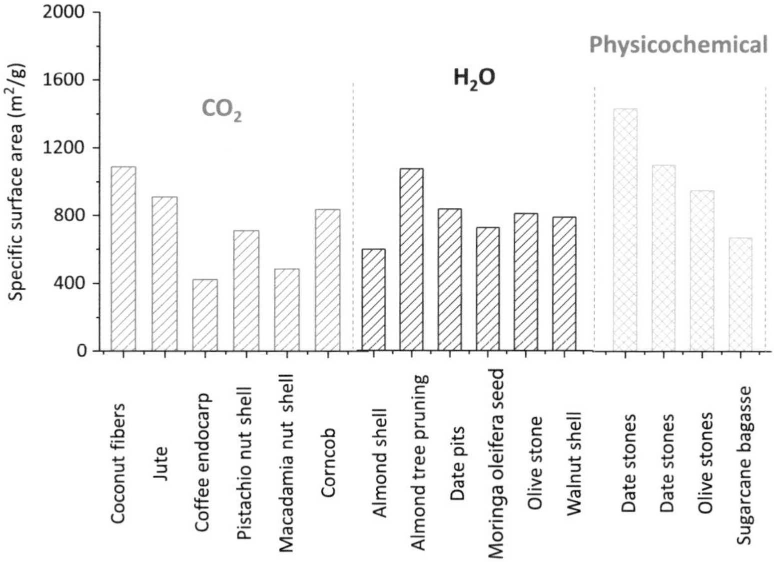

Fig. 8 Comparison of the specific surface area of the carbons derived from different bioprecursors, activated either physically with CO_2 (Aworn, Thiravetyan, & Nakbanpote, 2008; Lua & Yang, 2004; Nabais et al., 2008; Phan et al., 2006), or H_2O (Bouchelta et al., 2012; González, Román, Encinar, & Martínez, 2009; Warhurst, Fowler, McConnachie, & Pollard, 1997), or physicochemically with H_3PO_4 and steam (Hazourli, Ziati, & Hazourli, 2009), HNO_3 and steam (Hazourli et al., 2009), and H_2SO_4 and air (Valix, Cheung, & McKay, 2004). Based on Hernández-Montoya, V., García-Servin, J., & Bueno-López, J. I. (2012). Thermal treatments and activation procedures used in the preparation of activated carbons. In V. H. Montoya (Ed.), *Lignocellulosic precursors used in the synthesis of activated carbon—Characterization techniques and applications in the wastewater treatment*. InTech. (Chapter 2).

biomass rich and low in lignin. The physical and chemical properties of the end products, ACs, are to a large extent dependent on the concentration of the activating agent and its ratio with respect to the material subjected to activation (Fig. 7, top), the temperature (also the heating rate) of the carbonization and its dwell time, and to a smaller extent the carbonization atmosphere (Hernández-Montoya et al., 2012).

Carbonization temperature and dwell time are also important processing variables for physical activation during which a carbon-source material no longer reacts with the chemical agent but the oxidizing agent, CO_2, air, steam, or a mixture of these (Hernández-Montoya et al., 2012) (Fig. 8). The selection of the oxidizing agent is non-negligible for attaining application-desired SSA, pore volume, and pore size distribution. Usually, the use of CO_2 endows the end-products with narrow pore size distribution and relatively high micropore volume (Fałtynowicz, Kaczmarczyk, & Kułażyński, 2015), whereas the steam ACs are characterized by lower micropore volume and overall wider micropores (Molina-Sabio, Gonzalez, Rodriguez-Reinoso, & Sepúlveda-Escribano, 1996).

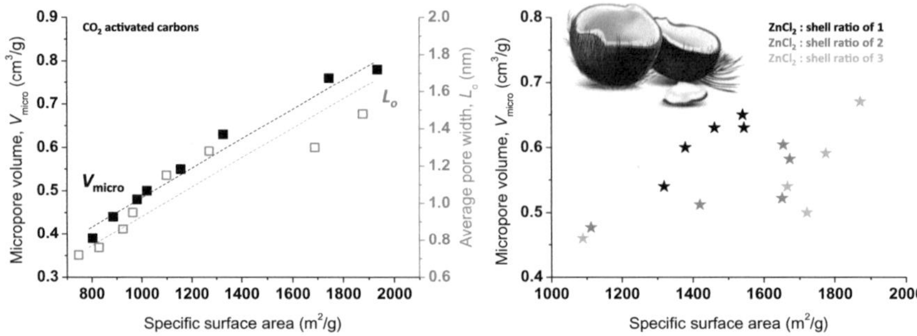

Fig. 9 Evolution of the micropore volume, V_{micro}, for the coconut shell-derived carbons prepared by (*left*) classical pyrolysis accompanied by CO_2 activation and (*right*) HTC-$ZnCl_2$ activation. The coconut graphics was designed by BSG Studio (All-free-download.com, 2018a). For the latter figure the data points were extracted from Jain, A., Balasubramanian, R., & Srinivasan, M. P. (2015). Tuning hydrochar properties for enhanced mesopore development in activated carbon by hydrothermal carbonization. *Microporous and Mesoporous Materials, 203*, 178–185; Decaux, C., Matei Ghimbeu, C., Dahbi, M., Anouti, M., Lemordant, D., Béguin, F., et al. (2014). Influence of electrolyte ion–solvent interactions on the performances of supercapacitors porous carbon electrodes. *Journal of Power Sources, 263*, 130–140.

Thus, in pursuit of high capacitance a preference would be given to CO_2 activation because higher micropore volume results in larger SSA and thus more active sites available for interactions with the electrolyte ions (Molina-Sabio et al., 1996). This micropore volume/SSA relationship initially established for olive stone-based carbons (Molina-Sabio et al., 1996) is, however, not always a rule of thumb and depends on the interplay between all processing parameters (Fig. 9).

Even if physical activation is considered as a cost-effective waste-free process, high activation temperature (800–1000°C), low carbon yield (Decaux et al., 2014), and limited control over the developing porosity (mainly microporosity with the average pore size <2nm, Fig. 9, right) (Decaux et al., 2014) do not speak in favor of implementing it as a part of the industrial process. In addition, with the capacitor application in mind, the ideal carbon electrode material should have both meso- and micropores, which, respectively, facilitate mobility of electrolyte ions and store the transferred charge.

To improve the development of meso- and microporosity, and render the ACs with tailor-made textural and chemical properties, sometimes chemical and physical methods are combined into one as, for example, in the case of coconut shell-derived carbon prepared by Jain and coworkers (Jain et al., 2015). Through careful analysis of the HTC-activation conditions and the interplay between the annealing temperature and $ZnCl_2$:shell ratio not only did they enhance the SSA and the mesoporosity of the coconut shell-derived carbon, but they also defined the optimal conditions for hydrochar upgrading (Jain et al., 2015). The latter included the optimization of the material in terms of micropore volume, a very important supercapacitor application-driven property (Fig. 9, right). Certainly, the work of Jain et al. can serve as a roadmap for preparing carbons with well-developed meso/microporous structures from biomass other than coconut shells. Among the examples of physicochemically ACs with highly developed porosity are also ACs derived from date(Hazourli et al., 2009) and olive stones with SSAs largely exceeding 900 m^2/g (Juárez-Galán, Silvestre-Albero, Silvestre-Albero, & Rodríguez-Reinoso, 2009).

Despite the wide range of activation methods and considerable knowledge acquired from studying the effect of activation conditions on the properties of the materials obtained from different carbon-rich precursors, extensive porosity development and relatively wide pore size distribution remain a sore point of the "waste to wealth" conversion process described in this chapter. Recently, it has been shown that porous high-surface-area carbon can be obtained also without the activation step. This was accomplished by Béguin et al.(Raymundo-Piñero, Leroux, & Béguin, 2006), who took advantage of Na, K, Ca, Mg, and their derivatives (earlier referred to as "impurities") naturally present in dead neem leaves, known as efficient porogens (Xu et al., 2013) promoting dehydration of the parent materials. As a result, the dead leaves-derived carbon prepared in this way outperformed other biomass-derived ACs in terms of SSA (1230 m^2/g) and a well-organized porous ion-conducting network allowing more efficient charge storage. This network was built from micropores matching the size of solvated ions. Similar work has been carried out on seaweeds (Raymundo-Piñero, Cadek, & Béguin, 2009). Although in this case the SSA of the carbons prepared without the activation step was on average lower than that of the dead neem leaves-based materials (from 416 to $<1300\,m^2/g$), impurity-induced engineering of the pore structure led to a competitive performance characteristics (high capacitance at low SSA).

Certainly, high SSA easily accessible to ions is important for efficient charge storage at the electrode/electrolyte interface and thus for attaining high specific capacitance but it is not a "must-have." A well-engineered meso/microporous ion-conducting network is equally or even more important when aiming at high-performance capacitor electrodes.

10.3.4 Doping of the biomass-derived carbons

Another strategy to enhance the performance of the biomass-derived carbon in an EC is to functionalize the surface of the material by introducing heteroatoms such as boron (B), oxygen (O), nitrogen (N), phosphorus (P), and sulfur (S). The functional groups containing these heteroatoms alter the wettability and polarity of the carbon surface, as well as its electronic conductivity. Consequently, they enhance capacitance through faradaic reactions with electrolyte ions and, in some cases, enlarge the potential window and extend the lifetime of the capacitor.

There are two main strategies to incorporate heteroatoms into the carbon framework, namely in situ and ex situ doping. The first is based on a simple pyrolysis of the biomass either containing heteroatoms or mixed with a heteroatom-containing compound. Many bioprecursors were explored for this purpose and it was observed that the nature and quantity of heteroatoms and resulting functional groups depend strongly on both biomass composition and synthesis procedure (pyrolysis temperature and time, nature and amount of activating agent, and posttreatment procedure, if any).

Antolini compared the content of nitrogen in botanic plants, algae, and animals (Table 1 in Antolini, 2016) and suggested to use the latter biomass rich in nitrogen (12%–17.5%) for the preparation of N-doped carbons. Zhao et al. (2016) pyrolyzed tobacco rods at 800°C in the presence of KOH activating agent and obtained N-doped carbon with low nitrogen content (1.1–1.4 at%). Xu et al. (2015) reported on the synthesis of N- and S-codoped carbon from shell beans by pyrolysis/activation with KOH at 650°C. *Enteromorpha prolifera* (Gao et al., 2014), natural silk (Hou, Cao, Idrees, & Ma, 2015), and willow catkins (Wang et al., 2015) were used to prepare materials doped with N, O, and S. The second approach to introduce heteroatoms in the carbon structure, ex situ doping, involves pyrolysis of biomass followed by the thermal treatment at high temperature using either "doping" gases (N_2, NH_3, H_2S) or chemical agents (melamine, urea). For the latter the examples might be N-doped porous carbons derived from wood, coconut shell, or potato wastes mixed with melamine and/or urea (Hulicova-Jurcakova, Seredych, Lu, & Bandosz, 2009; Ma et al., 2015; Seredych, Hulicova-Jurcakova, Lu, & Bandosz, 2008). The amount of nitrogen and oxygen introduced into these materials reached 6 and 10 at%, respectively. Higher concentration of oxygen functionalities on the surface of carbon was a sum of those developed during the activation process and those inherited from the parent material. One more example herein might be P-doped AC prepared from fruit stones and shiitake mushrooms mixed with phosphoric acid (H_3PO_4). Compared to undoped carbon materials, the use of P-doped AC in ECs resulted in higher capacitance and better stability within the wider voltage window (Cheng et al., 2015; Hulicova-Jurcakova et al., 2009).

The activation of the doped carbon that is necessary to develop the porosity and to achieve optimal SSA unfortunately always requires a high temperature. High temperature leads to the improvement of the carbon structure (especially its conductivity) but also risks the elimination of some of the functional groups, restricting pseudocapacitive properties of these materials. As mentioned before, the most common dopants naturally occurring in biomass are oxygen and nitrogen. Oxygen is present on the carbon surface in the form of carboxyl, ether, phenol, quinone, or ketone functional groups (Fig. 10A). While it contributes to the wettability of the carbon, a high amount of this heteroatom might be detrimental to electronic conductivity and induce side reactions leading to electrolyte decomposition. Nitrogen is usually found on the carbon surface in four different forms, namely graphitic (quaternary N), pyridinic, pyrrolic N, and in the form of oxidized species (Chen, Lin, Chen, Chen, & Yang, 2016; Tan et al., 2013) (Fig. 10B). The strong electron donor behavior of nitrogen enhances π bonding and the basic property of the carbon and thus its specific interactions with polar species via either electrostatic forces or hydrogen bonding. Additionally, the pyridine and pyrrolic groups cause the increase in the capacitance through redox reactions (Hulicova-Jurcakova, Seredych, et al., 2009; Lee, Chang, & Hu, 2013).

Fig. 11, left, shows the evolution of the amount of oxygen and nitrogen as a function of pyrolysis temperature for different selected biomass feedstocks, including cabbage, corn gluten, *Gigeriae galli*, and prawn shells. As can be seen, the amount of both elements depends on the source of the biomass and the conditions used for the preparation of the doped carbons. In addition, the amount of oxygen is always much higher

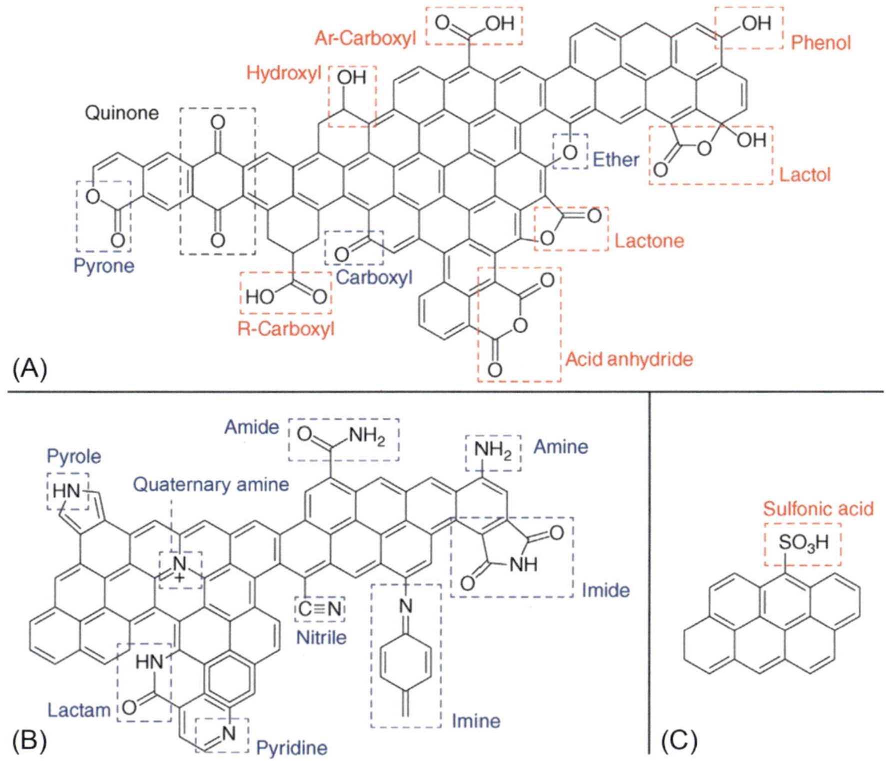

Fig. 10 Possible functional groups on a pyrolytic carbon surface as related to the presence of a heteroatom: (A) oxygen, (B) nitrogen, and (C) sulfur. Basic and acidic functionalities are indicated in *blue* and *red*, respectively.
Reproduced with permission from The Royal Society of Chemistry, de Clippel, F., Dusselier, M., Van de Vyver, S., Peng, L., Jacobs, P. A., & Sels, B. F. (2013). Tailoring nanohybrids and nanocomposites for catalytic applications. *Green Chemistry*, *15*(6), 1398–1430.

than that of nitrogen due to the higher abundance of the former in the parent material. As soon as the pyrolysis temperature increases, however, the amount of the two heteroatoms decreases, which is driven by the decomposition of O- and N-functional groups accompanied by the release of NH_3/HCN and CO/CO_2.

In the case of nitrogen, not only does the amount of N-functional groups decrease but also the ratio between the different N-groups changes as a function of temperature (Fig. 11, right) (Gao, Qu, Zhao, Wang, & Qiu, 2016; Xuan et al., 2017; Yan et al., 2018). Generally, at low carbonization temperatures (600°C) most of the nitrogen is present in pyrrolic form. The amount of pyridinic and quaternary N increases with the pyrolysis temperature, in particular at 800°C. Also, at low temperatures, nitrogen tends to appear in functional groups external to the aromatic ring (with localized charge), whereas at higher temperatures it is mainly found within the aromatic ring either delocalized or with no charge (Fig. 10). As can be seen in Fig. 11, right, high

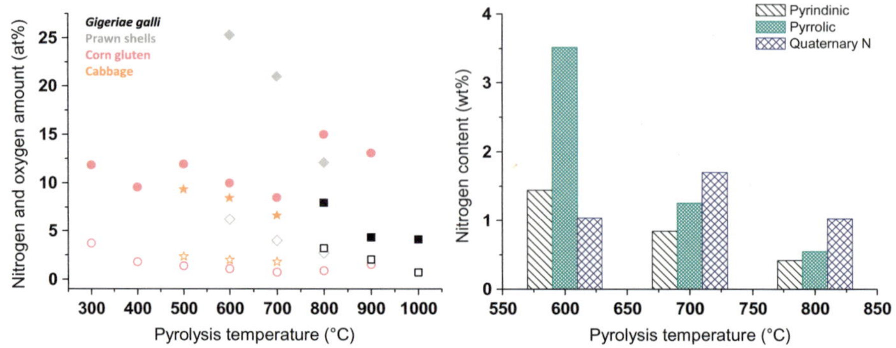

Fig. 11 (*Left*) Evolution of the amount of nitrogen (*open symbols*) and oxygen (*closed symbols*) as a function of the pyrolysis temperature for different biomass-derived carbons. (*Right*) Evolution of the pyridinic, pyrrolic, and quaternary N as a function of the pyrolysis temperature. (*Left*) Data points extracted from Gao, X., Xing, W., Zhou, J., Wang, G., Zhuo, S., Liu, Z., et al. (2014). Superior capacitive performance of active carbons derived from *Enteromorpha prolifera*. *Electrochimica Acta, 133*, 459–466; Cai, T., Wang, H., Jin, C., Sun, Q., & Nie, Y. (2018). Fabrication of nitrogen-doped porous electrically conductive carbon aerogel from waste cabbage for supercapacitors and oil/water separation. *Journal of Materials Science: Materials in Electronics, 29*(5), 4334–4344; Cheng, B.-H., Tian, K., Zeng, R. J., & Jiang, H. (2017). Preparation of high performance supercapacitor materials by fast pyrolysis of corn gluten meal waste. *Sustainable Energy & Fuels, 1*(4), 891–898; Hong, X., Hui, K. S., Zeng, Z., Hui, K. N., Zhang, L., Mo, M., et al. (2014). Hierarchical nitrogen-doped porous carbon with high surface area derived from endothelium corneum *Gigeriae galli* for high-performance supercapacitor. *Electrochimica Acta, 130*, 464–469; (*Right*) plotted using the data points from Gao, F., Qu, J., Zhao, Z., Wang, Z., & Qiu, J. (2016). Nitrogen-doped activated carbon derived from prawn shells for high-performance supercapacitors. *Electrochimica Acta, 190*, 1134–1141.

pyrolysis temperature favors the removal of heteroatom-containing groups from the carbon. One approach to preserve these functionalities might be, for example, to turn to microwave-assisted carbonization, a method used by Liang et al. (2018) in combination with KOH activation to prepare *Camellia oleifera* shell-derived ACs rich in oxygen (around 35 wt%).

Processing of the biomass into carbon with application-tuned properties and the performance of this "wealth" in capacitors are two different things. That is why the following sections are built upon concrete examples showing the influence of physicochemical properties of these materials on the performance characteristics of the actual device.

10.4 Electrical double-layer capacitors

The mechanism of energy storage in the most classical system, that is, EDLCs, is based on charge accumulation at the electrode/electrolyte interface in the form of the double layer. Usually, the average specific capacitance of commercial EDLCs with standard AC-based electrodes does not exceed 110 and 150 F/g in organic and aqueous (KOH) electrolyte, respectively (Brousse et al., 2017). According to Signorelli, Ku, Kassakian, and Schindall (2009), the electrode surface area, the gravimetric and volumetric specific capacitance (C_s), the frequency-dependent impedance response (the equivalent series resistance and the "knee" frequency), and the shape of the cyclic voltammetry (CV) curves are the basic metrics of EDLC performance. In addition, given the energy storage mechanism underlying the operation of EDLCs, all of these metrics are, to a variable extent, a function of the electrode material, its morphology, and physicochemical properties, as well as the function of the electrolyte (Pilon, Wang, & d'Entremont, 2015). The requirements for an ideal capacitor component are, however, not the same for the two. While the selection criteria for the electrolyte can be narrowed down to high ionic conductivity, high ion mobility, and wide potential and temperature operating range, many more criteria have to be considered when selecting a promising carbonaceous electrode material. In particular, the ideal candidate should feature high electrical conductivity and low electrical resistivity, high surface area easily accessible to electrolyte ions, optimal pore size, length and connectivity (hierarchical pore structures are preferred), low self-discharge, and relatively low cost (Pilon et al., 2015). Also, it should be designed and synthesized in accordance with the principles of green chemistry (i.e., through cost-effective and energetically efficient processes from low-risk, environmentally benign precursors). That is why biomass-derived (activated) carbons have entered the "energy storage war" and quickly become a strong competitor to commercial products.

Although the specific capacitance of the "real" EDLC electrodes is not directly proportional to their SSA, tuning the SSA of the biochar is often the first intuitive step toward high-performance energy carriers. As can be seen in the examples of corn waste-derived carbons given in Fig. 12, it is a reasonable approach because higher SSA is directly translated into higher C_s of the symmetric capacitor.

In this way, however, EDLC capacitance cannot be increased infinitely. Starting from 2000 m^2/g the gain in C_s is usually smaller, if any (Raymundo-Piñero, Leroux, & Béguin, 2006), because the large surface area is not supported by a well-organized pore network, enabling efficient charge transfer and storage.

Cellulose-rich biomass has in general a wide appeal for manufacturing of (micro) porous carbons for EDLCs. The reason for this lies in (1) an adequate carbon content (around 40%) (Viswanathan, Neel, & Varadarajan, 2009), (2) the presence of natural activating agents (described as impurities in Section 10.3.1), and (3) relatively narrow thermal decomposition range, all of which render this biopolymer a fairly good carbon precursor (Darmawan, Wistara, Pari, Maddu, & Syafii, 2016). Adding well-thought-out thermal processing conditions and a carefully selected activation method, it is almost a ready recipe for high-SSA EDLC carbon with pore size distribution matching that of commercial ACs, e.g., Norit Super 50. This has been demonstrated by Wang and coworkers (Wang, Geng, et al., 2015), who studied the influence of the activation temperature of biomass-derived carbon on its surface area, morphology, and microstructure. Through pyrolysis followed by chemical activation with KOH, they prepared a series of corncob-based ACs and assessed their charge storage abilities in a two-electrode symmetric system with either acidic or basic electrolyte (Fig. 13).

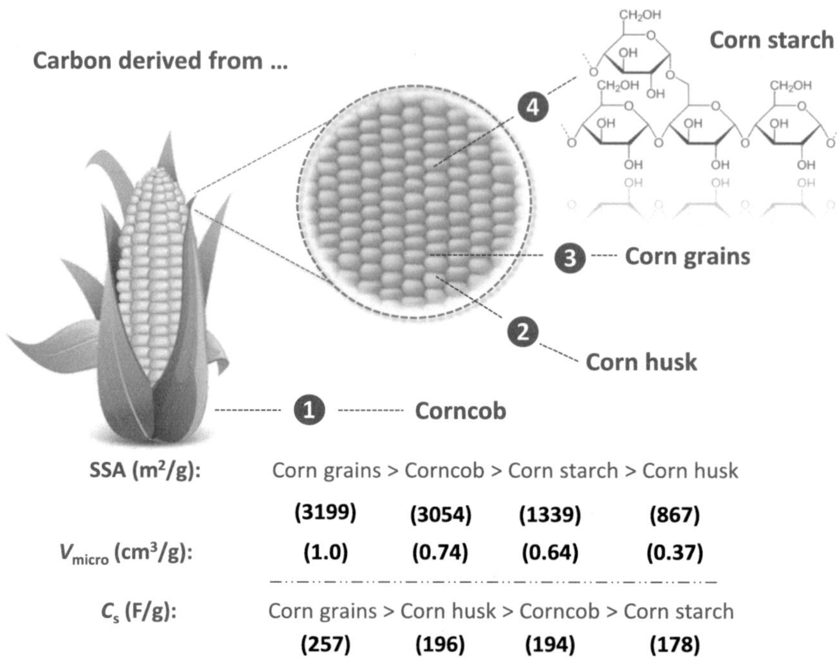

Fig. 12 Carbons derived from different corn wastes (Balathanigaimani et al., 2008; Li, Wang, Dai, Yang, & Zhong, 2008; Song et al., 2015; Wang, Geng, Li, & Zhang, 2015) and their corresponding SSA, V_{micro}, and C_s. The corncob graphic was designed by BSG Studio (All-free-download.com, 2018b). The corn grains were designed by Omelapics (Freepik, 2018b).

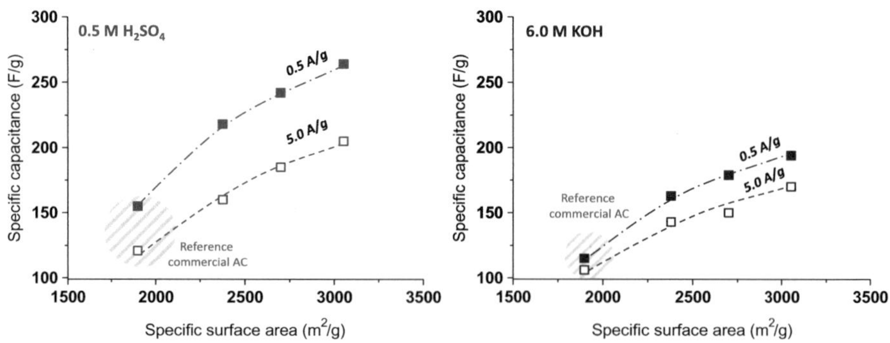

Fig. 13 Comparison of the electrochemical performance of different corncob-derived carbons in a two-electrode system (*left*) in 0.5 M H_2SO_4 and (*right*) 6 M KOH electrolyte.
Data points extracted from Table 4 from Wang, D., Geng, Z., Li, B., & Zhang, C. (2015). High performance electrode materials for electric double-layer capacitors based on biomass-derived activated carbons. *Electrochimica Acta*, *173*, 377–384.

Contrary to the previous example given in Fig. 12, herein the specific capacitance increased proportionally to the SSA of the corn waste-derived carbons. Moreover, C_s values were on average 50 F/g higher when H_2SO_4 was used as the electrolyte, which might be due to the presence of quinone groups on the surface of the carbon, bringing pseudocapacitance into the system (Wang, Geng, et al., 2015).

Regardless of the electrolyte used, the best-performing corncob-based carbon was that activated at 850°C (C850) with the highest proportion of meso- to micropores (0.651 to 0.492 cm^3/g). At low current density, the use of a pair of C850-based electrodes in EDLCs instead of that based on Norit A Supra resulted in almost two times higher capacitance. At high current density, at which the specific capacitance is controlled mainly by the average micropore width (Lota, Centeno, Frackowiak, & Stoeckli, 2008) and the number of groups generating CO_2 (Bichat et al., 2010), the difference between the two electrode materials was not as tremendous; however, the corncob-derived carbon was still able to store more charge. This was possible not only due to the well-developed surface area but also because of the great variety of functional groups present on the surface of corncob-derived carbon that improved its wettability and electrical conductivity.

Other interesting bioprecursors containing cellulose in their structure (in the cell walls) are seaweeds (Fig. 14). Their representative, *Lessonia nigrescens* (LN), was used by Raymundo-Piñero et al. (2009) to prepare a series of biomass-derived carbons.

Even if the surface area of these materials did not significantly exceed 1300 m^2/g, the capacitor assembled with a pair of LN900-based electrodes (900 denotes the carbonization temperature) outperformed that of commercial AC, Norit Super 50, in both aqueous (175 vs 119 F/g, respectively, in 1 M H_2SO_4) and organic electrolytes (94 vs 79 F/g, respectively, in 1 M tetraethylammonium tetrafluoroborate, $TEABF_4$, in

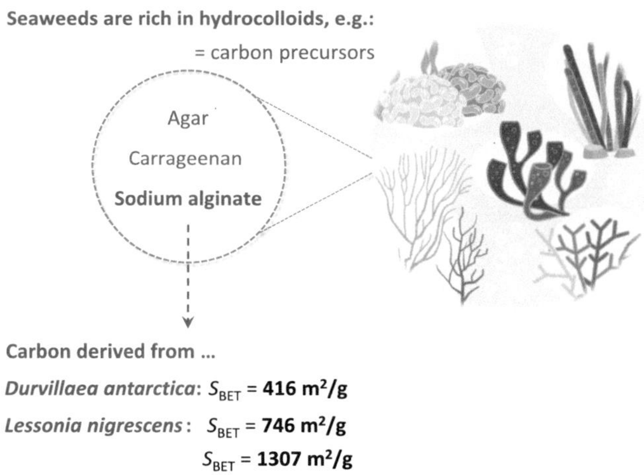

Fig. 14 Carbons derived from different seaweeds and their corresponding SSA. Seaweeds graphic was designed by Artsybunnies (Freepik, 2018c).

ACN). These results are even more remarkable given that the LN900 carbon was "activated" taking advantage of natural porogens present in the seaweed, mainly Na- (sodium alginate) and K-derivatives (carrageenan), homogeneously distributed in the bulk of this biomass (Raymundo-Piñero et al., 2009). The autoactivation of the LN seaweeds during the pyrolysis resulted in hierarchically porous material with meso- and micropores in the range of 3.0–4.0 nm and 0.7–0.8 nm, very beneficial for double-layer formation (Raymundo-Piñero, Leroux, & Béguin, 2006).

In commercial EDLCs, pairing the AC with H_2SO_4 electrolyte, in which the capacitance values for seaweed-derived carbons were the highest (Fig. 15, left), is not recommended. H_2SO_4 is corrosive to current collectors, cable connections, and other metallic components of the capacitor (Bichat et al., 2010). For this reason, the pursuit of "green carbon" electrode material for capacitor application is often accompanied by the pursuit of high-performing electrolyte (Fig. 15, right).

Organic electrolytes, e.g., $TEABF_4$ in ACN, are used to study and understand the relationship between the capacitance and the pore size (Chmiola et al., 2006; Raymundo-Piñero, Kierzek, Machnikowski, & Béguin, 2006). In general, the capacitance increases with the pore size approaching that of the bare electrolyte ions, i.e., 0.7 nm. The latter value varies, however, depending on the electrolyte/solvent(s) pair used. It remains at 0.7 nm for lithium bis(trifluoromethanesulfonyl)imide (LiTFSI) in ACN (Fig. 16, left, b and c) but increases to almost 1.0 nm when ACN is replaced by ethylene carbonate/PC/3 dimethyl carbonate (DMC) (Fig. 16, left, a). This might be due to the stronger interactions between the LiTFSI salt and the alkyl carbonate solvents, leading to difficulty in desolvating the LiTFSI ions and thus requiring wider

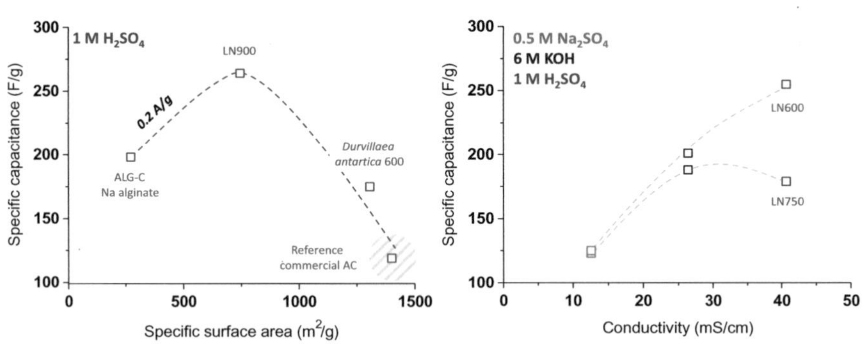

Fig. 15 (*Left*) Comparison of the electrochemical performance of different seaweed-derived carbons in symmetric two-electrode capacitors in 1 M H_2SO_4 electrolyte. (*Right*) Electrochemical performance of LN600- and LN750-based capacitors in three different electrolytes: 0.5 M Na_2SO_4, 6 M KOH, and 1 M H_2SO_4.
(*Left*) Data points extracted from Table 1 from Raymundo-Piñero, E., Cadek, M., & Béguin, F. (2009). Tuning carbon materials for supercapacitors by direct pyrolysis of seaweeds. *Advanced Functional Materials, 19*(7), 1032–1039; (*Right*) Data points extracted from Table 5 from Bichat, M. P., Raymundo-Piñero, E., & Béguin, F. (2010). High voltage supercapacitor built with seaweed carbons in neutral aqueous electrolyte. *Carbon, 48*(15), 4351–4361.

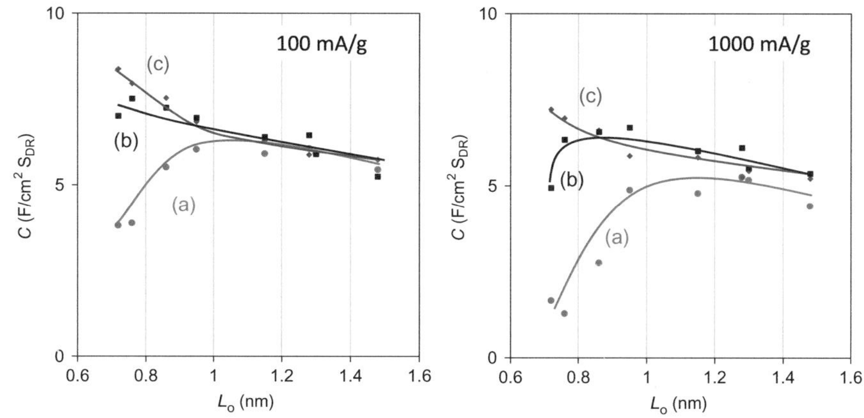

Fig. 16 Relation between the specific capacitance normalized by specific surface area and the pore size in 1 mol/L LiTFSI in ethylene carbonate/PC/3 DMC (a), $TEABF_4$ in ACN (b), and LiTFSI in ACN (c) for activated carbon derived from coconut shells activated with CO_2 gas, at (*left*) 100 mA/g and (*right*) 1000 mA/g.
Reproduced with permission from Elsevier, Decaux, C., Matei Ghimbeu, C., Dahbi, M., Anouti, M., Lemordant, D., Béguin, F., et al. (2014). Influence of electrolyte ion–solvent interactions on the performances of supercapacitors porous carbon electrodes. *Journal of Power Sources, 263*, 130–140.

pores to accommodate such ions. This is not the case for the ACN solvent, the ions of which are more easily desolvated and thus able to access smaller pores. The penetration of the electrolyte into the pores becomes especially important at higher current densities (Fig. 16).

The choice of the electrolyte/solvent(s) pair is, however, not solely limited to the ease of ion-pair formation, high ionic conductivity, and high ion mobility, but also to the affinity of the functional groups present on the surface of carbon to the electrolyte ions (Bichat et al., 2010). The latter brings about the pseudofaradaic reactions and thus is a source of additional capacitance in the system.

10.5 Carbon-based capacitors with pseudocapacitive effects

A common approach to trigger the pseudofaradaic reactions in the system is to employ electrodes based on doped carbons. Nitrogen is the most reported dopant (Cai, Wang, Jin, Sun, & Nie, 2018; Cheng, Tian, Zeng, & Jiang, 2017; Elmouwahidi et al., 2018; Fan et al., 2015; Li et al., 2012, 2014). It does not, however, stand alone, and is always accompanied by the oxygen naturally present in the biomass. Sometimes besides nitrogen (and oxygen), carbon is also doped with S (Elmouwahidi et al., 2018;

Lei et al., 2017), P (Seredych et al., 2008), or a mixture of heteroatoms (Lei et al., 2017). The chemical composition of the various oxygen and nitrogen groups formed during pyrolysis (see Fig. 10 in Section 10.3.4) depends mainly on the biomass source, pyrolysis temperature, and activation conditions. In addition, the acidity of these groups (Ghimbeu, Gadiou, Dentzer, Schwartz, & Vix-Guterl, 2010) defines their interactions with the electrolyte. In alkaline electrolyte (KOH), acidic O-groups (anhydrate, carboxyl) will react with hydroxyl ions (OH^-) (R1 and R2), whereas in acidic electrolyte (H_2SO_4), basic O-groups (ether, phenol, quinone) will interact with protons (H^+) (R3) (Chen et al., 2012; Oh et al., 2014; Tang, Liu, Yang, Yang, & Li, 2017). These (and other) redox reactions originating from the interactions of the surface functionalities with the active species in the electrolyte are usually tracked by means of CV performed in three-electrode cells.

$$-COOH + OH^- \leftrightarrow -COO^- + H_2O + e^- \tag{R1}$$

$$-COOH + OH^- \leftrightarrow > C = O + H_2O + e^- \tag{R2}$$

$$> C = O + H^+ + e^- \leftrightarrow > CH - O \tag{R3}$$

The pseudocapacitance triggered by N-functional groups is less understood and is still the subject of intense debate in the community. Some authors associate it with the pyrrolic and pyridinic N-groups, which are negatively charged and located at the edges of carbon. Also, although quaternary-N groups are known to mainly enhance electron transport, some studies showed that they may bring pseudocapacitance into the system, in both acid and basic electrolytes (Gao et al., 2016; Ornelas et al., 2014). Possible redox reactions involving pyridinic, pyridone, and pyrrolic N-groups are given here (Deng, Xie, Zou, & Ji, 2016; Seredych et al., 2008):

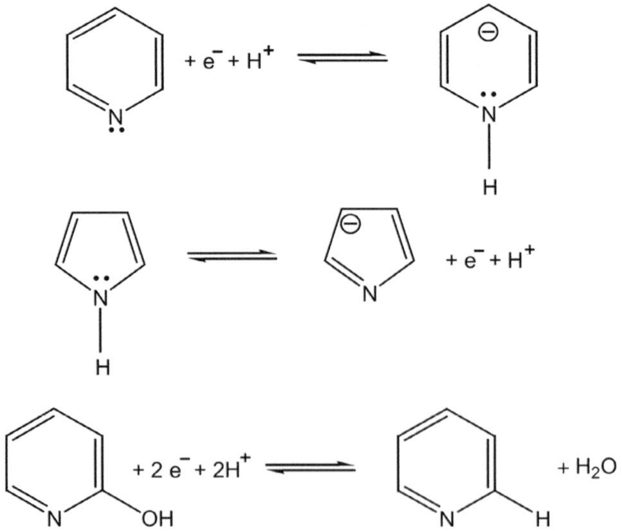

(R4)

$$\begin{aligned} &C^* - NH_2 + 2OH^- \rightleftharpoons C^* - NHOH + H_2O + 2e^- \\ &C^*H - NH_2 + 2OH^- \rightleftharpoons C^* = NH + 2H_2O + 2e^- \end{aligned} \tag{R5}$$

Fig. 17, left, shows the relation between specific capacitance, C_s, of different biomass-derived carbons in alkaline electrolyte, KOH, as a function of the atomic percentage of nitrogen and SSA. As can be seen, there is no clear correlation between these parameters, which might be related, among others, to: (1) the material characteristics, (2) the interactions of the doped carbon with the electrolyte, and (3) the electrode formulation and testing conditions. Each of these parameters has a strong impact on electrochemical performance. As explained before, the presence of nitrogen in the carbon structure is always accompanied by the presence of oxygen, which in turn leads to a large variety of possible pseudofaradaic interactions with the electrolyte (R1–R5). In addition, the nature and the amount of N- and O-groups directly influence the wettability and electronic conductivity of the material and thus the diffusion of the electrolyte ions and the charge propagation. Carbon texture (e.g., SSA), important for double-layer formation, also has its large share in defining the performance of the capacitor and if composed of very small pores inaccessible to the electrolyte ions it might as well limit C_s.

Therefore, it is very difficult to unambiguously evaluate the role of N- and O-functional groups in enhancing the performance of the capacitor due to their coexistence at different amounts and in different types. In an attempt to do so, N-free carbon materials with similar characteristics (SSA, pore size distribution, morphology, and structure) to those doped with nitrogen should be prepared. This is, however, a very difficult task. In addition, the variety of the electrode preparation recipes (the active material loading, the type of binder used, and the formulation as such), electrode testing conditions, and capacitance calculation (from two- or three-electrode cell configuration, from different current densities) contributes to the difficulty of obtaining reliable correlations in most cases. Despite that, Seredych et al. (2008) demonstrated that in the acidic medium a linear correlation between the number of basic N- (pyridinic and pyrrolic) and O-surface groups (quinone) can be obtained (Seredych et al., 2008) (Fig. 11, right). Interestingly, in the given example the correlation coefficient increases together with the current density, reaching almost linear dependence at 1 A/g. This, as explained by the authors, is because of a strong contribution of the electrochemically active basic functionalities to the capacitance at high current densities. At lower current densities, however, SSA and the formation of double-layers are the main contributors to the C_s. On the contrary, no correlation was found between the number of acidic surface groups and the capacitance.

10.6 Asymmetric and hybrid capacitors

Today's capacitors are expected not only to provide high power and thus to support the batteries in successfully fulfilling their tasks but also to store more energy. For this purpose, according to Eq. (3), not only should the specific capacitance, C_s, be enhanced but also the operational voltage window, U. This could be achieved by either

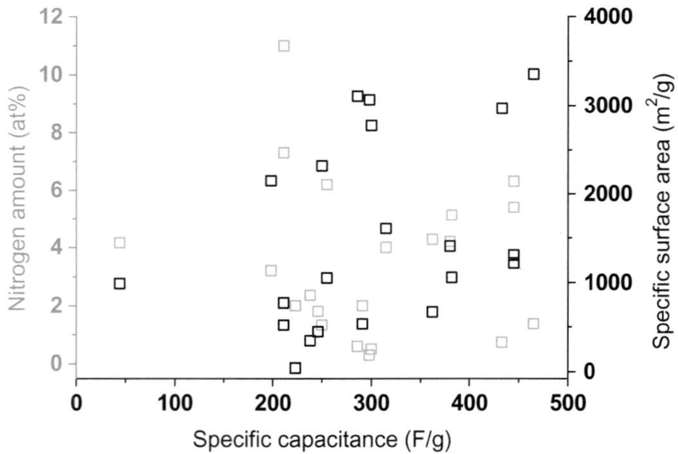

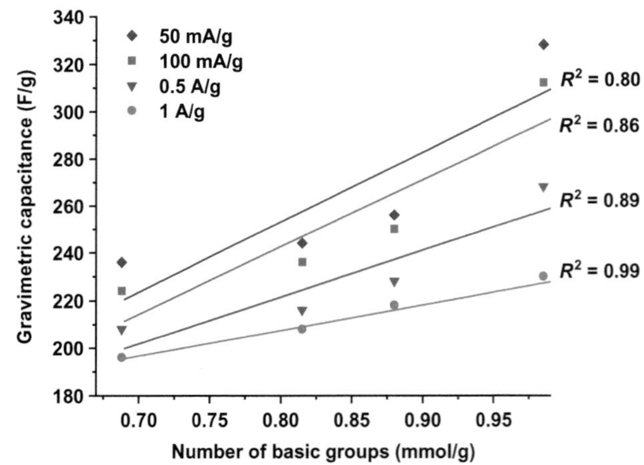

Fig. 17 (*Left*) Specific capacitance of different biomass-derived carbons tested in 6M KOH electrolyte as the function of the amount of nitrogen (at%) and SSA. (*Right*) Gravimetric capacitance versus number of basic groups at different current loads.
(*Left*) Plotted using data points extracted from Ma, G., Yang, Q., Sun, K., Peng, H., Ran, F., Zhao, X., et al. (2015). Nitrogen-doped porous carbon derived from biomass waste for high-performance supercapacitor. *Bioresource Technology*, *197*, 137–142; Cai, T., Wang, H., Jin, C., Sun, Q., & Nie, Y. (2018). Fabrication of nitrogen-doped porous electrically conductive carbon aerogel from waste cabbage for supercapacitors and oil/water separation. *Journal of Materials Science: Materials in Electronics*, *29*(5), 4334–4344; Cheng, B.-H., Tian, K., Zeng, R. J., & Jiang, H. (2017). Preparation of high performance supercapacitor materials by fast pyrolysis of corn gluten meal waste. *Sustainable Energy & Fuels*, *1*(4), 891–898; Hong, X., Hui, K. S., Zeng, Z., Hui, K. N., Zhang, L., Mo, M., et al. (2014). Hierarchical nitrogen-doped porous carbon with high surface area derived from endothelium corneum *Gigeriae galli* for high-performance supercapacitor. *Electrochimica Acta*, *130*, 464–469; Gao, F., Qu, J., Zhao, Z., Wang, Z., & Qiu, J. (2016). Nitrogen-doped activated carbon derived from prawn shells for high-performance supercapacitors. *Electrochimica Acta*, *190*, 1134–1141; Liang, J., Qu, T., Kun, X., Zhang, Y., Chen, S., Cao, Y.-C., et al. (2018). Microwave assisted synthesis of Camellia oleifera shell-derived porous carbon with rich oxygen functionalities and superior supercapacitor performance. *Applied Surface Science*, *436*, 934–940; Elmouwahidi, A., Castelo-Quibén, J., Vivo-Vilches, J. F., Pérez-Cadenas, A. F., Maldonado-Hódar, F. J., & Carrasco-Marín, F. (2018). Activated carbons from agricultural waste solvothermally doped with sulphur as electrodes for supercapacitors. *Chemical Engineering Journal*, *334*, 1835–1841; Li, Z., Zhang, L., Amirkhiz, B. S., Tan, X., Xu, Z., Wang, H., et al. (2012). Carbonized chicken eggshell membranes with

3D architectures as high-performance electrode materials for supercapacitors. *Advanced Energy Materials*, *2*(4), 431–437; Fan, X., Yu, C., Yang, J., Ling, Z., Hu, C., Zhang, M., et al. (2015). A layered-nanospace-confinement strategy for the synthesis of two-dimensional porous carbon nanosheets for high-rate performance supercapacitors. *Advanced Energy Materials*, *5*(7), 1401761; Li, Z., Xu, Z., Wang, H., Ding, J., Zahiri, B., Holt, C. M. B., et al. (2014). Colossal pseudocapacitance in a high functionality-high surface area carbon anode doubles the energy of an asymmetric supercapacitor. *Energy & Environmental Science*, *7*(5), 1708–1718; Lei, W., Guo, J., Wu, Z., Xuan, C., Xiao, W., & Wang, D. (2017). Highly nitrogen and sulfur dual-doped carbon microspheres for supercapacitors. *Science Bulletin*, *62*(14), 1011–1017; Han, X., Jiang, H., Zhou, Y., Hong, W., Zhou, Y., Gao, P., et al. (2018). A high performance nitrogen-doped porous activated carbon for supercapacitor derived from pueraria. *Journal of Alloys and Compounds*, *744*, 544–551; Li, J., Liu, W., Xiao, D., & Wang, X. (2017). Oxygen-rich hierarchical porous carbon made from pomelo peel fiber as electrode material for supercapacitor. *Applied Surface Science*, *416*, 918–924. (*Right*) Adapted from Seredych, M., Hulicova-Jurcakova, D., Lu, G. Q., & Bandosz, T. J. (2008). Surface functional groups of carbons and the effects of their chemical character, density and accessibility to ions on electrochemical performance. *Carbon*, *46*(11), 1475–1488 with permission from Elsevier.

employing a wide-voltage-window electrolyte (organic electrolyte or ionic liquid) or further elaborating on the concept of the supercapacitor. For the latter, ASCs and HECs are the "returning newbies" in the field. Although both are built up from a pair of dissimilar electrodes and thus offer significantly wider operational voltage windows compared to the aqueous and organic electrolytes (Ruiz, Santamaría, Granda, & Blanco, 2009), and enhanced energy density, they do not have the same working principle (Choudhary et al., 2017). ASCs can be considered as an adduct of the pseudocapacitive and EDL C_s (Fig. 18, left). HECs also utilize capacitor-like electrodes, either pseudocapacitive or EDLC electrode, but paired with a battery-like counterpart (Zuo et al., 2017), with the aim of simultaneously gaining in energy and power densities. It is therefore obvious that the requirements for the biomass-derived carbon for both applications are not exactly the same.

Corn (Xiaoqing, Chengfei, & Yue, 2017), potatoes (Ruibin et al., 2015), and other starch-rich biomasses are among the good candidates for ASC negative electrode material. In general, the structure of starch built up from glucose-based repeat units resembles that of cellulose. However, unlike cellulose, it contains not one but two glucose units, linear amylose and branched amylopectin, which tend to self-assemble into organized lamellar structures (Budarin et al., 2006). This intrinsic ability is used for the synthesis of pore size-controlled mesoporous materials, the preparation of which usually requires a templating agent, e.g., mesoporous silica (Budarin et al., 2006), and is associated with aggressive and potentially hazardous strong acid/alkali reactions upon removal of the template (Budarin et al., 2006). It was also the seed of the bio-based family of mesoporous materials known under the tradename Starbon. The SSA

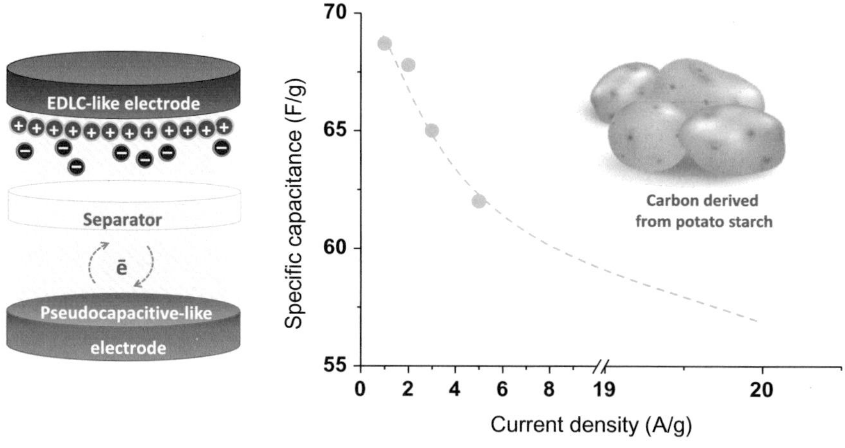

Fig. 18 (*Left*) Configuration of the asymmetric capacitor assembly. (*Right*) Specific capacitances calculated from galvanostatic charge/discharge curves of the asymmetric capacitor with 1 M KOH electrolyte as a function of current density. The potatoes graphic was designed by BSG Studio (All-free-download.com, 2018c).
(*Right*) Data points extracted from Ruibin, Q., Zhongai, H., Yuying, Y., Zhimin, L., Ning, A., Xiaoying, R., et al. (2015). Monodisperse carbon microspheres derived from potato starch for asymmetric supercapacitors. *Electrochimica Acta*, *167*, 303–310.

of the starch-based carbons prepared through classical pyrolysis followed by chemical activation typically exceeds 1300 m^2/g. It drops, however, sharply to only one-third of this value when pyrolysis is preceded by HTC and the resultant biochar is not activated (Li et al., 2008). The latter was the case of potato starch-derived carbon prepared by Rubinin and coworkers (Ruibin et al., 2015). Bound to acetylene black with Nafion, and paired with a reduced graphene oxide (rGO) positive electrode, this biomass-derived carbon showed very promising and highly reversible capacitive characteristics with less than 12 F/g difference between the extreme values of applied current densities, that is, 1 and 20 A/g (69 and 57 F/g, respectively) (Fig. 18, right). The practical potential window of potato starch derived-carbon and rGO capacitor utilizing aqueous electrolyte (1 M KOH) was up to 1.5 V, leading to 21.5 Wh/kg and thus rendering the concept of ASC an interesting solution to store more energy.

Besides the precursors rich in starch, other raw materials containing less of this polymeric carbohydrate (Krishnan, 2017) were also considered for the preparation of high-performance ASC negative electrodes. The representatives of this group, banana (Chaitra et al., 2017) and hemp stem fibers (Yang et al., 2017), have proven to be promising precursors for high-performance porous carbon materials. Interestingly, although both biowastes were not processed in the same conditions, the morphology of the end product was almost alike, with well-arranged arrays of mesopores forming a 3D honeycomb (Fig. 19). Even though this highly ordered

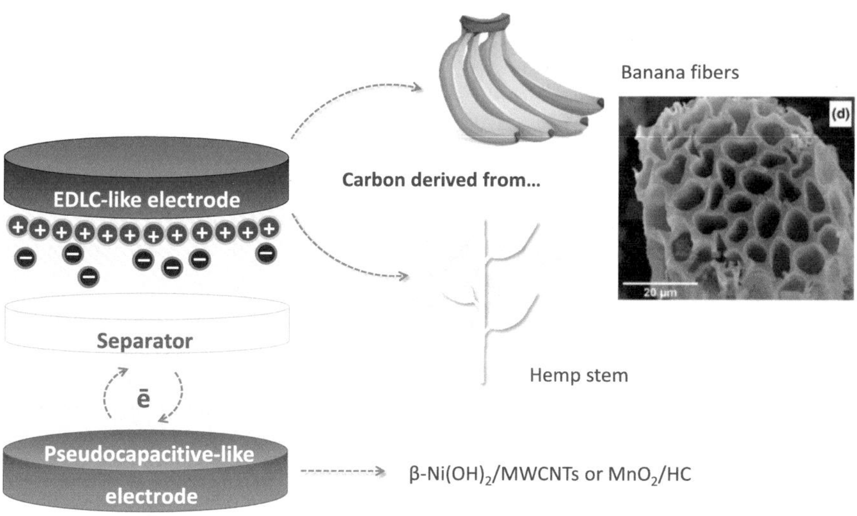

Fig. 19 Configuration of the asymmetric capacitor assembly in which the EDLC-like electrode is derived from either banana or hemp stem fibers. *EDLC*, electric double-layer capacitor; *MWCNTs*, multiwalled carbon nanotubes.
The SEM picture showing the 3D honeycomb-like morphology of the derived carbons is reproduced from Chaitra, K., Vinny, R. T., Sivaraman, P., Reddy, Narendra, Hu, Chunyan, Venkatesh, Krishna, et al. (2017). KOH activated carbon derived from biomass-banana fibers as an efficient negative electrode in high performance asymmetric supercapacitor. *Journal of Energy Chemistry*, *26*(1), 56–62 with the permission of Elsevier.

open-pore structure does not have a high SSA, it certainly offers high electrical conductivity, which in turn facilitates the electron and ion transport at the electrode/electrolyte interface (Yang et al., 2017). Paired with a carefully selected transition metal-based counter-electrode, bringing in pseudocapacitance and significantly adding to structural stability during cycling, these materials can cycle up to 2.0 V (in 1 M KOH), without apparent distortions in nearly rectangular CV profiles. Moreover, a synergy between the unique structural and electrochemical properties of each material used to assemble given ASC resulted in an energy density higher than 30 Wh/kg, leaving other biomass-derived competitors, such as jackfruit peel-based carbon (Sennu, Aravindan, & Lee, 2016), behind.

In all the examples of ASCs given thus far, the negative electrode was based on biomass-derived material storing the energy through the formation of a double-layer. Although the results were promising, the capacitance of the entire system was always limited by carbon, whose charge storage ability is certainly lower than that of TMO. One idea to tackle this limitation is to employ biomass-derived carbons with highly developed surface chemistry, in particular those containing N-functional groups, which trigger the redox reactions and thus boost the specific capacitance remarkably (a few orders of magnitude) (Yang & Zhou, 2017). To this end, Li and coworkers (Li et al., 2014) converted an N-rich bioprecursor, chicken egg whites, into a high functionality-high surface area capacitor electrode. Knowing that an activation temperature higher than 600°C risks the loss of nitrogen functionalities, they proposed an alternative processing of the denatured proteins, which does not require the use of a template. Low-temperature activation under the flow of argon resulted in highly microporous material (SSA and total pore volume up to 1405 m^2/g and 0.73 cm^3/g, respectively) containing a high amount of both N- and O-functional groups. As expected, the ASC assembled with this AC and an $NiCo_2O_4$/graphene positive electrode delivered twice the energy density of the same capacitor with commercial AC (48 vs 23 Wh/kg).

With their pseudocapacitive properties, N- and other heteroatom-doped carbons can also be used as positive electrodes. This is the case of the HEC combining a carbon-based electrode with an Li-ion battery-like electrode (Fig. 20). For the latter, the choice usually falls on silicon offering high theoretical energy (and power) density. There are also, however, examples of lithium titanate ($Li_4Ti_5O_{12}$) and metallic lithium (Jain et al., 2013) used in an HEC. Similar to the case of conventional positive electrodes in Li-ion batteries, the capacitor-like electrode also struggles to match the specific capacity (or capacitance) with that of the counter-electrode. Fortunately, high diversity of biomass precursors with different, often unique, structural properties and a wide range of processing methods can obtain materials with performance tailor-made characteristics.

Li et al. (2016a) used N-doped AC and Si/C composite to explore the benefits of multiple energy storage mechanisms, double-layer formation and pseudocapacitance, and alloying/dealloying reactions with Li in one system. The carbonaceous material was prepared from corncob by mixing the predried biomass with KOH and subsequently annealing it at either 400 or 600°C under the flow of an NH_3/N_2 mixture. A cycled versus Si/C electrode attained a high energy density of 230 Wh/kg without

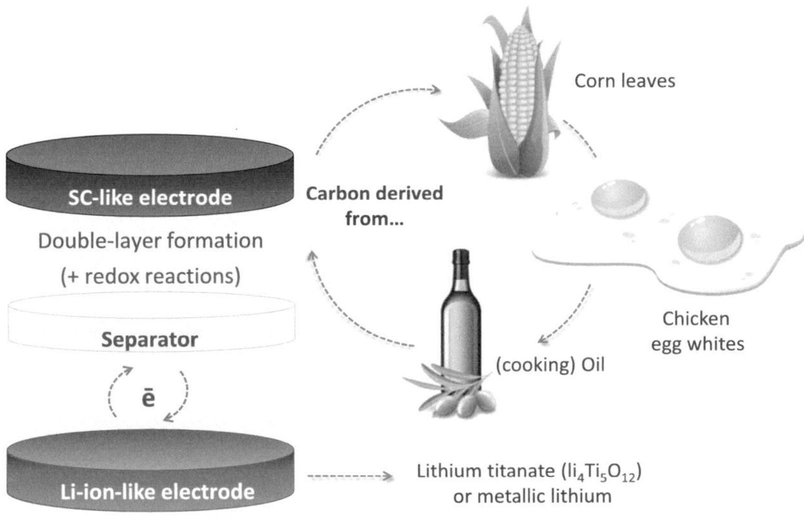

Fig. 20 Configuration of the hybrid capacitor assembly in which the SC-like electrode is derived from either corn leaves, or chicken egg whites, or (cooking) oil. The corncob graphic was designed by BSG Studio (All-free-download.com, 2018b).

scarifying the power density of the device. Interestingly, when the corncob-derived carbon was replaced by that obtained from egg white (Li et al., 2016b), the energy density of the HEC increased almost to 260 Wh/kg (at a similar high power density). The latter result was ascribed to the synergetic effect of the presence of oxygenated surface groups, partial graphitization of the biocarbon, high SSA, highly developed microporosity, and good electronic conductivity.

Despite the promising C_s and energy densities demonstrated at the cell level, today's green capacitors do not have an "easy life." Sustainable high-capacitance performance is often expected to go along with flexibility and an ease of fabrication in different shapes and sizes (Shi et al., 2013). Also in this case the granular self-assembly of starch into various shapes (Man et al., 2014) is an attribute, particularly in shortening the diffusion lengths (Ouyang et al., 2017), increasing packing density, and thus enhancing the rate capability and the amount of energy stored in a capacitor at a given mass (Wang, 2017). However, despite the promising performance characteristics described earlier in this chapter, the 3D spherical shape of the potato starch-derived carbon does not seem to be the most optimal geometry for increased charge accumulation and rapid electron transfer (Ouyang et al., 2017; Park, Li, Lu, & Sastry, 2016). The latter can be successfully met when using 1D (nano)structures. As shown by Ouyang et al.(Ouyang et al., 2017), 1D (nano)structures can be produced from biomass not necessarily predisposed to self-assembly into a particular shape, as is tofu with its irregular particle-like morphology. The $ZnCl_2$-molten-salt-assisted calcination of the soybean curd led to the formation of 1D carbon nanobelts (CNBs) with SSAs between 938 and 1208 m^2/g, depending on the residence time and interconnected pores network. The latter was built up mostly from mesopores, which occupied more than 80% of the total pore volume, lowering the density of the CNBs, but accelerating the ion transport and reducing the contact resistance. In addition, the quaternary, pyridinic, and pyrrolic nitrogen doping inherited from parent material significantly enhanced the electronic conductivity of 1D CNBs and brought in pseudocapacitance (Ouyang et al., 2017), further increasing the amount of charge stored in the capacitor. Indeed, the ASC assembled with a pair of nanobelt carbon-based electrodes, negative CNBs, and positive MnO_2-CNB electrode exhibited outstanding electrochemical performance (29 Wh/kg) with 96% capacity retention at a high current density of 3 A/g. Moreover, it could have been operated stably up to 2.0 V in aqueous electrolyte. A multiplication of the two, the capacitance and the voltage, resulted in nearly 30 Wh/kg, leaving behind most of the previously reported ASC with MnO_2-based electrodes (Ning et al., 2016; Sun et al., 2017).

10.7 Conclusions and prospects

As shown in this chapter, ACs with "environmental origin" have a broad spectrum of interesting features for ECs. They offer high specific surface area and well-developed porosity, and often contain heteroatoms altering electrode conductivity and its accessibility for electrolyte ions. Moreover, they could be formulated into electrodes without binders. The latter is an important step toward the development of sustainable (super)capacitors free from fluorine-based polymers. All solid-state (super)capacitors employing this kind of material have recently been developed for portable electronics.

Unfortunately, sometimes the term biomass is misused, and the parent materials considered as "wastes" and thus good candidates for carbonization serve better in other, often more important applications than energy storage. Also, the reproducibility of the carbons prepared from biomass with regard to the morphological and structural characteristics comes into question when upscaling these materials for industrial applications. For fundamental research, biomass brings in many opportunities for further development of EC technology, especially hybrid devices bringing together the capacitive and redox-based charge storage mechanisms. To properly assess the promise of these biomass-derived materials, however, the specific and/or volumetric capacitance values often reported for the three-electrode system should always be verified in two-electrode cells, the configuration used for commercial (super)capacitors. The same applies to specific and/or volumetric energy density, cycling stability, and cycle-life investigations.

Finally, despite the numerous efforts and significant progress in the "waste to wealth" conversion of biomass to (super)capacitor electrode materials, there is still room for improvement in specific capacitance and efficient charge propagation. These issues might be to a large extent solved by introducing pseudocapacitance into the system through the use of either heteroatom-doped carbons or organic electrolytes.

References

Abbas, Q., Babuchowska, P., Frąckowiak, E., & Béguin, F. (2016). Sustainable AC/AC hybrid electrochemical capacitors in aqueous electrolyte approaching the performance of organic systems. *Journal of Power Sources*, *326*, 652–659.

Ahiduzzaman, M., & Sadrul Islam, A. K. M. (2016). Preparation of porous bio-char and activated carbon from rice husk by leaching ash and chemical activation. *SpringerPlus*, *5*(1), 1248.

All-free-download.com. (2018a). http://all-free-download.com/free-vector/download/coconuts_310437.html.

All-free-download.com. (2018b). http://all-free-download.com/free-vector/download/corn_310464.html.

All-free-download.com. (2018c). http://all-free-download.com/free-vector/download/four-potatoes_310457.html.

Andreas, H. A. (2015). Self-discharge in electrochemical capacitors: a perspective article. *Journal of The Electrochemical Society*, *162*(5), A5047–A5053.

Antolini, E. (2016). Nitrogen-doped carbons by sustainable N- and C-containing natural resources as nonprecious catalysts and catalyst supports for low temperature fuel cells. *Renewable and Sustainable Energy Reviews*, *58*, 34–51.

Arami-Niya, A., Daud, W. M. A. W., & Mjalli, F. S. (2011). Comparative study of the textural characteristics of oil palm shell activated carbon produced by chemical and physical activation for methane adsorption. *Chemical Engineering Research and Design*, *89*(6), 657–664.

Aworn, A., Thiravetyan, P., & Nakbanpote, W. (2008). Preparation and characteristics of agricultural waste activated carbon by physical activation having micro- and mesopores. *Journal of Analytical and Applied Pyrolysis*, *82*(2), 279–285.

Azaïs, P., Duclaux, L., Florian, P., Massiot, D., Lillo-Rodenas, M.-A., Linares-Solano, A., et al. (2007). Causes of supercapacitors ageing in organic electrolyte. *Journal of Power Sources*, *171*(2), 1046–1053.

Azevedo, D. C. S., Araújo, J. C. S., Bastos-Neto, M., Torres, A. E. B., Jaguaribe, E. F., & Cavalcante, C. L. (2007). Microporous activated carbon prepared from coconut shells using chemical activation with zinc chloride. *Microporous and Mesoporous Materials*, *100*(1), 361–364.

Balathanigaimani, M. S., Shim, W.-G., Lee, M.-J., Kim, C., Lee, J.-W., & Moon, H. (2008). Highly porous electrodes from novel corn grains-based activated carbons for electrical double layer capacitors. *Electrochemistry Communications*, *10*(6), 868–871.

Balducci, A. (2016). Electrolytes for high voltage electrochemical double layer capacitors: a perspective article. *Journal of Power Sources*, *326*, 534–540.

Banks, S. W., Nowakowski, D. J., & Bridgwater, A. V. (2014). Fast pyrolysis processing of surfactant washed Miscanthus. *Fuel Processing Technology*, *128*, 94–103.

Bauer, L. (2017). *Biomass pyrolysis comes of age*. http://www.biofuelsdigest.com/bdigest/2017/06/08/biomass-pyrolysis-comes-of-age/.

Béguin, F., Presser, V., Balducci, A., & Frackowiak, E. (2014). Carbons and electrolytes for advanced supercapacitors. *Advanced Materials*, *26*(14), 2219–2251.

BGU. (2017). *Poultry excrement could partially replace coal as a renewable energy source according to BGU researchers*. http://in.bgu.ac.il/en/bidr/pages/news/Poultry%20Excrement%20Could%20Partially%20Replace%20Coal%20as%20a%20Renewable%20Energy%20Source%20According%20to%20BGU%20Researchers.aspx.

Bichat, M. P., Raymundo-Piñero, E., & Béguin, F. (2010). High voltage supercapacitor built with seaweed carbons in neutral aqueous electrolyte. *Carbon*, *48*(15), 4351–4361.

Bouchelta, C., Medjram, M. S., Zoubida, M., Chekkat, F. A., Ramdane, N., & Bellat, J.-P. (2012). Effects of pyrolysis conditions on the porous structure development of date pits activated carbon. *Journal of Analytical and Applied Pyrolysis*, *94*, 215–222.

Boudrahem, F., Aissani-Benissad, F., & Aït-Amar, H. (2009). Batch sorption dynamics and equilibrium for the removal of lead ions from aqueous phase using activated carbon developed from coffee residue activated with zinc chloride. *Journal of Environmental Management*, *90*(10), 3031–3039.

Brousse, T., Bélanger, D., Chiba, K., Egashira, M., Favier, F., Long, J., et al. (2017). Materials for electrochemical capacitors. In C. Breitkopf & K. Swider-Lyons (Eds.), *Springer handbook of electrochemical energy* (pp. 495–561). Berlin, Heidelberg: Springer Berlin Heidelberg.

Brousse, T., Bélanger, D., & Long, J. W. (2015). To be or not to be pseudocapacitive? *Journal of The Electrochemical Society*, *162*(5), A5185–A5189.

Budarin, V., Clark, J. H., Hardy, J. J. E., Luque, R., Milkowski, K., Tavener, S. J., et al. (2006). Starbons: new starch-derived mesoporous carbonaceous materials with tunable properties. *Angewandte Chemie International Edition*, *45*(23), 3782–3786.

Burke, A. (2007). R&D considerations for the performance and application of electrochemical capacitors. *Electrochimica Acta*, *53*(3), 1083–1091.

Cai, T., Wang, H., Jin, C., Sun, Q., & Nie, Y. (2018). Fabrication of nitrogen-doped porous electrically conductive carbon aerogel from waste cabbage for supercapacitors and oil/water separation. *Journal of Materials Science: Materials in Electronics*, *29*(5), 4334–4344.

Caturla, F., Molina-Sabio, M., & Rodríguez-Reinoso, F. (1991). Preparation of activated carbon by chemical activation with ZnCl2. *Carbon*, *29*(7), 999–1007.

Centeno, T. A., Sereda, O., & Stoeckli, F. (2011). Capacitance in carbon pores of 0.7 to 15 nm: a regular pattern. *Physical Chemistry Chemical Physics*, *13*(27), 12403–12406.

Chae, J. H., & Chen, G. Z. (2012). 1.9V aqueous carbon–carbon supercapacitors with unequal electrode capacitances. *Electrochimica Acta*, *86*, 248–254.

Chaitra, K., Vinny, R. T., Sivaraman, P., Reddy, N., Hu, C., Venkatesh, K., et al. (2017). KOH activated carbon derived from biomass-banana fibers as an efficient negative electrode in high performance asymmetric supercapacitor. *Journal of Energy Chemistry*, *26*(1), 56–62.

Chen, H. (2014). Chemical composition and structure of natural lignocellulose. In *Biotechnology of lignocellulose: Theory and practice* (pp. 25–71). Netherlands: Springer.

Chen, L., Bai, H., Huang, Z., & Li, L. (2014). Mechanism investigation and suppression of self-discharge in active electrolyte enhanced supercapacitors. *Energy & Environmental Science*, *7*(5), 1750–1759.

Chen, H., Lin, G., Chen, Y., Chen, W., & Yang, H. (2016). Biomass pyrolytic polygeneration of tobacco waste: product characteristics and nitrogen transformation. *Energy & Fuels*, *30*(3), 1579–1588.

Chen, C.-M., Zhang, Q., Zhao, X.-C., Zhang, B., Kong, Q.-Q., Yang, M.-G., et al. (2012). Hierarchically aminated graphene honeycombs for electrochemical capacitive energy storage. *Journal of Materials Chemistry*, *22*(28), 14076–14084.

Cheng, P., Gao, S., Zang, P., Yang, X., Bai, Y., Xu, H., et al. (2015). Hierarchically porous carbon by activation of shiitake mushroom for capacitive energy storage. *Carbon*, *93*, 315–324.

Cheng, B.-H., Tian, K., Zeng, R. J., & Jiang, H. (2017). Preparation of high performance supercapacitor materials by fast pyrolysis of corn gluten meal waste. *Sustainable Energy & Fuels*, *1*(4), 891–898.

Chien, H.-C., Wu, T.-H., Rajkumar, M., & Hu, C.-C. (2016). Effects of buffer agents on hydrogen adsorption and desorption at/within activated carbon for the negative electrode of aqueous asymmetric supercapacitors. *Electrochimica Acta*, *205*, 1–7.

Chmiola, J., Yushin, G., Dash, R., & Gogotsi, Y. (2006). Effect of pore size and surface area of carbide derived carbons on specific capacitance. *Journal of Power Sources*, *158*(1), 765–772.

Chmiola, J., Yushin, G., Gogotsi, Y., Portet, C., Simon, P., & Taberna, P. L. (2006). Anomalous increase in carbon capacitance at pore sizes less than 1 nanometer. *Science*, *313*(5794), 1760–1763.

Choudhary, N., Li, C., Moore, J., Nagaiah, N., Zhai, L., Jung, Y., et al. (2017). Asymmetric supercapacitor electrodes and devices. *Advanced Materials*, *29*(21), 1605336.

Conway, B. E., Birss, V., & Wojtowicz, J. (1997). The role and utilization of pseudocapacitance for energy storage by supercapacitors. *Journal of Power Sources*, *66*, 1), 1–14.

Conway, B. E., & Pell, W. G. (2003). Double-layer and pseudocapacitance types of electrochemical capacitors and their applications to the development of hybrid devices. *Journal of Solid State Electrochemistry*, *7*(9), 637–644.

Darmawan, S., Wistara, N. J., Pari, G., Maddu, A., & Syafii, W. (2016). Characterization of lignocellulosic biomass as raw material for the production of porous carbon-based materials. *BioResources*, *11*, 3561–3574.

Decaux, C., Matei Ghimbeu, C., Dahbi, M., Anouti, M., Lemordant, D., Béguin, F., et al. (2014). Influence of electrolyte ion–solvent interactions on the performances of supercapacitors porous carbon electrodes. *Journal of Power Sources*, *263*, 130–140.

Demirbas, A. (2004). Effects of temperature and particle size on bio-char yield from pyrolysis of agricultural residues. *Journal of Analytical and Applied Pyrolysis*, *72*(2), 243–248.

Deng, J., Li, M., & Wang, Y. (2016). Biomass-derived carbon: synthesis and applications in energy storage and conversion. *Green Chemistry*, *18*(18), 4824–4854.

Deng, Y., Xie, Y., Zou, K., & Ji, X. (2016). Review on recent advances in nitrogen-doped carbons: preparations and applications in supercapacitors. *Journal of Materials Chemistry A*, *4*(4), 1144–1173.

Dubal, D. P., Ayyad, O., Ruiz, V., & Gomez-Romero, P. (2015). Hybrid energy storage: the merging of battery and supercapacitor chemistries. *Chemical Society Reviews*, *44*(7), 1777–1790.

Elizalde-González, M. P., Mattusch, J., Peláez-Cid, A. A., & Wennrich, R. (2007). Characterization of adsorbent materials prepared from avocado kernel seeds: natural, activated and carbonized forms. *Journal of Analytical and Applied Pyrolysis*, *78*(1), 185–193.

Elmouwahidi, A., Castelo-Quibén, J., Vivo-Vilches, J. F., Pérez-Cadenas, A. F., Maldonado-Hódar, F. J., & Carrasco-Marín, F. (2018). Activated carbons from agricultural waste solvothermally doped with sulphur as electrodes for supercapacitors. *Chemical Engineering Journal*, *334*, 1835–1841.

Fałtynowicz, H., Kaczmarczyk, J., & Kułażyński, M. (2015). Preparation and characterization of activated carbons from biomass material—giant knotweed (*Reynoutria sachalinensis*). *Open Chemistry*, *13*(1), 2391–5420.

Fan, X., Yu, C., Yang, J., Ling, Z., Hu, C., Zhang, M., et al. (2015). A layered-nanospace-confinement strategy for the synthesis of two-dimensional porous carbon nanosheets for high-rate performance supercapacitors. *Advanced Energy Materials*, *5*(7), 1401761.

Fic, K., Lota, G., Meller, M., & Frackowiak, E. (2012). Novel insight into neutral medium as electrolyte for high-voltage supercapacitors. *Energy & Environmental Science*, *5*(2), 5842–5850.

Fic, K., Meller, M., & Frackowiak, E. (2015). Interfacial redox phenomena for enhanced aqueous supercapacitors. *Journal of The Electrochemical Society*, *162*(5), A5140–A5147.

Fic, K., Meller, M., Menzel, J., & Frackowiak, E. (2016). Around the thermodynamic limitations of supercapacitors operating in aqueous electrolytes. *Electrochimica Acta*, *206*, 496–503.

Freepik. (2018a). https://www.freepik.com/free-vector/bamboo-reeds_794180.htm#termbamboo&page=1&position=4.

Freepik. (2018b). https://www.freepik.com/free-vector/corn-background_1062541.htm#term=corn%20background&page=1&position=0.

Freepik. (2018c). https://www.freepik.com/free-vector/seaweed-elements-collection_1195378.htm#term=seaweed&page=1&position=27.

Funke, A., & Ziegler, F. (2010). Hydrothermal carbonization of biomass: a summary and discussion of chemical mechanisms for process engineering. *Biofuels, Bioproducts and Biorefining*, *4*(2), 160–177.

Gallifuoco, A., Taglieri, L., Scimia, F., Papa, A. A., & Di Giacomo, G. (2017). Hydrothermal carbonization of biomass: new experimental procedures for improving the industrial Processes. *Bioresource Technology*, *244*, 160–165.

Gao, Q., Demarconnay, L., Raymundo-Pinero, E., & Béguin, F. (2012). Exploring the large voltage range of carbon/carbon supercapacitors in aqueous lithium sulfate electrolyte. *Energy & Environmental Science*, *5*(11), 9611–9617.

Gao, F., Qu, J., Zhao, Z., Wang, Z., & Qiu, J. (2016). Nitrogen-doped activated carbon derived from prawn shells for high-performance supercapacitors. *Electrochimica Acta*, *190*, 1134–1141.

Gao, X., Xing, W., Zhou, J., Wang, G., Zhuo, S., Liu, Z., et al. (2014). Superior capacitive performance of active carbons derived from *Enteromorpha prolifera*. *Electrochimica Acta*, *133*, 459–466.

García-Cruz, L., Ratajczak, P., Iniesta, J., Montiel, V., & Béguin, F. (2016). Self-discharge of AC/AC electrochemical capacitors in salt aqueous electrolyte. *Electrochimica Acta*, *202*, 66–72.

Garcia-Gomez, A., Moreno-Fernandez, G., Lobato, B., & Centeno, T. A. (2015). Constant capacitance in nanopores of carbon monoliths. *Physical Chemistry Chemical Physics*, *17*(24), 15687–15690.

Ghimbeu, C. M., Gadiou, R., Dentzer, J., Schwartz, D., & Vix-Guterl, C. (2010). Influence of surface chemistry on the adsorption of oxygenated hydrocarbons on activated carbons. *Langmuir*, *26*(24), 18824–18833.

Gogotsi, Y., & Simon, P. (2011). True performance metrics in electrochemical energy storage. *Science*, *334*(6058), 917–918.

González, J. F., Román, S., Encinar, J. M., & Martínez, G. (2009). Pyrolysis of various biomass residues and char utilization for the production of activated carbons. *Journal of Analytical and Applied Pyrolysis*, *85*(1), 134–141.

Guan, L., Yu, L., & Chen, G. Z. (2016). Capacitive and non-capacitive faradaic charge storage. *Electrochimica Acta*, *206*, 464–478.

Haddad, K., Jeguirim, M., Jellali, S., Guizani, C., Delmotte, L., Bennici, S., et al. (2017). Combined NMR structural characterization and thermogravimetric analyses for the assessment of the AAEM effect during lignocellulosic biomass pyrolysis. *Energy*, *134*, 10–23.

Hayashi, J. I., Horikawa, T., Muroyama, K., & Gomes, V. G. (2002). Activated carbon from chickpea husk by chemical activation with K2CO3: preparation and characterization. *Microporous and Mesoporous Materials*, *55*(1), 63–68.

Hayashi, J. I., Horikawa, T., Takeda, I., Muroyama, K., & Nasir Ani, F. (2002). Preparing activated carbon from various nutshells by chemical activation with K2CO3. *Carbon*, *40*(13), 2381–2386.

Hazourli, S., Ziati, M., & Hazourli, A. (2009). Characterization of activated carbon prepared from lignocellulosic natural residue: example of date stones. *Physics Procedia*, *2*(3), 1039–1043.

He, M., Fic, K., Frckowiak, E., Novak, P., & Berg, E. J. (2016). Ageing phenomena in high-voltage aqueous supercapacitors investigated by in situ gas analysis. *Energy & Environmental Science*, *9*(2), 623–633.

Hernández-Montoya, V., García-Servin, J., & Bueno-López, J. I. (2012). Thermal treatments and activation procedures used in the preparation of activated carbons. In V. H. Montoya (Ed.), *Lignocellulosic precursors used in the synthesis of activated carbon—Characterization techniques and applications in the wastewater treatment*: InTech (Chapter 2).

Hou, J., Cao, C., Idrees, F., & Ma, X. (2015). Hierarchical porous nitrogen-doped carbon nanosheets derived from silk for ultrahigh-capacity battery anodes and supercapacitors. *ACS Nano*, *9*(3), 2556–2564.

Huang, W., Zhang, H., Huang, Y., Wang, W., & Wei, S. (2011). Hierarchical porous carbon obtained from animal bone and evaluation in electric double-layer capacitors. *Carbon*, *49*(3), 838–843.

Hulicova-Jurcakova, D., Puziy, A. M., Poddubnaya, O. I., Suárez-García, F., Tascón, J. M. D., & Lu, G. Q. (2009). Highly stable performance of supercapacitors from phosphorus-enriched carbons. *Journal of the American Chemical Society*, *131*(14), 5026–5027.

Hulicova-Jurcakova, D., Seredych, M., Lu, G. Q., & Bandosz, T. J. (2009). Combined effect of nitrogen- and oxygen-containing functional groups of microporous activated carbon on its electrochemical performance in supercapacitors. *Advanced Functional Materials*, *19*(3), 438–447.

Jagiello, J., & Olivier, J. P. (2013). 2D-NLDFT adsorption models for carbon slit-shaped pores with surface energetical heterogeneity and geometrical corrugation. *Carbon*, *55*, 70–80.

Jain, A., Aravindan, V., Jayaraman, S., Kumar, P. S., Balasubramanian, R., Ramakrishna, S., et al. (2013). Activated carbons derived from coconut shells as high energy density cathode material for Li-ion capacitors. *Scientific Reports*, *3*, 3002.

Jain, A., Balasubramanian, R., & Srinivasan, M. P. (2015). Tuning hydrochar properties for enhanced mesopore development in activated carbon by hydrothermal carbonization. *Microporous and Mesoporous Materials*, *203*, 178–185.

Jenkins, B. M., Bakker, R. R., & Wei, J. B. (1996). On the properties of washed straw. *Biomass and Bioenergy*, *10*(4), 177–200.

Johns, P. A., Roberts, M. R., Wakizaka, Y., Sanders, J. H., & Owen, J. R. (2009). How the electrolyte limits fast discharge in nanostructured batteries and supercapacitors. *Electrochemistry Communications*, *11*(11), 2089–2092.

Juárez-Galán, J. M., Silvestre-Albero, A., Silvestre-Albero, J., & Rodríguez-Reinoso, F. (2009). Synthesis of activated carbon with highly developed "mesoporosity". *Microporous and Mesoporous Materials*, *117*(1), 519–521.

Kaghazchi, T., Kolur, N. A., & Soleimani, M. (2010). Licorice residue and pistachio-nut shell mixture: a promising precursor for activated carbon. *Journal of Industrial and Engineering Chemistry*, *16*(3), 368–374.

Karthikeyan, K., Amaresh, S., Lee, S. N., Sun, X., Aravindan, V., Lee, Y.-G., et al. (2014). Construction of high-energy-density supercapacitors from pine-cone-derived high-surface-area carbons. *ChemSusChem*, *7*(5), 1435–1442.

Kötz, R., & Carlen, M. (2000). Principles and applications of electrochemical capacitors. *Electrochimica Acta*, *45*(15), 2483–2498.

Kowal, J., Avaroglu, E., Chamekh, F., Šenfelds, A., Thien, T., Wijaya, D., et al. (2011). Detailed analysis of the self-discharge of supercapacitors. *Journal of Power Sources*, *196*(1), 573–579.

Krishnan, P. (2017). *Fracture and failure analyses of plastics and reinforced plastics*. https://fr.slideshare.net/PadmanabhanKrishnan2/fracture-and-failure-analyses-of-plastics-and-reinforced-plastics.

Kurig, H., Russina, M., Tallo, I., Siebenbürger, M., Romann, T., & Lust, E. (2016). The suitability of infinite slit-shaped pore model to describe the pores in highly porous carbon materials. *Carbon*, *100*, 617–624.

Laheäär, A., Przygocki, P., Abbas, Q., & Béguin, F. (2015). Appropriate methods for evaluating the efficiency and capacitive behavior of different types of supercapacitors. *Electrochemistry Communications*, *60*, 21–25.

Le Bideau, J., Viau, L., & Vioux, A. (2011). Ionogels, ionic liquid based hybrid materials. *Chemical Society reviews*, *40*(2), 907–925.

Lee, Y.-H., Chang, K.-H., & Hu, C.-C. (2013). Differentiate the pseudocapacitance and double-layer capacitance contributions for nitrogen-doped reduced graphene oxide in acidic and alkaline electrolytes. *Journal of Power Sources*, *227*, 300–308.

Lei, W., Guo, J., Wu, Z., Xuan, C., Xiao, W., & Wang, D. (2017). Highly nitrogen and sulfur dual-doped carbon microspheres for supercapacitors. *Science Bulletin*, *62*(14), 1011–1017.

Li, B., Dai, F., Xiao, Q., Yang, L., Shen, J., Zhang, C., et al. (2016a). Nitrogen-doped activated carbon for a high energy hybrid supercapacitor. *Energy & Environmental Science*, *9*(1), 102–106.

Li, B., Dai, F., Xiao, Q., Yang, L., Shen, J., Zhang, C., et al. (2016b). Activated carbon from biomass transfer for high-energy density lithium-ion supercapacitors. *Advanced Energy Materials*, *6*(18), 1600802.

Li, Q.-Y., Wang, H.-Q., Dai, Q.-F., Yang, J.-H., & Zhong, Y.-L. (2008). Novel activated carbons as electrode materials for electrochemical capacitors from a series of starch. *Solid State Ionics*, *179*(7), 269–273.

Li, Z., Xu, Z., Wang, H., Ding, J., Zahiri, B., Holt, C. M. B., et al. (2014). Colossal pseudocapacitance in a high functionality-high surface area carbon anode doubles the energy of an asymmetric supercapacitor. *Energy & Environmental Science*, *7*(5), 1708–1718.

Li, Z., Zhang, L., Amirkhiz, B. S., Tan, X., Xu, Z., Wang, H., et al. (2012). Carbonized chicken eggshell membranes with 3D architectures as high-performance electrode materials for supercapacitors. *Advanced Energy Materials*, *2*(4), 431–437.

Liang, J., Qu, T., Kun, X., Zhang, Y., Chen, S., Cao, Y.-C., et al. (2018). Microwave assisted synthesis of *Camellia oleifera* shell-derived porous carbon with rich oxygen functionalities and superior supercapacitor performance. *Applied Surface Science*, *436*, 934–940.

Liu, F., & Guo, M. (2015). Comparison of the characteristics of hydrothermal carbons derived from holocellulose and crude biomass. *Journal of Materials Science*, *50*(4), 1624–1631.

Lobato, B., Suárez, L., Guardia, L., & Centeno, T. A. (2017). Capacitance and surface of carbons in supercapacitors. *Carbon*, *122*, 434–445.

Long, J. W., Bélanger, D., Brousse, T., Sugimoto, W., Sassin, M. B., & Crosnier, O. (2011). Asymmetric electrochemical capacitors—stretching the limits of aqueous electrolytes. *MRS Bulletin*, *36*(7), 513–522.

Lota, G., Centeno, T. A., Frackowiak, E., & Stoeckli, F. (2008). Improvement of the structural and chemical properties of a commercial activated carbon for its application in electrochemical capacitors. *Electrochimica Acta*, *53*(5), 2210–2216.

Lota, G., Fic, K., & Frackowiak, E. (2011). Alkali metal iodide/carbon interface as a source of pseudocapacitance. *Electrochemistry Communications*, *13*(1), 38–41.

Lota, G., & Frackowiak, E. (2009). Striking capacitance of carbon/iodide interface. *Electrochemistry Communications*, *11*(1), 87–90.

Lu, M., Béguin, F., & Frackowiak, E. (2013). *Supercapacitors: Materials, systems, and applications*. Wiley.

Lu, B., Hu, L., Yin, H., Xiao, W., & Wang, D. (2016). One-step molten salt carbonization (MSC) of firwood biomass for capacitive carbon. *RSC Advances*, *6*(108), 106485–106490.

Lu, H., & Zhao, X. S. (2017). Biomass-derived carbon electrode materials for supercapacitors. *Sustainable Energy & Fuels*, *1*(6), 1265–1281.

Lua, A. C., Lau, F. Y., & Guo, J. (2006). Influence of pyrolysis conditions on pore development of oil-palm-shell activated carbons. *Journal of Analytical and Applied Pyrolysis*, *76*(1), 96–102.

Lua, A. C., & Yang, T. (2004). Effect of activation temperature on the textural and chemical properties of potassium hydroxide activated carbon prepared from pistachio-nut shell. *Journal of Colloid and Interface Science*, *274*(2), 594–601.

Ma, G., Yang, Q., Sun, K., Peng, H., Ran, F., Zhao, X., et al. (2015). Nitrogen-doped porous carbon derived from biomass waste for high-performance supercapacitor. *Bioresource Technology*, *197*, 137–142.

Man, J., Lin, L., Wang, Z., Wang, Y., Liu, Q., & Wei, C. (2014). Different structures of heterogeneous starch granules from high-amylose rice. *Journal of Agricultural and Food Chemistry*, *62*(46), 11254–11263.

Mayer, Z. A., Apfelbacher, A., & Hornung, A. (2012). Effect of sample preparation on the thermal degradation of metal-added biomass. *Journal of Analytical and Applied Pyrolysis*, *94*, 170–176.

Menzel, J., Fic, K., Meller, M., & Frackowiak, E. (2014). The effect of halide ion concentration on capacitor performance. *Journal of Applied Electrochemistry*, *44*(4), 439–445.

Miller, J. R. (2010). Introduction to electrochemical capacitor technology. *IEEE Electrical Insulation Magazine*, *26*(4), 40–47.

Molina-Sabio, M., Gonzalez, M. T., Rodriguez-Reinoso, F., & Sepúlveda-Escribano, A. (1996). Effect of steam and carbon dioxide activation in the micropore size distribution of activated carbon. *Carbon*, *34*(4), 505–509.

Morita, M., Kaigaishi, T., Yoshimoto, N., Egashira, M., & Aida, T. (2006). Effects of the electrolyte composition on the electric double-layer capacitance at carbon electrodes. *Electrochemical and Solid-State Letters*, *9*(8), A386.

Nabais, J. V., Carrott, P., Ribeiro Carrott, M. M. L., Luz, V., & Ortiz, A. L. (2008). Influence of preparation conditions in the textural and chemical properties of activated carbons from a novel biomass precursor: the coffee endocarp. *Bioresource Technology*, *99*(15), 7224–7231.

Naoi, K., Nagano, Y., & Naoi, W. (2013). New hybrid supercapacitors and their prospects. *Carbon*, *57*, 539.

Ning, P., Duan, X., Ju, X., Lin, X., Tong, X., Pan, X., et al. (2016). Facile synthesis of carbon nanofibers/MnO2 nanosheets as high-performance electrodes for asymmetric supercapacitors. *Electrochimica Acta*, *210*, 754–761.

Nygård, H. S., & Olsen, E. (2012). Review of thermal processing of biomass and waste in molten salts for production of renewable fuels and chemicals. *International Journal of Low-Carbon Technologies*, *7*(4), 318–324.

Oh, Y. J., Yoo, J. J., Kim, Y. I., Yoon, J. K., Yoon, H. N., Kim, J.-H., et al. (2014). Oxygen functional groups and electrochemical capacitive behavior of incompletely reduced graphene oxides as a thin-film electrode of supercapacitor. *Electrochimica Acta*, *116*, 118–128.

Olivares-Marín, M., Fernández-González, C., Macías-García, A., & Gómez-Serrano, V. (2006). Preparation of activated carbon from cherry stones by chemical activation with ZnCl2. *Applied Surface Science*, *252*(17), 5967–5971.

Ornelas, O., Sieben, J. M., Ruiz-Rosas, R., Morallon, E., Cazorla-Amoros, D., Geng, J., et al. (2014). On the origin of the high capacitance of nitrogen-containing carbon nanotubes in acidic and alkaline electrolytes. *Chemical Communications*, *50*(77), 11343–11346.

Ouyang, T., Cheng, K., Yang, F., Zhou, L., Zhu, K., Ye, K., et al. (2017). From biomass with irregular structures to 1D carbon nanobelts: a stripping and cutting strategy to fabricate high performance supercapacitor materials. *Journal of Materials Chemistry A*, *5*(28), 14551–14561.

Park, J., Li, J., Lu, W., & Sastry, A. M. (2016). Geometric consideration of nanostructures for energy storage systems. *Journal of Applied Physics*, *119*(2), 025101.

Park, D., Yun, Y.-S., & Park, J. M. (2005). Studies on hexavalent chromium biosorption by chemically-treated biomass of Ecklonia sp. *Chemosphere*, *60*(10), 1356–1364.

Phan, N. H., Rio, S., Faur, C., Le Coq, L., Le Cloirec, P., & Nguyen, T. H. (2006). Production of fibrous activated carbons from natural cellulose (jute, coconut) fibers for water treatment applications. *Carbon*, *44*(12), 2569–2577.

Pilon, L., Wang, H., & d'Entremont, A. (2015). Recent advances in continuum modeling of interfacial and transport phenomena in electric double layer capacitors. *Journal of The Electrochemical Society*, *162*(5), A5158–A5178.

Pognon, G., Brousse, T., & Bélanger, D. (2011). Effect of molecular grafting on the pore size distribution and the double layer capacitance of activated carbon for electrochemical double layer capacitors. *Carbon*, *49*(4), 1340–1348.

R3Water. *Hydrothermal carbonisation—HTC*. http://r3water.eu/wp-content/uploads/2014/04/R3Water-technology-fact-sheet_HTC_AvaCO2.pdf. Accessed 3 March 2018.

Ratajczak, P., Jurewicz, K., & Béguin, F. (2014). Factors contributing to ageing of high voltage carbon/carbon supercapacitors in salt aqueous electrolyte. *Journal of Applied Electrochemistry*, *44*(4), 475–480.

Raymundo-Piñero, E., Cadek, M., & Béguin, F. (2009). Tuning carbon materials for supercapacitors by direct pyrolysis of seaweeds. *Advanced Functional Materials*, *19*(7), 1032–1039.

Raymundo-Piñero, E., Kierzek, K., Machnikowski, J., & Béguin, F. (2006). Relationship between the nanoporous texture of activated carbons and their capacitance properties in different electrolytes. *Carbon*, *44*(12), 2498–2507.

Raymundo-Piñero, E., Leroux, F., & Béguin, F. (2006). A high-performance carbon for supercapacitors obtained by carbonization of a seaweed biopolymer. *Advanced Materials*, *18*(14), 1877–1882.

Redondo, E., Carretero-González, J., Goikolea, E., Ségalini, J., & Mysyk, R. (2015). Effect of pore texture on performance of activated carbon supercapacitor electrodes derived from olive pits. *Electrochimica Acta*, *160*, 178–184.

Roldán, S., González, Z., Blanco, C., Granda, M., Menéndez, R., & Santamaría, R. (2011). Redox-active electrolyte for carbon nanotube-based electric double layer capacitors. *Electrochimica Acta*, *56*(9), 3401–3405.

Roldán, S., Granda, M., Menéndez, R., Santamaría, R., & Blanco, C. (2012). Supercapacitor modified with methylene blue as redox active electrolyte. *Electrochimica Acta*, *83*, 241–246.

Ruibin, Q., Zhongai, H., Yuying, Y., Zhimin, L., Ning, A., Xiaoying, R., et al. (2015). Monodisperse carbon microspheres derived from potato starch for asymmetric supercapacitors. *Electrochimica Acta*, *167*, 303–310.

Ruiz, V., Santamaría, R., Granda, M., & Blanco, C. (2009). Long-term cycling of carbon-based supercapacitors in aqueous media. *Electrochimica Acta*, *54*(19), 4481–4486.
Sarkar, A., & Praveen, G. (2017). *Utilization of waste biomass into useful forms of energy* (pp. 117–132). Cham: Springer International Publishing.
Sennu, P., Aravindan, V., & Lee, Y.-S. (2016). High energy asymmetric supercapacitor with 1D@2D structured NiCo2O4@Co3O4 and jackfruit derived high surface area porous carbon. *Journal of Power Sources*, *306*, 248–257.
Seredych, M., Hulicova-Jurcakova, D., Lu, G. Q., & Bandosz, T. J. (2008). Surface functional groups of carbons and the effects of their chemical character, density and accessibility to ions on electrochemical performance. *Carbon*, *46*(11), 1475–1488.
Sharma, P., & Bhatti, T. S. (2010). A review on electrochemical double-layer capacitors. *Energy Conversion and Management*, *51*(12), 2901–2912.
Shi, S., Xu, C., Yang, C., Chen, Y., Liu, J., & Kang, F. (2013). Flexible asymmetric supercapacitors based on ultrathin two-dimensional nanosheets with outstanding electrochemical performance and aesthetic property. *Scientific Reports*, *3*, 2598.
Shul, G., & Belanger, D. (2016). Self-discharge of electrochemical capacitors based on soluble or grafted quinone. *Physical Chemistry Chemical Physics*, *18*(28), 19137–19145.
Signorelli, R., Ku, D. C., Kassakian, J. G., & Schindall, J. E. (2009). Electrochemical double-layer capacitors using carbon nanotube electrode structures. *Proceedings of the IEEE*, *97*(11), 1837–1847.
Simon, P., & Gogotsi, Y. (2008). *Materials for electrochemical capacitors. Vol. 7* (pp. 845–854).
Simon, P., Gogotsi, Y., & Dunn, B. (2014). Where do batteries end and supercapacitors begin? *Science*, *343*(6176), 1210–1211.
Singh, L., & Kalia, V. C. (2017). *Waste biomass management—A holistic approach*. Springer International Publishing.
Song, S., Ma, F., Wu, G., Ma, D., Geng, W., & Wan, J. (2015). Facile self-templating large scale preparation of biomass-derived 3D hierarchical porous carbon for advanced supercapacitors. *Journal of Materials Chemistry A*, *3*(35), 18154–18162.
Stoeckli, F., & Centeno, T. A. (2012). Pore size distribution and capacitance in microporous carbons. *Physical Chemistry Chemical Physics*, *14*(33), 11589–11591.
Sun, S., Jiang, G., Liu, Y., Zhang, Y., Zhou, J., & Xu, B. (2017). Growth of MnO2 nanoparticles on hybrid carbon nanofibers for flexible symmetrical supercapacitors. *Materials Letters*, *197*, 35–37.
Tan, H., & Wang, S.-R. (2009). Experimental study of the effect of acid-washing pretreatment on biomass pyrolysis. *Journal of Fuel Chemistry and Technology*, *37*(6), 668–672.
Tan, Y., Xu, C., Chen, G., Liu, Z., Ma, M., Xie, Q., et al. (2013). Synthesis of ultrathin nitrogen-doped graphitic carbon nanocages as advanced electrode materials for supercapacitor. *ACS Applied Materials & Interfaces*, *5*(6), 2241–2248.
Tang, C., Liu, Y., Yang, D., Yang, M., & Li, H. (2017). Oxygen and nitrogen co-doped porous carbons with finely-layered schistose structure for high-rate-performance supercapacitors. *Carbon*, *122*, 538–546.
Tay, T., Ucar, S., & Karagöz, S. (2009). Preparation and characterization of activated carbon from waste biomass. *Journal of Hazardous Materials*, *165*(1), 481–485.
Thomberg, T., Tooming, T., Liivand, K., Siinor, L., Jänes, A., & Lust, E. (2015). Supercapacitors based on mixture of room temperature ionic liquids containing specifically adsorbed iodide anions. *ECS Transactions*, *64*(20), 1–11.

Valix, M., Cheung, W. H., & McKay, G. (2004). Preparation of activated carbon using low temperature carbonisation and physical activation of high ash raw bagasse for acid dye adsorption. *Chemosphere*, *56*(5), 493–501.

Viswanathan, B., Neel, P. I., & Varadarajan, T. (2009). *Methods of activation and specific applications of carbon materials*. India, Chennai: National Centre for Catalysis Research-IIT Madras.

Vix-Guterl, C., Frackowiak, E., Jurewicz, K., Friebe, M., Parmentier, J., & Béguin, F. (2005). Electrochemical energy storage in ordered porous carbon materials. *Carbon*, *43*(6), 1293–1302.

Wang, S. (2017). Biosynthesis of carbon spheres for supercapacitor electrode application. *International journal of electrochemical science*, *12*, 11496–11505.

Wang, D., Geng, Z., Li, B., & Zhang, C. (2015). High performance electrode materials for electric double-layer capacitors based on biomass-derived activated carbons. *Electrochimica Acta*, *173*, 377–384.

Wang, D.-W., Li, F., Liu, M., Lu, G. Q., & Cheng, H.-M. (2008). 3D aperiodic hierarchical porous graphitic carbon material for high-rate electrochemical capacitive energy storage. *Angewandte Chemie*, *120*(2), 379–382.

Wang, X., Li, Y., Lou, F., Melandso Buan, M. E., Sheridan, E., & Chen, D. (2017). Enhancing capacitance of supercapacitor with both organic electrolyte and ionic liquid electrolyte on a biomass-derived carbon. *RSC Advances*, *7*(38), 23859–23865.

Wang, K., Zhao, N., Lei, S., Yan, R., Tian, X., Wang, J., et al. (2015). Promising biomass-based activated carbons derived from willow catkins for high performance supercapacitors. *Electrochimica Acta*, *166*, 1–11.

Warhurst, A. M., Fowler, G. D., McConnachie, G. L., & Pollard, S. J. T. (1997). Pore structure and adsorption characteristics of steam pyrolysis carbons from *Moringa oleifera*. *Carbon*, *35*(8), 1039–1045.

Wasiński, K., Walkowiak, M., & Lota, G. (2014). Humic acids as pseudocapacitive electrolyte additive for electrochemical double layer capacitors. *Journal of Power Sources*, *255*, 230–234.

Williams, P. T., & Onwudili, J. (2005). Composition of products from the supercritical water gasification of glucose: a model biomass compound. *Industrial & Engineering Chemistry Research*, *44*(23), 8739–8749.

Xiaoqing, Y., Chengfei, L., & Yue, C. (2017). Hierarchical porous carbon with ultrahigh surface area from corn leaf for high-performance supercapacitors application. *Journal of Physics D: Applied Physics*, *50*(5), 055501.

Xu, G., Han, J., Ding, B., Nie, P., Pan, J., Dou, H., et al. (2015). Biomass-derived porous carbon materials with sulfur and nitrogen dual-doping for energy storage. *Green Chemistry*, *17*(3), 1668–1674.

Xu, C., Xu, B., Gu, Y., Xiong, Z., Sun, J., & Zhao, X. S. (2013). Graphene-based electrodes for electrochemical energy storage. *Energy & Environmental Science*, *6*(5), 1388–1414.

Xuan, C., Peng, Z., Wang, J., Lei, W., Xia, K., Wu, Z., et al. (2017). Biomass derived nitrogen doped carbon with porous architecture as efficient electrode materials for supercapacitors. *Chinese Chemical Letters*, *28*(12), 2227–2230.

Yagmur, E., Ozmak, M., & Aktas, Z. (2008). A novel method for production of activated carbon from waste tea by chemical activation with microwave energy. *Fuel*, *87*(15), 3278–3285.

Yamazaki, S., Ito, T., Yamagata, M., & Ishikawa, M. (2012). Non-aqueous electrochemical capacitor utilizing electrolytic redox reactions of bromide species in ionic liquid. *Electrochimica Acta*, *86*, 294–297.

Yan, S., Lin, J., Liu, P., Zhao, Z., Lian, J., Chang, W., et al. (2018). Preparation of nitrogen-doped porous carbons for high-performance supercapacitor using biomass of waste lotus stems. *RSC Advances*, *8*(13), 6806–6813.

Yang, M., Kim, D. S., Hong, S. B., Sim, J.-W., Kim, J., Kim, S.-S., et al. (2017). MnO2 nanowire/biomass-derived carbon from hemp stem for high-performance supercapacitors. *Langmuir*, *33*(21), 5140–5147.

Yang, H., Yan, R., Chen, H., Lee, D. H., & Zheng, C. (2007). Characteristics of hemicellulose, cellulose and lignin pyrolysis. *Fuel*, *86*(12), 1781–1788.

Yang, M., & Zhou, Z. (2017). Recent breakthroughs in supercapacitors boosted by nitrogen-rich porous carbon materials. *Advanced Science*, *4*(8), 1600408.

Yu, Z., Tetard, L., Zhai, L., & Thomas, J. (2015). Supercapacitor electrode materials: nanostructures from 0 to 3 dimensions. *Energy & Environmental Science*, *8*(3), 702–730.

Yuliansyah, A. T., & Hirajima, T. (2012). Efficacy of hydrothermal treatment for production of solid fuel from oil palm wastes. In V. Abrol (Ed.), *Resource management for sustainable agriculture*: InTech (Chapter 1).

Zhao, Y.-Q., Lu, M., Tao, P.-Y., Zhang, Y.-J., Gong, X.-T., Yang, Z., et al. (2016). Hierarchically porous and heteroatom doped carbon derived from tobacco rods for supercapacitors. *Journal of Power Sources*, *307*, 391–400.

Zuo, W., Li, R., Zhou, C., Li, Y., Xia, J., & Liu, J. (2017). Battery-supercapacitor hybrid devices: recent progress and future prospects. *Advanced Science*, *4*(7), 1600539.

Further reading

Adinaveen, T., Kennedy, L. J., Vijaya, J. J., & Sekaran, G. (2013). Studies on structural, morphological, electrical and electrochemical properties of activated carbon prepared from sugarcane bagasse. *Journal of Industrial and Engineering Chemistry*, *19*(5), 1470–1476.

Arnold, S., Moss, K., Henkel, M., & Hausmann, R. (2017). Biotechnological perspectives of pyrolysis oil for a bio-based economy. *Trends in Biotechnology*, *35*(10), 925–936.

de Clippel, F., Dusselier, M., Van de Vyver, S., Peng, L., Jacobs, P. A., & Sels, B. F. (2013). Tailoring nanohybrids and nanocomposites for catalytic applications. *Green Chemistry*, *15* (6), 1398–1430.

Han, X., Jiang, H., Zhou, Y., Hong, W., Zhou, Y., Gao, P., et al. (2018). A high performance nitrogen-doped porous activated carbon for supercapacitor derived from pueraria. *Journal of Alloys and Compounds*, *744*, 544–551.

Hong, X., Hui, K. S., Zeng, Z., Hui, K. N., Zhang, L., Mo, M., et al. (2014). Hierarchical nitrogen-doped porous carbon with high surface area derived from endothelium corneum *Gigeriae galli* for high-performance supercapacitor. *Electrochimica Acta*, *130*, 464–469.

Li, J., Liu, W., Xiao, D., & Wang, X. (2017). Oxygen-rich hierarchical porous carbon made from pomelo peel fiber as electrode material for supercapacitor. *Applied Surface Science*, *416*, 918–924.

Wahid, M., Puthusseri, D., Phase, D., & Ogale, S. (2014). Enhanced capacitance retention in a supercapacitor made of carbon from sugarcane bagasse by hydrothermal pretreatment. *Energy & Fuels*, *28*(6), 4233–4240.

Zhang, L., Zhang, F., Yang, X., Leng, K., Huang, Y., & Chen, Y. (2013). High-performance supercapacitor electrode materials prepared from various pollens. *Small*, *9*(8), 1342–1347.

Zhang, L. L., & Zhao, X. S. (2009). Carbon-based materials as supercapacitor electrodes. *Chemical Society Reviews*, *38*(9), 2520–2531.

Biochars in iron and steel industries

11

Rita Khanna[*], *Kejiang Li*[†], *Ziming Wang*[†], *Minmin Sun*[†], *Jianliang Zhang*[†], *Partha Sarathi Mukherjee*[‡]
[*]School of Materials Science and Engineering, The University of New South Wales, Sydney, NSW, Australia, [†]School of Metallurgical and Ecological Engineering, University of Science and Technology Beijing, Beijing, PR China, [‡]Institute of Minerals and Materials Technology, Advanced Materials Technology Department, Bhubaneshwar, India

Chapter Outline

11.1 Introduction

The iron and steel industrial sector plays a significant role in enhancing industrial output, the national economy, and overall strength of a given country. Iron, steel, and ferroalloys find extensive applications in major industries such as energy, transport, construction, agriculture, light and heavy industries, healthcare, and several other enterprises. The global demand for steel has been growing steadily with total production in small-, medium-, and large-sized units reaching 493 million tonnes in 1967, 850 million tonnes in 2000, and a massive 1603 million tonnes in 2017 (World Steel Association, 2017a, 2017b). Overall steel usage is projected to increase to over 2.5 billion tonnes by 2050 (World Steel Association, 2018). Due to such large volumes

Char and Carbon Materials Derived from Biomass. https://doi.org/10.1016/B978-0-12-814893-8.00011-0

being produced, high levels of energy consumption in the steel industry, associated costs, and CO_2 emissions have become key challenges for the sector.

Steel is primarily produced through one of two manufacturing routes: the integrated iron and steelmaking route and the electric arc furnace (EAF) steelmaking route. In the integrated route, iron ore is reduced by cokes/coals to produce carbon-rich pig iron in blast furnaces (BFs), decarburized in basic oxygen furnaces (BOF), and refined further to produce steel (Peacy, 1979). This route approximately uses 1400 kg of iron ore, 800 kg of coal, 120 kg of recycled steel, along with 300 kg of limestone to produce 1000 kg of crude steel (World Steel Association, 2012). Ironmaking, presently dominated by the BF, is a highly energy-intensive process and accounts for nearly 75%–80% of total energy consumed while generating most CO_2 emissions. A possible changeover to alternative ironmaking technologies (COREX and FINMET) is only likely over several decades due to huge capital requirements and very long average plant lives (Delport & Holaschke, 1990). The BF is likely to remain the primary ironmaking route for the foreseeable future.

The EAF steelmaking route uses steel scrap and other iron-bearing materials for producing steel. Due to low capital and operational costs, small sizes, fast turnovers, and high productivity, the EAF route has grown considerably in recent decades. EAF-based operations have gradually started to replace production areas previously dominated by the integrated BF route. Approximately 40% of steel in North America is presently made via the EAF route (Jones, Bowman, & Lefrank, 1998). For an EAF-based operation, the capital cost per tonne of annual installed capacity is typically in the range of \$140–200/t, whereas the corresponding costs for BF operations are approximately \$1000/t. The EAF route uses approximately 880 kg of recycled steel, granulated iron, direct reduced iron (DRI), and hot metal, along with 16 kg of coal and 64 kg of limestone to produce 1000 kg of crude steel (World Steel Association, 2012).

With electricity providing up to 44% of primary energy consumed (1.56×10^{18} J), direct fuels such as natural gas, coke and breeze, coke oven gas, coal, and residual fuel oils are also used to provide up to 42% of total energy requirements of the iron and steel sector. While EAFs use only small amounts of coal as a raw material, the electricity used is generally generated by a coal-fired power plant. The greenhouse gas of most relevance to the world steel industry is carbon dioxide (CO_2); $\sim$1.8 t of emissions are generated by producing a tonne of steel. Several new technologies are being developed to reduce CO_2 emissions toward lowering the environmental impact of the steel industry (see Hasanbeigi, Arens, & Price, 2014 and Quader, Ahmed, Ghazilla, Ahmed, & Dahari, 2015 for extensive reviews). These include, among others, Ultra-Low CO_2 Steelmaking (ULCOS) (Quader, Ahmed, Dawal, & Nukman, 2016; ULCOS, 2004), American Iron and Steel Institute (AISI) technology roadmap program, CO_2 Breakthrough Framework (POSCO, 2014), and CO_2 Ultimate Reduction in Steelmaking Process (Pardo & Moya, 2013). ULCOS has identified three carbon-lean routes: (1) replacement of coal/coke with hydrogen or electricity in hydrogen reduction or the electrolysis of iron ore processes, (2) carbon capture and storage technology, and (3) use of biomass chars in the industry in a sustainable manner.

In this chapter, we focus our attention on the partial or complete replacement of coals and cokes with biomass chars in iron and steelmaking processes. The utilization of biochars in ironmaking is expected to lower carbon and energy consumption and reduce operating costs as well as greenhouse gas (GHG) emissions. Kumar and Gupta (1998) critically reviewed the use of wood chars in the manufacture of iron and some ferroalloys based on chemical composition, reactivity, heating value, capacity, and strength. Because the basic characteristics of biomass-based carbons differ significantly from those of fossil-based carbons, biomass char utilization in iron and steelmaking processes is expected to face several technological challenges, constraints and limitations, availability, environmental issues, and economic considerations. This chapter presents an overview of the utilization of biomass chars in iron and steelmaking industries with specific focus on the influence of biochar replacement during cokemaking, sintering, agglomeration, BF fuel injection, EAF slag foaming and reduction reactions, economic and supply issues, along with directions toward industrial-scale applications in various stages of iron and steelmaking.

11.2 Basic characteristics of biobased carbons, biochars, and fossil-based carbons

Biomass, a generic term used for organic matter derived from plants (crops, grasses, wood, trees, etc.), has carbon, hydrogen, oxygen, and nitrogen as its key constituents along with some mineral matter. Being naturally abundant and available in a wide variety, biomass represents a valuable resource of renewable carbons and energy, while containing low levels of harmful impurities (S, P, and ash impurities). Biochars are also expected to produce less CO_2 as compared to coals during steelmaking reactions. A brief comparison of fundamental characteristics of biomass from a variety of sources, biobased chars and charcoals, coals, and cokes, is presented in Table 1.

The following were some of the key differences between various carbon resources:

- The fixed carbon contents of biobased charcoal and coke were quite high and typically ranged between 85% and 88%; whereas the corresponding values for coal ranged between 50% and 55%. The fixed carbon in biomass/biochar was much lower and ranged between 16% and 35% only.
- Biomass also had a much higher volatile matter (VM) (55%–84%) as compared to coke (1%–1.5%), coal (15%–30%), and biocharcoal (10%–12%).
- The calorific values of biomass (7%–24%) were much less than the corresponding values for fossil-based carbon resources (23%–32%). However, the concentrations of mineral impurities present in biomass (0.2%–2%) were much smaller than the typical ash content of cokes and coals (10%–12%).

Coals generally have a densely packed structure, whereas biochars typically have a highly porous structure caused by the release of volatiles during the charring process. The Brunauer-Emmett-Teller surface area of biochars was determined to be 228–950 m^2/g and was found to be much higher than coke (16–17 m^2/g) and coal (66 m^2/g) (Alakangas et al., 2016). The utilization of the raw biomass (woods, etc.) in metallurgical processes is currently not feasible due to large volumes, high oxygen and moisture contents, low heating values, extensive volatile release at high

Table 1 A comparison of fundamental characteristics of a range of biobased and fossil-based carbon resources (Alakangas, Hurskainen, Laatikainen-Luntama, & Korhonen, 2016; Carbo, Leiser, Kiel, Threan, & Witt, 2014; Loison, Foch, & Boyer, 1989; Suopajärvi, Kemppainen, Haapakangas, & Fabritius, 2017)

Basic fuel characteristics	Woodchip	Wood pellet	Torrefied wood pellet	Charcoal	Coal	Coke
Moisture content (wt%)	30–55	7–10	1–5	1–5	10–15	1–10
Calorific value (MJ/kg)	7–12	15–17	18–24	30–32	23–28	30
Fixed carbon (wt%)	16–25	16–25	22–35	85–87	50–55	85–88
Mineral matter (wt%)	0.4–2	0.2–0.4	0.5	0.8–1.4	10	8–12
Volatile content (wt%)	75–84	75–84	55–65	10–12	15–30	1–1.5
Bulk density (kg/m^3)	200–300	550–650	650–800	180–240	800–850	400–500
Milling requirements	Special	Special	Standard	Standard	Standard	NA
Transport costs	High	Medium	Low	Medium	Medium	Low

temperatures, associated volume changes, and their detrimental influence on furnace operations (Lovel, Vining, & Dell'Amico, 2009; Suopajärvi, Pongrácz, & Fabritius, 2013). To overcome some of these key issues, biomass-based chars such as torrefied woods or wood charcoals are being used in various applications. Several different types of heat treatments can be used to produce biochars and to enhance their characteristics. Torrefaction, i.e., mild carbonization, in the temperature range 225–300°C produces chars with carbon contents in the range 47%–66% (dry basis) and heating values of 19.5–26.4 MJ/kg (Phanphanich & Mani, 2011). Slow pyrolysis can enhance the carbon content of charcoals to as high as 95%; heating values of 34–35 MJ/kg have been achieved. While increasing carbonization temperatures can enhance biochar properties further for utilization in iron and steelmaking, the use of higher temperatures may significantly reduce the volumes/yields of charcoal produced.

11.3 Biochar utilization in various aspects of iron and steelmaking

11.3.1 Biochar usage in cokemaking

Metallurgical coke has several significant roles in BF ironmaking (Li et al., 2014). Its key roles in an operating BF are as a fuel providing heat energy for the endothermic chemical reactions and for the melting of slag and metal; as a chemical reducing agent providing reducing gases; as carburizer for the hot metal in the hearth; and as the only solid support medium for the iron-bearing burden providing a permeable matrix for liquid slag and hot metal. Metallurgical coke is produced from low-ash, low-sulfur coking coals in the coking process involving devolatilization and carbonization at ~1000–1100°C in airless ovens over 18–20h. Volatile hydrocarbon by-products are later used to generate energy in separate combustion processes. Due to the limited supply and high costs of high-grade coking coals, blending of coals with acceptable levels of volatility and mineral content is a common practice for producing cokes of appropriate strength, ensuring limited volatile release, swelling, and wall pressure during baking in coke ovens.

To produce metallurgical coke with adequate quality, up to 15% of charcoals may be added to the coals being used (MacPhee, Gransden, Giroux, & Price, 2009). Fluidity, a key requirement for producing high-quality metallurgical cokes determines the capacity of coals/coal blends to form a plastic phase. The fluidity of coals is known to range from 1 ddpm (dial division per minute) for noncoking coals and up to 5000 ddpm for hard coking coals. The production of good-quality coke requires fluidities in the range 400–1000 ddpm (Kumar, Barman, Singh, & Ranjan, 2008). The influence of biomass char addition on the fluidity of coal blends has been investigated for a range of biomasses such as Kraft lignin from the pulp/paper industry, woody components, and cellulose, lignin, xylan, torrefied biomass, biochars, bio-oils, etc. (Castro-Díaz, Vega, Barriocanal, & Snape, 2015; Castro-Díaz, Zhao, Kokonya, Dufour, & Snape, 2012; Kokonya, Castro-Díaz, Barriocanal, & Snape, 2013). Because biochars do not undergo transformation to the plastic phase during the coking process, the addition of most types of biomaterials generally lowers the maximum fluidity achieved in the coal-biochar blend.

Thermal degradation behavior of biobased materials depends significantly on the nature/type of biomass being investigated. The decomposition of raw biomasses takes place at low temperatures, and most of the volatiles are released before reaching the plastic stage of the coking process (300–350°C), whereas thermal decomposition temperatures for biochars are similar to those of coal, occurring during the transformation of semicoke to coke. Biochar remains relatively inert during the softening and melting processes but plays an important role for binding various plastic phases present. The addition of biochars lowers the fluidity of the coal-biochar blend, which in turn could affect the formation and stability of the coke matrix.

Table 2 summarizes changes in fluidity caused by the addition of different biochar fractions to various coal blends. Experiments were carried out on coking coals (CC1) with particle sizes ranging from 20 to 212 μm, and their blends with sawdust char, CC2, and lignin. Additional amounts of blending constituents varied from 0 to 15 wt%, and their influence on maximum fluidity F_{max} (ddpm) and percentage reduction in F_{max} were determined. Very high levels of F_{max} reductions were observed for sawdust chars ranging from 53.7 to 97.4 wt% with the concentration of sawdust char increasing from 5 to 10 wt%. A similar trend was observed for lignin as well.

Coal rank and the fluidities of vitrinite and inertinite phases affect the optical texture of coke and play an important role in determining the overall fluidity in the blend (Montiano, Díaz-Faes, Barriocanal, & Alvarez, 2014). Because the relative

Table 2 Influence of biomass char additions on the fluidity of coal blends (Diez, Alvarez, & Fernández, 2012; Guerrero, Diez, & Borrego, 2015)

Coal fluidity	Additives (μm)	F_{max} (additions wt%)				% reduction F_{max} (additions wt%)		
		0	5	10	15	5	10	15
Low	CC1, 212 μm	373	247	145	80	34	61	79
	CC1, 80 μm	373	210	110	45	44	71	88
	CC1, 20 μm	373	150	55	10	60	85	97
Medium	CC1, 212 μm	541	400	350	275	26	35	49
	CC1, 80 μm	541	390	300	185	28	45	66
	CC1, 20 μm	541	325	160	90	40	70	83
High	CC1, 212 μm	1891	1040	700	550	45	63	71
	CC1, 80 μm	1891	910	680	375	52	64	80
	CC1, 20 μm	1891	740	370	140	61	80	93
Low	Sawdust	389	71	10	NA	82	97	NA
	CC2	389	250	146	NA	36	63	NA
	Lignin	389	192	133	NA	51	66	NA

proportions of reactive and inert components present can impact coke behavior, the addition of inert biochars could have an influence on blend fluidity as well as texture. The addition of sawdust char to the coking coal blends decreased the anisotropic texture with a corresponding increase in the isotropic texture.

Volume changes (expansion or contraction) in the coal blends are an important factor during cokemaking, because they can impact oven pressure and the sticking of coke charges. Only a few studies have been reported on the influence of biomass chars during cokemaking. MacPhee et al. (2009) reported negligible contraction for adding up to 10% coarse biochar particles to coal blends; negligible contractions were reported for fine particles as well. Thomas et al. (2011) reported a smaller contraction for coarse torrefied biomass; however, no contraction was reported for particle sizes <0.25 mm. Ng, Giroux, MacPhee, and Todoschuk (2012) determined the coking pressure with different concentrations of charcoal added to coal blends; the wall pressure was found to decrease significantly especially for smaller charcoal particles as compared to pure coal blends. The reduction in pressure was attributed to the gases escaping from the plastic layer in the presence of inert materials. Coarse torrefied biomass particles increased the wall pressure, whereas the oven pressure tended to decrease with finer particles.

The strength of metallurgical coke is measured with shatter tests and/or drum tests. Postreaction strength (coke strength after reaction, CSR) of coke is measured with the CSR index, which is coupled to the coke reactivity index (CRI) (Li et al., 2014). Nearly 200 g of a sized fraction of coke is exposed to 100% carbon dioxide at 1100°C for 2 h and the weight loss is measured as CRI. The product is then subjected to a drum test (600 revolutions in an I-drum) and the mass of sample remaining (>10 mm) is expressed to give the CSR. A high CSR value helps prevent coke from breaking down, improves the permeability of gas and liquid, increases productivity, and reduces specific coke consumption in the blast furnace. Various BFs require cokes with CRI values typically ranging between 20 and 30, and CSR values ranging between 58 and 65 (Alvarez, Diez, Barriocanal, Diaz-Faes, & Cimadevilla, 2007).

The addition of biochars to coal blends affects the reactivity of the resulting coke as well. MacPhee et al. (2009) added up to 10 wt% highly carbonized biochars to three different coal blends. Reactivity was found to increase with increasing levels of biochars in the blend; this increase was attributed to a higher presence of calcium in the blend. Qin et al. (2014) found that the addition of pine sawdust char (up to 5%) to a coal blend kept the CRI value of biocoke below the requisite level of 30. Montiano et al. (2014) used chestnut and pine sawdust chars (1–5 wt%) to produce biocokes with two different coal blends. The CRI and CSR measurements were found to agree well with other studies, indicating a CRI increase and CSR decrease with increasing additions of biochars. To produce industrially viable CRI and CSR ranges, addition of sawdust chars is presently feasible only below 2 wt%.

11.3.2 Biochar usage in iron ore sintering

Sintering of iron ore with returned sinter fines and plant dusts is a commonly used pretreatment approach in BF ironmaking. Other sinter constituents include up to 3–5 wt% coke breeze to provide fuel and heat in situ, and dolomite, limestone, and silica fluxes for forming suitable agglomerates with appropriate compositions. All these materials are blended together and formed into micro-sized pellets in the granulation drum with small amounts of water. Raw sinter pellets are heat treated (1300–1480°C) on a continuously moving sinter belt. Size distribution, cold strength, reduction behavior, and disintegration characteristics determine the product quality of sinters (Geerdes, Chaigneau, & Kurunov, 2015). Biobased fuels can be used to partially replace coke breeze, differences in their basic characteristics may impact the quality of the sinter product. While several studies have been reported on the replacement of coke breeze with up to 60% biofuels, only a few studies have been reported on 100% biofuel replacement with mildly or highly carbonized biochars. Due mainly to their higher porosity and surface area, the reactivity of biobased fuels is significantly higher than coke breeze reactivity; it can therefore have a detrimental impact on the overall sintering process and sinter quality (Gan et al., 2015).

Ooi et al. (2008) evaluated the partial replacement (20%–50%) of coke with chars from sunflower husks. The sintering time decreased from 440 (0% husk) to 415 s (20% husk) and 387 s (50% husk); however, not much change was observed in the combustion efficiency for up to 20% husk replacement. The reductions in sintering time were attributed to low heat capacity and higher reactivity of the sunflower husk chars. They also investigated the use of biochars from pine wood flour, sawdust, red gum charcoal, and charcoal prepared from prickly acacia as fuels for sintering. Preliminary tests showed poor granulation and low bed densities with wood flour and sawdust, while the wood char produced a satisfactory sinter (Lovel, Vining, & Dell'Amico, 2007).

Substituting coke breeze with biocharcoals was found to increase the vertical sintering speed; however, increasing substitution rates resulted in lower sinter yields and productivity (Mousa, Babich, & Senk, 2015). While the maximum temperature and the holding time were found to decrease due to higher reactivity of biomass-based chars, the yield as well as tumble index of sinters were found to worsen. As charcoals provided better results than straw chars, up to 40% biocharcoals (and only 20% straw chars) could be used to replace coke breeze without affecting product quality. Increasing levels of biochars reduced the bulk density of the sinters; combustion temperatures were achieved relatively quickly in these sinters due to rapid heating and lower thermal loads (Lovel et al., 2009). According to modified tumble index results, the addition of biochars as fuel lowers sinter strength along with reductions in bulk density, thinner combustion and sintering zones within operating BFs. The share of charcoal in iron ore sinters is presently limited to less than 25 wt% (Naglaa, Khalifa, El-Midany, Ahmed, & Shalabi, 2015).

11.3.3 Biochar use in agglomeration: carbon composites

Carbon composite agglomerates (CCAs) represent a class of materials produced by agglomerating fine powders of carbonaceous materials such as coal and coke fines, waste carbons, biobased charcoals, and chars, with iron ore fines with or without the use of binder. These blends are either cold pressed to form agglomerated pellets or alternatively composite briquettes are produced through hot or cold pressing for applications in BFs, DRI production, and EAF steelmaking (Yokoyama, Higuchi, Ito, & Oshio, 2012). Biomass chars have also been investigated as a carbon resource in CCA production. Most of these studies were focused on the reduction behavior, the metallization degree of DRI-CCAs, and the mechanical strength of CCAs. Several different types of biomass such as wood, straw fiber, sawdust and rice crust chars, bamboo char, and charcoal have been investigated (Ahmed, Viswanathan, & Bjorkman, 2014; Guo et al., 2015). Bentonite is typically used as a binder; it is also a common practice to use no binder at all. While the typical concentration of biomass chars in CCAs ranges between 2 and 14 wt%, up to 46 wt% biochars have been used in agglomerates for DRI applications (Ueki et al., 2013).

Iron oxide reduction rates were found to be higher in biochar-based CCAs as compared to corresponding rates for coal or coke-based CCAs. Biochar-based composites achieved a higher degree of reduction in 60 min; this was attributed to the higher rate of gasification for biochars (Konishi, Ichikawa, & Usui, 2010). The particle sizes of biochars had a small effect on the reduction rate; a small enhancement in reaction kinetics observed for smaller char particles can be attributed to their higher surface area (Ueda et al., 2009; Ueda, Watanabe, Yanagiya, Inoue, & Ariyama, 2009). Luo, Yi, and Zhou (2011) carried out isothermic reduction of composite briquettes to metallic iron in the presence of H_2–CO gas mixtures; the presence of coke in briquettes produced higher amounts of reduced iron as compared to biomass chars. Iron concentration in biochar-based composite pellets was found to reach 95%, when the temperature of the direct reduction was increased to 1100°C. Increasing the relative proportion of CO in the gas mixture had a detrimental influence on the overall reduction behavior and metal quality.

The presence of biochars increases pellet porosity and the contact area between iron oxides and reducing gas, which in turn enhances the kinetics of interfacial reactions. Kowitwarangkul, Babich, and Senk (2014) investigated the reduction behavior of self-reducing pellets (SRPs) using isothermal as well as nonisothermal tests in an attempt to simulate the thermal reserve zone in the BF. The overall reaction rate of composites was affected by composition (relative proportions of iron oxide and carbon), physical form and shape (pellet, briquette), and agglomerate size, temperature, gas atmosphere, and pressure (Bagatini, Zymla, Osório, & Vilela, 2014). The reduction of composite briquettes is controlled by the gasification of carbon and the reduction of iron ore, which promotes the use of highly reactive biomass char. High metallization degrees have been reported while using biochars from wood, charcoal, and straw fiber. High metallization degrees of reduced DRI pellets are important for their utilization in BF, BOF, and EAF steelmaking. Mechanical strength of CCAs is also of crucial importance during utilization in a variety of applications. It is, however, difficult to produce biochar-based self-reducing pellets with sufficient crushing strength for modern, large BFs with working volumes up to $6000\,m^3$; higher charging ratios of SRP might be possible in smaller ($<1000\,m^3$) BFs (Fu, Zhang, Hwang, Liau, & Lin, 2012).

11.3.4 Biochar use as an injectant in the blast furnace

A major advancement in BF technology has been the injection of supplementary fuels for reducing coke consumption, minimizing fuel costs, better control, and for extending coke oven life. Several injectants, including pulverized coal, natural gas, coke oven gas, waste plastics, etc., have been trialed. Key requirements for carbonaceous injectants include high carbon contents, low ash and alkali contents, low sulfur and phosphorus contents, and high heating values. VM also needs to be low because it can affect gasification in the raceway and the coke replacement ratio in the BF. Other parameters include reactivity, ignition, chemical composition, impurities, particle size, and particle size distribution (Konishi et al., 2010). Unsustainably high rates of injection can seriously undermine furnace stability with an increasing amount of unburnt char entering the blast furnace burden (Yellishetty, Mudd, Ranjith, & Tharumarajah, 2011). A number of studies have also been carried out to evaluate biochar injection in BFs; biochar injection rates can reach up to 200 kg for producing 1000 kg of hot metal (Mathieson, Truelove, & Rogers, 2005).

Material handling facilities in integrated steel plants use a single machine for simultaneously grinding, drying, and classifying coal. As the coal is milled to a very small size (60% under 75 mm size), particles with lower density are likely to leave the mill in larger particle sizes (Ueki et al., 2013). The grinding behavior of different biochars therefore needs to be investigated carefully in comparison to coal. Gil, García, Pevida, and Rubiera (2015) compared the grinding behavior of a number of biochars from pine, chestnut chips, black poplar, etc. torrefied in the temperature range 240–300°C with 11–43 min residence time. Particle size distribution profiles were used as a measure of grindability; biochar with the highest proportion of smallest size was considered to have best grindability. Grindability was found to improve with torrefaction temperature as well as with residence time. Grinding behavior, storage, as well as conveying properties of biochars and biochar fossil coal blends were investigated. Simultaneous grinding of coal and chestnut chips (5–55 wt%) torrefied at 280°C with 22 min residence time did not show much interaction. The flow behavior of pine and birch charcoals produced at 520°C was found to be comparable to that of coals. The relative proportion of biocharcoals in PCI blends could range between 10 and 50 wt% without any detrimental influence.

Machado, Osório, and Vilela (2010) compared the CO_2 reactivities of a number of carbonaceous materials, including several coals, biochars, and their blends. The reactivity of biochars (VM 21.9%) was found to be higher than that of a high volatile (38.7%) coal and was attributed to the high porosity and disordered carbon structure of biochars. Using a drop-tube furnace, Pohlmann, Borrego, Osório, Diez, and Vilela (2016) compared the combustibility of eucalyptus chars with injection coals at 1300°C in a range of oxygen levels. Materials were chosen with similar levels of VM ranging from low to high; biochars showed a higher level of burnout in all cases. Biomass chars have an isotropic structure, higher micro- as well as macroporosity, and higher specific surface area than coals. Industrial trials as well as theoretical models have indicated that the injection of pulverized biochar particles ($<75\,\mu m$) into a BF could reduce CO_2 emissions between 18% and 40% (Feliciano & Mathews, 2012; Ueda, Watanabe, et al., 2009; Ueda, Yanagiya, et al., 2009). However, a complete replacement of coals with biochars could increase operating costs by 5%–16% (Bohlin, 1998).

11.3.5 Biochar use in EAF steelmaking: slag foaming and reducing agents

The utilization of biochars has proved to be much easier in EAF steelmaking as compared to the BF. EAF steelmaking is a batch process with fast turnovers and there are no long-term disturbances, whereas BFs have been known to run continuously over several years (Li et al., 2014). Biochar-based DRIs can partially replace feedstock during the charging process, and biochars could also be used as a slag foaming agent either by itself or as coke–biochar blends.

Fidalgo, Berrueco, and Millan (2015) investigated chemical and thermal characteristics of biochars from agricultural waste for utilization in EAF steelmaking; fixed carbon, ash content, and VM of biochars were found to be the key parameters. Echterhof and Pfeifer (2011) observed that composition, calorific value, and reactivity also played an important role in their utilization. The CO_2 reactivity of biochars affects the kinetics of iron oxide reduction, gas generation, and slag foaming; high reactivity biochars produce higher levels of slag foaming than coke. Mineral impurities present in biochars can affect slag chemistry, slag viscosity, and surface tension. For adequate carbon transfer into slag, the low density of biochars may necessitate briquetting or pelletizing. Yunos et al. (2011) reported that the injection of palm shell chars produced significantly higher and sustained slag foaming in EAF slags as compared to metallurgical coke.

Yunos (2012) reported in-depth investigations of blends of agricultural waste (coconut shells and palm shells) chars with metallurgical coke in EAF steelmaking for its fuel and carbon requirements. The addition of agricultural waste chars had a strong influence on the combustion performance and carbon-slag interactions, including FeO reduction, gas generation, and its entrapment within the slag as well as enhancement of the slag volumes. Results showed extensive reactions in the gas phase and carbon-slag interactions for 100% palm chars. The combustion burnout of coke-agricultural blends was seen to be higher than the burnout of coke and raw agricultural waste alone. The rate of volatile release was found to affect the physical structure of the resulting char. Whereas palm shell's cell structures opened with increases in surface area, coconut shells retained their cylindrical cell structures with an initial decrease in surface area. The aliphatic structures present in the agricultural waste were consumed during combustion, leaving behind a more ordered aromatic structure. Metallurgical coke showed high rates of iron oxide reduction, very high initial rates of gas generation, but poor foaming behavior. Palm chars showed high slag volumes but slow rates for iron oxide reduction. Slower rates of gas generation made it easier for slag to trap gases and to generate sustained levels of slag foaming. A balance between iron oxide reduction and foaming needs to be established to optimize furnace efficiency and operations.

During plant trials, charging with biochar fines did not have a negative impact on steel quality, slag chemistry, or slag foaming (Demus et al., 2012); however, the carburization of molten iron was found to be below acceptable levels. Due to high combustion rates of biochars, only a limited time was available for carburization of molten iron. It may be possible to control reactivity and the combustion behavior of biochars by briquetting with water and molasses. Industrial-scale trials have been reported on injection as well as charging of biochars in three different EAFs; no significant differences in metal and slag quality were observed as compared to corresponding results from coal. Handling, charging, dispersion of low-density powders, and intense flame emissions were some of the major issues during biochar trials. Partially successful injection trials showed that inadequate and limited slag foaming was affected by the penetration of biochar in the slag (Bianco et al., 2013).

11.4 Economic and supply issues

11.4.1 Economic issues

Commercial-scale utilization of biomass chars in iron and steelmaking is presently limited by the high costs of biobased carbons in comparison with fossil-based carbons. The production costs of biochars are impacted by several factors such as location and costs of raw biomass, processing methods, labor costs, location of iron and steel plants, distances and associated costs involved in transportation and requisite infrastructure upgrades, etc. Jahanshahi et al. (2015) estimated the capital costs of a proposed charcoal plant to be $111 million, including costs for pyrolysis unit (39%), engineering and construction (21%), storage (18%), etc. Total operating costs were computed to be $446/t and average charcoal cost was $243/t.

Table 3 Production costs of a range of biomass-based chars (Suopajärvi et al., 2017)

Biochars	Biomass used	Technology	Capacity (dry t/day)	Location	Production cost €/t
Torrefied biomass	Logging residue	Torrefaction	230	Finland	147
	Small diameter wood	Torrefaction	230	Finland	188
	Debarked roundwood	Torrefaction/ pelletization	350	United States	150
	Sawdust chips	Torrefaction/ pelletization	380	Canada	113
Charcoal	Pine forest residue	Slow pyrolysis	230	Finland	337
	Small diameter wood	Slow pyrolysis	230	Finland	513
	Eucalyptus forest residue	Slow pyrolysis	910	Australia	303

Estimated production costs of torrefied biomass and charcoals are summarized in Table 3. These studies are from Europe, Australia, and North America with plant capacities ranging from 50 to 5000 dry biomass tonnes per day. Wood being the most abundant resource, most of the biochar production was wood based. Details of the feedstock used, plant capacity, production technology, location, and the base year of various studies are available elsewhere (Suopajärvi et al., 2017). These costs are generally quite high, show a strong dependence on the feedstock, and range from 113 to 188 €/t for torrefied biomass and 223 to 513 €/t for charcoal.

Using thermodynamic and mathematical modeling, Helle, Helle, Saxén, and Pettersson (2009) investigated the economy of biomass injection, raw material costs, and emissions. This study suggested the need for preprocessing biomass prior to its utilization in BFs because high oxygen content and low heating value of the injectant can have a detrimental influence on the productivity of BFs. Using a lifecycle assessment approach, Norgate and Langberg (2009) estimated potential reductions in CO_2 emissions from charcoal substitution; emissions during all intermediate process steps were taken into account. For charcoal to be economically competitive would require benefits under the carbon trading scheme. Using a cost objective function, Bruzual (2014) calculated the impact of biochar substitution during BF injection on CO_2 abatement to be ~28%–40%. Based on the value-in-use approach, Mathieson estimated rational purchasing prices of a variety of reductants such as torrefied softwood, lignite softwood, polychars, etc. as compared to coal for injection into BFs. Various charcoals were found to have appropriate thermochemistry and were comparable in costs (Mathieson, 2007).

11.4.2 Supply issues

Because iron and steel industries use large amounts of raw materials, a regular supply of biochars can be a significant issue for biochar utilization in the sector. Through species selection, silvicultural, and management practices, forest productivity has been increased multifold toward sustainable production of biofuels and charcoal. For example, Brazil has been well known for producing high-quality iron and steel using wood charcoal. Brazil has the largest abundance of biomass because of its suitable landscape, climate, rainfall, and soils. Because Brazil did not have sufficient reserves of coking coal, nearly 8 million tonnes per annum of wood charcoal were produced specifically for iron and steel industries (Kammen & Lew, 2005). Limiting factors associated with deforestation and subsequent replanting include labor costs, long (14–21 years) cycles of forest regeneration, environmental licenses, land management, etc. To reduce associated CO_2 emissions by 62 Mt by 2030, it has been proposed to use renewable charcoals in BFs to produce up to 46% of pig iron.

Australia is a relatively small producer of charcoal for predominant use in cooking and metallurgy. Charcoal supplies depend on supplies of appropriate biomass feedstock as well as imports. These include mallee trees grown in Western Australia's southwest wheatbelt as well as mangrove wood charcoal from Indonesia, oak charcoal from Argentina, and coconut shell charcoal dust briquettes and briquetted sawdust charcoal from Malaysia. Iron and steel industries require ~5.7 Mt/year of raw biomass or 1.7 Mt/year of charcoal a current hot metal production of 8.4 Mt/year (Fung, 1988). Some short-term targets include 5%–10% replacement of coal with biochars during cokemaking; 100% utilization of biochars during injection; 100% biochar utilization as injection; and charged carbon in EAF steelmaking for slag foaming and supplementary energy. These changes are expected to reduce GHG emissions by 25%–50% in these processes. With the steel industry expected to emerge as a major consumer of biocarbons, efforts are being made toward expanding biochar production capacity in Canada. While China has extensive biomass resources, the utilization of biochars is primarily focused on producing chemicals, energy, and power generation, and not so much on the iron and steel sector.

11.5 Concluding remarks

A critical overview of biochar utilization in iron and steelmaking has been presented along with economic and supply issues facing biochar utilization and replacement. Beneficial effects of charcoal substitution in BF ironmaking include reductions in CO_2 and SO_2 generation, improvements in the quality of hot metal, and reduction in slag volumes. Extensive research is being carried out on several key processes. We conclude this chapter by providing the current status and future directions of various biochar applications in the iron and steel sector.

- For biochar replacement in cokemaking, optimal charcoal characteristics have been identified as: VM <10%, size <1 mm, density >700 kg/m^3, and low alkali content (Todoschek, Giroux, & Wing, 2016). With net CO_2 emissions savings of 1%–5%, this research is still at the laboratory stage and on its way toward industrial trials.
- For BF nut coke replacement, optical charcoal parameters include VM <7%, high density, and sizes <20–25 mm. With CO_2 emissions savings estimated to be ~3%–7%, this biochar substitution is already being tested in commercial mini BFs and industrial trials in large BFs (Lu et al., 2012). Lump charcoal for BF charging requires high strength, VM 0%–25%, and sizes in the range 30–60 mm. Insufficient strength of biochars can be a limiting issue for coke substitution.
- For iron ore sintering, the replacement charcoal needs to have low VM <3%, high densities >700 kg/m^3, small sizes <0.3–3 mm, and could achieve net CO_2 emissions reductions of 5%–15%. This research is presently at pilot-scale test stage (Rodd, Koehler, Walker, & Voermann, 2010). For carbon composites, biochars need to have high strength >500 N, low VM <5%, and sizes typically ~75 μm and could achieve net CO_2 emissions reductions of 3%–7%. In addition to bench-scale R&D, industrial trials are being conducted.
- As BF injectants, biochars need to have a high VM 10%–20%, low ash content <5%, and low level of alkalis. This replacement can produce high CO_2 emissions savings of up to 19%–25%. Physical parameters of biochars such as porosity, grain size distribution, and specific surface within a certain range do not have a detrimental effect on charcoal injection rates. Currently, these are being investigated theoretically as well as undergoing combustion testing in mini BFs and industrial trials in large BFs.
- For EAF slag foaming and injectants, biochars require a low VM (2%–7%), low moisture content (1%–7%), particle size ~0.5–5 mm, and low alkalis, and are at the industrial trial stage (Wibberley et al., 2001).

The basic characteristics of solid biomass-based chars are significantly different from those of fossil-based coals and cokes. These differences are the main barriers as well as opportunities for the utilization of biochars in different metallurgical processes. While highly reactive biomass-based reducing agents provide high burnout during BF injection, their high reactivity has a detrimental influence on their sintering capacity. The chief barrier to the utilization of chars in the iron and steelmaking sector continues to be high prices and limited supplies of biochars as compared to fossil-based carbonaceous resources.

References

Ahmed, H. M., Viswanathan, N., & Bjorkman, B. (2014). Composite pellets—a potential raw material for iron-making. *Steel Research International*, *85*(3), 293–306.

Alakangas, E., Hurskainen, M., Laatikainen-Luntama, J., & Korhonen, J. (2016). *Properties of Fuels Used in Finland (in Finnish)* (p. 258). Technical Research Centre of Finland, VTT Technology.

Alvarez, R., Diez, M., Barriocanal, C., Diaz-Faes, E., & Cimadevilla, J. (2007). An approach to blast furnace coke quality prediction. *Fuel*, *86*, 2159–2166.

Bagatini, M. C., Zymla, V., Osório, E., & Vilela, A. C. F. (2014). Carbon gasification in self-reducing mixtures. *ISIJ International*, *54*(12), 2687–2696.

Bianco, L., Baracchini, G., Cirilli, F., Di Sante, L., Moriconi, A., & Moriconi, E. (2013). Sustainable electric arc furnace steel production: GREENEAF. *BHM Berg-und Hüttenmännische Monatshefte*, *158*(1), 17–23.

Bohlin, F. (1998). The Swedish carbon dioxide tax: effects on biofuel use and carbon dioxide emissions. *Biomass and Bioenergy*, *15*, 283–291.

Bruzual, C. F. (2014). Charcoal injection in blast furnaces (Bio-PCI): CO_2 reduction potential and economic prospects. *Journal of Materials Research and Technology*, *3*(3), 233–243.

Carbo, M., Leiser, S., Kiel, J. H. A., Threan, D., & Witt, J. (2014). New results of the SECTOR project: Production of solid sustainable energy carriers from biomass by means of torrefaction. In: *39th Clearwater Clean Coal Conference, Clearwater, Florida, US, 1–5 June, 2014*.

Castro-Díaz, M., Vega, M. F., Barriocanal, C., & Snape, C. E. (2015). Utilization of carbonaceous materials to restore the coking properties of weathered coals. *Energy & Fuels*, *29*(9), 5744–5749.

Castro-Díaz, M., Zhao, H., Kokonya, S., Dufour, A., & Snape, C. E. (2012). The effect of biomass on fluidity development in coking blends using high-temperature SAOS rheometry. *Energy & Fuels*, *26*(3), 1767–1775.

Delport, H. M. W., & Holaschke, P. J. (1990). *COREX Symposium, South Africa* (pp. 21–34).

Demus, T., Echterhof, T., Pfeifer, H., Schulten, M., Noel, Y., & Quicker, P. (2012). Investigations on the use of biogenic residues as a substitute for fossil coal in the EAF steelmaking process. In: *10th European Electric Steelmaking Conference (EEC), Graz, 25*.

Diez, M., Alvarez, R., & Fernández, M. (2012). Biomass derived products as modifiers of the rheological properties of coking coals. *Fuel*, *96*, 306–313.

Echterhof, T., & Pfeifer, H. (2011). Potential of biomass usage in electric steelmaking. In: *Proceedings of 1st International Conference on Energy Efficiency and CO2 Reduction in the Steel Industry. 28.06–30.06*.

Feliciano, C., & Mathews, J. A. (2012). Bio-PCI a renewable reductant for blast furnaces: CO2 mitigation potential and economical assessment. In *Proceedings of the ICSTI12. 2012* (pp. 1914–1927).

Fidalgo, B., Berrueco, C., & Millan, M. (2015). Chars from agricultural wastes as greener fuels for electric arc furnaces. *Journal of Analytical and Applied Pyrolysis*, *113*, 274–280.

Fu, J.-X., Zhang, C., Hwang, W.-S., Liau, Y.-T., & Lin, Y.-T. (2012). Exploration of biomass char for CO_2 reduction in RHF process for steel production. *International Journal of Greenhouse Gas Control*, *8*, 143–149.

Fung, P. (1988). Eucalypts in Brazil for charcoal, fibres and energy and the IUFRO All Division 5 Conference. In: *Report to the Australian Academy of Science, Maxwell Ralph Jacobs Memorial Fund in cooperation With the Australian Institute of Foresters Inc 12 May—6 June 1988*.

Gan, M., Fan, X., Ji, Z., Jiang, T., Chen, X., & Yu, Z. (2015). Application of biomass fuel in iron ore sintering: influencing mechanism and emission reduction. *Ironmaking and Steelmaking*, *42*(1), 27–33.

Geerdes, M., Chaigneau, R., & Kurunov, I. (2015). *Modern Blast Furnace Ironmaking: An Introduction*. Ios Press.

Gil, M., García, R., Pevida, C., & Rubiera, F. (2015). Grindability and combustion behavior of coal and torrefied biomass blends. *Bioresource Technology*, *191*, 205–212.

Guerrero, A., Diez, M. A., & Borrego, A. G. (2015). Influence of charcoal fines on the thermoplastic properties of coking coals and the optical properties of the semicoke. *International Journal of Coal Geology*, *147*, 105–114.

Guo, D., Hu, M., Pu, C., Xiao, B., Hu, Z., & Liu, S. (2015). Kinetics and mechanisms of direct reduction of iron ore-biomass composite pellets with hydrogen gas. *International Journal of Hydrogen Energy*, *40*(14), 4733–4740.

Hasanbeigi, A., Arens, M., & Price, L. (2014). Alternative emerging ironmaking technologies for energy-efficiency and carbon dioxide emissions reduction: a technical review. *Renewable and Sustainable Energy Reviews*, *33*, 645–658.

Helle, H., Helle, M., Saxén, H., & Pettersson, F. (2009). Mathematical optimization of ironmaking with biomass as auxiliary reductant in the blast furnace. *ISIJ International*, *49*(9), 1316–1324.

Jahanshahi, S., Mathieson, J. G., Somerville, M. A., Haque, N., Norgate, T. E., & Deev, A. (2015). Development of low-emission integrated steelmaking process. *Journal of Sustainable Metallurgy*, *1*(1), 94–114.

Jones, J. A. T., Bowman, B., & Lefrank, P. A. (1998). Electric arc furnace steelmaking. In *Steelmaking and Refining Volume* (pp. 525–560). Pittsburg, PA: The AISI Steel Foundation.

Kammen, D., & Lew, D. (2005). *Review of technologies for the production and use of biochar. Energy and Resources group and Goldman school of public policy UC Berkeley and NREL.*

Kokonya, S., Castro-Díaz, M., Barriocanal, C., & Snape, C. E. (2013). An investigation into the effect of fast heating on fluidity development and coke quality for blends of coal and biomass. *Biomass and Bioenergy*, *56*, 295–306.

Konishi, H., Ichikawa, K., & Usui, T. (2010). Effect of residual volatile matter on reduction of iron oxide in semi-charcoal composite pellets. *ISIJ International*, *50*(3), 386–389.

Kowitwarangkul, P., Babich, A., & Senk, D. (2014). Reduction behavior of self-reducing pellet (SRP) for low height blast furnace. *Steel Research International*, *85*(11), 1501–1509.

Kumar, P. P., Barman, S., Singh, S., & Ranjan, M. (2008). Influence of coal fluidity on coal blend and coke quality. *Ironmaking and Steelmaking*, *35*(6), 416–420.

Kumar, M., & Gupta, R. C. (1998). Industrial uses of wood char. *Energy Sources*, *20*, 575–589.

Li, K., Khanna, R., Zhang, J., Liu, Z., Sahajwalla, V., & Yang, T. (2014). The evolution of structural order, microstructure and mineral matter of metallurgical coke in a blast furnace: a review. *Fuel*, *133*, 194–215.

Loison, R., Foch, P., & Boyer, A. (1989). *Coke: Quality and Production*. London: Butterworths.

Lovel, R., Vining, K., & Dell'Amico, M. (2007). Iron ore sintering with charcoal. *Minerals Processing and Extractive Metallurgy*, *116*(2), 85–92.

Lovel, R. R., Vining, K. R., & Dell'Amico, M. (2009). The influence of fuel reactivity on iron ore sintering. *ISIJ International*, *49*(2), 195–202.

Lu, L., Adam, M., Somerville, M., Hapugoda, S., Jahanshahi, S., & Mathieson, J. G. (2012). Iron ore sintering with charcoal. In *6th International Congress on the Science and Technology of Ironmaking* (pp. 1121–1131).

Luo, S., Yi, C., & Zhou, Y. (2011). Direct reduction of mixed biomass-Fe_2O_3 briquettes using biomass-generated syngas. *Renewable Energy*, *36*(12), 3332–3336.

Machado, J. G. M. S., Osório, E., & Vilela, A. C. F. (2010). Reactivity of brazilian coal, charcoal, imported coal and blends aiming to their injection into blast furnaces. *Materials Research*, *13*(3), 287–292.

MacPhee, J., Gransden, J., Giroux, L., & Price, J. (2009). Possible CO_2 mitigation via addition of charcoal to coking coal blends. *Fuel Processing Technology*, *90*(1), 16–20.

Mathieson, J. (2007). *The value-in-use of some biomass-derived blast furnace injectants. Technote 2007. BSR/N/2007/071.*

Mathieson, J. G., Truelove, J. S., & Rogers, H. (2005). Toward an understanding of coal combustion in blast furnace tuyere injection. *Fuel, 84*(10), 1229–1237.

Montiano, M., Díaz-Faes, E., Barriocanal, C., & Alvarez, R. (2014). Influence of biomass on metallurgical coke quality. *Fuel, 116*, 175–182.

Mousa, E., Babich, A., & Senk, D. (2015). Iron ore sintering process with biomass utilization. In: *Proc. of METEC & 2nd ESTAT, Düsseldorf, Gemany, 15–19 June 2015.*

Naglaa, A. E. H., Khalifa, A. A., El-Midany, A. A., Ahmed, A., & Shalabi, M. E. H. (2015). Effect of replacement coke breeze by charcoal on technical operation of iron ore sintering. *International Journal of Scientific and Engineering Research, 6*(2), 681–687.

Ng, K. W., Giroux, L., MacPhee, T., & Todoschuk, T. (2012). Incorporation of charcoal in coking coal blend–a study of the effects on carbonization conditions and coke quality. In: *AISTech 2012: Proceedings of the iron & steel technology conference* (pp. 225–236).

Norgate, T., & Langberg, D. (2009). Environmental and economic aspects of charcoal use in steelmaking. *ISIJ International, 49*, 587–595.

Ooi, T. C., Aries, E., Ewan, B. C., Thompson, D., Anderson, D. R., & Fisher, R. (2008). The study of sunflower seed husks as a fuel in the iron ore sintering process. *Minerals Engineering, 21*(2), 167–177.

Pardo, N., & Moya, J. A. (2013). Prospective scenarios on energy efficiency and CO_2 emissions in the European iron & steel industry. *Energy, 54*, 113–128.

Peacy, J. G. (1979). *The Iron Blast Furnace-Theory and Practice* (p. 13). Pergamon Press.

Phanphanich, M., & Mani, S. (2011). Impact of torrefaction on the grindability and fuel characteristics of forest biomass. *Bioresource Technology, 102*(2), 1246–1253.

Pohlmann, J. G., Borrego, A. G., Osório, E., Diez, M. A., & Vilela, A. C. (2016). Combustion of eucalyptus charcoals and coals of similar volatile yields aiming at blast furnace injection in a CO_2 mitigation environment. *Journal of Cleaner Production, 129*, 1–11.

POSCO. (2014). http://www.posco.kr/homepage/docs/eng2/html/sustain/environ/s91d1050150c.jsp.

Qin, L., Han, J., Ye, W., Zhang, S., Yan, Q., & Yu, F. (2014). Characteristics of coal and pine sawdust co-carbonization. *Energy & Fuels, 28*(2), 848–857.

Quader, M. A., Ahmed, S., Dawal, S., & Nukman, Y. (2016). Present needs, recent progress and future trends of energy-efficient Ultra-Low Carbon Dioxide (CO2) Steelmaking (ULCOS) program. *Renewable and Sustainable Energy Reviews, 55*, 537–549.

Quader, M. A., Ahmed, S., Ghazilla, R. A. R., Ahmed, S., & Dahari, M. A. (2015). comprehensive review on energy efficient CO_2 breakthrough technologies for sustainable green iron and steel manufacturing. *Renewable and Sustainable Energy Reviews, 50*, 594–614.

Rodd, L., Koehler, T., Walker, C., & Voermann, N. (2010). *Economics of slag heat recovery from ferronickel slags. In Sustainability for Profit. Conference of Metallurgists (COM2010)* (pp. 3–17). https://www.steel.org/~/media/Files/AISI/Making%20Steel/TechReportResearchProgramFINAL.pdf.

Suopajärvi, H., Kemppainen, A., Haapakangas, J., & Fabritius, T. (2017). Extensive review of the opportunities to use biomass-based fuels in iron and steelmaking processes. *Journal of Cleaner Production, 148*, 709–734.

Suopajärvi, H., Pongrácz, E., & Fabritius, T. (2013). The potential of using biomass-based reducing agents in the blast furnace: a review of thermochemical conversion technologies and assessments related to sustainability. *Renewable and Sustainable Energy Reviews, 25*, 511–528.

Thomas, S., McKnight, S., Serrano, E., Mašlejová, A., Želinský, R., & Tomáš, J. (2011). *Laboratory evaluation of biomass usage for coke and sinter production.* Düsseldorf (Germany): Proceedings of METEC InSteelCon.

Todoschek, T., Giroux, L., & Wing, N. K. W. (2016). Development of biocarbons for Canadian steel production. In: Biocleantech Forum *Nov. 2-3, 2016.*

Ueda, S., Watanabe, K., Yanagiya, K., Inoue, R., & Ariyama, T. (2009). Optimization of biomass utilization for reducing CO_2 in ironmaking process. *Journal of Iron and Steel Research International, 16*, 593–599.

Ueda, S., Yanagiya, K., Watanabe, K., Murakami, T., Inoue, R., & Ariyama, T. (2009). Reaction model and reduction behavior of carbon iron ore composite in blast furnace. *ISIJ International, 49*(6), 827–836.

Ueki, Y., Yoshiie, R., Naruse, I., Ohno, K.-i., Maeda, T., & Nishioka, K. (2013). Reaction behavior during heating biomass materials and iron oxide composites. *Fuel, 104*, 58–61.

ULCOS. (2004). https://cordis.europa.eu/project/rcn/80229_en.html.

Wibberley, L., Nunn, J., Scaife, P., Zivlas, A., Andrews, L., & Coomber, B. (2001). *Large scale use of forest biomass for iron and steelmaking. NSW Government Sustainable Energy Research Development Fund (SERDF) Report by BHP Minerals Technology and NSW State Forests 2001.*

World Steel Association. (2012). *Sustainable Steel: At the Core of the Green Economy, Worldsteel.*

World Steel Association. (2017a). https://www.worldsteel.org/en/dam/jcr:0474d208-9108-4927-ace8-4ac5445c5df8/World+Steel+in+Figures+2017.pdf.

World Steel Association. (2017b). https://www.worldsteel.org/en/dam/jcr:938bf06f-764e-441c-874a-057932e06dba/Sust_Steel_2017_vfinal_web.pdf.

World Steel Association. (2018). https://www.worldsteel.org/en/dam/jcr:f07b864c-908e-4229-9f92-669f1c3abf4c/fact_energy_2018_.pdf.

Yellishetty, M., Mudd, G. M., Ranjith, P. G., & Tharumarajah, A. (2011). Environmental lifecycle comparisons of steel production and recycling: sustainability issues, problems and prospects. *Environmental Science and Policy, 14*(6), 650–663.

Yokoyama, H., Higuchi, K., Ito, T., & Oshio, A. (2012). Decrease in carbon consumption of a commercial blast furnace by using carbon composite iron ore. *ISIJ International, 52*(11), 2000–2006.

Yunos, N. F. M. (2012). *High Temperature Phenomena Occurring During Reactions of Agricultural Wastes in Electric Arc Furnace Steelmaking: Interactions With Gas and Slag Phases.* PhD Thesis Sydney: University of New South Wales.

Yunos, N. F. M., Zaharia, M., Idris, M. A., Nath, D., Khanna, R., & Sahajwalla, V. (2011). Recycling agricultural waste from palm shells during electric arc furnace steelmaking. *Energy & Fuels, 26*(1), 278–286.

Further reading

Han, H., Duan, D., & Yuan, P. (2015). Exploration of straw fiber as reducing agent utilization in rotary hearth furnace process for direct reduced iron production. *Steel Research International, 86*(11), 1361–1369.

Han, H., Duan, D., Yuan, P., & Li, D. (2015). Biomass reducing agent utilisation in rotary hearth furnace process for DRI production. *Ironmaking and Steelmaking, 42*(8), 579–584.

Sustainability assessment for biomass-derived char production and applications

12

*Mejdi Jeguirim**, *Antonis A. Zorpas*‡, *Jose Navarro Pedreno*§, *Lionel Limousy**, *Pantelitsa Loizia*‡,¶, *Marinos Stylianou*†,‖, *Agapios Agapiou*†
*University of Strasbourg, University of Upper Alsace, Institute of Materials Science of Mulhouse (IS2M—UMR CNRS 7361), Mulhouse, France, †University of Cyprus, Department of Chemistry, Nicosia, Cyprus, ‡Open University of Cyprus, Faculty of Pure and Applied Sciences, Environmental Conservation and Management, Laboratory of Chemical Engineering and Engineering Sustainability, Nicosia, Cyprus, §Department of Agrochemistry and Environment, University Miguel Hernández of Elche, Alicante, Spain, ¶I.E.S.T—EnviTech Ltd, (Institute of Environmental Technology and Sustainable Development), Department of Research and Development, Paralimni, Cyprus, ‖University of Cyprus, Department of Civil & Environmental Engineering, NIREAS-International Water Research Center, Subsurface Research Laboratory, Nicosia, Cyprus

Chapter Outline

Char and Carbon Materials Derived from Biomass. https://doi.org/10.1016/B978-0-12-814893-8.00012-2

12.1 Biomass-derived chars: Circularity, classification, definitions, and brief history

12.1.1 Circularity: Carbon cycle

As a systemic description, living systems are considered and characterized by a stable organizational structure if rapidly developed and if the structure undergoes change through a continuous flow of energy and matter (Masullo, 2014). Generally, these are considered to be open systems related by massive ecological and natural nets where the wastes of one are nourishment for the others. In the ecological nets, plants and systems have a vital role in connecting not-living with living systems. Plants absorb water and minerals from the soils, use solar energy, and absorb CO_2 from the atmosphere, and as a result create sugars, proteins, and other organic compounds. Plants are the real producers of the bricks of life, being the links between living and lifeless organized systems. The stream of nutrients that declare an ecosystem is not always consistent and systematic, but usually indicates repeated movements of material and energy among its parts in a typical circular movement.

A typical example of circularity is the cycle of carbon (circularity as a definition is used to describe how close an "object" should be to a true circle). The carbon cycle is represented by big tanks, atmospheres, oceans, carbonate rocks, soils, and biomass. The continual movement of carbon from one area to another is the basis for the cycle because carbon is crucial for all life on earth, considering that all living organisms are made up of carbon. Carbon is found in the atmosphere as CO_2, in the lithosphere in the form of carbonate rocks, in the biosphere stored in plants and trees, and finally in the hydrosphere dissolved in oceans and lakes. Moreover, carbon is also found in fossil fuels and soils from dead organisms and wastes (http://www.ei.lehigh.edu/learners/cc/pdf/CarbonCycle_PrintVersion.pdf).

12.1.2 Biomass conversion

Organic materials originating from plant and animal resources as well as wastes generated from their transformations are referred as biomass. Biomass covers lignocellulosic materials, including waste wood, agricultural residues, as well as non-lignocellulosic materials such as food processing residues, animal excrement, municipal solid wastes, and sewage sludge. These biomass feedstocks could be transformed into high value-added products, including fuels, materials, and chemical compounds.

Among the conversion techniques, pyrolysis is a thermochemical process converting biomass into liquid (bio-oil), gaseous (syngas), and solid (char) products. Syngas and bio-oil are considered as alternative fuels. An extensive research is currently conducted on their formation, upgrading, and applications. The solid product could be used for a variety of purposes, such as soil amendment, water and gas treatment, and energy production. Currently, several investigations are being conducted to identify new sustainable applications such as gas storage, catalyst, and supercapacitor synthesis.

12.1.3 Biomass char classification

Several designations of biomass-derived chars are available in the literature. The International Biochar Initiative (IBI, http://www.biochar-international.org/biochar) called the solid product obtained during biomass thermochemical decomposition *biochar*. The IBI generalized biochar as "a solid material obtained from the carbonization of biomass." Lehmann and Joseph (2009) used the term biochar and defined it as "the carbon-rich product obtained when biomass, such as wood, manure or leaves, is heated in a closed container with little or no available air. In more technical terms, biochar is produced by so-called thermal decomposition of organic material under limited supply of oxygen (O_2), and at relatively low temperatures (<700°C)." However, the authors stated that the term "biochar" should unambiguously designate biomass-derived chars that are used to improve soil characteristics. Brown, Wright, and Brown (2011) and Shackley, Hammond, Gaunt, and Ibarrola (2011) mentioned also the term biochar during pyrolytic char applications to soil. In particular, Brown et al. (2011) defined biochar as "a carbon-rich material capable of resisting chemical and microbial breakdown, allowing the carbon to be sequestered for periods of time approaching hundreds or thousands of years." Shackley et al. (2011) described biochar as the "porous carbonaceous solid produced by thermochemical conversion of organic materials in an oxygen-depleted atmosphere that has physiochemical properties suitable for the safe and long-term storage of carbon in the environment and, potentially, soil improvement."

For other applications (not soil application), Lehmann and Joseph (2009) used the term *charcoal* when pyrolytic char is used as fuel for heat, as a filter for aqueous and gaseous effluents treatment, and as a reductant in ironmaking. Hagemann et al. (2018) used the term *pyrogenic carbonaceous materials* (PCMs) to describe materials produced by different thermochemical conversion processes and containing some organic carbon. They have classified PCMs in biochar, charcoal, and activated carbon. These materials have similarity in terms of elemental composition and main chemical bonds, but each one has specific properties and applications as shown in Fig. 1. This latter aims (1) to highlight various options for the combination of the different thermochemical conversion processes with each other as well as with nonthermal pre- and post-treatments; (2) to deliver a synopsis on feedstock and applications; and (3) to see whether a PCM can be considered as a biochar based on the cycle of carbon circularity. Such classification depends on the envisioned application (nonoxidative for carbon sequestration), sustainable production (sustainably sourced biomass, no fossil

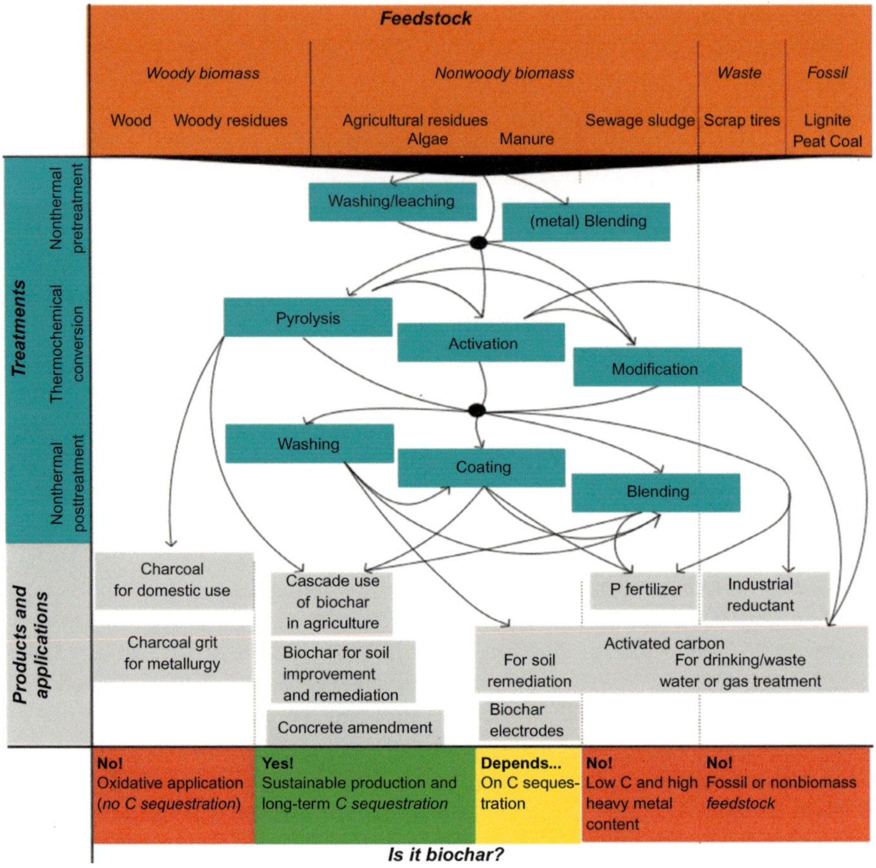

Fig. 1 Overview of feedstock *(orange)*, treatments for production *(blue)*, and applications *(gray)* of pyrogenic carbonaceous materials (Hagemann et al., 2018).

feedstock), and chemical properties (carbon content, e.g., >50% or >10%, and a low content of pollutants as detailed in the guidelines of the European Biochar Certificate or the IBI).

In addition to the three above materials, Hagemann et al. (2018) proposed two PCMs, including natural pyrogenic matter and other solid products obtained by pyrolysis. However, the authors do not include in their classification carbonaceous materials produced in the liquid phase, such as hydrochar (hydrothermal carbonization (HTC) product) and in the gas phase, such as carbon black and soot.

12.1.4 Biochar

The discussion on biochar comes from the investigation of highly fertile anthropogenic soils rich in pyrogenic carbon such as the Amazonian dark earth (Fig. 2). "The evolution of terra preta into biochar is a bizarre and intriguing story."

Fig. 2 Black carbon presented in a paleosoil of about 5 million years ago.
From Navarro Pedreño, J., & Mataix, J. (2014). An approach to knowledge of the paleosols of the province of Alicante. *Cidaris*, *32*, 21–31.

In 1495 Jaume Ferrer de Blanes sent a letter to inform Christopher Columbus about his findings in the New World: the relation between "great and valuable things" and "hot regions inhabited by darker skinned peoples." In 1542 Francisco de Orellana referred to the term "terra preta" indicating that "there were many roads entering the interior of the land…the land is a fertile and as normal in appearance as our Spain." In 1670 John Philip Bettendorff, a German Jesuit missionary, provided the earliest known use of the term "terra preta." In Brazil in 1867 Ballard S. Dunn talked about the quality of the lands indicating that "The character of the soils is usually called with us 'mulatto' and its depth from eight inches to five feet." Also, in 1879 *Scribner's Monthly* mentioned the "rich terra preta, black land, the best on the Amazons. It is a fine dark loam a foot and often two feet thick" (Bates, 2011). In the same year a naturalist called Herbert H. Smith stated that "the bluff land owes its reaches to the refuge of a thousand kitchens for maybe a thousand years." This result, reinforced by geologist William Katzer's early 20th century analysis of soil composition—a blend of mineral residuum, charred plant materials, and decomposed organics—started turning heads. Archeological investigations have confirmed the connection between the situation of the terra preta sites and the civilizations Orellana described back in the 16th century (Glaser, Haumaier, Guggenberger, & Zech, 2001).

This brief history confirms that biochar is a strategic agent converting poor soil into highly productive agroecosystems (Glaser et al., 2001). Nowadays, biochar use is not limited to the direct application of biochar to soil and several promising issues have been examined (Fig. 1). In particular, biochar can be used as a feed supplement to improve ruminant productivity and to reduce odors and nutrient losses from manure. In addition, biochar serves as a slow-release fertilizer when mixed with manure or compost applied to soil (Joseph et al., 2015).

Several applications of chars derived from biomass are emerging nowadays. These applications include their use for syngas production through the gasification process, as a catalyst support, and for supercapacitor production. However, we prefer to use the terms "chars" and "biomass-derived chars" keeping the term biochar for agricultural applications.

12.1.5 Charcoal

Charcoal is defined as carbonized wood used mainly as fuel for cooking and heating or as a reductant in the iron and steel industry. Charcoal is the oldest chemical product: the Tyrolean iceman was tattooed using charcoal as a pigment, probably for medical purposes, more than 5000 years ago (Spindler, 2013). During the last few centuries, the main motivation for research into charcoal was gunpowder production, which contains 15% charcoal (Gray, Marsh, & McLaren, 1982). The main investigations concerned the search for suitable production technologies allowing a clean combustion process.

Charcoal is mainly produced using traditional methods such as earth mounds. These techniques do not cover the generated exhaust gases that contain carbon monoxide, methane, volatile organic compounds, and fine particulate matter (PM). During the last few decades, charcoal has been produced using modern technologies such as industrial retorts that valorize all the generated products.

According to the United Nations Food and Agriculture Organization, in 2015 50 million tons of charcoal were produced from wood globally. Two-thirds of worldwide production is situated in Africa where charcoal is still used as cooking and heating fuel. However, a significant part is also exported to Europe where it is used mainly for barbecues. In fact, industrial processes use fossil coal because of its lower costs. However, in Brazil, charcoal is also used in iron and steel industries due lower local price and low ash content.

12.1.6 Activated carbon

Activated carbon was defined as "any form of carbon capable of adsorption" (Schanz & Parry, 1962). Historically, the use of charcoal as a sorbent for water purification dates back to the Roman Empire. However, charcoal optimization for specific contaminants removal occurred many years later. In 1862, Smith detected that charcoal adsorbs oxygen from air after a long period. However, the adsorption capacity was different between different tested charcoals. Such observation was the starting point to correlate charcoal characteristics to their sorption capacities. Therefore, investigations started in the 19th century to improve charcoal properties. In particular, activation strategies were implemented to develop new surface areas. Hence, activated carbon science began and the first patents were obtained in the 1920s (Chaney, 1924). However, at this stage, there was no clear explanation for the increased sorption capacities after the activation process. The main investigations correlated the adsorption capacities to activation conditions and assumed an increased surface area. In the 1950s, activation was defined as "any process which selectively removes the hydrogen or hydrogen-rich fractions from a carbonaceous raw material in such a manner as to produce an open, porous residue" (Lewis & Metzner, 1954). Chemical and structural modifications leading to increased surface areas and tailored surface chemistry of the activated carbon are the key factors affecting the adsorption capacities (Belhachemi, Jeguirim, Limousy, & Addoun, 2014).

12.2 Char production pathways

Char production and applications involve three main steps, presented previously in Fig. 1: (1) feedstock collection and preparation, (2) feedstock conversion, and (3) product management and applications. Several combinations of biomass resources, conversion processes, and applications were proposed in the literature. These investigations were mainly performed at the laboratory scale producing small amounts of chars. In particular, different feedstocks were converted to chars under different operating conditions such as temperature, heating rate, and residence time. This research resulted in interesting information on the effect of these operating conditions on char yield and properties. However, due to the small scale, these studies are not very useful to evaluate the sustainability of char production pathways through life-cycle and cost analysis.

Recently, larger pilot-dimensioned facilities have been developed to produce significant amounts through different conversion techniques. These facilities could be more useful for the evaluation of environmental and economic issues associated with char production. However, these investigations do not focus on feedstock collection and pretreatment or char transport and utilization. Furthermore, the char properties are variable and suitable applications depend on soil type or available technologies. Therefore, each combination feedstock, conversion process, and application is a unique situation inducing complexity in the selection of a sustainable strategy.

Due to such complexity, sustainability assessment, in the following sections, considers biomass resources, conversion techniques, and char applications separately. The main purposes are to identify the most promising feedstocks for char production, the optimal conversion scheme, and suitable applications from an environmental and economic standpoint.

12.2.1 Biochar feedstocks

Various feedstocks have been used for char production from different types and sectors, including agricultural and industrial waste, household and garden waste, livestock waste, and algae (Table 1). Three main categories can be identified, according to the production method, as well as transportation and environmental and economic costs:

1. *Energy crops and algae* (such as poplar, black locust, eucalyptus, *Miscanthus*, giant reed, kenaf, sorghum, microalgae). These are specifically cultivated for bioenergy purposes. The cost of energy and the environmental issues associated with this category are important considerations. In particular, significant energy is consumed for the production (plantation,

Table 1 Examples of biochar feedstock material.

Category	Origin	Biomass feedstock	References
Energy crops and cultivated algae	Agricultural biomass	*Miscanthus*, *Arundo donax*, kenaf	Jeguirim, Dorge, Loth, and Trouvé (2010)
	Algae	Marine macroalgae (*Saccharina japonica* and *Sargassum fusiforme*)	Poo, Son, Chang, et al. (2018)
		Algae biomass of *Spirulina platensis*	Nautiyal, Subramanian, and Dastidar (2016)
		Freshwater microalga *Chlorella* sp.	Shen, Li, Zhu, et al. (2017)
Residual biomass	Agricultural waste	Waste wood, corn cob and corn stalk, wheat residues, rice straw, sugar beet tailing	Xie, Reddy, Wang, et al. (2015)
		Date palm residues	Jeguirim, Elmay, Limousy, Lajili, and Said (2014)
		Corn cobs, woodchips	Micháleková-Richveisová, Frišták, Pipíška, et al. (2017)
		Oak wood, bamboo woods, maize residue, soybean stover, peanut shells	Jung, Hwang, Ahn, and Ok (2015)
		Luffa cylindrica sponge	Liatsou, Pashalidis, Oezaslan, and Dosche (2017)
		Opuntia ficus cladodes, *Opuntia ficus indica* (cactus plant)	Hadjittofi, Prodromou, and Pashalidis (2014), Hadjittofi, Charalambous, and Pashalidis (2016)
		Pine needles	Chen, Zhou, and Zhu (2008)
Industrial and farming residues	Household and garden waste	Garden wood waste	Micháleková-Richveisová et al. (2017)
	Industrial waste	Oak sawdust	Wang, Guo, Shen, et al. (2015)
		Sugar cane bagasse	Abdelhafez and Li (2016)
		Orange peel	Abdelhafez and Li (2016), Chen, Chen, and Lv (2011)
		Giant reed straw	Hou, Huang, Yang, et al. (2016)
		Tire rubber	Lian, Huang, Chen, et al. (2011)

Table 1 Continued

	Livestock	Composted swine manures	Rosales, Meijide, Pazos, and Sanromán (2017)
		Pig manure	Gascó, Paz-Ferreiro, Cely, et al. (2016)
		Poultry manure	Batool, Idrees, Hussain, and Kong (2017)
		Cattle manure Chicken manure Pig slurry	Cely, Gascó, Paz-Ferreiro, and Méndez (2015)
		Swine-separated solids, paved feedlot manure, dairy manure, poultry litter, turkey litter	Cantrell, Hunt, Uchimiya, et al. (2012)

harvesting) of perennial grasses, tree crops, and microalgae. Furthermore, for perennial grasses and tree crops, the agricultural land requires plowing, which my lead to the loss of stored soil carbon. Such carbon lost should be considered in the carbon footprint calculation. In contrast, for algae cultivation, significant water consumption should be taken into account for carbon balance.

2. *Residual biomasses* (such as olive and grape prunings, wheat straw, corn and tobacco stovers, forestry residues, etc.). This category requires only transportation cost since production costs are assigned to the raw product. However, a fundamental distinction exists between geographically concentrated feedstocks and those that request a collection plan. In fact, dispersed feedstocks are costly and lead to an increase in gaseous emissions related to transport and therefore to the decrease in carbon benefit for char production and applications.
3. *Industrial and farming residues* (pomace and vinasse, whey, meat and bone meal, tallow, skins, shells and stones, rice and grain husk, cooking oils, swine and cow manure, poultry litter and feathers, etc.). These are usually available as a free cost biomass directly at the production site (or even as an avoided cost, when considered as waste to be disposed of). Therefore, this category is the most interesting biomass since it does not generate either production or transportation costs. In some cases, producers pay handling and disposal costs and the conversion of the wastes to char may generate an economic advantage. As for biomass residues, concentrated industrial and farming residues will have more economic advantages compared to dispersed residues.

In addition to the economic and energetic advantages, the redirection of agricultural and industrial residues from land spreading to char production could decrease their decomposition into methane. In fact, food processing residues, improperly composted or land spread, emit huge amounts of methane. These undesirable emissions could be prevented when these wastes are converted to char. In this context, Table 2 displays a preliminary assessment of the environmental and energetic issues related to the various biomass resources used for char production.

Table 2 Environmental and energetic issues with biomass feedstocks.

	Land-use conversion involved	**Cultivation and harvest required**	**Already concentrated**	**Drying needed**	**Preparation required**	**Other**
Energy crops and algae						
Perennial grasses	Y	Y	N	Y	Y	Biodiversity loss to monoculture
Tree crops	Y	Y	N	Y	Y	Biodiversity loss to monoculture
Algae	N	Y	Y	Y	N	Water use
Residual biomass						
Logging residue	N	Y	N	Y	Y	Habitat reduction, avoided methane
Forest products processing	N	N	Y	Y	Y	Avoided methane
Crop residue	N	Y	N	Y	Y	Increase soil erosion, decrease tilth
Industrial and farming residues						
Urban wood residue	N	N	N	Y	Y	Avoided methane
Animal manure	N	N	Y	Y	N	Avoided methane Avoided nutrients
Municipal waste	N	N	Y	N	Y	Avoided methane
Food processing waste	N	N	Y	Y	N	Avoided methane
Sewage sludge	N	N	Y	N	Y	Avoided methane

12.2.2 Char conversion techniques and environmental impact

Conversion systems are applied to transform biomass feedstocks into solid residue (char). Conversion techniques can be categorized into four types: (1) slow pyrolysis, (2) intermediate/fast pyrolysis, (3) gasification, and (4) HTC, as detailed in Table 3. These techniques are different in terms feedstock suitability and operating conditions (temperature, pressure, heating rates, residence time) resulting in different product yields and properties.

Among the different processes for char production, slow pyrolysis is the preferred one for biomass feedstock carbonization since it is characterized by slow heating rates, long residence time, and low temperature (300–600°C). These operating conditions maximize the char yield (Guizani, Jeguirim, Valin, Limousy, & Salvador, 2017). In contrast, fast pyrolysis aims to maximize the production of liquid fuels called bio-oil. The pyrolysis process also produces fuel gas (mostly CO and CO_2) and char, which could provide process heat and generate electricity if a combined heat and power system is installed. In addition to pyrolysis, undesired char could be produced during the gasification process. In fact, during this conversion technique, an oxidizing atmosphere (air, O_2, H_2O, CO_2, or their mixtures) is applied leading to the partial combustion of biomass feedstock. The product gases include a mixture of CO, CO_2, CH_4, and H_2. These gaseous products can be used directly as a fuel or converted to synthetic natural gas (syngas). However, in addition to a gas yield of approximately 85%, a small char yield is obtained, which could also be used for various applications.

Recently, there has been a growing interest in char production using HTC. This technique is a chemical process more suited to biomass feedstocks with higher moisture content, e.g., food processing residues (vinasse, olive mill waste water), animal manures, and algae. Recent investigations show that hydrochars with high yields could be produced through HTC at low temperatures (~200°C) and short processing times (<12h). During the dry thermochemical conversion processes (pyrolysis or gasification), energy is consumed for biomass drying leading to an increase in the operating cost. Furthermore, gaseous and PM emissions are generated from the pyrolysis

Table 3 Comparison of reaction conditions and typical product yields for thermochemical conversion processes with char as a product.

Process	Reaction conditions (temperature (°C); vapor residence time)	Production distribution (wt%)		
		Char	Liquid	Gas
Slow pyrolysis	~400; hours to weeks	35	30	35
Intermediate pyrolysis	~500; ~10–20 s	20	50	30
Fast pyrolysis	~500, ~1 s	12	75	13
Gasification	~800; ~10–20 s	10	5	85
Hydrothermal carbonization	~180–250; no vapor residence time; ~1–12 h processing time	50–80	5–20 (dissolved)	2–5

and gasification system leading to negative impact and a request for a costly effluent treatment system. Therefore, HTC has important advantages because it decreases drying and pollutant treatment costs. However, HTC has not significantly progressed to the pilot-scale stage and the performance of hydrochars produced using the HTC technique has not been widely evaluated for several applications such as soil amendments. The current available technology for char production as well as its suitability is analyzed in the following section based on the work of Lohri, Rajabu, Sweeney, and Zurbrügg (2016).

12.2.2.1 Overview of available technologies for char production

During the last century, char production was realized mainly using traditional methods. Char production occurs in earth pits or kilns where wood is ignited and covered allowing its carbonization. These kilns are low cost, simple to construct, and can be applied anywhere. However, the char yield is low (10%) and the gaseous products are emitted to the atmosphere. Therefore, modern technologies for char production emerged in the 20th century. A classification of the available carbonization technology ranging from small-scale, low-cost pyrolyzers to more modern systems is presented in Table 4. This classification is based on the reactor type, operation type, scale, construction material, conversion efficiency, emissions, and auxiliary requirements.

12.2.2.2 Sustainability assessment

To assess the sustainability of char production, different technologies were evaluated by Lohri et al. (2016) using technical, financial, and environmental criteria. The technical criterion is based on biomass feedstock suitability (moisture, particle size), feedstock pretreatment (drying, grinding), carbonization capacity, system portability, labor intensity, carbonization controllability (temperature, residence time, energy consumption), lifespan, char yield, and demonstrated use (technical functionality validation). The financial criterion is based on capital cost, operating cost, and gas recovery (heat or other applications). The environmental criterion is based on pollutant emission, tar recovery, and water requirement. Further details on these criteria are found in Lohri et al. (2016). The authors assessed the different carbonization technologies based on the literature review and their experience, and have proposed a five-point scale (+2: much better than, +1: better than, 0: equal to, −1: worse than, −2: much worse than) to evaluate the different technologies with the baseline one. Drum reactors were chosen as the baseline technology since it is the carbonization technology widely used for agricultural waste. Weights were ascribed to designate the importance of each criterion for overall technology sustainability (3: high importance, 2: medium importance, 1: low importance). Total scores were obtained by multiplying weights and scores. The obtained values allowed the different technologies to be ranked. Table 5 shows the overall sustainability assessment matrix.

The low-tech retorts obtained the highest weighted score. Therefore, this technology was ranked first in terms of overall sustainability for biomass feedstock

Table 4 Classification and important characteristics of carbonization technologies (Lohri et al., 2016).

Reactor type	Process type	Capacity	Construction materials	Conversion efficiency (mass%)	Energy source	Residence time	Emissions (g/kg char)	Auxiliary	Portability/ permanence	Capital cost
Earthen kilns Earth pit, mound	Batch	10–330 m^3	Soil, sod	90 kg char/ m^3 wood	Partial oxidation	20 days/ 180 m^3	CO_2: 1058–3027; CO: 143–333; CH_4: 32–62; TSP: 13–411	None	Impermanent	\$27/ton charcoal
Brick kilns Brazilian beehive	Batch	50–330 m^3	Brick, mortar	90 kg char/ m^3 wood	Partial oxidation	20–30 days/ 270 m^3	CO_2: 1533; CO: 373; CH_4: 57	None	Stationary	\$150–1500
Metal kilns Missouri	Batch	350 m^3	Steel, brick/ concrete	25%–36%	Partial oxidation	80 h	CO_2: 543–560; CO: 140–162; CH_4: 37–54; TSP: 160	Tar recovery	Stationary	\$15,000
Drum reactors Vertical (D-Lab, ARTI)	Batch	200 L	Mild steel	19%–30%	Partial oxidation	0.5–4 h, 1 day	CO_2: 1517; CO: 336; CH_4: 58; TSP: 4.2	None	Portable	\$13–61/ ton charcoal
Low-tech retorts Adam (Improved Charcoal Production System)	Batch	3 m^3	Brick or earth blocks	30–42 (dry basis)	Partial oxidation and volatile combustion	12 h	n/a	None	Stationary and portable version	€300

Continued

Table 4 Continued

High-tech retorts Carbo twin retort	Batch	$2 \times 5\,m^3$	Steel	30–33	External heat and volatile combustion	32–36h	Complies with Dutch emission standards	Oil burner, fork lift, hoist and rail, sand lock, exhaust gas recirculation	Stationary	€1 million
Flash carbonization HNEI flash carbonization	Batch	594 tons/year	Steel vessel and piping components	30–50	Partial combustion	20min	n/a	Compressed air source, electronic ignition	Stationary	€180/ton charcoal
Hydrothermal carbonization by AVA-CO_2	Batch	2664 tons/year char produced	Steel vessels and piping components	37–60	Steam	5–10h	n/a	Mixing tank high-pressure reactors, buffer tank	Stationary	€10–12 million

n/a, not available; *TSP*, total suspended particles.

Table 5 Technology assessment matrix (drum reactor as baseline technology).

Assessment criteria	***Weight***	**Reactor**							
		Earthen pit/ mound	**Brick kiln**	**Metal kiln**	**Drum reactor (baseline)**	**Low-tech retort**	**High-tech retort**	**Flash carbonizer**	**Hydrothermal carbonization reactor**
Technical aspects									
Suitability for biowaste	*3*	−1	−1	−1	0	0	0	0	+2
Feedstock pretreatment	*2*	−1	−1	−1	0	0	−1	−1	+2
Throughput	*2*	−2	−1	−1	0	0	+2	+1	+1
Portability	*2*	−2	−2	−1	0	−1	−2	−1	−1
Labor intensity	*2*	−1	+1	+1	0	0	−1	−1	−1
Controllability	*2*	−2	−1	−1	0	+1	+2	+2	+2
Conversion efficiency	*2*	0	0	0	0	+1	+1	+2	+2
Lifespan	*2*	−2	+2	0	0	0	+2	+2	+1
Demonstrated use	*2*	+1	0	0	0	−1	−1	−2	−2
Financial aspects									
Capital cost	*3*	+1	−1	−1	0	−1	−2	−2	−2
Operating cost	*3*	+1	0	0	0	0	−2	−2	−2
Gas recovery	*2*	0	0	0	0	+1	+1	+1	+1

Continued

Table 5 Continued

Environmental and health aspects									
Pollutant emissions	*3*	0	0	0	0	+1	+1	+1	+1
Tar recovery	*1*	−1	0	0	0	+1	+1	+2	+2
Water requirement	*2*	0	0	0	0	0	0	0	−2
Safety	*3*	−1	+1	0	0	+1	+1	−2	−2
Total weighted score		**−19**	**−7**	**−12**	**0**	**+6**	**+1**	**−7**	**−1**
(Overall ranking)		*(7)*	*(5)*	*(8)*	*(3)*	*(1)*	*(2)*	*(5)*	*(4)*

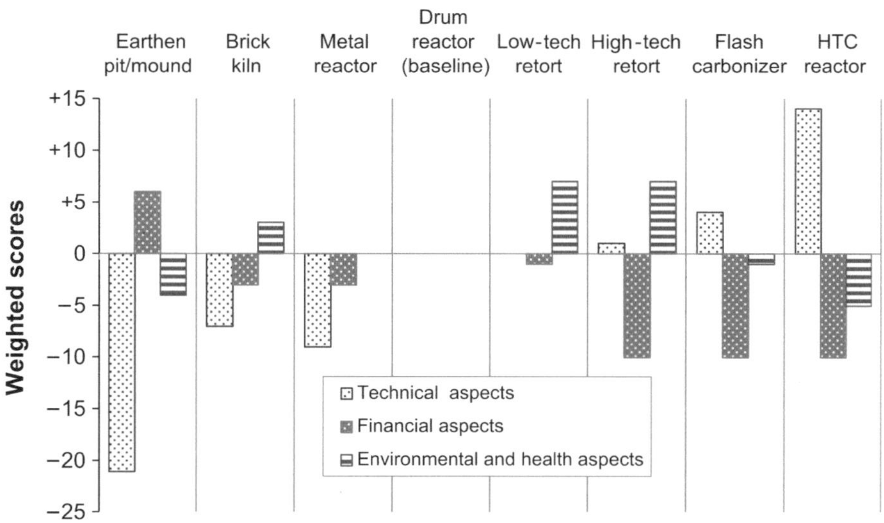

Fig. 3 Assessment results by criteria categories: Technology suitability compared to drum reactor.
Elsevier, Copyright 2018.

carbonization, followed by high-tech retorts. Drum reactors were ranked third, while metal kilns and earthen pit/mounds obtained the lowest weighted score. Furthermore, Fig. 3 shows the sustainability of the different technologies from the technical, financial, and environmental criteria. The summed results of each criterion are expressed as weighted score differences in comparison to the baseline (drum reactor). This comparison indicates that the high-tech systems (retorts, flash pyrolysis, HTC) score positively from a technical point of view. However, these technologies receive remarkably negative scores in the financial criterion. In contrast, earthen pit/mounds have a low technical score followed by metal reactors.

12.3 Biochar for agricultural application

12.3.1 Soil properties improvement and combating climate change

Biochar has been proven to increase soil organic carbon (SOC), water retention, and nutrients availability over a longer time period and to sequester carbon. However, biochar affects soils and crops differently. Both of these factors have hampered biochar's marketability for widespread use. Biochar can be added to a soil as a possible strategy to store carbon in soils. However, this addition is a nonnatural increment of soil organic matter and exceeds the soil potential for carbon sequestration from plant biomass. This means that this artificial increment of SOC can affect the soil properties and vary the natural process of organic carbon sink. Even more, soil properties can determine the effectiveness of this addition and the persistence or not along time of this organic carbon forms added as biochar.

The use of biochar as a soil additive has been proposed as a means to simultaneously mitigate anthropogenic climate change while improving agricultural soil fertility (Woolf, 2008). Both effects should be considered positive nowadays, especially due to the growing scarcity of organic matter in all the agricultural soils and the negative effects associated with the presence of greenhouse gases (GHGs) in the atmosphere. However, some works indicated that there was no evidence of changes in crop yield or quality in the short term (Jay, Fitzgerald, Hipps, & Atkinson, 2015), therefore long-term studies with repeated biochar applications are needed with commercial crops.

Soil properties are changed when adding biochar. Many authors indicated that biochar acts as a soil conditioner enhancing soil physical, chemical, and biological properties. Most of them documented that biochar can perform as a soil conditioner helping plant growth by retaining nutrients and by providing other services such as soil physical and biological properties (Glaser, Lehmann, & Zech, 2002; Haddad et al., 2017; Lehmann et al., 2003; Lehmann & Rondon, 2005). Biochar could improve soil physical and physicochemical properties by increasing soil pH and cation exchange capacity, and enhancing nutrient retention (Shin et al., 2014). The chemical surface properties of the biochar determine the possibilities of retaining cations and nutrients like Ca or Mg. Moreover, biochar can facilitate the liberation of N, S, and P. However, decomposition is not a good requirement if the aim is to maintain organic carbon storage in the soils.

Other possible benefits include reduction of nitrate leaching (Singh, Hatton, Singh, Cowie, & Kathuria, 2010) although nitrate has a high solubility and pH plays an important role. In this sense, the adsorption of contaminants, such as As and Cu from soils, was also reported (Beesley, Moreno-Jimenez, & Gomez-Eyles, 2010) as was the reduction of trace gas emissions from soils (nitrous oxide and methane) (Clough, Bertram, Ray, & Condron, 2010). Table 6 indicates the major properties affected by the addition of biochar in soils.

Table 6 Soil properties affected by the application of biochar.

Properties	Expected result
Retaining nutrients	The possibility of a chemical surface interaction with cations and avoiding their leachate
Increasing cation exchange capacity	An increment of the specific surface of soils to interact with dissolved substances
Source of nutrients	Decomposition of biochar can facilitate the bioavailability of several nutrients, including N, S, and P
Modifying soil pH	The soil pH can be altered and acidified by using biochar. This can also be reflected in other soil properties
Retaining of pollutants	Nitrate or trace elements can also interact with the biochar. This interaction depends on the soil pH

12.3.2 Assessment of biochar using lifecycle analysis and SWOT analysis

The Society for Environmental Toxicology and Chemistry (SETAC) defined the concept of lifecycle analysis (LCA) during 1999 (Azapagic, 1999). LCA methodology, as defined by SETAC and from ISO (International Organization for Standardization), includes four specific steps (De Feo & Malvano, 2009; Goudouva and Zorpas, 2017), which include the goal and scope definition, inventory analysis, impact assessment, and improvement/monitoring assessment. LCA is a very useful tool to measure the environmental impact of any processing like solid waste management (De Feo & Malvano, 2009), mining activities (Goudouva & Zorpas, 2017), etc. LCA (also known as ecobalance) is a standard technique (ISO 14040: 2006 series) to assess environmental impacts associated with all stages of a product's life from cradle to grave (i.e., from raw material extraction through materials processing, manufacturing, distribution, use, repair and maintenance, and disposal or recycling). LCA is also applied to measure the impacts of biochars in the environment as indicated from Biederman and Harpole (2013). The LCA approach measures the aquatic system of the application of biochars. Also, Homagain, Shahi, Luckai, and Sharma (2014) used LCA methodology to evaluate the energy consumption of the production of bioenergy from biochar land application. They found that the application of biochars to land consumed 4847.6 MJ/t of dry feedstock, which was more energy than conventional systems, but the GHG emissions reduced by almost 69 kg CO_2.

Roberts, Loy, Joseph, Scott, and Lehmann (2010) estimated the energy economy and climate potential of biochars through LCA (Fig. 4). They found that in each feedstock that was evaluated, the net energy of the system was positive, which meant that more energy is generated than consumed and the net energies of 1 dry ton of late stover, early stover, switchgrass, and yard waste were +4116, +3044, +4899, and +4043 MJ, respectively. The additional syngas heat energies produced per ton of feedstock were +4859, +4002, +5787, and +3507 MJ for late stover, early stover, switchgrass, and yard waste, respectively. Furthermore, for climate change impacts, net negative GHG emissions imply more CO_2e reductions than emissions. The net GHG emissions for late stover, early stover, and yard waste were −864, −793, and −885 kg CO_2e/t dry biomass. Of all of the biomass sources, the yard waste system resulted in most GHG emission reductions per functional unit, primarily because there were no emissions associated with yard waste production or collection but only for transport. Moreover, the economic analysis shows that the uncertainty in the value of sequestered CO_2e creates a large variability in net profitability. Each feedstock shown has a high and low revenue scenario, according to an \$80/t CO_2e versus \$20/t CO_2e GHG offset value. The high revenue of late stover (+\$35/t stover) indicates a moderate potential for economic viability.

On the other hand, Homagain et al. (2014) using LCA techniques (Fig. 5) evaluated the cost of biochar for bioenergy production using as feedstock forest harvest residue, saw mill residue, and underutilized trees. The results indicated that a 25-year average annual cost inventory of the biochar-based bioenergy system of a 1 MWh plant showed that the cost of pyrolysis (\$381,536 per year) was the most expensive stage

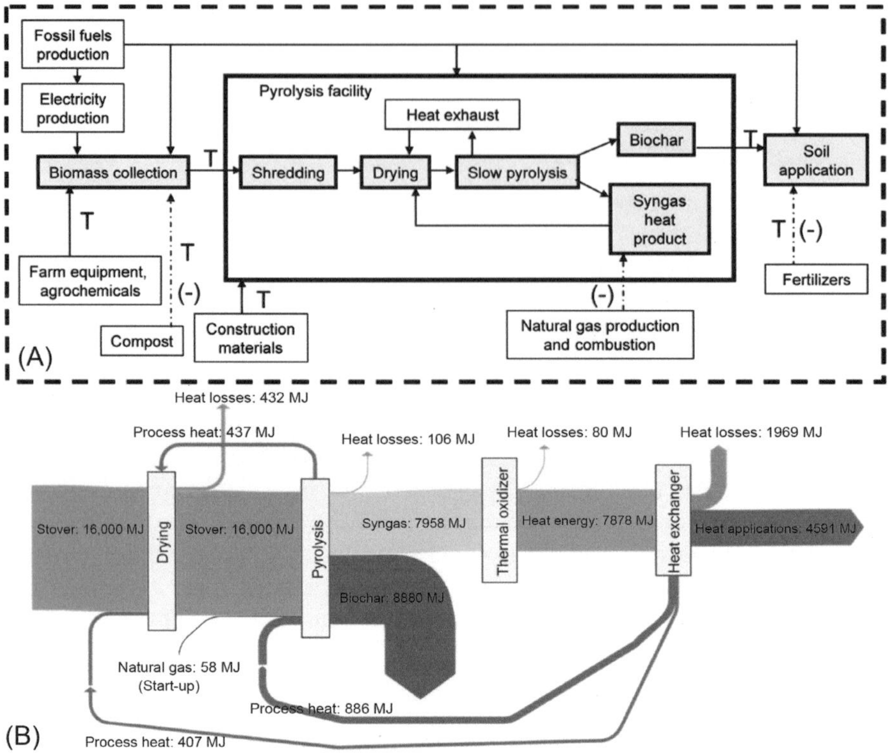

Fig. 4 (A) System boundaries for the LCA of a biochar system with bioenergy production are denoted by the *dashed box*. *Dashed arrows with (-)* indicate avoided processes. The "T" represents transportation. The avoided compost process applies to the yard waste scenario only. (B) Energy flows (MJ/t dry feedstock) of a pyrolysis system for biochar with bioenergy production using the late stover functional unit.
American Chemical Society, Copyright 2018.

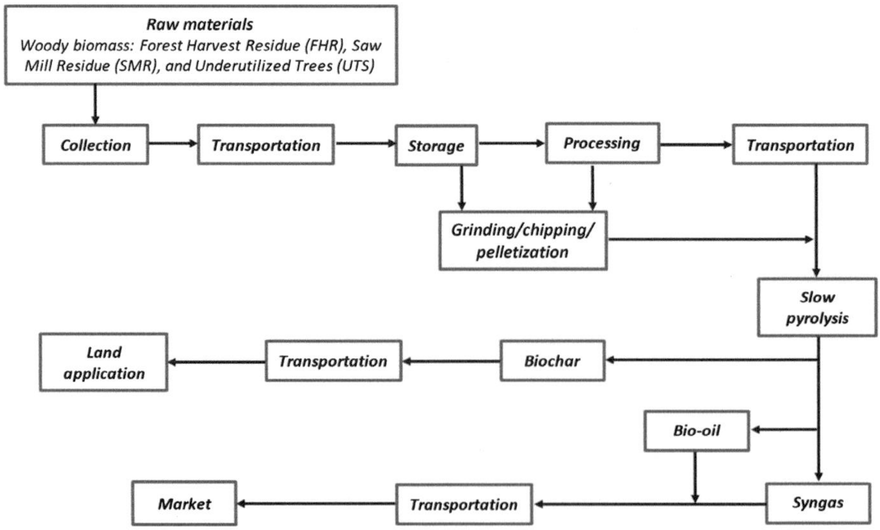

Fig. 5 System boundary for Life Cycle Cost Analysis (LCCA) of biochar-based bioenergy. Springer Nature, Copyright 2018.

of production followed by storage/processing ($237,171 per year), which includes palletization. The transportation costs for low and high availability were $97,962 per year and $83,268 per year, respectively. Additionally, the total inventory cost was added to estimate the total annual cost of plant operation and was approximately $875,000 for biochar (low), $850,000 for biochar (high), $1,100,000 for land application (low), and $1,050,000 for land application (high). Total annual plant operation cost was high in land application with low feedstock availability followed by land application with high feedstock availability, biochar with low feedstock availability, and biochar with high feedstock availability. The average annual cost of operation from all scenarios was $988,550 with a present value of $532,816.

SWOT methodology is a strategic analysis tool that combines the strengths and weaknesses study of an organization, territory, or sector with the study of opportunities and threats in its environment (Fertel, Bahn, Vaillancourt, & Waaub, 2013). SWOT analysis derives its name from the words strengths (S), weaknesses (W), opportunities (O), and threats (T). Strengths are positives and weaknesses are negatives related to system internal factors. Opportunities are external factors that have a positive interaction with the system and the negative effects of the system environment represent threats to the system (Srdjevic, Bajcetic, & Srdjevic, 2012). According to Islam and Mamun (2017) SWOT analysis is an effective tool, able to support the policymakers. SWOT analysis has been used by Zabaniotou, Rovas, Libutti, and Monteleone (2015) to indicate the main opportunities and strengths as well as the main threats and weaknesses of the internal and external environment of the production method (pyrolysis) of biochars. They indicated that the main strengths of the methods are that enhanced bioeconomy, the production of bio-oil, is in line with 20-20-20 EU strategy and minimized CO_2 emissions. The most relevant opportunities include the application of biochar as soil amendment, as carbon sequester, as increased sustainability and the market of biochars, as well as new green job positions. On the other hand, the main weaknesses are the requirement of storage capacity of feedstocks due to seasonality, while the main treats include the use of pomace as animal feed and the competition that exists from other renewable sources, biorefineries, and other waste management practices.

12.4 Biomass-derived chars for energy production

12.4.1 Combustion systems

Biomass-derived chars could be used for energy purposes as an alternative to charcoal or fossil coal. In particular, pyrolytic chars (or hydrochars) could be used in domestic applications such as biomass stoves, biomass boilers, and barbecues. These chars could also be used for industrial applications through cofiring in coal power plants for heat and electricity production.

Peters, Iribarren, and Dufour (2015) compared different pathways for the recovery of biomass-derived chars obtained by slow pyrolysis. The evaluation of the environmental impact of each pathway was performed using the Centrum voor

Milieuwetenschappen Leiden (CML) method. In particular, the authors evaluated different impact potentials, including eutrophication, acidification, abiotic depletion, and global warming (100-year perspective). Furthermore, the cumulative nonrenewable (fossil and nuclear; CEDnr) and total (renewable and nonrenewable; CEDt) energy demands are estimated.

12.4.1.1 Domestic applications

Peters et al. (2015) showed that the replacement of charcoal by biomass char is an option with low beneficial impact compared to its use for agriculture application. In particular, the use of biochar is more favorable in terms of global warming (Table 7). However, net energy savings (CEDt) were found for charcoal. This result is attributed to the fact that conventional charcoal is produced from forest wood in inefficient traditional kilns. The replacement of this production by residues originating from resources other than wood saves significant amounts of wood primary energy, resulting in a negative CEDt result. In contrast, the nonrenewable energy savings and the reduction in abiotic depletion are not highly affected since biomass char is a substitute for other renewable energy resources (charcoal from wood).

12.4.1.2 Industrial applications. Cofiring for electricity production

The substitution of fossil coal by biomass char in coal power plants for heat and electricity production shows the highest reduction of environmental impacts in the major assessed categories (Table 7). Such a result is explained by the high environmental impacts associated with mining and coal combustion. The replacement of fossil coal by biomass-derived chars prevents these impacts and gives a net reduction for the cofiring system.

Huang, Syu, Chiueh, and Lo (2013) also evaluated the environmental impact and benefits of the use of biomass-derived chars in the cofiring supply chain used for electricity production. The authors analyzed 15 impact categories (Table 8). They

Table 7 Characterization results of the alternative systems.

Category	Unit	BC-1k	DBC	CC	NG	FC
ADP	kg Sb eq	−38.19	−121.80	−29.75	−108.92	−140.21
AP	kg SO_2 eq	29.27	21.55	27.53	29.47	−122.29
EP	kg PO_4^{3-} eq	10.30	8.74	9.41	9.37	−18.21
GWP	t CO_2 eq	−17.73	−15.74	−7.96	−13.95	−19.09
CED_{nr}	GJ	−48.17	−221.48	−30.94	−193.23	−190.46
CED_t	GJ	238.47	45.59	−67.31	75.33	77.27

ADP, abiotic depletion potential; *AP*, acidification potential; *BC-1k*, base-case system (1000-year stable biochar); *CC*, char for charcoal substitution; *CEDnr*, cumulative nonrenewable energy demand; *CEDt*, cumulative total energy demand; *DBC*, direct biomass combustion; *EP*, eutrophication potential; *FC*, char for fossil coal substitution; *GWP*, global warming potential; *NG*, char for natural gas substitution.

Table 8 The potential environmental impact and impact reduction per kWh of electricity generated by 10% and 20% cofiring.

Impact category	Unit	10%	Variation[a] (%)	20%	Variation[a] (%)	Reference system
Carcinogens	kg C_2H_3Cl eq	2.86E−03	5.27	3.01E−03	10.79	2.72E−03
Noncarcinogens	kg C_2H_3Cl eq	1.58E−02	3.11	1.63E−02	6.37	1.53E−02
Respiratory inorganics	kg $PM_{2.5}$ eq	6.18E−04	2.63	6.34E−04	5.29	6.02E−04
Ionizing radiations	Bq C-14 eq	1.94E+00	55.49	2.68E+00	114.81	1.25E+00
Ozone layer depletion	kg CFC-11 eq	8.76E−09	7.37	9.40E−09	15.21	8.16E−09
Respiratory organics	kg C_2H_4 eq	5.93E−05	2.87	6.10E−05	5.82	5.76E−05
Aquatic ecotoxicity	kg TEG water eq[b]	4.21E+01	−2.35	4.11E+01	−4.67	4.31E+01
Terrestrial ecotoxicity	kg TEG soil eq[b]	9.20E+00	−2.52	8.94E+00	−5.28	9.44E+00
Terrestrial acidification/ nutrification	kg SO_2 eq	1.67E−02	4.43	1.75E−02	9.43	1.60E−02
Land occupation	m^b org.arable[c]	3.85E−03	−3.85	3.68E−03	−8.10	4.00E−03
Aquatic acidification	kg SO_2 eq	4.37E−03	3.66	4.54E−03	7.69	4.22E−03
Aquatic eutrophication	kg PO_4 P-lim[d]	2.82E−05	792.31	5.46E−05	1627.66	3.16E−06
Global warming	kg CO_2 eq	1.04E+00	−4.10	9.90E−01	−8.71	1.08E+00
Nonrenewable energy	MJ primary	9.92E+00	−3.89	9.50E+00	−7.96	1.03E+01
Mineral extraction	MJ surplus	1.21E−03	12.01	1.35E−03	24.97	1.08E−03

[a]The increase or reduction of environmental impact compared with the reference system (electricity generation only by coal).
[b]TEG water/soil: Triethylene glycol into water/soil.
[c]Org.arable: Organic arable land.
[d]P-lim: Into a phosphorus-limited land.

concluded that the introduction of biomass char in the supply chain leads to impact reduction benefits in aquatic ecotoxicity, land occupation, terrestrial ecotoxicity, global warming, and nonrenewable energy. However, this substitution generates high negative impacts on human health (Table 8) and to a lower extent on ecosystem quality, climate change, and resources. Carbon reduction could be 4.32 and 4.68 metric tons CO_2 eq/ha/year at 10% and 20% cofiring ratios, respectively. The improvement of cofiring efficiency and the increase in char density may be major factors for reducing its environmental impact.

12.4.2 Syngas production (gasification)

In spite of the main advantages that waste/biomass gasification could provide, it has so far not been able to fully compete with traditional energy production methods such as combustion, anaerobic digestion, and fermentation. In fact, gasification technologies are not capable of achieving successful commercialization due to a number of challenges and barriers. Among these barriers, high capital and operational costs of gasification plants are limiting their development. In particular, gasification facility costs include site construction, the biomass pretreatment unit (drying, chips, or briquetting), the gasifier reactor, the producer gas conditioning system (tar and ash removal, filtration, purification, etc.), the engine or gas/steam turbine, and an electricity distribution network. Moreover, often new technologies require technological upgrade based on research and development for technology advancements and sustainability, which is also costly and not always successfully achieved. Therefore, the economic viability of the gasification plant depends strongly on the revenue stream, which depends on policy aspects. In fact, the generated streams such as gate fees and produced syngas utilization for electricity/fuel production are not, in general, sufficient for the viability of the gasification facility. The latter should be supported by renewable energy credits for electricity produced by means of renewable technologies as well as carbon credits for every ton of CO_2 prevented from entering the atmosphere.

Few investigations have proven the environmental benefit of biomass/char gasification. The main investigations examined the LCA of different technologies for converting energy to waste. In particular, Evangelisti et al. (2015) showed that conventional (incineration with energy recovery and landfill with electricity recovery) and advanced energy-from-waste technologies (gasification-plasma (G-PI), fast pyrolysis with combustor (FP-C), air blown gasification (G-SC)) have a better environmental performance over conventional municipal solid waste treatment technologies in terms of carbon footprint, global warming potential, acidification potential, eutrophication potential, photochemical ozone creation potential, and human toxicity potential. In addition, it was also reported that net electrical efficiency was higher for the case of a combination of fluidized-bed gasification with plasma processes over the other two advanced methods used. The net electrical efficiency of the G-PI process was 28%, whereas that of the G-SC was only 8%. The net electrical efficiency of the FP-C process amounted to 26% and was close to that of G-PI. However, this value does not take into consideration the energy content of the oil used in the producer gas cleaning part. If it is included in the LCA, the net electrical efficiency will be lower.

Maya, Sarmiento, de Sales Oliveira, Lora, and Andrade (2016) compared gasification and incineration technologies for the disposal of municipal solid waste. It was found that gasification led to four times fewer emissions of SO_2 than incineration processes, leading to less acidification of the air. Further, 33% fewer NO_x emissions were emitted for gasification than incineration leading to less smog production. Watson, Zhang, Si, Chen, and de Souza (2018) reported that $PM_{2.5}$ emissions were three times lower for the gasification process than for the combustion process due to lower NO_x emissions. Such difference was equal to −0.05 and −0.15 kg $PM_{2.5}$ eq for the moving-grate combustor and the gasifier, respectively. To sum up, the gasification process leads to net environmental and social benefits over the combustion process.

Although there are environmental advantages of gasification technology, it still faces technical barriers limiting its development. In particular, the presence of tar in syngas is a bottleneck if high-value chemicals are planned to be produced from syngas. In such a case, the tar content should be less than 1 mg/m^3. For electricity production in gas engines or turbines, the syngas should have a tar level lower than 30 mg/m^3. The tar content can be reduced by using catalysts during the gasification process. De Diego et al. (2016) reported that the use of catalytic fibers, made of MgO and NiO, in a dual fluidized-bed reactor used for biomass gasification reduced the tar content by up to 95%, which amounted to as little as 0.2 g/m^3 of the total gas. However, the use of catalysts creates another issue, such as catalyst poisoning and deactivation. Gasification products of biowaste usually contain sulfur and other compounds, which are responsible for the poisoning of catalysts even at low concentrations (<10 ppm). Another major challenge for gasification technology development is the separation of gaseous products. A pure stream of syngas is required if high-value chemical products (methanol, hydrogen) or liquid fuels via Fischer-Tropsch synthesis are planned to be produced. Generally, there are still constructional and design problems associated with gasifiers as well as scale-up, periodic replacement of refractory materials, and even more unique issues if plasma gasification is applied for, etc. Due to this difficulty, conventional incineration combined with the use of fossil fuels for energy/heat or chemicals production is more flexible, simpler, and cheaper to use.

12.4.3 Char as reductant in iron and steel industries

Because of renewable energy, development of the large-scale utilization of biomass and charcoal as reducing agents in iron and steel industries is an economic barrier compared to fossil-based reducing agents. In fact, the production cost of biomass-based reducing agents is affected by several variables: applied processing method, location of the plant (transportation costs and cost of infrastructure and labor), and raw material cost. The capital costs of a proposed charcoal plant were estimated to be $111 million, with costs for the pyrolysis unit (39%), engineering and construction (21%), storage (18%), etc. (Jahanshahi et al., 2015). Total operating costs were computed to be $446/ton and average charcoal cost was $243/ton.

Several studies were carried out in Europe, Australia, and North America where the capacities of the biomass-based reducing agent plants ranged from 50 to 5000 dry biomass tons per day. The feedstock in most cases was wood based, wood being the most suitable raw material for biomass-based reducing agent production. These costs are generally quite high, show a strong dependence on the feedstock, and range from 113 to 188€/t for torrefied biomass, 223 to 513€/t for charcoal, 722 to 974€/t for biosyngas, 164 to 189€/t for bio-oils, 890 to 2002€/t for biohydrogen, and 223 to 399€/t for biomethanol.

The major environmental challenge for the use of charcoal in the iron and steel industries is char production. In fact, because these industries use large amounts of raw materials, a regular supply of chars can be a significant issue for char utilization in the sector. Through species selection, silvicultural, and management practices, forest productivity has increased multifold toward sustainable production of biofuels and charcoal. For example, Brazil is well known for producing high-quality iron and steel using wood charcoal. Brazil has the largest abundance of biomass because of its suitable landscape, climate, rainfall, and soils. Because Brazil does not have sufficient reserves of coking coal, nearly 8 million tonnes per annum of wood charcoal were produced for the iron and steel industries. Higher transaction costs for deforestation as compared to coal production can be a serious issue, because planted forest activity is labor intensive, has a cycle of 14–21 years of production, and high costs of land management and environmental licensing. The displacement of nonrenewable charcoal by renewable charcoal by 2017 and the use of charcoal to produce up to 46% of pig iron and steel have been proposed by 2030, which would potentially mitigate 62 Mt of CO_2 between 2010 and 2030. It is currently the world's largest producer of charcoal with an annual production of 13 million tonnes accounting for ~30% of global production. This would represent 31% of all emissions reductions expected from the steel industry and Brazil's effort to reduce its GHG emissions by 39% by 2020.

12.5 Biomass-derived char as green environmental adsorbent for pollutants removal

Several investigations have shown that biomass-derived chars could be efficient materials for the removal of pollutants from gaseous and aqueous effluents. Such pollutants removal could be insured by the raw chars or after chemical and physical treatments by activated carbons. Numerous investigations have examined the sustainability of the use of biomass chars and activated carbons as green adsorbents. Moreira, Noya, and Feijoo (2017) compared the environmental impacts of activated carbon production with different chars elaborated from different biomass feedstocks. The authors selected four potential impact categories, including climate change, terrestrial acidification, freshwater eutrophication, and fossil depletion. Table 9 shows a comparison of the environmental impacts of activated carbon biomass-derived char production for each impact category. It is clearly observed that char production and its application for pollutants removal has lower environmental impacts than activated carbon (Table 9). Moreira et al. (2017) indicated that woodchips, corn stover, and poplar are the most environmentally friendly biochars in terms of climate change, terrestrial acidification, and fossil depletion. In contrast, forest residues and palm oil wastes have higher

Table 9 Comparative environmental results between activated carbon production and systems assessed in this study assuming biochar as a substitute for activated carbon.

Feedstock	Activated carbon production	Sewage sludge	Yard waste	Woodchips	Corn stover	Forest residue	Poultry litter	Poplar	Oil palm empty fruit bunch
CC (kg CO_2 eq/kg adsorption material)	1.44	1.26	−1.34	−3.57	−3.42	4.26	−0.41	−3.89	11.1
TA (kg SO_2 eq/kg adsorption material)	2.50×10^{-3}	-9.36×10^{-3}	-2.42×10^{-2}	-6.85×10^{-2}	-4.44×10^{-2}	-9.76×10^{-3}	-2.40×10^{-2}	-1.16×10^{-2}	5.33×10^{-2}
FE (kg P eq/kg adsorption material)	9.99×10^{-5}	3.71×10^{-4}	1.14×10^{-3}	1.32×10^{-5}	7.21×10^{-4}	1.11×10^{-3}	2.37×10^{-5}	1.66×10^{-3}	5.91×10^{-4}
FD (kg oil eq/kg adsorption material)	0.17	−0.02	−0.53	−2.01	−1.19	−5.21	−0.76	−0.26	−0.06

Negative values imply environmental credits while positive ones account for environmental impacts.
CC, climate change; *FD*, fossil depletion; *FE*, freshwater eutrophication; *TA*, terrestrial acidification.

environmental impacts in terms of climate change. In addition, higher negative impacts were obtained in freshwater eutrophication for all the tested biomass feedstocks. Such results are attributed to the greater energy requirements during the pyrolysis process. Such negative impacts could be avoided by the recovery of the gaseous (syngas) and liquid (bio-oil) products during char production.

The environmental benefits of the use of biomass chars for environmental application are confirmed through the analysis of several investigations of LCA of activated carbons. As an example, Hjaila, Baccar, Sarrà, Gasol, and Blánquez (2013) evaluated the potential environmental impacts of the use of olive waste cakes in the production of activated carbon using H_3PO_4 chemical activation. The authors obtained higher negative environmental impacts for the main categories, including eutrophication (96%), terrestrial ecotoxicity (92%), freshwater aquatic ecotoxicity (90%), human toxicity (64%), acidification potential (62%), and ozone depletion potential (44%). Furthermore, warming potential was one of the highest impacts (11.096 kg CO_2 eq/kg activated carbon). In fact, the cumulative energy demand of the production process of activated carbon using olive waste cakes precursor was 167.63 MJ issued from the impregnation, carbonization, drying, and washing steps. Modification of certain procedures and the recovery of the pyrolysis products during the carbonization step may help to lower the negative impacts of activated carbon production. In a similar investigation, Arena, Lee, and Clift (2016) indicated the requirement of the use of a renewable energy source for electricity production to decrease the negative environmental impacts of the production of activated carbons from coconut shell.

12.6 Catalyst, supercapacitor, and energy storage applications

12.6.1 Catalyst

Biomass-derived chars could be used as a precursor for the synthesis of different catalysts. These catalysts could be used in several applications such as syngas cleaning, syngas conversion into liquid hydrocarbons through Fischer-Tropsch synthesis, and biodiesel production. The use of biomass char as a catalyst precursor is motivated by its low cost and easy recyclability of the supported metal. However, the development of this application is limited by relatively low efficiency and low abrasive resistance compared with the available commercial catalysts.

12.6.2 Supercapacitor

Biomass-derived chars could be used as a precursor for the elaboration of supercapacitors. The attractiveness of these carbonaceous materials is motivated by their availability and low environmental impact. In addition, their activation process allows materials to be obtained with high surface area and rich porous structure at low cost. Such properties and advantages are critical for supercapacitor development at the industrial scale.

12.6.3 Energy storage

Biomass-based carbon adsorbents were used for gas storage, including CO_2, H_2, and CH_4. Such application is encouraged by low cost, availability, sustainable resource, and easy regeneration. However, specific surface treatments are required to reach an interesting selectivity and adsorption capacity.

12.7 Conclusion

The performance of char production and application systems is strongly influenced by feedstock characteristics, conversion techniques, char logistics (handling, densification, transport), and utilization methods. Several combinations of biomass resources, conversion technologies, and utilization systems have been reported in the literature. However, few studies have analyzed the sustainability of the whole process chain. Therefore, an attempt was made to analyze separately the different steps based on the available data in the literature.

The sustainability assessment was performed through environmental and economic issues. It seems that biomass residues are preferable as char feedstocks compared to energy crops and algae. In fact, biomass residues as well as farming and industrial wastes are concentrated, minimizing the energy cost and the environmental impact of collection. Furthermore, these biomass wastes do not incur the energetic cost and environmental impact associated with dedicated land use and water consumption for energy crops and algae production.

Concerning the conversion systems, slow pyrolysis seems to be the optimal process for maximizing char yield. Large production systems, uniform feedstocks, and tightly controlled application regimes are apt to be more reliable from a monitoring and verification standpoint. However, the performance of small char production systems should be improved due to their utility in the developing world. Such improvement is necessary to insure their role for combating climate change.

Among the various char applications, soil amendment seems a promising issue for carbon sequestration, soil fertility improvement, GHG emissions reduction, and soil remediation. The development of large-scale pilot studies for the emerging technology (electrodes, syngas production, etc.) is requested to analyze their sustainability.

References

Abdelhafez, A. A., & Li, J. (2016). Removal of Pb(II) from aqueous solution by using biochars derived from sugar cane bagasse and orange peel. *Journal of the Taiwan Institute of Chemical Engineers*, *61*, 367–375.

Arena, N., Lee, J., & Clift, R. (2016). Life Cycle Assessment of activated carbon production from coconut shells. *Journal of Cleaner Production*, *125*, 68–77.

Azapagic, A. (1999). Life cycle assessment and its application to process selection, design and optimisation. *Chemical Engineering Journal*, *73*, 1–21.

Bates, A. (2011). *Beyond nero; Biochar and the carbon cycle*. Curso de Diseno de Permacultura en Eco Convivencias en sporles, Mallorca.

Batool, S., Idrees, M., Hussain, Q., & Kong, J. (2017). Adsorption of copper (II) by using derived-farmyard and poultry manure biochars: efficiency and mechanism. *Chemical Physics Letters*, *689*, 190–198.

Beesley, L., Moreno-Jimenez, E., & Gomez-Eyles, J. (2010). Effects of biochar and green-waste compost amendments on mobility, bioavailability and toxicity of inorganic and organic contaminants in a multi-element polluted soil. *Environmental Pollution*, *158*(1), 155–160.

Belhachemi, M., Jeguirim, M., Limousy, L., & Addoun, F. (2014). Comparison of NO_2 removal using date pits activated carbon and modified commercialized activated carbon via different preparation methods: effect of porosity and surface chemistry. *Chemical Engineering Journal*, *253*, 121–129.

Biederman, L. A., & Harpole, W. S. (2013). Biochar and its effects on plant productivity and nutrient cycling: a meta-analysis. *GCB Bioenergy*, *5*, 202–214.

Brown, T. R., Wright, M. M., & Brown, R. C. (2011). Estimating profitability of two biochar production scenarios: slow pyrolysis vs. fast pyrolysis. *Biofuels, Bioproducts and Biorefining*, *5*, 54–68.

Cantrell, K. B., Hunt, P. G., Uchimiya, M., et al. (2012). Impact of pyrolysis temperature and manure source on physicochemical characteristics of biochar. *Bioresource Technology*, *107*, 419–428.

Cely, P., Gascó, G., Paz-Ferreiro, J., & Méndez, A. (2015). Agronomic properties of biochars from different manure wastes. *Journal of Analytical and Applied Pyrolysis*, *111*, 173–182.

Chaney, N. K. (1924). Adsorptive carbon and process of making the same. U.S. Patent 1497543, 10 June 1924.

Chen, B., Zhou, D., & Zhu, L. (2008). Transitional adsorption and partition of nonpolar and polar aromatic contaminants by biochars of pine needles with different pyrolytic temperatures. *Environmental Science & Technology*, *42*, 5137–5143.

Chen, B., Chen, Z., & Lv, S. (2011). A novel magnetic biochar efficiently sorbs organic pollutants and phosphate. *Bioresource Technology*, *102*, 716–723.

Clough, T. J., Bertram, J. E., Ray, J. L., & Condron, L. M. (2010). Unweathered wood biochar impact on nitrous oxide emissions from a bovine-urine-amended pasture soil. *Soil Science Society of America Journal*, *74*(3), 852–860.

De Diego, L. F., García-Labiano, F., Gayán, P., Abad, A., Mendiara, T., Adánez, J., et al. (2016). Tar abatement for clean syngas production during biomass gasification in a dual fluidized bed. *Fuel Processing Technology*, *152*, 116–123.

De Feo, G., & Malvano, C. (2009). The use of LCA in selecting the best MSW management system. *Waste Management*, *29*, 1901–1915.

Evangelisti, S., Tagliaferri, C., Clift, R., Lettieri, P., Taylor, R., & Chapman, C. (2015). Life cycle assessment of conventional and two-stage advanced energy-from-waste technologies for municipal solid waste treatment. *Journal of Cleaner Production*, *100*, 212–223.

Fertel, C., Bahn, O., Vaillancourt, K., & Waaub, P. J. (2013). Canadian energy and climate policies: a SWOT analysis in search of federal/provincial coherence. *Energy Policy*, *63*, 1139–1150.

Goudouva, G. T., & Zorpas, A. A. (2017). Water Footprint determination by quarry operation in island regions. *Desalination and Water Treatment*, *86*, 271–276.

Gascó, G., Paz-Ferreiro, J., Cely, P., et al. (2016). Influence of pig manure and its biochar on soil CO_2 emissions and soil enzymes. *Ecological Engineering*, *95*, 19–24.

Glaser, B., Haumaier, L., Guggenberger, G., & Zech, W. (2001). The 'terra preta' phenomenon: a model for sustainable agriculture in the humid tropics. *Naturwissenschaften*, *88*, 37–41.

Glaser, B., Lehmann, J., & Zech, W. (2002). Ameliorating physical and chemical properties of highly weathered soils in the tropics with charcoal—a review. *Biology and Fertility of Soils*, *35*, 219–230.

Gray, E., Marsh, H., & McLaren, M. (1982). A short history of gunpowder and the role of charcoal in its manufacture. *Journal of Materials Science*, *17*, 3385–3400.

Guizani, C., Jeguirim, M., Valin, S., Limousy, L., & Salvador, S. (2017). Biomass chars: the effects of pyrolysis conditions on their morphology, structure, chemical properties and reactivity. *Energies*, *10*(6), 796.

Haddad, K., Jeguirim, M., Jerbi, B., Chouchene, A., Dutournié, P., Thevenin, N., et al. (2017). Olive mill wastewater: from a pollutant to green fuels, agricultural water source and biofertilizer. *ACS Sustainable Chemistry & Engineering*, *5*, 8988–8996.

Hadjittofi, L., Charalambous, S., & Pashalidis, I. (2016). Removal of trivalent samarium from aqueous solutions by activated biochar derived from cactus fibres. *Journal of Rare Earths*, *34*, 99–104.

Hadjittofi, L., Prodromou, M., & Pashalidis, I. (2014). Activated biochar derived from cactus fibres—preparation, characterization and application on Cu(II) removal from aqueous solutions. *Bioresource Technology*, *159*, 460–464.

Hagemann, N., Spokas, K., Schmidt, H.-P., Kägi, R., Böhler, M. A., & Bucheli, T. D. (2018). Activated carbon, biochar and charcoal: linkages and synergies across pyrogenic carbon's ABCs. *Water*, *10*, 182.

Hjaila, K., Baccar, R., Sarrà, M., Gasol, C. M., & Blánquez, P. (2013). Environmental impact associated with activated carbon preparation from olive-waste cake via life cycle assessment. *Journal of Environmental Management*, *130*, 242–247.

Homagain, K., Shahi, C., Luckai, N., & Sharma, M. (2014). Biochar-based bioenergy and its environmental impact in northwestern Ontario Canada: a review. *Journal of Forest Research*, *25*(4), 737–748.

Hou, J., Huang, L., Yang, Z., et al. (2016). Adsorption of ammonium on biochar prepared from giant reed. *Environmental Science and Pollution Research*, *23*, 19107–19115.

Huang, Y. F., Syu, F. S., Chiueh, P. T., & Lo, S. H. (2013). Life cycle assessment of biochar cofiring with coal. *Bioresource Technology*, *131*, 166–171.

Islam, R. F., & Mamun, A. K. (2017). Possibilities and challenges of implementing renewable energy in the light of PESTLE & SWOT analyses for island countries. In F. Islam, K. Mamun, & M. Amanullah (Eds.), *Smart energy grid design for island countries, green energy and technology* (pp. 1–19). Cham: Springer.

Jahanshahi, S., Mathieson, J. G., Somerville, M. A., Haque, N., Norgate, T. E., Deev, A., et al. (2015). Development of low-emission integrated steelmaking process. *Journal of Sustainable Metallurgy*, *1*, 94–114.

Jay, C. N., Fitzgerald, J. D., Hipps, N. A., & Atkinson, C. J. (2015). Why short-term biochar application has no yield benefits: evidence from three field-grown crops. *Soil Use and Management*, *31*, 241–250.

Jeguirim, M., Dorge, S., Loth, A., & Trouvé, G. (2010). Devolatilization kinetics of Miscanthus straw from thermogravimetric analysis. *International Journal of Green Energy*, *7*, 164–173.

Jeguirim, M., Elmay, Y., Limousy, L., Lajili, M., & Said, R. (2014). Devolatilization behavior and pyrolysis kinetics of potential Tunisian biomass fuels. *Environmental Progress & Sustainable Energy*, *33*, 1452–1458.

Joseph, S., Doug, P., Dawson, K., Mitchell, D. R., Rawal, A., James, H., et al. (2015). Feeding biochar to cows: an innovative solution for improving soil fertility and farm productivity. *Pedosphere*, *25*, 666–679.

Jung, K.-W., Hwang, M.-J., Ahn, K.-H., & Ok, Y.-S. (2015). Kinetic study on phosphate removal from aqueous solution by biochar derived from peanut shell as renewable adsorptive media. *International journal of Environmental Science and Technology*, *12*, 3363–3372.

Lehmann, J., da Silva Jr, J. P., Steiner, C., Nehls, T., Zech, W., & Glaser, B. (2003). Nutrient availability and leaching in an archaeological Anthrosol and a Ferralsol of the Central Amazon basin: fertilizer, manure and charcoal amendments. *Plant and Soil*, *249*, 343–357.

Lehmann, J., & Joseph, S. (2009). *Biochar for environmental management: Science and technology*. London: Earthscan.

Lehmann, J., & Rondon, M. (2005). Bio-char soil management on highly-weathered soils in the humid tropics. In N. Uphoff (Ed.), *Biological approaches to sustainable soil systems*. Boca Raton, FL: CRC Press.

Lewis, W. K., & Metzner, A. B. (1954). Engineering, design and process development section: activation of carbons. *Industrial and Engineering Chemistry*, *46*, 849–858.

Liatsou, I., Pashalidis, I., Oezaslan, M., & Dosche, C. (2017). Surface characterization of oxidized biochar fibers derived from *Luffa cylindrica* and lanthanide binding. *Journal of Environmental Chemical Engineering*, *5*, 4069–4074.

Lian, F., Huang, F., Chen, W., et al. (2011). Sorption of apolar and polar organic contaminants by waste tire rubber and its chars in single- and bi-solute systems. *Environmental Pollution*, *159*, 850–857.

Lohri, C. R., Rajabu, H. M., Sweeney, D. J., & Zurbrügg, C. (2016). Char fuel production in developing countries—a review of urban biowaste carbonization. *Renewable and Sustainable Energy Reviews*, *59*, 1514–1530.

Masullo, A. (2014). *Human economy and natural economy, issue E3S web of conferences*, *2*, 04002.

Maya, D. M. Y., Sarmiento, A. L. E., de Sales Oliveira, C. A. V. B., Lora, E. D. S., & Andrade, R. V. (2016). Gasification of municipal solid waste for power generation in Brazil, a review of available technologies and their environmental benefits. *Journal of Chemistry and Chemical Engineering*, *10*, 249–255.

Micháleková-Richveisová, B., Frišták, V., Pipíška, M., et al. (2017). Iron-impregnated biochars as effective phosphate sorption materials. *Environmental Science and Pollution Research*, *24*, 463–475.

Moreira, M. T., Noya, I., & Feijoo, G. (2017). The prospective use of biochar as adsorption matrix—a review from a Life cycle perspective. *Bioresource Technology*, *246*, 135–141.

Nautiyal, P., Subramanian, K. A., & Dastidar, M. G. (2016). Adsorptive removal of dye using biochar derived from residual algae after in-situ transesterification: alternate use of waste of biodiesel industry. *Journal of Environmental Management*, *182*, 187–197.

Peters, J. F., Iribarren, D., & Dufour, J. (2015). Biomass pyrolysis for biochar or energy applications? A life cycle assessment. *Environmental Science & Technology*, *49*, 5195–5202.

Poo, K.-M., Son, E.-B., Chang, J.-S., et al. (2018). Biochars derived from wasted marine macroalgae (*Saccharina japonica* and *Sargassum fusiforme*) and their potential for heavy metal removal in aqueous solution. *Journal of Environmental Management*, *206*, 364–372.

Roberts, K. G., Loy, B. G., Joseph, S., Scott, N. R., & Lehmann, J. (2010). Life cycle assessment of biochar systems: Estimating the energetic, economic, and climate change potential. *Environmental Science & Technology*, *44*, 827–833.

Rosales, E., Meijide, J., Pazos, M., & Sanromán, M. A. (2017). Challenges and recent advances in biochar as low-cost biosorbent: from batch assays to continuous-flow systems. *Bioresource Technology*, *246*, 176–192.

Schanz, J. J., & Parry, R. H. (1962). The activated carbon industry. *Industrial and Engineering Chemistry*, *54*, 24–28.

Singh, B. P., Hatton, B. J., Singh, B., Cowie, A. L., & Kathuria, A. (2010). Influence of biochars on nitrous oxide emission and nitrogen leaching from two contrasting soils. *Journal of Environmental Quality*, *39*(4), 1224–1235.

Shackley, S., Hammond, J., Gaunt, J., & Ibarrola, R. (2011). The feasibility and costs of biochar deployment in the UK. *Carbon Management*, *2*, 335–356.

Shen, Y., Li, H., Zhu, W., et al. (2017). Microalgal-biochar immobilized complex: a novel efficient biosorbent for cadmium removal from aqueous solution. *Bioresource Technology*, *244*, 1031–1038.

Shin, J., Lee, S., Park, W., Choi, Y., Hong, S., & Park, S. (2014). Carbon sequestration in soil cooperated with organic composts and bio-char during corn (*Zea mays*) cultivation. *Journal of Agricultural Chemistry and Environment*, *3*, 151–155.

Spindler, K. (2013). *The man in the ice*. London: Hachette UK.

Srdjevic, Z., Bajcetic, R., & Srdjevic, B. (2012). Identifying the criteria set for multicriteria decision making based on SWOT/PESTLE analysis: a case study of reconstructing a water intake structure. *Water Resources Management*, *26*, 3379–3393.

Xie, T., Reddy, K. R., Wang, C., et al. (2015). Characteristics and applications of biochar for environmental remediation: a review. *Critical Reviews in Environmental Science and Technology*, *45*, 939–969.

Watson, J., Zhang, Y., Si, B., Chen, W. T., & de Souza, R. (2018). Gasification of biowaste: a critical review and outlooks. *Renewable and Sustainable Energy Reviews*, *83*, 1–17.

Wang, Z., Guo, H., Shen, F., et al. (2015). Biochar produced from oak sawdust by Lanthanum (La)-involved pyrolysis for adsorption of ammonium (NH4+), nitrate (NO3 –), and phosphate (PO43 –). *Chemosphere*, *119*, 646–653.

Woolf, D. (2008). *Biochar as a soil amendment: A review of the environmental implications*. Swansea University, School of the Environment and Society. http://orgprints.org/13268/.

Zabaniotou, A., Rovas, D., Libutti, A., & Monteleone, M. (2015). Boosting circular economy and closing the loop in agriculture: case study of as mall-scale pyrolysis–biochar based system integrated in an olive farm in symbiosis with an olive mill. *Environmental Development*, *14*, 22–36.

Further reading

Navarro Pedreño, J., & Mataix, J. (2014). An approach to knowledge of the paleosols of the province of Alicante. *Cidaris*, *32*, 21–31.

Smith, A. (1862). On the absorption of gases by charcoal.—No. I. *Proceedings of the Royal Society of London*, *12*, 424–426.

Index

Note: Page numbers followed by *f* indicate figures, and *t* indicate tables.

V

W

X

Made in the USA
Monee, IL
12 September 2026

62977623R00302